Zeneb Luo

Biomathematics & statistics scotland

POTATO GENETICS

POTATO GENETICS

Edited by

J.E. Bradshaw

and

G.R. Mackay

Scottish Crop Research Institute
Invergowrie
Dundee
DD2 5DA
UK

CAB INTERNATIONAL

CAB INTERNATIONAL Tel: Wallingford (0491) 832111
Wallingford Telex: 847964 (COMAGG G)
Oxon OX10 8DE Telecom Gold/Dialcom: 84: CAU001
UK Fax: (0491) 833508

© CAB INTERNATIONAL 1994. All rights reserved. No part of this publication may be reproduced in any form or by any means, electronically, mechanically, by photocopying, recording or otherwise, without the prior permission of the copyright owners

A catalogue entry for this book is available from the British Library

ISBN 0 85198 869 5

Typeset by Colset Pte Ltd, Singapore
Printed and bound in the UK at the University Press, Cambridge

Contents

Contributors ix

Preface xi

Part One: Origins, Species and Cytology

1. Origins of Cultivated Potatoes and Species Relationships 3
   J.G. HAWKES

2. Genome Evolution in Potatoes 43
   M.J. WILKINSON

Part Two: Theory and Methods of Genetical Analysis

3. Quantitative Genetics Theory for Tetrasomic Inheritance 71
   J.E. BRADSHAW

4. Theory for Locating Quantitative Trait Loci 101
   J.E. BRADSHAW

5. Use of 2n Gametes 109
   G.C.C. TAI

6. Use of 24-Chromosome Potatoes (Diploids and Dihaploids) 133
   for Genetical Analysis
   R. ORTIZ and S.J. PELOQUIN

7. Production of Monohaploids of *Solanum tuberosum* L. and 155
   Their Use in Genetics, Molecular Biology and Breeding
   E. JACOBSEN and M.S. RAMANNA

Part Three: Cellular and Molecular Genetics

8.	Tissue Culture G. Wenzel	173
9.	Somaclonal Variation A. Kumar	197
10.	Molecular Genetics K.N. Watanabe	213

Part Four: Environmental Stress, Morphology and Quality

11.	Environmental Stress and Its Impact on Potato Yield M.E. Vayda	239
12.	Inheritance of Morphological and Tuber Characteristics R. Ortiz and Z. Huaman	263
13.	Inheritance of Table and Processing Quality M.F.B. Dale and G.R. Mackay	285

Part Five: Inheritance of Resistance to Pests and Diseases

14.	Inheritance of Resistance to Nematodes M.S. Phillips	319
15.	Inheritance of Resistance to Viruses K.M. Swieżyński	339
16.	Inheritance of Resistance to Late Blight V. Umaerus and M. Umaerus	365
17.	Inheritance of Resistance to Warm-growing-season Fungal Diseases J.J. Pavek and D.L. Corsini	403
18.	Inheritance of Resistance to Fungal Diseases of Tubers R.L. Wastie	411
19.	Inheritance of Resistance to Bacterial Diseases J.G. Elphinstone	429
20.	Inheritance of Resistance to Insects and Mites K.V. Raman, A.M. Golmirzaie, M. Palacios and J. Tenorio	447

Part Six: Potato Breeding

21. Breeding Strategies for Clonally Propagated Potatoes 467
    J.E. BRADSHAW and G.R. MACKAY

22. Breeding Potatoes Based on True Seed Propagation 499
    A.M. GOLMIRZAIE, P. MALAGAMBA and N. PALLAIS

23. Introgression of Genes from Wild Species, Including Molecular 515
    and Cellular Approaches
    J.G.Th. HERMSEN

Index 539

Contributors

J.E. Bradshaw, *Crop Genetics Department, Scottish Crop Research Institute, Invergowrie, Dundee DD2 5DA, UK.*

D.L. Corsini, *United States Department of Agriculture, Agricultural Research Service, Pacific West Area, PO Box AA, Aberdeen, Idaho 83210, USA.*

M.F.B. Dale, *Crop Genetics Department, Scottish Crop Research Institute, Invergowrie, Dundee DD2 5DA, UK.*

J.G. Elphinstone, *Department of Plant Pathology, Rothamsted Experimental Station, ARFC Institute of Arable Crops Research (IACR), Harpenden, Herts AL5 2JQ, UK.*

A.M. Golmirzaie, *International Potato Center (CIP), Apartado Postal 5969, Lima, Peru.*

J.G. Hawkes, *Emeritus Professor, School of Continuing Studies, University of Birmingham, PO Box 363, Edgbaston, Birmingham B15 2TT, UK.*

J.G.Th. Hermsen, *Department of Plant Breeding (IVP), Wageningen Agricultural University, PO Box 386, 6700 AJ Wageningen, The Netherlands.*

Z. Huaman, *Department of Genetic Resources, International Potato Center (CIP), Apartado Postal 5969, Lima, Peru.*

E. Jacobsen, *The Graduate School of Experimental Plant Sciences, Agricultural University, PO Box 386, 6700 AJ Wageningen, The Netherlands.*

A. Kumar, *Cell and Molecular Genetics Department, Scottish Crop Research Institute, Invergowrie, Dundee DD2 5DA, UK.*

G.R. Mackay, *Crop Genetics Department, Scottish Crop Research Institute, Invergowrie, Dundee DD2 5DA, UK.*

P. Malagamba, *International Potato Center (CIP), Apartado Postal 5969, Lima, Peru.*

R. Ortiz, *Department of Genetics, University of Wisconsin-Madison, 1575 Linden Drive, Madison, Wisconsin 53706, USA.*

M. Palacios, *International Potato Center (CIP), Apartado Postal 5969, Lima, Peru.*

N. Pallais, *International Potato Center (CIP), Apartado Postal 5969, Lima, Peru.*

J.J. Pavek, *United States Department of Agriculture, Agricultural Research Service, Pacific West Area, PO Box AA, Aberdeen, Idaho 83210, USA.*

S.J. Peloquin, *Department of Horticulture, University of Wisconsin-Madison, 1575 Linden Drive, Madison, Wisconsin 53706, USA.*

M.S. Phillips, *Zoology Department, Scottish Crop Research Institute, Invergowrie, Dundee DD2 5DA, UK.*

K.V. Raman, *International Potato Center (CIP), Apartado Postal 5969, Lima, Peru.*

M.S. Ramanna, *Department of Plant Breeding (IVP), Agricultural University, PO Box 386, 6700 AJ Wageningen, The Netherlands.*

K.M. Swieżyński, *Potato Research Institute, Research Center for Genetics, Breeding and Virology, Młochów, 05-832 Rozalin, Poland.*

G.C.C. Tai, *Agriculture Canada Research Station, PO Box 20280, Fredericton, New Brunswick, Canada, E3B 4Z7.*

J. Tenorio, *International Potato Center (CIP), Apartado Postal 5969, Lima, Peru.*

M. Umaerus, *Department of Plant Breeding, Swedish University of Agricultural Sciences, PO Box 7003, S-750 07 Uppsala, Sweden.*

V. Umaerus, *Department of Plant and Forest Protection, Swedish University of Agricultural Sciences, PO Box 7044, S-750 07 Uppsala, Sweden.*

M.E. Vayda, *Department of Biochemistry, Microbiology and Molecular Biology, University of Maine, Orono, Maine 04469, USA.*

R.L. Wastie, *Crop Genetics Department, Scottish Crop Research Institute, Invergowrie, Dundee DD2 5DA, UK.*

K.N. Watanabe, *Department of Plant Breeding and Biometry, 252 Emerson Hall, Cornell University, Ithaca, New York 14853-1902, USA.*

G. Wenzel, *Federal Centre for Breeding Research on Cultivated Plants, Institute for Resistance Genetics, Graf-Seinsheimstrasse 23, D-85461 Grünbach, Germany.*

M.J. Wilkinson, *Crop Genetics Department, Scottish Crop Research Institute, Invergowrie, Dundee DD2 5DA, UK.*

Preface

Whilst the rediscovery in 1900 of Mendel's published work of 1865 marked the birth of modern genetics and opened the way to crop improvement by scientific breeding methods, the principal cultivated potato (*S. tuberosum* subsp. *tuberosum*) proved a difficult species for genetic research because of its complex inheritance patterns, and for cytological investigations because of its small and relatively numerous chromosomes. Indeed, not until the end of the 1930s did geneticists recognize that it was a tetraploid ($2n = 4x = 48$) which displays tetrasomic inheritance. Potato breeding in Europe and North America in the early 20th century therefore remained empirical and genetically unsophisticated. Furthermore, the rate of progress was impeded because much of the genetical variation originally introduced from South America had been utilized and not replenished. As the 20th century progressed, this was remedied as new introductions of primitive cultivated species and wild species became available from the collecting expeditions to Central and South America, pioneered by the Russians in the 1920s. In contrast, detailed genetical knowledge remained elusive, as did a detailed understanding of the factors which governed the crossability of species. Furthermore, the empirical introgression of genes from other species into the Tuberosum gene pool greatly complicated genetic interpretation of the heritable variation of all but a few simply inherited traits.

Some of the problems and complexities of working with a tetraploid were overcome by breeding and genetic studies at the diploid level. These became feasible from 1958 onwards, following the production of relatively large numbers of dihaploids from tetraploid Tuberosum, and the discovery that they could be crossed with many wild species which were also diploid. However, the dihaploids were usually male sterile, and most dihaploids, diploid species, and their hybrids displayed gametophytic self-incompatibility. Therefore, true breeding lines which could readily be selfed and crossed and which displayed disomic inheritance could not be produced; thus inheritance studies remained difficult. This was particularly so for economically important traits, most of which displayed continuous variation. These required the biometrical approach pioneered by Fisher in 1918, in which genetic information is inferred from measurements on related individuals, even though individual genes can not be recognized by Mendelian analysis because discrete classes can not be discerned.

Perhaps, then, it is not that surprising that so little was known about the genetics of the potato when Howard published his short treatise in 1970. Indeed, the complexity of the potato genome continues to deny detailed formal genetic interpretation and for many important economic traits it has only been possible to partition variation into general and specific combining ability – concepts which aid plant breeders but explain little of the underlying genetics, that is the number of genes involved and the nature of their action. There have nevertheless been a number of exciting developments in recent years which have led to major advances in potato genetics and breeding. The potato, or at least some of its genotypes, has proved particularly amenable to tissue culture techniques, including rapid methods of micro-propagation, anther culture and protoplast fusion. The use of *Agrobacterium tumifaciens* Ti plasmid mediated transformation for the transfer of specific genes is readily available. A reasonably high density linkage map of molecular markers has been obtained from diploids and aligned with the homoeologous tomato genome, and a start has been made to locating the genes underlying continuous variation through their linkage to these molecular markers. It should therefore soon prove possible to manipulate quantitative trait loci in the same way as Mendelian factors, by following the segregation patterns of closely linked markers, as well as determining their function and even cloning them. Finally, there is now a better understanding of species relationships, their ploidy, effective ploidy and crossability, which allows the great ecogeographical range of adaptation of wild species to be exploited in breeding potatoes for an ever wider range of environments.

We therefore considered it timely to ask a number of experts to help us review the current state of potato genetics, from theories on the origins of cultivated potatoes, species relationships and genome evolution; through the theory and limitations of biometrical genetical analysis at the tetraploid and diploid levels to the potential of the new techniques in cellular and molecular genetics; on to a consideration of what is already known about the inheritance of economically important traits; and finally to how this information can be used to develop new and more efficient breeding strategies.

We gave authors as much freedom as possible over the content and style of their chapters, consistent with the subject matter forming a reasonably comprehensive and coherent book without unnecessary duplication. We did our best to help authors make their contributions as readable and as free from errors as is possible in a human endeavour. We would like to thank our colleague, Dr Roger Wastie, and our typist, Mrs Sheena Forsyth, for their help in this respect. We are also indebted to many other members of the Scottish Crop Research Institute for their support.

We hope that the finished product will be of value to geneticists and breeders throughout the world as they strive to make real progress with this genetically complex but extremely important food crop. Genetics is the science of heredity; a more comprehensive understanding of the genetics of the potato and its application are essential if the potato is to continue to make a substantial contribution to the feeding of mankind, and permit the evolution of sustainable agricultural systems with increased reliance on intrinsic disease and pest resistance to maintain quality and productivity.

J.E. Bradshaw
G.R. Mackay
July 1993

I Origins, Species and Cytology

1 Origins of Cultivated Potatoes and Species Relationships

J.G. HAWKES

Emeritus Professor, School of Continuing Studies, University of Birmingham, PO Box 363, Edgbaston, Birmingham B15 2TT, UK.

The potato is one of the world's most important crops, exceeded only by wheat, maize and rice in total production. Yet until the 16th century it was unknown to the peoples of Europe, Asia, Africa and North America. In South America, however, it was probably the most productive source of food for the communities of the high Andes and southern Chile. The Spanish conquerors found potato cultivation already very widely adopted in what are now Colombia, Ecuador, Peru and Bolivia, in addition to the Araucanian region of Chile.

Archaeological Evidence for Potato Domestication

The centre of origin of potato cultivation may well have been the Andes of southern Peru and northern Bolivia, where likely wild prototypes still exist. Archaeological remains of potatoes and an unrelated tuber crop, *Ulluco*, have been radiocarbon-dated to 7000 years before present (Hawkes, 1990: 18). It is highly likely that the potato was domesticated even before that date, but we have no information at present to support that assumption.

There is much later evidence from rubbish heaps, graves and food stores of potato cultivation at 4500 to 3500 before present (Martins-Farias, 1976; Ugent *et al.*, 1982).

There are also spectacular ceramics dating from the Moche cultures in northern Peru (*c.* AD 1–600) and the Chimu peoples (*c.* AD 900–1450), as well as Huari or Pacheco urns from the Nazca valley in southern Peru, dated to *c.* AD 650–700. All these ceramics are from coastal areas, and it is assumed that the potters obtained potatoes by barter or other means from farmers in the highlands where potatoes were actually cultivated (see Hawkes, 1990, for illustrations).

Historical Evidence – South America

The conqueror of Peru, Francisco Pizarro, may well have been the first European to see potatoes in 1533, but there is no historical record of this event. The first historical record was in 1537 (Juan de Castellanos, 1601 (Castellanos, 1886), when a band of Spaniards led by Jiménez de Quesada penetrated into the highlands of what is now Colombia (see Hawkes, 1990: 22, for full description). This was followed by accounts of López de Gómara (1552) for potatoes in southern Peru and by Pedro Cieza de León (1553) in the area of what is now southern Colombia and northern Ecuador. Potatoes in Chile received first mention by Sir Francis Drake for 1578 (Drake, 1628). No cultural practices were mentioned in these accounts, except for that of Felipe Guamán Poma de Ayala, written in 1583 to 1613. This work was never published by the author and remained relatively unknown until a facsimile edition was published by P. Rivet in 1936.

The Potato in 16th- and 17th-Century Europe

Although the Spanish chroniclers wrote detailed accounts of the history of their conquests and the life and customs of the peoples, no contemporary accounts exist (and perhaps none were ever written) of the first introduction of the potato into Europe. However, much thought and sifting of such evidence that exists have been accorded to this subject (see, particularly, Salaman, 1949, revised edition, 1985), with the following results.

Apparently there were two introductions into Europe, one into Spain in *c.* 1570 and the second into England in *c.* 1590 (Hawkes, 1990: 31). The Spanish introduction rests on market records from the Hospital de la Sangre in Seville (Hawkes and Francisco-Ortega, 1992), whilst the English records are extremely complex; nevertheless, we know from Gerard that he grew the potato in his garden and he described and figured it in his *Herbal* of 1597. This was certainly a distinct introduction from the Spanish one.

European herbalists obtained their potatoes from the Flemish herbalist Carol Clusius (1601), his specimens being derived from Spain, via Italy. Particularly interesting accounts are given also by C. Bauhin (1596, 1620), J. Bauhin (1651) and Parkinson (1629).

It is interesting to note that these first European potatoes were adapted to develop tubers under the short 12-hour day of the Andes and not, as our present potatoes do, under the long 16–18-hour day of Europe. Evidence from the Spanish archives shows clearly that the first European potatoes were short-day-adapted, tuberizing in December to January in the apparently frost-free zones of southern Spain and Italy (Hawkes and Francisco-Ortega, 1992), as Salaman (1946, 1954; Salaman and Hawkes, 1949) had correctly surmised.

These potatoes were the Andean form of the tetraploid potato (*Solanum tuberosum* L. subsp. *andigena* Hawkes), which needed several centuries of 'unconscious' selection in Europe to adapt it to the long summer days of northern Europe. Hence it was not until the late 18th and early 19th centuries that this new day-length adaptation was complete, thus allowing potato cultivation on a field scale to spread into central and eastern Europe also.

Potatoes in the North American colonies were received first from Bermuda in 1691, where they had been grown from an earlier importation from England in 1613. British missionaries took potatoes to India and China in the 17th century and they were introduced into Japan and parts of Africa at about the same time. They appeared in New Zealand in 1769 and were grown by the Maoris by 1840 (Hawkes, 1990).

Thus, the potato, which was restricted to the Americas until the 16th century, became one of the most important world crops in no more than some 300 years.

Cultivated Potatoes in South America

As is generally well known, potato species belong to the large and very diversified genus, *Solanum*. However, they constitute a very small part of this worldwide genus, the wild potatoes occurring only in the Americas. The cultivated species were at one time confined to the Andes of South America and the lowlands of southern Chile, in both cases being adapted to the cool temperate climates of these regions, either at high altitudes or at high latitudes. The related wild species are much more widespread, as we shall see later.

There are seven cultivated species of potato, occurring in a polyploid series with a base number of 12 and ranging from diploid to pentaploid. Several of them are fairly similar to each other and for that reason were classified by Dodds (1962) as 'groups' of *S. tuberosum* rather than distinct species. Their probable evolutionary relationships are shown in Figure 1.1.

The diploid species, *S. stenotomum* is grown from central Peru to central Bolivia and is believed to be the most primitive, probably having been derived from the diploid wild species, *S. leptophyes*, or possibly *S. canasense*, both of which still occur in the central part of its distribution area. However, cultivated potato evolution did not stop at *S. stenotomum* but really just began there.

At least four wild potato species were involved in the process. Evidence indicates that hybridization of *S. stenotomum* with the weedy species *S. sparsipilum* and subsequent chromosome doubling produced the tetraploid *S. tuberosum* subsp. *andigena* in the central Andes (Cribb and Hawkes, 1986). Some workers, however, consider that the tetraploid Andean potatoes are derived from *S. stenotomum* by simple chromosome doubling. The evidence for this will be discussed later. This tetraploid subspecies was carried by ancient peoples into southern Chile, where it became adapted to the long day length, there to evolve into subsp. *tuberosum*. A similar process in Europe caused the same development to take place under the long day conditions mentioned in an earlier section. However, it should also be stated that certain authors (Grun, 1979, 1990) believe that subsp. *tuberosum* from Chile and

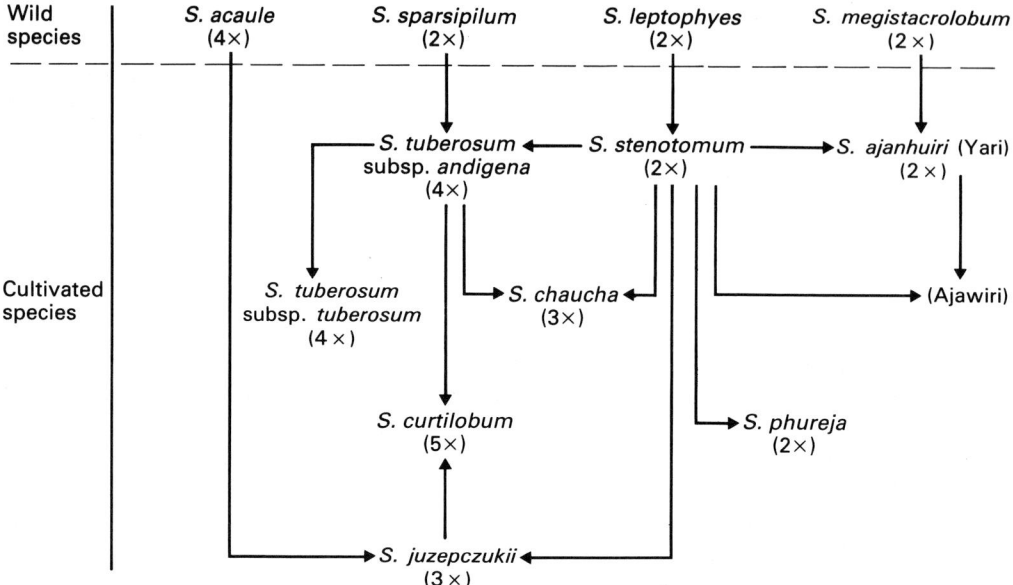

Fig. 1.1. Evolutionary relationships of cultivated potatoes and their ploidy levels (adapted from Hawkes, 1990, with kind permission of Belhaven Press).

Europe differs from subspecies *andigena* by certain cytoplasmic factors that it may have acquired from some wild diploid species, such as *S. chacoense*.

S. stenotomum had been taken in pre-conquest days into the lower, warmer eastern valleys, where the tubers lost their dormancy period and speeded up tuber development, so that three crops a year could be grown. This was clearly a process of artificial selection by the Andean farmers. These non-dormant eastern valley potatoes have been given the name *S. phureja*.

Natural hybridization of *S. stenotomum* with the wild frost-resistant species *S. megistacrolobum* also took place, to form the diploid *S. ajanhuiri* (Huamán *et al.*, 1982, 1983; Johns and Keen, 1986). The F_1 hybrid produced the 'Yari' group of varieties and a probable backcross to the cultivated parent gave rise to the 'Ajawiri' group of varieties.

A further series of hybridizations took place between *S. stenotomum* and the wild tetraploid species *S. acaule*. The F_1 cross gave rise to a highly sterile triploid, *S. juzepczukii* (Hawkes, 1962a; Schmiediche *et al.*, 1980, 1982), which incorporated the strong frost resistance of *S. acaule*. A further natural cross took place between *S. juzepczukii* and *S. tuberosum* subsp. *andigena* to form the only slightly less frost-resistant pentaploid species *S. curtilobum*. This evidently involved a $2n$ gamete from *S. juzepczukii* and a normal gamete from *S. tuberosum* subsp. *andigena* (see above references).

A series of crosses between *S. stenotomum* and subsp. *andigena* have given rise to the triploid hybrids named *S. chaucha*. The processes involved in this formation were worked out by Jackson *et al.* (1977, 1978).

We thus have a network of cultivated species or species groups which evolved chiefly in the central Andes of Peru and Bolivia, involving four original wild species (Figure 1.1). All but two of these cultivated potatoes have always been confined to that central area. However, the diploid *S. phureja* has extended northwards into Ecuador, Colombia and Venezuela, whilst the tetraploid *S. tuberosum* spread into southern Chile also (Figures 1.2 and 1.3).

Wild Potatoes and Their Adaptive Ranges

The cultivated potato possesses more related wild species than probably any other crop. Nearly all of these are in the same gene pool and will thus hybridize with it, though some difficulties occur, which will be discussed later.

Some 228 wild potato species are now recognized, but not all of these have been investigated fully. They possess the same base number ($x = 12$) of chromosomes as the cultivated potatoes, and range from diploid ($2n = 2x = 24$) to hexaploid ($2n = 6x = 72$).

To facilitate the understanding of their suggested taxonomic and evolutionary relationships potato species have been divided into Series. Two subsections are recognized in section Petota of subgenus Potatoe. The first is subsection Estolonifera, which indicates the absence of stolons or tubers. The species in one series of this subsection, namely, Etuberosa, can be induced to form hybrids with the tuber-bearing species. The other series, Juglandifolia, cannot. Species in series Juglandifolia possess yellow flowers and appear to have closer connections with the tomato genus, *Lycopersicon*, than with *Solanum*. The series Juglandifolia is listed here for completeness, but need not be considered further.

The species in all 19 series of subsection Potatoe bear tubers and are clearly of potential interest to breeders. The better-known species in both subsections are listed in Table 1.1 in columns according to their chromosome numbers. The taxonomic system used here is that of the author (Hawkes, 1990).

An important point to bear in mind when assessing wild potatoes for plant breeding purposes is their wide geographical distribution and very great range of ecological adaptation. Their total distribution ranges from the southwestern states of the USA (Arizona, Colorado, New Mexico and Texas) through Mexico and into Guatemala, Honduras, Costa Rica and western Panamá, all generally at medium to fairly high altitudes (Figure 1.4).

In South America, wild species are found along the Andes from Venezuela through Colombia, Ecuador, Peru, Bolivia and northwest Argentina. They also occur in the lowlands of Chile, Argentina, Uruguay, Paraguay and southeastern Brazil (Figure 1.5). Again, the adaptative range among different species is very great.

Thus, some species are adapted to growth in the cold, very high Andean regions from 3000 to 4500 m, where frosts are very common (*S. acaule* and *S. megistacrolobum*, for example), whilst others occur in dry semidesert conditions (e.g. *S. berthaultii*, *S. tarijense* and *S. neocardenasii*). Others are

Fig. 1.2. Distribution of diploid cultivated potato species (from Hawkes, 1990, with kind permission of Belhaven Press).

Fig. 1.3. Distribution of triploid, tetraploid and pentaploid potato species (from Hawkes, 1990, with kind permission of Belhaven Press).

Table 1.1. Classification and chromosome numbers of the more important wild and cultivated potato species and their close relatives.

Subsections and series	Species arranged by chromosome number ($x = 12$)				
	2x	3x	4x	5x	6x
Subsection Estolonifera					
Series					
I. Etuberosa	S. brevidens S. etuberosum				
II. Juglandifolia	S. lycopersicoides				
Subsection Potatoe					
Series					
I. Morelliformia	S. morelliforme				
II. Bulbocastana	S. bulbocastanum S. clarum	S. bulbocastanum			
III. Pinnatisecta	S. brachistotrichum S. cardiophyllum S. jamesii S. pinnatisectum S. trifidum	S. cardiophyllum S. jamesii			
IV. Polyadenia	S. polyadenium S. lesteri				
V. Commersoniana	S. commersonii	S. commersonii			
VI. Circaeifolia	S. capsicibaccatum S. circaeifolium				
VII. Lignicaulia	S. lignicaule				

Series	Species
VIII. Olmosiana	S. olmosense
IX. Yungasensa	S. chacoense, S. tarijense, S. yungasense
X. Megistacroloba	S. boliviense, S. megistacrolobum, S. sanctae-rosae, S. toralapanum
XI. Cuneoalata	S. infundibuliforme
XII. Conicibaccata	S. chomatophilum, S. santolallae, S. violaceimarmoratum, S. agrimonifolium, S. colombianum, S. longiconicum, S. oxycarpum, S. moscopanum
XIII. Piurana	S. piurae, S. tuquerrense
XIV. Ingifolia	S. ingifolium
XV. Maglia	S. maglia
XVI. Tuberosa (wild)	S. alandiae, S. berthaultii, S. brevicaule, S. bukasovii, S. canasense, S. gandarillasii, S. gourlayi, S. hondelmannii, S. kurtzianum, S. leptophyes, S. marinasense, S. microdontum, S. multidissectum, S. neocardenasii, S. oplocense

Table 1.1. (cont.)

Subsections and series	Species arranged by chromosome number ($x = 12$)				
	2x	3x	4x	5x	6x
XVI. Tuberosa (cultivated)	S. sparsipilum S. spegazzinii S. vernei S. verrucosum S. × ajanhuiri S. phureja S. stenotomum	S. × chaucha S. × juzepczukii	S. sucrense S. tuberosum subsp. tuberosum S. tuberosum subsp. andigena	S. × curtilobum	
XVII. Acaulia			S. acaule		S. albicans
XVIII. Longipedicellata		S. × vallis-mexici	S. fendleri S. hjertingii S. papita S. polytrichon S. stoloniferum		
XIX. Demissa				S. × semidemissum S. × edinense	S. brachycarpum S. demissum S. guerreroense S. hougasii S. iopetalum S. schenckii

Origins of Cultivated Potatoes and Species Relationships

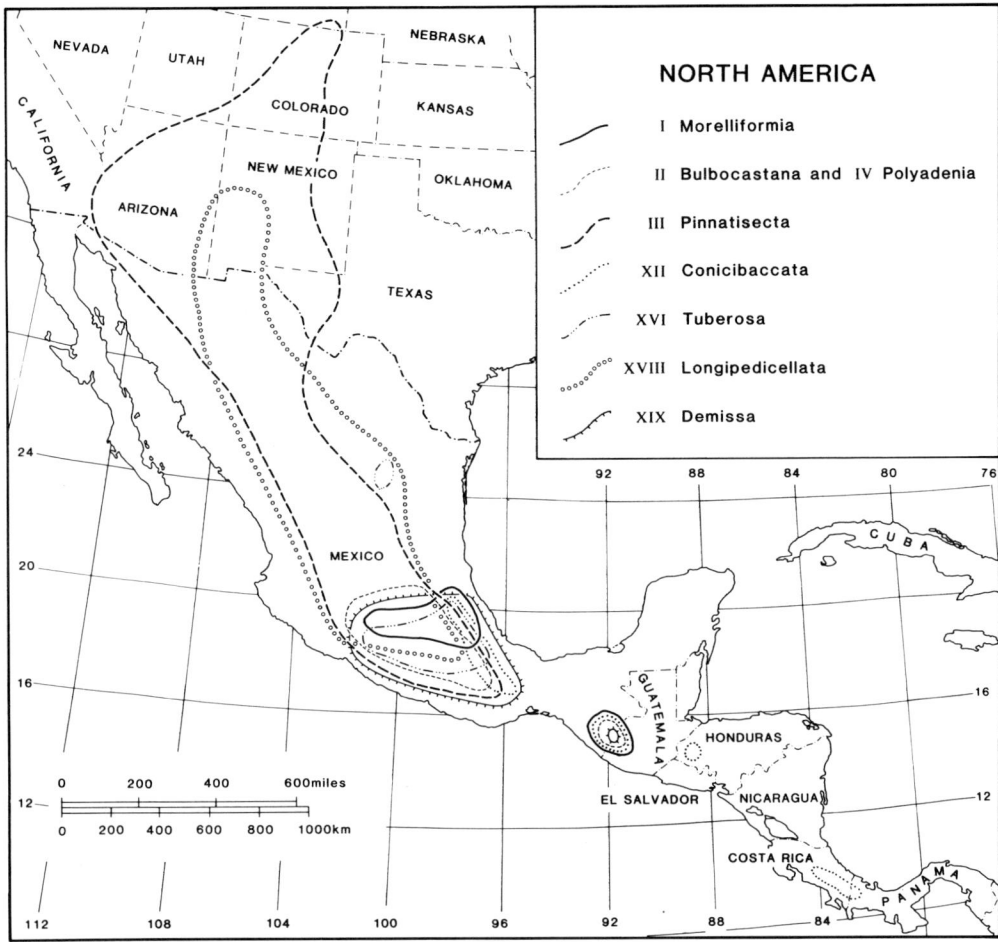

Fig. 1.4. Distribution of wild potato series in North and Central America (from Hawkes, 1990, with kind permission of Belhaven Press).

found in cool temperate rain forest (e.g. *S. violaceimarmoratum* and *S. colombianum*), and others on the coastal plains of Argentina and surrounding countries (*S. commersonii* and *S. chacoense*).

In Mexico and the USA wild potatoes are commonly found in scrub and cactus deserts, as well as juniper and scrub pine regions (e.g. *S. stoloniferum* and *S. jamesii*). Higher up, they inhabit mesic to cool temperate pine and *Abies* forests (e.g. *S. brachycarpum*, *S. demissum* and *S. verrucosum*), whilst in South America, where pines and *Abies* are not native, wild potatoes often inhabit *Podocarpus–Alnus–Polylepis* woodlands (e.g. *S. vernei* and *S. microdontum*).

There is thus an extraordinarily wide range of habitats in which wild potatoes are found, and this underlines the way in which they have become adapted to stress environments and have developed strong resistances to a

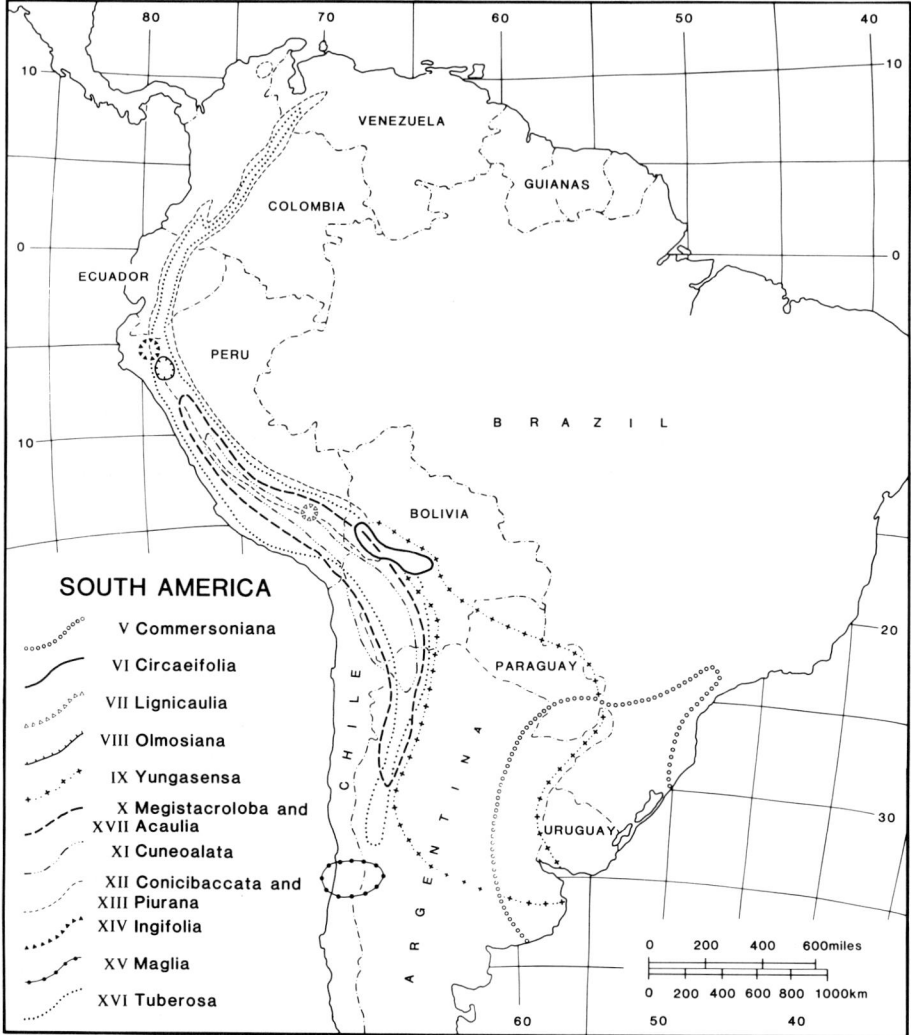

Fig. 1.5. Distribution of wild potato series in South America (from Hawkes, 1990, with kind permission of Belhaven Press).

wide range of pests and diseases. On the other hand, cultivated potatoes have evolved under a very limited range of environmental conditions in cool temperate regions and are thus often unable to resist the attacks of pests and diseases occurring over the much wider range of conditions in which they are now cultivated. Thus a knowledge of the great ecogeographical range of wild species, contrasted with the originally very narrow range of cultivated ones, can assist in an understanding of the need for the use of wild species in modern potato breeding.

The Systematics of Wild Potatoes and Their Close Relatives

Wild and cultivated plants are traditionally described and named by means of a study of their morphological diversity. This is normally referred to as 'alpha taxonomy'. Other features, such as interspecific crossability, chromosome number, chromosome morphology, chromosome pairing in natural and artificial hybrids, biochemical markers such as protein electrophoresis, isozyme analysis, serological relationships, presence of species-specific compounds such as alkaloids, etc., have also been widely used to distinguish species or highlight intraspecific differences. Recently, molecular techniques, such as RFLP analysis and DNA fingerprinting, have been found to be of great importance. All such methods, which work towards what is termed 'omega taxonomy', have often been found to be of immense value in underlining or casting doubt on the results of the more traditional techniques mentioned above and in suggesting relationships of value in choosing initial plant breeding materials.

The taxonomic system used in this chapter is based primarily on morphological characters, as well as to a great extent on cytological and crossability features.

Clearly, there are far too many wild potato species to be considered here, and, in any event, a large proportion are not well known and are of limited interest to breeders at present, though they might well be important in the future. To take a particular example, the potatoes of Bolivia are generally very well known in gene banks and other living collections (Hawkes and Hjerting, 1989). Of the 32 species recognized by these authors, 23 are of great to very great value to breeders, whilst nine are at present insufficiently known in the living state for much value to be ascribed to them yet. This remarkably high proportion may not be typical of the wild potatoes of other countries, but it certainly also applies to those of Argentina (Hawkes and Hjerting, 1969), and of Mexico and elsewhere (Flanders *et al.*, 1992).

For economy of space and bearing in mind the interests of potato breeders, I have set out in Table 1.1 those species most likely to be referred to in the literature, adding some lesser-known ones here and there because of their value in establishing probable evolutionary sequences.

It will be seen that most species in this list are diploids ($2n = 2x = 24$) but autotriploids have been discovered in some species (Tarn and Hawkes, 1986), formed presumably by the fusion of n and $2n$ gametes. The $2n$ gametes have been found in nearly every potato species examined (den Nijs and Peloquin, 1977a, b), but it would seem that they are generally eliminated. However, the present knowledge of triploid occurrences may be false. In the days when the quarantine authorities allowed us to import tubers, triploids were often found. Now that tubers are not allowed and since the seeds from obviously fertile plants are even-numbered polyploids, the sterile triploids and pentaploids are left behind and remain undiscovered.

In view of the fact that $2n$ gametes are quite frequent, it is surprising that more tetraploids are not found. There is clearly a partial $2n$ gamete block somewhere in the system, even though some tetraploids occur. Those in series Conicibaccata, Piurana, Acaulia and Longipedicellata are clearly disomic allotetraploids, with regular bivalent pairing. Some of those in series Tuberosa behave as tetrasomic autotetraploids, such as *S. tuberosum*, tetraploid races of *S. gourlayi* and *S. sucrense* (but see Astley and Hawkes, 1979), whilst others, such as *S. oplocense*, function as allotetraploids (Hawkes and Jackson, 1992).

The hexaploids in series Conicibaccata, Tuberosa, Acaulia and Demissa are disomic allohexaploids, though in no instance has it yet been possible to discover their exact ancestry. Triploids and pentaploids with '×' placed before the specific epithet are naturally occurring hybrids and their origins are known with some certainty.

Evolutionary Relationships

Morphological and other considerations

There is much to be said for regarding the Mexican and Guatemalan *S. morelliforme* in series Morelliformia as the most primitive of the tuber-bearing species, with its simple leaves, delicate habit and small stellate flowers (but see Spooner *et al.*, 1991a; later in this chapter). Its epiphytic habit is possibly a refuge from competition with more successful terrestrial species. It possesses the simplest immunological spectrum of any species against antiserum with *S. tuberosum*, consisting of only one line, whereas other species show two lines (series Pinnatisecta), four lines (series Tuberosa, Demissa, Longipedicellata and Polyadenia) or five lines (series Commersoniana).

The species most similar in flower type to *S. morelliforme* are those in series Bulbocastana and Pinnatisecta, with a small white star-shaped (stellate) corolla, also occurring in Mexico. Not far off morphologically is the Mexican series Polyadenia, although in this series the corolla often becomes expanded to pentagonal and the immunological spectrum is four lines against *S. tuberosum* antiserum rather than two (see section on comparative immunology later). All these species possess an endosperm balance number (EBN) of 1, the nature and significance of which will be discussed below.

In addition to these species from Mexico, parts of Central America and the southwest USA, there are other groups in South America with the white star-shaped corolla and an endosperm balance number of 1. These are series Commersoniana, Circaeifolia and Lignicaulia. A fourth series, Olmosiana, possesses the same corolla type but has not yet been examined for its endosperm balance number.

It has been suggested elsewhere (Hawkes, 1988, 1990) that wild potatoes evolved first in Mexico and later migrated to South America in the early Pliocene, when the Panama isthmus was formed, some 3½ million years ago. The series Commersoniana, Circaeifolia, Lignicaulia and Olmosiana are considered to be remnants of this southwards migration, all with EBN = 1.

In Argentina and adjoining countries we find a group of white stellate-flowered species, series Yungasensa, possibly derived from series Commersoniana, but now with an EBN of 2. Why or how this change occurred we do not know, but it has the effect that crosses are extremely difficult between diploid species with EBN of 1 and diploid species with an EBN of 2 (Johnston and Hanneman, 1978, 1980a, b, 1982; Hanneman, 1989). However, if 1 EBN species are tetraploidized, giving rise to 2 EBN progeny, the crosses can be effected with comparative ease.

Not only do series Yungasensa diploid species possess an EBN of 2, but all diploids in series Megistacroloba, Cuneoalata, Conicibaccata, Piurana and Tuberosa so far investigated (with two or three exceptions only) also possess an EBN of 2 (see Tables 1.2 and 1.3).

Allotetraploid species which possess bivalent pairing also possess an EBN of 2, such as those in series Conicibaccata, Piurana, Acaulia and Longipedicellata. Those which act as autotetraploids, such as *S. tuberosum*, *S. gourlayi* (4*x*) and *S. sucrense*, possess an EBN of 4, and can cross readily with the allohexaploid species in series Demissa – hence the successes obtained by breeders in crossing *S. tuberosum*, by *S. demissum* when transferring *Phytophthora* resistance to the cultivated potato. Results with the tetraploid and hexaploid *S. oplocense* and the hexaploid *S. albicans* are not so clear.

The nature of the EBN mechanism is not well understood at present, but Hanneman *et al.* (1990) found that it seemed to be under genetic control and was dosage-dependent. Hawkes and Jackson (1992) showed that the evolution of a rotate corolla seemed to be correlated with 2 EBN and that tetraploids with 2 EBN were allotetraploids, but those with 4 EBN were cytological autotetraploids. It is further suggested that the 2 EBN condition evolved as an isolating mechanism in the evolution of Series Yungasensa species from series Commersoniana species in what is now Argentina.

Returning to the main theme of wild potato evolution, it has been suggested that series X (Megistacroloba) up to and including series XVI (Tuberosa) and XVII (Acaulia) were derived ultimately from series Yungasensa, some of them developing allotetraploids, either sporadically (series Conicibaccata, Piurana and Tuberosa) or entirely (series Acaulia).

A possible scenario for wild potato evolution is suggested in Figure 1.6. Series Yungasensa is considered to be the 2 EBN 'advanced Stellata' (star-shaped flowers), derived from the 'primitive Stellata' with 1 EBN. In turn the 'advanced Stellata' gave rise in Argentina, Bolivia and southern Peru to the 'primitive Rotata' with pentagonal corolla and broadly triangular lobes (series Maglia, Megistacroloba, Cuneoalata, Conicibaccata (in part) and Tuberosa (southern species)).

Series Conicibaccata possesses diploid species in the south (northern Bolivia and Peru), but has become entirely tetraploidized in Colombia and Venezuela. This series evidently effected a return migration through Central America (*S. longiconicum*) into Mexico (*S. agrimonifolium* and *S. oxycarpum*) of wholly tetraploid species. When this took place we have no idea.

Another return migration, or perhaps two, of South American species into Mexico also took place. Thus the allotetraploid series, Longipedicellata species

Table 1.2. Potato (*Solanum*) species grouped by endosperm balance number (EBN) and ploidy levels (from Hawkes and Jackson, 1992 (based on Hanneman and Bamberg, 1986, and Chávez *et al.*, 1988a, b), with kind permission of the authors and *Theoretical and Applied Genetics*).

Geographical distribution	1 EBN	2 EBN		4 EBN	
	2x	2x	4x	4x	6x
United States	jamesii		fendleri		
Mexico	brachistotrichum bulbocastanum cardiophyllum jamesii pinnatisectum trifidum	verrucosum	agrimonifolium fendleri hjertingii oxycarpum papita polytrichon stoloniferum		brachycarpum demissum guerreroense hougasii iopetalum
South America	brevidens capsicibaccatum circaeifolium chancayense commersonii etuberosum fernandezianum lignicaule mochiquense	abancayense amabile acroglossum ambosinum berthaultii boliviense brevicaule bukasovii canasense chacoense chomatophilum gandarillasii gourlayi (2x) huancabambense infundibuliforme kurtzianum laxissimum leptophyes marinasense medians megistacrolobum microdontum multidissectum multiinterruptum pampasense pascoense phureja raphanifolium sanctae-rosae sogarandinum sparsipilum spegazzinii stenotomum tarijense venturii violaceimarmoratum weberbaueri	acaule colombianum sucrense (?) tuquerrense	gourlayi (4x) sucrense (?) andigena tuberosum	albicans moscopanum oplocense (6x)

Table 1.3. Taxonomic groupings of Solanum section Petota, showing ploidy levels, endosperm balance number (EBN) and geographical distribution (from Hawkes and Jackson, 1992, with kind permission of Theoretical and Applied Genetics).

Subsections	Superseries (corolla groups)	Series	Ploidy[1]	EBN	Geographical distribution
Estolonifera		Etuberosa	2x	1	South America
		Juglandifolia	2x	?	
Potatoe	Stellata (primitive)	Morelliformia	2x	1	Southwestern USA, Mexico, Central America
		Bulbocastana	2x	1	
		Pinnatisecta	2x	1	
		Polyadenia	2x	1	
		Lignicaulia	2x	1	South America
		Circaeifolia	2x	1	
		Commersoniana	2x	1	
		Olmosiana	2x	?	
	Stellata (advanced)	Yungasensa	2x	2	South America
	Rotata (primitive)	Cuneoalata	2x	2	Southern to central regions of South America
		Megistacroloba	2x	2	
		Southern forms of Tuberosa	2x, 4x, 6x	2, 4	
		Southern forms of Conicibaccata	2x	2	
	Rotata (advanced)	Piurana	2x, 4x	2	Central to northern regions of South America
		Ingifolia	2x	?	
		Acaulia	4x, 6x	2, 4	
		Central and northern forms of Tuberosa and Conicibaccata	2x, 4x, 6x	(1), 2, 4	
		Longipedicellata	4x	2	Southwestern USA, Mexico, Central America
		Demissa	6x	4	
		Mexican forms of Conicibaccata	4x	2	

[1] Not including odd polyploid cytotypes found in some species.

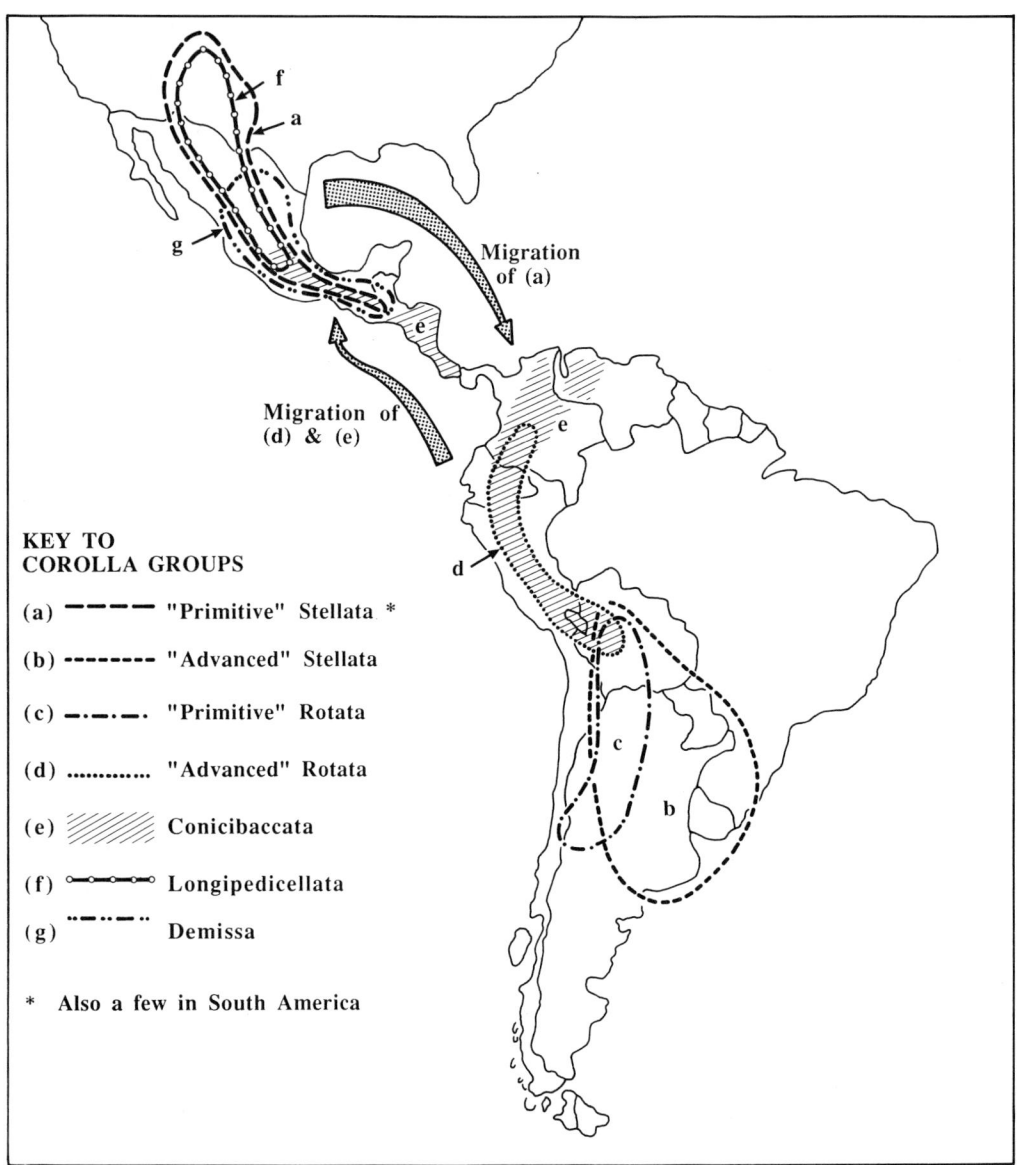

Fig. 1.6. Probable migration routes and geographical distribution of corolla groups in potato species (modified from Hawkes, 1988).

possesses the A genome of *S. chacoense* (series Yungasense), according to Matsubayashi (1955), and all species in this series are closely similar genetically (Hawkes, 1966). The origin of the other genome in these species is unknown, but one or more of the Mexican 1 EBN diploids could be considered as B genome donors.

Finally, we come to the Mexican series Demissa allohexaploids, of which the best known is *S. demissum* itself. Hawkes (1956) found that the F_1 hybrids of all Mexican hexaploids so far crossed with each other were sterile, and this was explained by Marks (1955), who showed that the hybrids contained two sets of homologous chromosomes from each parent, the third set being non-homologous. One of the homologous sets may have been *S. verrucosum*, but what the others were has not been answered.

We have not yet mentioned in this discussion the position of the series Etuberosa species so far investigated, namely *S. etuberosum*, *S. brevidens* and *S. fernandezianum*. These have been shown to be 1 EBN species (Johnston and Hanneman, 1978, 1980a, b, 1982), which will cross to some extent with other 1 EBN species, but rather rarely with any others. Since they possess rotate pigmented corollas they must be somewhat far removed from all the Mexican species, which are generally thought of as nearer the ancestral wild potatoes. It is possible that series Etuberosa species may have been derived from section Basarthrum, which also has an articulated (jointed) pedicel as have all the section Petota species. The non-tuber bearing section Basarthrum (not mentioned in Table 1.1) possesses a basal articulation, and it is interesting to note that the pedicel in series Etuberosa is also very low. These two groups may possibly possess a common ancestor, but, if so, this must have been a very long time ago.

Crossability and compatibility

Most wild and cultivated potato species can be hybridized together, though with some exceptions. When hybrids can be made, they are generally fertile at the same level of ploidy. Tetraploids with disomic inheritance can be crossed together also, but not so easily with *S. tuberosum*, which possesses tetrasomic inheritance. The Mexican hexaploids can be crossed, but their F_1 hybrids are sterile, evidently because they do not possess at least one genome in common (Marks, 1955). Diploid × tetraploid crosses are easy if the species are both disomic (as with *S. chacoense* × *S. stoloniferum*), although the hybrids are of reduced fertility. Diploid (disomic) × tetraploid (tetrasomic) crosses are more difficult, sometimes producing tetraploid progeny through the functioning of $2n$ gametes. This phenomenon will be discussed in other chapters.

By and large, potato breeders have at their disposal a very wide range of germplasm, and this is partly due to the rather low level of genome differentiation in most species. Potato species seem to have evolved by means of geographical and ecological isolation rather than by genetic incompatibility, though some exceptions are known. With the occurrence of man-made changes in habitat, many previously isolated species have come together and formed natural hybrids. There is some evidence for the introgression of genes from one species to another by this means (Hawkes, 1962b), but the hybrids themselves tend to disappear in a few years, evidently because they lack an integrated set of adaptive complexes from one or other of the parent species (Hawkes and Hjerting 1969: 459–460).

Most diploid species are self-incompatible, due mainly to S alleles, which cause arrested pollen growth. These species are therefore obligate outbreeders. Exceptions to this rule are seen in *S. brevidens* and *S. etuberusom* (series Etuberosa), *S. morelliforme* (series Morelliformia), *S. polyadenium* (series Polyadenia) and *S. verrucosum* (series Tuberosa). Tetraploid and hexaploid species seem all to be self-compatible, but there are indications of inbreeding depression after a few generations of selfing in those species investigated.

Genome relationships

On the whole, there seems to be a somewhat poorly developed genome system in potatoes. Many years ago it was suggested (Hawkes, 1958) that series Tuberosa and Yungasensa (then included in series Commersoniana) could be allocated a genome formula of A_1A_1 for diploid species, and this formula could probably be extended to series Megistacroloba and Cuneoalata, as well as at least the diploid species of series Conicibaccata and Piurana. Series Acaulia was allotted a formula of $A_2A_2A_3A_3$, noting, however, that the differences between A_1, A_2 and A_3 were rather slight. Thus, even though *S. acaule* behaves as an allotetraploid with regular bivalent pairing, its triploid hybrid with diploid A_1 species behaves like an autotriploid. It is assumed that when homologues are present they pair normally at the expense of homoeologues, but when absent they associate at random. In the same publication (Hawkes, 1958), the Mexican diploids were given a B genome formula and the tetraploids in series Longipedicellata an A_4B formula. This was because hybrids between $4x$ *S. stoloniferum* (series Longipedicellata) and the A_1A_1 *S. chacoense* (now in series Yungasensa) give approximately $12_{II} + 12_I$ at meiotic metaphase. The situation in the series Demissa hexaploids of Mexico is not so clear, but the AB genome still seems to exist in them, as we shall see later in this chapter.

Matsubayashi (1981) proposed a more complex system for South American species: all A genome but each series with different superscripts. He agreed with the AB formula for the Mexican series Longipedicellata but gave an ACC formula for the hexaploid *S. demissum*. However, in a later publication, Matsubayashi (1991) gave an AACC formula to the tetraploid Mexican and Central American Conicibaccata species, *S. agrimonifolium*, *S. oxycarpum* and *S. longiconicum,* and an AADDDD formula for the Mexican hexaploids, the third D genome being distinguished by different superscripts. This replaces the earlier formula of AACCCC, because in his 1991 publication Matsubayashi reserves the C genome for the AACC Conicibaccata, even though it seems to the present writer that at least some of the series Demissa species contain Conicibaccata elements. The origin of the Matsubayashi D genome is not given, and there is thus still much doubt about the evolutionay origins of these hexaploids. Matsubayashi allocates an E genome formula to series Etuberosa. We shall see later that these results, based on meiotic pairing studies, do not agree very well with chloroplast DNA results (see, for instance, Spooner *et al.,* 1991a).

Finally, López and Hawkes (1991) suggested genome formulae for diploid, tetraploid and hexaploid species in series Conicibaccata, all variants of a hypothetical X genome (possibly equivalent to the A genome) but with a distinct M genome in the hexaploid *S. moscopanum*.

From a conventional cytological point of view this seems to be as far as one can go, but the genome debate will be taken up again in the section on RFLP analysis.

Biochemical systematics

Flavonoids

A systematic review of leaf flavonoids was carried out by Wietschel and Reznik (1980a, b). More complex patterns were noted in *S. bulbocastanum* than in *S. morelliforme* and *S. clarum*, whose patterns were almost identical, thus adding weight to the suggestion that *S. clarum* should be included in series Morelliformia. Complex tetraglycosides were observed in series Conicibaccata and one or two series Tuberosa species. The authors conclude that series Tuberosa can be placed at the end of a developmental line for South American species, but even series Pinnatisecta in Mexico shows considerable complexity of glycosides, whilst series Bulbocastana, Circaeifolia and Cuneoalata lack much of this diversity.

Protein electrophoresis

The investigation of affinities by protein electrophoresis has been attempted by several investigators. Thus Desborough and Peloquin (1969) analysed 22 species in acid gel systems. Apart from the fact that *S. lignicaule* and *S. marinasense* showed very distinctive patterns, no taxonomic deductions could be drawn from these results. However, interesting results were obtained by Rickeman and Desborough (1978), who compared the cultivated *S. tuberosum* subsp. *tuberosum* with *S. tuberosum* subsp. *andigena*, *S. stenotomum* and *S. phureja* by polyacrylamide gel electrophoresis. The results showed that: (i) there were close affinities between *S. stenotomum* and *S. phureja*, lending weight to the hypothesis that the latter may have been derived from the former; (ii) subsp. *andigena* was derived from a cross between *S. stenotomum* and a wild diploid species; (iii) subsp. *tuberosum* was derived from subsp. *andigena*. Unfortunately, they gave no opinion on what wild species might have been involved in subsp. *andigena* origins.

Polyacrylamide gel electrophoresis was also used by Cribb and Hawkes (1986) to investigate the origin of *S. tuberosum* subsp. *andigena*. They showed that two protein bands were unique to *S. stenotomum* and one to *S. sparsipilum* – the two putative ancestral species. None was unique to subsp. *andigena*. Four bands were common to *S. sparsipilum* and subsp. *andigena* but not to *S. stenotomum*, whereas one was found in *S. stenotomum* and subsp. *andigena* but not in *S. sparsipilum*. Ten artificial amplidiploid hybrids of the two diploid species were compared with natural subsp. *andigena*, of which six could not be distinguished from it. Three differed by one band only. This would strongly reinforce the hypothesis of subsp. *andigena* being an amphiploid hybrid of *S. stenotomum* and *S. sparsipilum*. However, one has to set against these conclusions the results of Gatenby and Cocking (1978), who examined the electrofocusing pattern of fraction 1 protein from *S. sparsipilum*, *S. stenotomum* and *S. tuberosum* subsp. *andigena*. They found that the fraction 1 protein of the two cultivated species was identical, but the small subunit of *S. sparsipilum* was not seen in either of them. There were differences also between subsp. *andigena* and subsp. *tuberosum*. This brings us to mention the hypothesis set out by Grun (1990) in which he thinks it possible that subsp. *tuberosum* evolved from a cross between subsp. *andigena* as staminate parent and an as yet unidentified wild female parent which contributed cytoplasmic sterility factors that do not occur in subsp. *andigena*. It is hypothesized that *S. maglia* or *S. chacoense* might have played a role here, though this has not yet been substantiated.

Clausen and Okada (1990) analysed by tuber protein electrophoresis a group of rather 'difficult' wild potato species in northern Argentina, including diploid and tetraploid *S. gourlayi*, *S. vidaurrei*, *S. incamayoense* and *S. infundibuliforme*. On the whole, the 4x and 2x forms of *S. gourlayi* formed distinct groups; these were close to but distinct from *S. vidaurrei*, which has been considered as a subspecies of *S. gourlayi* (Hawkes and Hjerting, 1989) in any case. *S. infundibuliforme* and *S. incamayoense* also separated quite nicely from the other two species. This seems to be a good vindication for the use of protein electrophoresis in separating the distinct components of a group of very similar wild potato species.

One can conclude from protein electrophoresis work that very broad surveys do not solve the whole problem of potato species relationships but that they are reasonably successful when groups of close similarities or affinities are investigated.

Isozyme and allozyme analyses

A study of esterase isozymes from potato tubers was reported by Desborough and Peloquin (1967) for 48 wild and four cultivated species, as well as several interspecific hybrids. Although the patterns were, in general, not species-specific, where the number of isozymes was five or less similar patterns were observed, as in *S. pinnatisectum*, *S. bulbocastanum*, *S. canasense*, *S. demissum*, *S. fendleri*, *S. megistacrolobum*, *S. sparsipilum* and *S. phureja*. No conclusions could be drawn, however, on species or series relationships.

Further work on isozymes was carried out by Hosaka and Matsubayashi (1983a), in which the peroxidase enzymes were studied in diploid species of series Commersoniana, Circaeifolia, Conicibaccata, Cuneoalata, Megistacroloba and Tuberosa in South America and in the Mexican series Morelliformia, Bulbocastana, Pinnatisecta and Polyadenia. The results, from a study of 22 bands, indicated that all South American species belonged to the A genome group and only series Conicibaccata could be distinguished from the rest by possessing band 8. Both *S. morelliforme* and *S. clarum* had very weak bands but neither coincided closely with *S. bulbocastanum*. In the scatter diagrams, *S. capsicibaccatum* was closely associated with *S. bulbocastanum*, *S. brachistotrichum* and *S. polyadenium*, an interesting fact which tends to strengthen the present writer's hypothesis of its evolutionary relationships.

In a second paper (Hosaka and Matsubayashi, 1983b) the same methods were used to classify diploid cultivated potatoes, leading them to form three subspecies (*stenotomum*, *phureja* and *goniocalyx*) of *S. stenotomum* and to place *S. ajanhuiri* outside this group – a reasonable conclusion that accords well with current views.

Another work of considerable interest deals with the hitherto rather neglected series Etuberosa (Spooner and Douches, 1991; Spooner *et al.*, 1992). Thirty-two populations of *S. brevidens*, *S. etuberosum* and *S. fernandezianum* were subjected to allozyme electophoresis. The first two species are from southern Chile (and Argentina) and the last from the Juan Fernandez Islands, 600 km west of the Chilean mainland. All are self-compatible and diploid. The results show similar mean genetic identities in *S. brevidens* and *S. etuberosum*. *S. fernandezianum* shows much less genetic diversity, possibly owing to founder effects and isolation. The series Etuberosa exhibits very low within-population diversity contrasted with other diploid species of wild potato, probably due to inbreeding. It is clear from these results that series Etuberosa represents a very distinct group of species, rather closely related to each other genetically.

Comparative immunology

A valuable understanding of species relationships was obtained from immunological studies of tuber proteins (Gell *et al.*, 1956, 1960). The authors reported that protein extracts from different species run by double diffusion against *S. tuberosum* antiserum formed three groups. A four-band spectrum included *S. tuberosum* itself and species in series Longipedicellata and Demissa (also Bulbocastana, later found to be an error). A two-band spectrum was formed by species in series Pinnatisecta and a one-band spectrum by series Morelliformia (*S. morelliforme*) (Figure 1.7). An antiserum raised from *S. cardiophyllum* subsp. *ehrengergii* showed up to six bands, but when absorbed with *S. tuberosum* extract it clearly differentiated *S. jamesii* and *S. pinnatisectum*, on the one hand, from subsp. *cardiophyllum* and its hybrid with *S. pinnatisectum* – *S.* × *sambucinum* – on the other. The results also demonstrated a closer relationship of *S. stoloniferum* and *S. polytrichon* to series Pinnatisecta, whilst series Demissa veered towards series Tuberosa, probably through the influence of *S. verrucosum*. Following this, Lester (1965) showed clear immunological differences between *S. commersonii* (series Commersoniana), on the one hand, and *S. chacoense* and *S. tarijense* (series Yungasensa), on the other, the latter being closer to *S. tuberosum*. Further work by Hawkes and Lester (1966, 1968), using double diffusion and

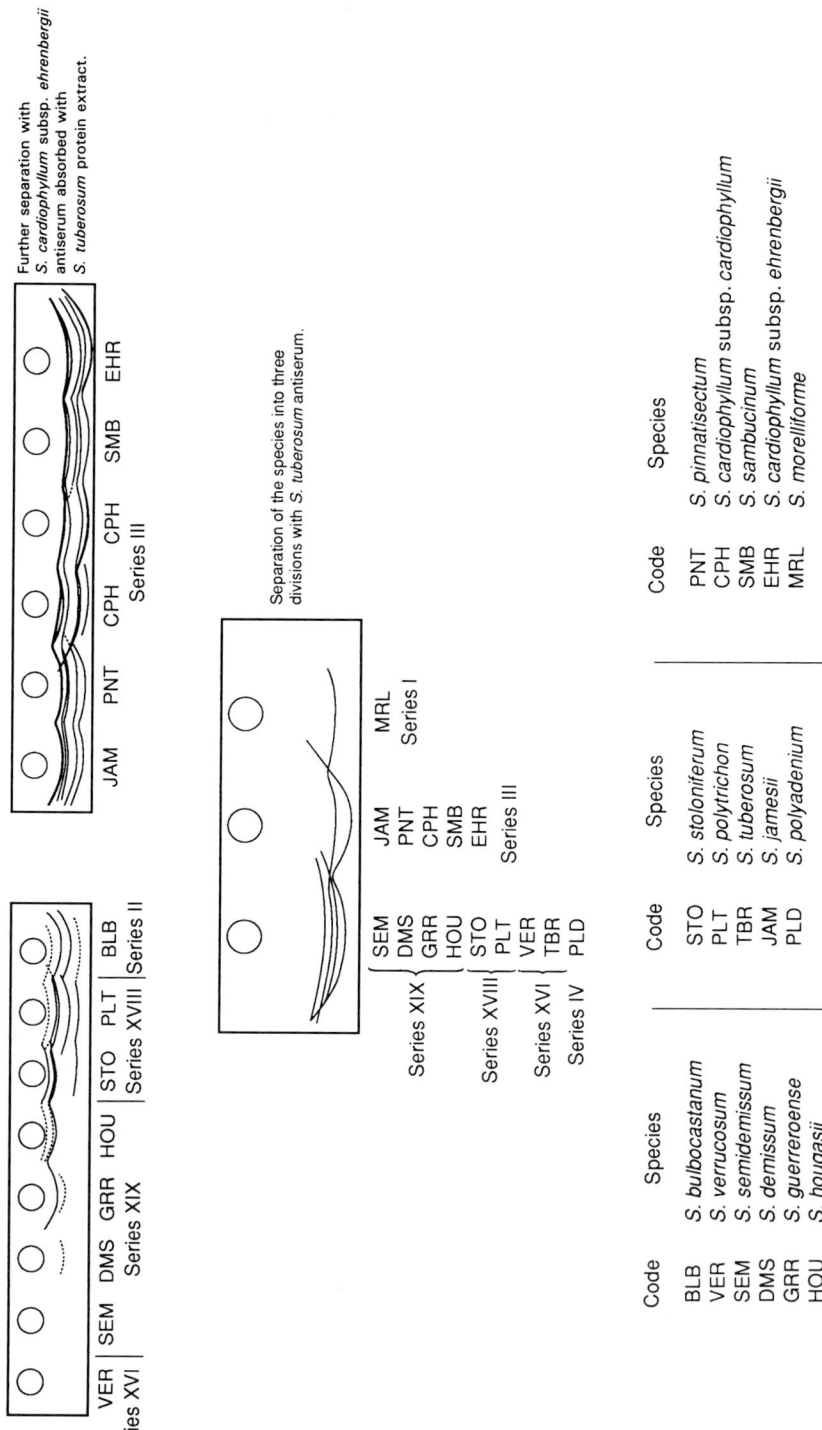

Fig. 1.7. Serological groupings of Mexican potato species obtained with *S. tuberosum* and *S. cardiophyllum* antisera (the latter absorbed with *S. tuberosum* protein extract). Series I: Morelliformia; series II: Bulbocastania; series III: Pinnatisecta; series IV: Polyadenia; series XVI: Tuberosa; series XVIII: Longipedicellata; series XIX: Demissa. (Modified from Gell *et al.*, 1960, with kind permission of the authors and the Royal Society.)

immunoelectrophoretic techniques indicated that the Mexican species *S. clarum* was more closely related to *S. bulbocastanum* and *S. cardiophyllum* than to *S. morelliforme*. *S. polyadenium* showed some relationships to series Bulbocastana and Pinnatisecta, on the one hand, and to series Longipedicellata and Demissa, on the other, seeming to be intermediate between the two groups (Figure 1.7). Further immunoelectrophoretic studies of *S. bulbocastanum* collections indicated clearly visible differences between subsp. *bulbocastanum*, subsp. *dolichophyllum* and subsp. *partitum*, with forms in southern Mexico veering towards the Guatemalan subsp. *partitum*. Finally, the nature of *S.* × *sambucinum* was investigated, showing by means of morphological and immunoelectrophoretic methods that this 'species' was in reality a natural hybrid of *S. pinnatisectum* × *S. cardiophyllum* subsp. *ehrenbergii*.

Molecular systematics

Much work has been published in the last ten years on species relationships as indicated by restriction fragment length polymorphisms (RFLPs) of chloroplast or nuclear DNA. The benefit of using chloroplast DNA is that it is easier to handle. On the other hand, it gives an idea of only those evolutionary pathways derived from female parents. This may not be of so much importance in broader perspectives but could be misleading when investigating possible species crosses, where, of course, the male parent cannot be identified.

Much of the early work in this field on potatoes was carried out by Hosaka and various colleagues. Particularly interesting results were obtained (Hosaka *et al.*, 1984) on 24 *Solanum* species in section Petota and two in *Lycopersicon*. These formed four distinct groups (Figure 1.8). One of these included South American species and Mexican polyploids. The second comprised the Mexican diploids. The third was represented by *S. etuberosum* and the fourth by *Solanum lycopersicoides* and two *Lycopersicon* species. It is quite clear from this that at least *S. lycopersicoides* in series Juglandifolia lies much closer to *Lycopersicon* (the tomato) than it does to the potato. Secondly, the Mexican diploids branch off from the same stock as *S. etuberosum* – though that species is genetically much further away. The rest are quite closely grouped, even including *S. commersonii* and *S. capsicibaccatum*. Thus the *S. etuberosum*/Mexican diploid group only partly coincides with the 1 EBN group discussed in an earlier part of this chapter. Nevertheless, it coincides remarkably well with many ideas on evolutionary relationships obtained from morphological, cytological and crossability studies.

Another very interesting study is that of Debener *et al.* (1990), using nuclear RFLPs, which might be more reliable than the chloroplast RFLP work. These authors used 34 accessions of 15 wild and two cultivated potatoes, including *S. etuberosum*, but unfortunately excluding series Morelliformia, Bulbocastana, Commersoniana (as at present defined), Circaeifolia, Lignicaulia, Cuneoalata, Conicibaccata, Piurana, Ingifolia and Maglia. However,

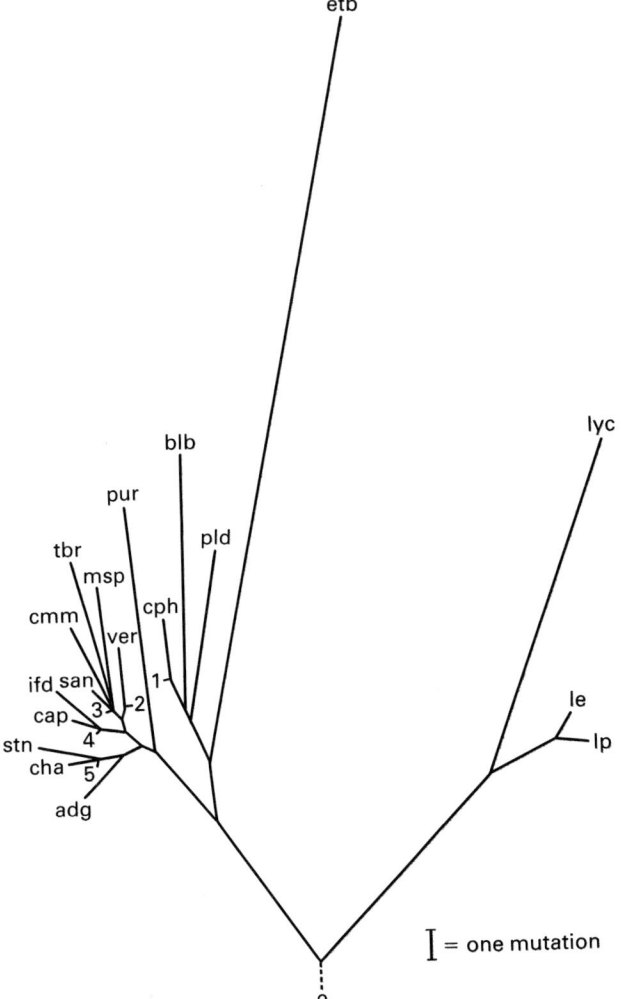

Fig. 1.8. Chloroplast genome relationships (from Hosaka et al., 1984, with kind permission of the authors and the *Japanese Journal of Genetics*).

they cannot be blamed too strictly since at that time none of the RFLP work had incorporated all these series, some of which were difficult to obtain. Two data sets were analysed, of which the second, in the form of an unrooted tree, is shown here (Figure 1.9). It is interesting to see again the relationships between series Etuberosa and the Mexican wild diploids (Pinnatisecta and Polyadenia series). The clustering of subsp. *andigena* with *S. tuberosum* (Bintje), *S. stenotomum* and *S. canasense* also agrees with current hypotheses. Unfortunately, the other possible ancestor of *S. stenotomum*, namely *S. leptophyes*, was not included. Unexpectedly, *S. sparsipilum* was quite isolated from the cultivated potato group – which calls into question its assumed partial ancestry of the tetraploid cultivated potato, *S. tuberosum* subsp. *andigena*.

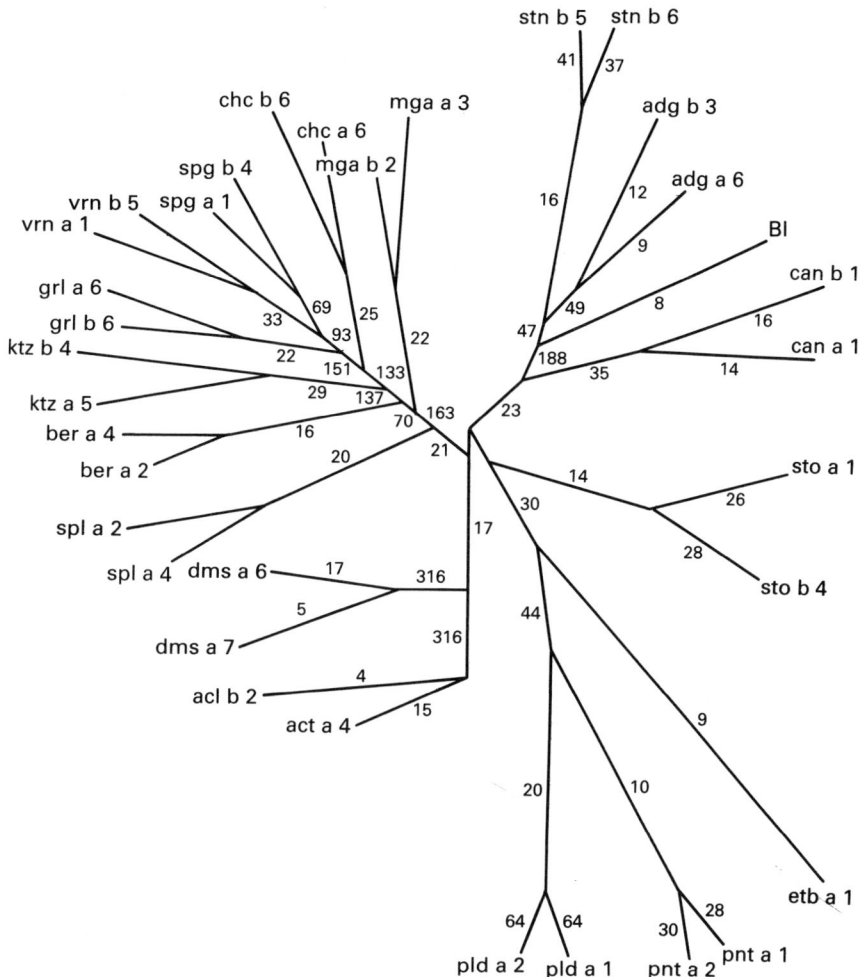

Fig. 1.9. Unrooted phylogenetic tree for certain wild and cultivated potato species. Key: acl = S. acaule; adg = S. tuberosum subsp. andigena; ber = S. berthaultii; Bl = Bintje; can = S. canasense; chc = S. chacoense; dms = S. demissum; etb = S. etuberosum; grl = S. gourlayi; ktz = S. kurtzianum; mga = S. megistacrolobum; pld = S. polyadenium; pnt = S. pinnatisectum; spg = S. spegazzinii; spl = S. sparsipilum; stn = S. stenotomum; sto = S. stoloniferum; vrn = S. vernii. a and b indicate different accessions. (From Debener et al., 1990, with kind permission of the authors and Theoretical and Applied Genetics.)

A good close clustering of the Argentinian and Bolivian diploid species in series Yungasensa, Megistacroloba and Tuberosa was to be expected. What stands out as very curious is the large distance between S. stoloniferum and S. demissum, since the latter is assumed to be derived partly from the former. Another curious feature is the closeness of series Acaulia to the geographically remote series Demissa – also the remoteness of series Acaulia from the Tuberosa group in spite of the evidence of the closely related A genomes of all of them.

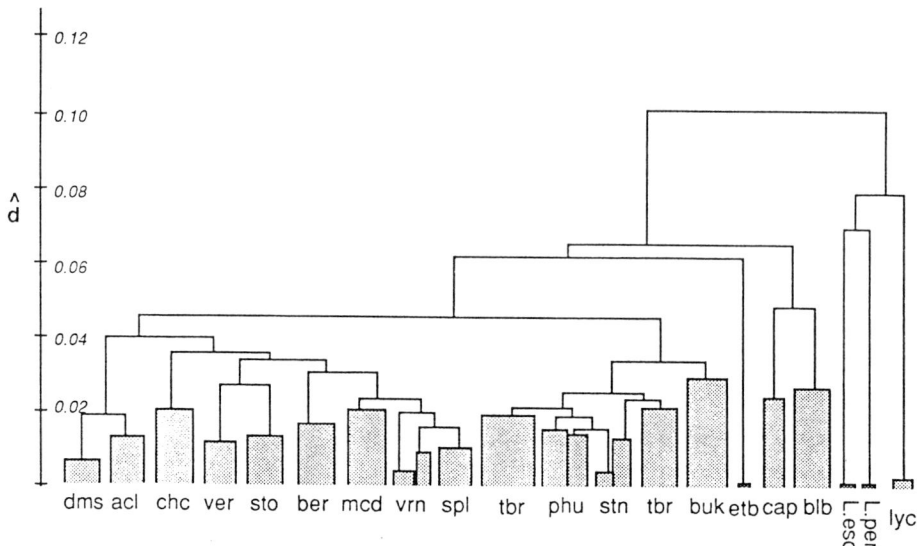

Fig. 1.10. Estimates of genetic distance (vertical axis) among 18 *Solanum* and two *Lycopersicon* species. Key: acl = *S. acaule*; ber = *S. berthaultii*; blb = *S. bulbocastanum*; buk = *S. bukasovii*; cap = *S. capsicibaccatum*; chc = *S. chacoense*; dms = *S. demissum*; etb = *S. etuberosum*; lyc = *S. lycopersicoides*; L. esc = *Lycopersicon esculentum*; L. pen = *Lycopersicon pennellii*; mcd = *S. microdontum*; phu = *S. phureja*; spl = *S. sparsipilum*; stn = *S. stenotomum*; sto = *S. stoloniferum*; tbr = *S. tuberosum*; ver = *S. verrucosum*; vrn = *S. vernei*. (From Bonierbale *et al.*, 1990, with kind permission of the authors and CAB International.)

Before we deal with the RFLP results concerning the cultivated potato, we shall continue to look at the wider spectrum of the wild species. In this context another nuclear DNA study by Bonierbale *et al.* (1990) is relevant (Figure 1.10). Again, the distant groupings of *Lycopersicon* and *Solanum lycopersicoides* show up clearly. Related rather closely are *S. capsicibaccatum* and *S. bulbocastanum* (both EBN = 1 species) and then *S. etuberosum*, which is also an EBN = 1 species. All the rest are clearly distinct from the above, and fall into two groups: (i) the cultivated species, including the wild *S. bukasovii*; and (ii) wild diploid series Tuberosa, Yungasensa, Acaulia, Longipedicellata and Demissa. Again, *S. sparsipilum* belongs to this group, and not, as one would have hoped, to the cultivated group.

The work of Spooner *et al.* (1991a), using chloroplast DNA RFLP methods, compares closely to the Bonierbale paper and very interestingly relates results to the genome formulae work described in an earlier section (see Figure 1.11). As usual, series Etuberosa (as clade 1) stands out clearly from the rest, and is given the genome formula D. Clade 2 includes Polyadenia, Pinnatisecta and Morelliformia, and the B genome formula for these species is confirmed. Clade 3 - Bulbocastana - also genome B - is somewhat separated from clade 2. Clade 4 includes series Tuberosa, Conicibaccata, Longipedicellata (AB) and Demissa (AAB or ABB). This is to be expected and confirms previous work using other techniques. What is curious, however, is

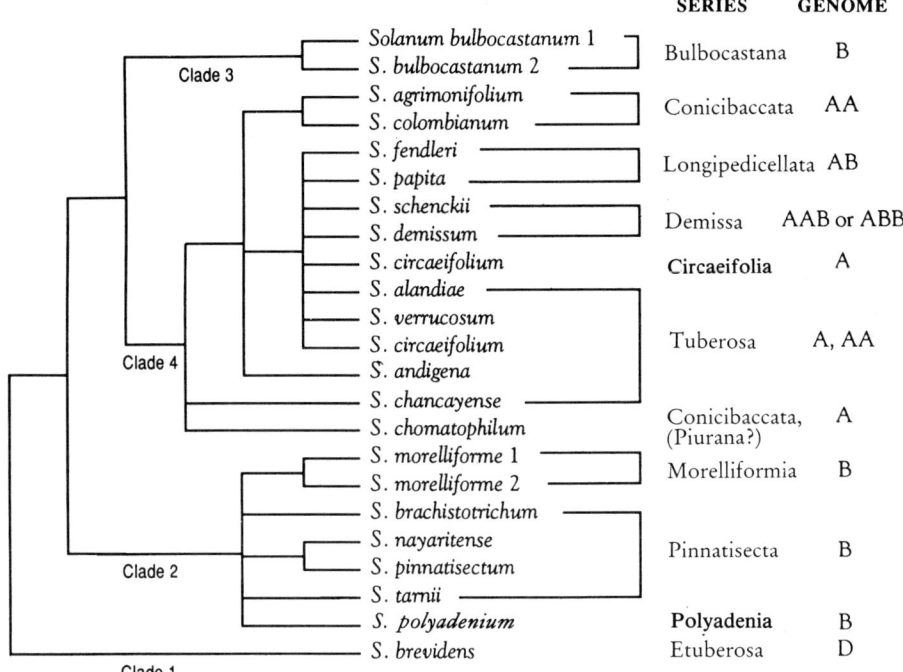

Fig. 1.11. Strict consensus tree of 20 equally parsimonious 124-step Wagner trees with genome designations following Hawkes (1990) (from Spooner et al., 1991a, with kind permission of the authors and the *American Journal of Botany*).

that *S. circaeifolium* is included in this clade and not (as Bonierbale's results indicate) related to *S. bulbocastanum* in a quite separate group. Further work on *S. circaeifolium* is therefore needed.

Spooner and Sytsma (1992) continue the work reported above, but add very many more species and accessions (see Figure 1.12). Series Etuberosa is again clearly separated from the other species at the bottom of the diagram. Clade 1 is similar to the Spooner *et al.* (1991a) paper except that more species are added and *S. clarum* appears close to *S. morelliforme*. Close similarities between certain species accord well with present taxonomic judgements. Clade 3 shows a distinct grouping of the tetraploid series Conicibaccata species and another grouping of the Peruvian *S. chancayense* with the anomalous *S. chomatophilum*. Very curiously, *S. stoloniferum* is separated from the rest of its series Longipedicellata. Because *S. stoloniferum* is very widespread and variable, the fact that only one accession is included in this study is unfortunate. Apart from this, the rest of the Longipedicellata and six series Demissa species are all closely grouped, though it seems most curious that the quite distinct *S. circaeifolium* species from Bolivia is included in the same assemblage. Finally, clade 2 includes the morphologically highly disparate species *S. bulbocastanum* and *S. cardiophyllum* – a real enigma at present.

To terminate this section on RFLP studies of wild potatoes, we should

32   J.G. Hawkes

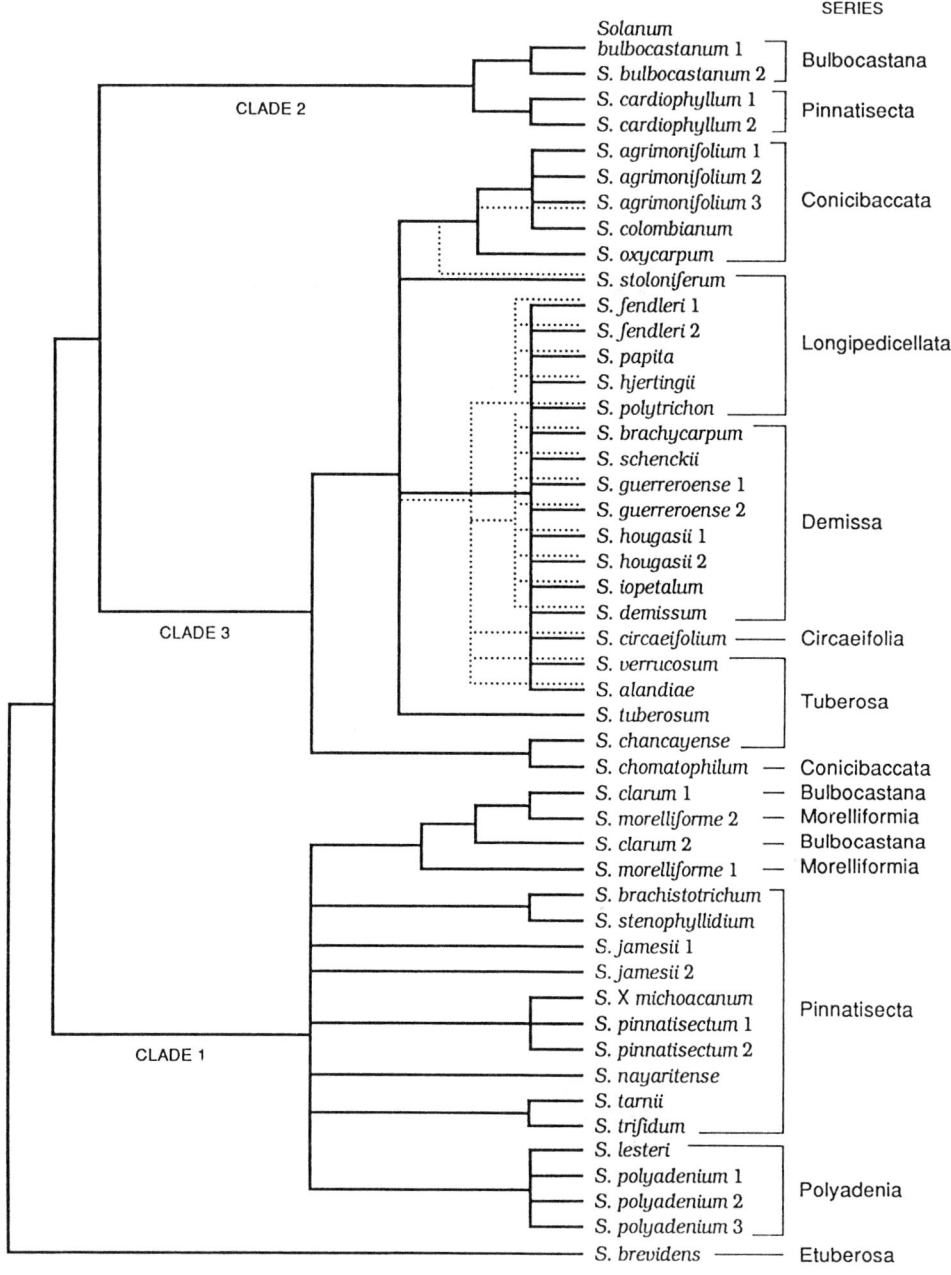

Fig. 1.12. Strict consensus tree (solid lines) of 2107 equally parsimonious 152-step Wagner trees, with genome designations following Hawkes (1990) (from Spooner and Sytsma, 1992, with kind permission of the authors and *Systematic Botany*). Dotted lines indicate the strict consensus tree where it differs from the Wagner tree.

mention a paper by Ugent (1970) in which the wild Peruvian potato *S. raphanifolium* is concluded to be a stabilized hybrid of *S. megistacrolobum* × *S. canasense*. The present author has unpublished work indicating that this hypothesis is incorrect. A recent publication by Spooner *et al.* (1991b) shows by means of chloroplast DNA hybridization techniques that Ugent's hypothesis is not correct and that *S. raphanifolium* is an independent species in its own right.

We must now turn to the very interesting work by Hosaka and colleagues concerning the relationships of *S. tuberosum* and its subspecies *andigena* to each other and to other cultivated species (Hosaka, 1986; Hosaka and Hanneman, 1988a, b; Hosaka *et al.*, 1988). These were all based on chloroplast DNA studies, mainly of cultivated species. The hypothesis clearly differentiates five main chloroplast DNA types (Figure 1.13), namely the A-type from subsp. *andigena* and *S. maglia*, the T-type from Chilean subsp. *tuberosum*, the S-type from *S. stenotomum* and other diploid cultivated species, as well as *S. curtilobum*, the C-type from wild Andean species, such as *S. acaule*, *S. bukasovii*, *S. multidissectum*, *S. canasense* and *S. juzepczukii*, and the W-type from a whole range of Argentinian and Bolivian species. As will be seen from Figure 1.13, the T-type preponderates in southern Chile and is found to a small extent in Argentina but nowhere else. The A-type is found to the exclusion of the others in Mexico, Costa Rica and Ecuador, frequently in Colombia and in more than 50% of cases in Peru and Bolivia. On the basis of this work, Hosaka and colleagues discount the hypothesis put forward by Ugent *et al.* (1987) that the Chilean *S. tuberosum* was derived from *S. maglia*, since the latter possesses an A-type cytoplasm similar to that of subsp. *andigena*. Thus, although the first European potatoes would have possessed A-type cytoplasm (as Myatt's Ashleaf still does), nearly all were decimated by the late blight epidemic of the 1840s, to be replaced by the T-type cytoplasm from Chile derived from Goodrich's 'Rough Purple Chile' and its derivatives. It is possible that the T-type cytoplasm of the Chilean potato and modern varieties may have been derived from *S. chacoense* or some related species (see also Grun, 1979).

To sum up these previous sections, it is gratifying to discover that, by and large, the results from experimental studies correlate remarkably well with those based on morphology. It must be admitted here that the taxonomic scheme now proposed (Hawkes, 1990) uses experimental data wherever possible but then leaps forward to propose a possible scenario for the evolution of the tuber-bearing species and series Etuberosa. Recent RFLP results sometimes support this scenario and sometimes do not. At times, also, the results obtained by different workers contradict each other, as with, for instance, the position of *S. clarum* (series Morelliformia or Bulbocastana) and the position of series Circaeifolia (linked to the Mexican 1 EBN diploids or not).

Nevertheless, the taxonomic and evolutionary system now in use is capable of modification where necessary and meanwhile can be used to form a basis for continuing experimental work.

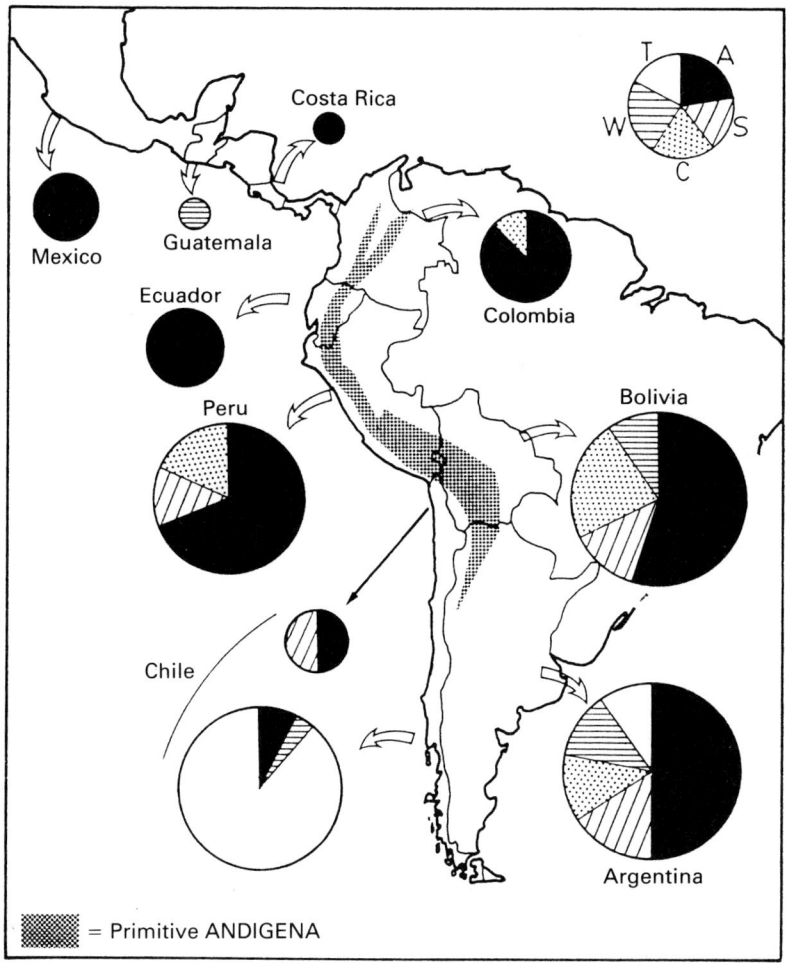

Fig. 1.13. Distribution of ctDNA types in *S. tuberosum*. The percentage of accessions of each ctDNA type are indicated by the area occupied in a given circle. The Chilean potatoes are shown by two groups, coastal, or southern Chilean, potatoes and those of the province of Arica, or northern Chile. For the former group subsp. *tuberosum* and subsp. *andigena* were combined and shown here. Key: T = *S. tuberosum*; A = *S. andigena/S. maglia*; S = *S. stenotomum, S. goniocalyx, S. phureja, S. chaucha, S. curtilobum*; C = *S. acaule, S. bukasovii, S. canasense, S. multidissectum, S. juzepczukii*; W = *S. chacoense, S. demissum, S. gourlayi, S. kurtzianum, S. leptophyes, S. microdontum, S. oplocense, S. sparsipilum, S. spegazzinii, S. sucrense, S. vernei*. (From Hosaka and Hanneman, 1988a, with kind permission of the authors and *Theoretical and Applied Genetics*.)

Value of Wild Species and Primitive Forms in Potato Breeding

The potato was probably the first crop plant in which breeding for disease resistance was attempted. This was largely due to the late blight disaster in Ireland during 1845 and 1846 and subsequent blight epidemics in Britain and Europe. It was thought in the 19th century that 'new blood' was needed to invigorate the crop, which was also subject to 'degeneration', due, as we now know, to virus infection. The well-known horticulturalist, John Lindley, introduced *S. demissum* from Mexico in 1846 but it seemed to show no *Phytophthora* resistance. The German scientist, Klotzsch, also introduced this species (as *S. utile*) in 1849, and some of his seedlings were resistant. These are thought to have given rise to the German 'W-races' through the F_1 hybrid of *S. tuberosum* × *S. demissum*, named *S. edinense* because it was first described from material in the Edinburgh botanic garden (Ross, 1958).

R.N. Salaman also found blight resistance in 1909 and 1910 in a similar hybrid (wrongly named '*S. etuberosum*'). From then until the late 1920s all potato breeding for resistance was directed towards *Phytophthora*, using *S. demissum* or its derivatives.

This was changed entirely in the 1920s and 1930s by the results of the N.I. Vavilov expeditions to the Americas, largely led by S.M. Bukasov and S.W. Juzepczuk (Bukasov, 1933) (for a more detailed account see Hawkes and Hjerting, 1989). The British Empire (later, Commonwealth) expedition in 1939, taking advantage of the pioneering Russian work, collected materials in Mexico and the Andean countries of South America. As a result of this expedition the Commonwealth Potato Collection (CPC) was established in Cambridge and is now situated at SCRI. One of the valuable results of this expedition was the discovery of eelworm (cyst nematode) resistance in the Argentinian wild species, *S. vernei*. The Russians had already discovered frost resistance in *S. acaule* and Colorado beetle resistance in *S. chacoense*, as well as resistance to various other pests and diseases (Bukasov and Lebedeva, 1955; Bukasov and Kameraz, 1959, 1973) and had examined species origins and relationships by means of hybridization studies (Bukasov, 1938, 1939a, b).

Potato gene banks, or germplasm collections, as they are often called, have been established, not only in Britain and Russia, as mentioned above, but also in Braunschweig, Germany (collaboratively with the Netherlands), and in the USA (inter-regional collections at Sturgeon Bay, Wisconsin). After the establishment of the International Potato Center in Lima, Peru, a concerted effort was successfully carried out to collect all the cultivated potato germplasm from the Andes. Important collections of wild species were established at Balcarce, Argentina, and of mainly *S. tuberosum* varieties at Valdivia, Chile.

To sum up, we can say that only about 130 out of the total known number of 235 potato species exist in gene banks, so much more effort to collect the others, which are known only as descriptions or dried specimens, is needed

Table 1.4. General summary of evaluation results on resistance to the major potato diseases and pests, and adaptation to environmental extremes. Only more important species included. (From Hawkes, 1990 (based on Ross, 1986, Hawkes and Hjerting, 1989, CIP reports and various inventories), with kind permission of Belhaven Press.).

Fungus resistance
  Phytophthora infestans (late blight)
    S. berthaultii, S. bulbocastanum, S. circaeifolium, S. demissum, S. microdontum, S. phureja, S. pinnatisectum, S. polyadenium, S. stoloniferum, S. tarijense, S. tuberosum subsp. andigena, S. vernei, S. verrucosum
  Synchytrium endobioticum (wart)
    S. tuberosum (both subspecies), also to R_2 and R_3 races in a range of wild species from Bolivia including S. acaule, S. sparsipilum (and S. spegazzinii from Argentina)
  Streptomyces scabies (common scab)
    S. chacoense, S. commersonii, S. yungasense and various cultivated species

Bacterial resistance
  Pseudomonas solanacearum (bacterial wilt)
    Promising species are, in particular, S. chacoense and S. sparsipilum. Resistance is also found in S. microdontum, S. phureja and S. stenotomum
  Erwinia carotovora (soft rot; blackleg)
    Resistance found in some accessions of S. bulbocastanum, S. chacoense, S. demissum, S. hjertingii, S. leptophyes, S. megistacrolobum, S. microdontum, S. phureja, S. pinnatisectum, S. tuberosum subsp. andigena, etc.

Virus resistance
  Potato virus X
    S. acaule, S. chacoense, S. curtilobum, S. phureja, S. sparsipilum, S. sucrense, S. tarijense, S. tuberosum subsp. andigena, and several other species (Hawkes and Hjerting, 1989)
  Potato virus Y
    S. chacoense, S. demissum, S. phureja, S. stoloniferum, S. tuberosum subsp. andigena (Ross, 1986: 70-72)
  Potato leaf roll virus
    S. acaule, S. brevidens, S. etuberosum, S. raphanifolium
  Spindle tuber viroid
    S. acaule from Peru (good resistance), S. berthaultii, S. guerreroense

Insect resistance
  Leptinotarsa decemlineata (Colorado beetle)
    S. berthaultii, S. chacoense, S. commersonii, S. demissum, S. polyadenium, S. tarijense
  Myzus persicae, Macrosiphum euphorbiae (aphids)
    S. berthaultii, S. bukasovii, S. bulbocastanum, S. chomatophilum, S. infundibuliforme, S. lignicaule, S. marinasense, S. medians, S. multidissectum, S. stoloniferum

Nematode resistance
  Globodera rostochiensis, G. pallida (potato cyst nematode)
    S. acaule, S. boliviense, S. bulbocastanum, S. capsicibaccatum, S. cardiophyllum, S. gourlayi, S. oplocense, S. sparsipilum, S. spegazzinii, S. sucrense, S. vernei and several other species from Bolivia and Argentina

Table 1.4. (cont.)

Meloidogyne incognita (root-knot nematode)
  S. chacoense, S. curtilobum, S. microdontum, S. phureja, S. sparsipilum and
  S. tuberosum subsp. andigena

Physiological characters
  Frost
    S. acaule, S. ajanhuiri, S. boliviense, S. brachistotrichum, S. brevicaule, S. brevidens,
    S. canasense, S. chomatophilum, S. commersonii, S. curtilobum, S. demissum,
    S. juzepczukii, S. megistacrolobum, S. multidissectum, S. raphanifolium, S. sanctae-rosae,
    S. toralapanum and S. vernei. By far the greatest frost resister is S. acaule
  Heat and drought
    S. acaule, S. bulbocastanum, S. chacoense, S. megistacrolobum, S. microdontum,
    S. papita, S. pinnatisectum and S. tarijense
  Lack of tuber blackening
    S. hjertingii

(Hawkes, 1990). This figure – 130 out of 235 – is indeed a very small proportion, and of course even fewer have been evaluated for various characters of resistance or adaptation. However, in a recent work (Hawkes and Hjerting, 1989), resistances of some sort have been shown to exist in the Bolivian potato species (some 40 in all) to all the pests and diseases so far known. Certain species are particularly useful in this respect and some, such as *S. chacoense, S. demissum, S. berthaultii, S. tarijense, S. stoloniferum, S. acaule*, etc., are well known in the literature (see Table 1.4). It comes as rather a shock, however, to learn (Ross, 1986) that germplasm from only six wild species is frequently incorporated into European cultivars, namely: *S. demissum* (late blight and PLRV); *S. acaule* (PVX, PLRV, PSTV, wart, *Globodera* and frost); *S. chacoense* (PVA, PVY, late blight, Colorado beetle, tuber moth); *S. spegazzinii* (*Fusarium*, wart, *Globodera*); *S. stoloniferum* (PVA, PVY); and *S. vernei* (*Globodera*). Genes from *S. microdontum, S. sparsipilum, S. verrucosum, S. phureja, S. tuberosum* subsp. *andigena, S. commersonii* and *S. maglia* have also been used on occasion. In the same publication, Ross (1986) notes that no less than 97 European cultivars contain genes from these 13 species.

Apart from this, the evaluation work has far outstripped the actual use to which wild potatoes have been put. Evaluation results are noted in the more recent inventories, e.g. Hanneman and Bamberg (1986), and details are still appearing in the literature, as, for instance, Flanders *et al.* (1992), who have evaluated 100 species of wild potato for resistance to Colorado potato beetle, green peach aphid, potato aphid, potato flea beetle and potato leaf hopper, with considerable success, finding an interesting range of within-species diversity. These results compare with those of other workers in respect of the fact that intraspecific resistance diversity is generally quite high. It is thus misleading to state, as sometimes occurs in the literature, that a particular species does or does not possess resistance to a particular pest or disease, based on one or only a few samples.

To sum up, it would be safe to say that, although probably a great deal of effort has been put into the assessment and use of wild potato species in breeding for resistance and adaptation, there is still a large potential for the incorporation of their useful genes into new varieties. Unfortunately, as is well known, it takes many generations of backcrossing to the cultivated parent species in order to eliminate the characters of poor yield and poor quality which are transferred together with the good resistance characters of the wild parents. It is to be hoped that new methods of gene transfer will be available in the near future to render these tasks easier to accomplish, so reducing the time necessary to breed varieties with selected genes or gene complexes transferred from wild species and primitive forms (see Chapter 23 by Hermsen).

References

Astley, D. and Hawkes, J.G. (1979) The nature of the Bolivian weed potato species *Solanum sucrense* Hawkes. *Euphytica* 28, 685–696.
Bauhin, C. (1596) *Phytopinax*. Sebastianum Henricpetri, Basle.
Bauhin, C. (1620) *Prodromus Theatri Botanici*. Paul Jacob and John Treudel, Frankfurt-on-Main.
Bauhin, J. (1651) *Historia Plantarum Universalis*. Graflenreid, Yverdon.
Bonierbale, M.W., Ganal, M.W. and Tanksley, S.D. (1990) Applications of restriction fragment length polymorphisms and genetic mapping in potato breeding and molecular genetics. In: Vayda, M.E. and Park, W.D. (eds) *The Molecular and Cellular Biology of the Potato*. CAB International, Wallingford, pp. 13–24.
Bukasov, S.M. (1933) The potatoes of South America and their breeding possibilities. Suppl. 58, *Bulletin of Applied Botany, Genetics and Plant Breeding (Leningrad)* 192.
Bukasov, S.M. (1938) Interspecific hybridization in the potato (in Russian). *Izvestia Akademii Navuk, CCCP*, 711–732.
Bukasov, S.M. (1939a) The origin of potato species. *Physis (Buenos Aires)*, 18, 41–46.
Bukasov, S.M. (1939b) Interspecific hybridization in the potato. *Physis (Buenos Aires)*, 18, 269–284.
Bukasov, S.M. and Kameraz, A.Y. (1959) *Ocnovi Selektsii Kartofelya (Bases of Potato Breeding)*. Gosudarstvennoe Isdatelstvo Selskokhozyaictvennoi Literaturi, Moscow and Leningrad.
Bukasov, S.M. and Kameraz, A.Y. (1973) *Genetics of the Potato*. Izdatelstvo 'Nauka', Moscow.
Bukasov, S.M. and Lebedeva, N.A. (1955) Breeding the potato for resistance to Colorado beetle (*Leptinotarsa decemlineata* Say). In: *Mezhduvedomstvennaya metodicheskaya Komissiya po koloradskomu zhuku*. Academy of Sciences of the USSR, Moscow.
Castellanos, J. de (1886) *Elegías de Varones Ilustres de Indias, Part 4* (manuscript completed 1601). Antonio Paz y Melia, Madrid.
Chávez, R., Jackson, M.T., Schmiediche, P.E. and Franco, J. (1988a) The importance of wild potato species resistant to the potato cyst nematode, *Globodera pallida*, pathotypes Pa4 and Pa5, in potato breeding. I. Resistance studies. *Euphytica* 36(3), 9–14.

Chávez, R., Schmiediche, P.E., Jackson, M.T. and Raman, K.V. (1988b) The breeding potential of wild potato species resistant to the potato tuber moth, *Phthorimaea operculella* (Zeller). *Euphytica* 38, 123-132.

Cieza de León, Pedro de (1553) La Crónica del Perú. Seville. English translation by Sir Clements R. Markham in Hakluyt Society 33 and 68, 1864 and 1883, London.

Clausen, A. and Okada, K.A. (1990) Electroforesis de proteinas de los tubérculos en *Solanum* tuberosos silvestres (Solanaceae). *Darwiniana* 30, 163-169.

Clusius, C. (1601) *Rariorum Plantarum Historia*. Moretus, Plantin, Antwerp.

Cribb, P.J. and Hawkes, J.G. (1986) Experimental evidence for the origin of *S. tuberosum* subsp. *andigena*. In: D'Arcy, W.G. (ed.) *Solanaceae: Biology and Systematics*. Columbia University Press, New York, pp. 383-404.

Debener, T., Salamini, F. and Gebhardt, C. (1990) Phylogeny of wild and cultivated *Solanum* species based on nuclear restriction fragment length polymorphisms (RFLPs). *Theoretical and Applied Genetics* 79, 360-368.

den Nijs, T.P.M. and Peloquin, S.J. (1977a) Polyploid evolution via 2*n* gametes. *American Potato Journal* 54, 377-386.

den Nijs, T.P.M. and Peloquin, S.J. (1977b) 2*n* gametes in potato species and their function in sexual polyploidization. *Euphytica* 26, 585-600.

Desborough, S. and Peloquin, S.J. (1967) Esterase isozymes from *Solanum* tubers. *Phytochemistry* 6, 989-994.

Desborough, S. and Peloquin, S.J. (1969) Acid gel disc electrophoresis of tuber proteins from *Solanum* species. *Phytochemistry* 8, 425-429.

Dodds, K.S. (1962) Classification of cultivated potatoes. In: Correll, D.S. (ed.) *The Potato and its Wild Relatives*. Texas Research Foundation, Renner, Texas, pp. 517-539.

Drake, Sir Francis (1628) *The World Encompassed*. Hakluyt Society 16, 1854, London.

Flanders, K.L., Hawkes, J.G., Radcliffe, E.B. and Lauer, F.I. (1992) Insect resistance in potatoes: sources, evolutionary relationships, morphological and chemical defenses and ecogeographical associations. *Euphytica* 61, 83-111.

Gatenby, A.A. and Cocking, E.C. (1978) Fraction 1 protein and the origin of the European potato. *Plant Science Letters* 12, 177-181.

Gell, P.G.H., Wright, S.T.C. and Hawkes, J.G. (1956) Immunological methods in plant taxonomy. *Nature* 177, 573.

Gell, P.G.H., Hawkes, J.G. and Wright, S.T.C. (1960) The application of imunological methods to the taxonomy of species within the genus *Solanum*. *Proceedings of the Royal Society, Series B* 151, 364-383.

Gerard, J. (1597) *The Herball or Generall Historie of Plantes*. John Norton, London.

Grun, P. (1979) Evolution of the cultivated potato: a cytoplasmic analysis. In: Hawkes, J.G., Lester, R.N. and Skelding, A.D. (eds) *The Biology and Taxonomy of the Solanaceae*. Academic Press, London, pp. 655-665.

Grun, P. (1990) The evolution of cultivated potatoes. *Economic Botany* 44 (Suppl. 3), 39-55.

Hanneman, R.E., Jr (1989) Interspecific crossability studies among 1 EBN species. *American Potato Journal* 66, 524 (Abstract).

Hanneman, R.E. and Bamberg, J.B. (1986) Inventory of tuber-bearing *Solanum* species. *USDA Bulletin* 533, 216.

Hanneman, R.E., Jr, Bamberg, J.B., Dodd, J.B., Ehlenfeldt, M.K. and Johnston, S.A. (1990) The genetics of the endospern balance number (EBN). In: MacKerron, D.K.L., Edmond, H.D., Hall, D., Kirkman, M.A., Lang, R.W., Mackay, G.R., McRae, D.C. and Oxley, S.J.P. (eds), *Abstract of the 11th Triennial Conference of the European Association for Potato Research*. Edinburgh, UK pp. 124-125.

Hawkes, J.G. (1956) Hybridization studies on four hexaploid *Solanum* species in series *Demissa* Buk. *New Phytologist* 55, 191–205.

Hawkes, J.G. (1958) Potatoes: taxonomy, cytology and crossability. In: Kappert, H. and Rudorf, W. (eds) *Handbuch der Pflanzenzüchtung*, 2nd edn. Paul Parey, Berlin and Hamburg, Chapter 1, Vol. 3, pp. 1–43.

Hawkes, J.G. (1962a) The origin of *Solanum juzepczukii* Buk. and *S. curtilobum* Juz. et Buk. *Zeitschrift für Pflanzenzüchtung* 47, 1–14.

Hawkes, J.G. (1962b) Introgression in certain wild potato species. *Euphytica* 11, 26–35.

Hawkes, J.G. (1966) Modern taxonomic work on the *Solanum* species of Mexico and adjacent countries. *American Potato Journal* 43, 81–103.

Hawkes, J.G. (1988) The evolution of cultivated potatoes and their tuber-bearing wild relatives. *Die Kulturpflanze* 36, 189–208.

Hawkes, J.G. (1990) *The Potato, Evolution, Biodiversity and Genetic Resources*. Belhaven Press, London.

Hawkes, J.G. and Francisco-Ortega, J. (1992) The potato in Spain during the late 16th century. *Economic Botany* 46, 89–97.

Hawkes, J.G. and Hjerting, J.P. (1969) *The Potatoes of Argentina, Brazil, Paraguay and Uruguay – a Biosystematic Study*. Oxford University Press, Oxford.

Hawkes, J.G. and Hjerting, J.P. (1989) *The Potatoes of Bolivia: Their Breeding Value and Evolutionary Relationships*. Oxford University Press, Oxford.

Hawkes, J.G. and Jackson, M.T. (1992) Taxonomic and evolutionary implications of the endosperm balance number hypothesis in potatoes. *Theoretical and Applied Genetics* 84, 180–185.

Hawkes, J.G. and Lester, R.N. (1966) Immunological studies on the tuber-bearing *Solanums*. II. Relationships of the North American species. *Annals of Botany* 30, 269–290.

Hawkes, J.G. and Lester, R.N. (1968) Immunological studies on the tuber-bearing *Solanums*. III. Variability within *S. bulbocastanum* and its hybrids with species in series *Pinnatisecta*. *Annals of Botany* 32, 165–186.

Hosaka, K. (1986) Who is the mother of the potato? Restriction endonuclease analysis of chloroplast DNA of cultivated potatoes. *Theoretical and Applied Genetics* 72, 606–618.

Hosaka, K. and Hanneman, R.E., Jr (1988a) The origin of the cultivated tetraploid potato based on chloroplast DNA. *Theoretical and Applied Genetics* 76, 172–176.

Hosaka, K. and Hanneman, R.E., Jr (1988b) Origin of chloroplast DNA diversity in the Andean potatoes. *Theoretical and Applied Genetics* 76, 333–340.

Hosaka, K. and Matsubayashi, M. (1983a) Studies on the phylogenetic relationships in tuberous *Solanums* by isozyme analysis. II. Phylogenetic relationships between Mexican and South American diploid species. *Science Reports, Faculty of Agriculture, Kobe University* 15, 217–228.

Hosaka, K. and Matsubayashi, M. (1983b) Studies on the phylogenetic relationships in tuberous *Solanums* by isozyme analysis. III. Interspecific differences between each of four cultivated diploid species (in Japanese). *Reports of the Society of Crop Science and Breeding, Kinki* 28, 28–32.

Hosaka, K., Ogihara, Y., Matsubayashi, M. and Tsunewaki, K. (1984) Phylogenetic relationship between the tuberous *Solanum* species as revealed by restriction endonuclease analysis of chloroplast DNA. *Japanese Journal of Genetics* 59, 349–369.

Hosaka, K., Zoeten, G.A. de and Hanneman, R.E. (1988) Cultivated potato chloroplast DNA differs from the wild type by one deletion: evidence and implications. *Theoretical and Applied Genetics* 75, 741–745.

Huamán, Z., Hawkes, J.G. and Rowe, P.R. (1982) A biosystematic study of the origin of the diploid potato, *Solanum ajanhuiri*. *Euphytica* 31, 665–675.

Huamán, Z., Hawkes, J.G. and Rowe, P.R. (1983) Chromatographic studies on the origin of the cultivated potato, *Solanum ajanhuiri*. *American Potato Journal* 60, 361–367.

Jackson, M.T., Hawkes, J.G. and Rowe, P.R. (1977) The nature of *Solanum chaucha*, Juz. et Buk., a triploid cultivated potato of the South American Andes. *Euphytica* 26, 775–783.

Jackson, M.T., Rowe, P.R. and Hawkes, J.G. (1978) Crossability relationships of Andean potato varieties of three ploidy levels. *Euphytica* 27, 541–555.

Johns, T. and Keen, S.L. (1986) Ongoing evolution of the potato on the altiplano of Bolivia. *Economic Botany* 40, 409–424.

Johnston, S.A. and Hanneman, R.E. (1978) Endosperm balance factors in some tuber-bearing *Solanum* species. *American Potato Journal* 55, 380.

Johnston, S.A. and Hanneman, R.E., Jr (1980a) Support of the endosperm balance number hypothesis utilizing some tuber-bearing *Solanum* species. *American Potato Journal* 57, 7–14.

Johnston, S.A. and Hanneman, R.E., Jr (1980b) The discovery of effective ploidy barriers between diploid *Solanums*. *American Potato Journal* 57, 484–485 (Abstract).

Johnston, S.A. and Hanneman, R.E. (1982) Manipulations of endosperm balance number overcome crossing barriers between diploid *Solanum* species. *Science* 217, 446–448.

Lester, R.N. (1965) Immunological studies on the tuber-bearing *Solanums*. I. Techniques and South American species. *Annals of Botany* 29, 609–624.

López, L.E. and Hawkes, J.G. (1991) Cytology and genome constitution of the tuber-bearing *Solanum* species in series *Conicibaccata*. In: Hawkes, J.G., Lester, R.N., Nee, M. and Estrada-R., N. (eds) *Solanaceae III, Taxonomy, Chemistry and Evolution*. Royal Botanic Gardens, Kew, England, pp. 327–346.

López de Gómara, F. (1552) *Historia General de las Indias*. Zaragoza (Madrid edition, 1922).

Marks, G.E. (1955) Cytogenetic studies in tuberous *Solanum* species. I. Genomic differentiation in the group *Demissa*. *Journal of Genetics* 53, 262–269.

Martins-Farias, R. (1976) New archaeological techniques for the study of ancient root crops in Peru. Unpublished PhD thesis, University of Birmingham.

Matsubayashi, M. (1955) Studies on the species differentiation in the section *Tuberarium* of *Solanum*. III. Behaviour of meiotic chromosomes in F_1 hybrid between *S. longipedicellatum* and *S. schickii* in relation to its parent species. *Science Reports Hyogo University of Agriculture* 2, 25–31.

Matsubayashi, M. (1981) Species differentiation in tuberous *Solanum* and the origin of cultivated potatoes (in Japanese). *Recent Advances in Breeding* 22, 86–106.

Matsubayashi, M. (1991) Phylogenetic relationships in the potato and its related species. In: Tsuchiya. T. and Gupta, P.K. (eds) *Chromosome Engineering in Plants: Genetics, Breeding and Evolution, Part B*. Elsevier, Amsterdam, pp. 93–118.

Parkinson, J. (1629) *Paradisi in Sole Paradisus Terrestris*. Humfrey Lownes and Robert Young, London.

Poma de Ayala, Felipe Guamán (1936) *Nueva Corónica y Buen Gobierno*. Ed. and publ. P. Rivet (1936). Traveaux et Mémoires de l'Institut d'Ethnologie, 23, Paris.

Rickeman, V.S. and Desborough, S.L. (1978) Elucidation of the evolution and taxonomy of cultivated potatoes with electrophoresis. I. Groups *Tuberosum*, *Andigena*, *Phureja* and *Stenotomum*. *Theoretical and Applied Genetics* 52, 187-190.

Ross, H. (1958) Ausgangsmaterial für die Züchtung. In: Kappert, H. and Rudorf, W. (eds) *Handbuch der Pflanzenzüchtung*, 2nd edn. Paul Parey, Berlin and Hamburg. Vol. 3, pp. 43-59.

Ross, H. (1986) Potato breeding – problems and perspectives. *Journal of Plant Breeding* Suppl. 13 (Horn, W. and Röbbelen, G., eds) p. 132.

Salaman, R.N. (1946) The early European potato: its character and place of origin. *Journal of the Linnean Society (Botany)* 53, 1-27.

Salaman, R.N. (1949) *The History and Social Influence of the Potato*. Cambridge University Press, Cambridge. Revised edition, Hawkes J.G. (ed.) (1985), Cambridge University Press, Cambridge.

Salaman, R.N. (1954) The origin of the early European potato. *Journal of the Linnean Society (Botany)* 55, 185-190.

Salaman, R.N. and Hawkes, J.G. (1949) The character of the early European potato. *Proceedings of the Linnean Society, London* 161, 71-84.

Schmiediche, P.E., Hawkes, J.G. and Ochoa, C.M. (1980) Breeding of the cultivated potato species *Solanum juzepczukii* Buk. and *S. curtilobum* Juz. et Buk. I. *Euphytica* 29, 685-704.

Schmiediche, P.E., Hawkes, J.G. and Ochoa, C.M. (1982) The breeding of the cultivated potato species *Solanum juzepczukii* Buk. and *S. curtilobum* Juz. et Buk. II. *Euphytica* 31, 695-707.

Spooner, D.M. and Douches D.S. (1991) Allozyme variation within *Solanum* Sect. *Petota* Ser. *Etuberosa*. *American Potato Journal* 68, 636 (Abstract).

Spooner, D.M. and Sytsma, K.J. (1992) Re-examination of Series relationships of Mexican and Central American wild potatoes (*Solanum* sect. *Petota*): evidence from chloroplast DNA restriction site variation. *Systematic Botany* 17, 432-448.

Spooner, D.M., Sytsma, K.J. and Conti, E. (1991a) Chloroplast DNA evidence for genome differentiation in wild potatoes (*Solanum* sect. *Petota*: Solanaceae). *American Journal of Botany* 78, 1354-1366.

Spooner, D.M., Sytsma, K.J. and Smith, J.F. (1991b) A molecular re-examination of diploid hybrid speciation of *Solanum raphanifolium*. *Evolution* 45, 757-764.

Spooner, D.M., Douches, D.S. and Contreras, A. (1992) Allozyme variation within *Solanum* Sect. *Petota*, Ser. *Etuberosa (Solanaceae)*. *American Journal of Botany* 79, 467-471.

Tarn, T.R. and Hawkes, J.G. (1986) Cytogenetic studies and the occurrence of triploidy in the wild potato species *Solanum commersonii* Dun. *Euphytica* 35, 293-302.

Ugent, D. (1970) *Solanum raphanifolium*, a Peruvian wild potato species of hybrid origin. *Botanical Gazette* 131, 225-233.

Ugent, D., Pozorski, S. and Pozorski, T. (1982) Archaeological potato tuber remains from the Casma valley of Peru. *Economic Botany* 36, 181-192.

Ugent, D., Dillehay, T. and Ramírez, C. (1987) Potato remains from a late Pleistocene settlement in south central Chile. *Economic Botany* 4, 17-27.

Wietschel, G. and Reznik, H. (1980a) Die Flavonoid-Muster der Knollentragenden *Solanum* – Arten II. Die Flavonoid-Glycoside der Arten aus Series I-XVI. *Zeitschrift für Pflanzenphysiologie* 97, 79-88.

Wietschel, G. and Reznik, H. (1980b) Die Flavonoid-Muster der Knollentragenden *Solanum* – Arten III. Die Flavonoid-Glycoside der Arten aus Series XVII. *Zeitschrift für Pflanzenphysiologie* 99, 149-158.

2 Genome Evolution in Potatoes

M.J. WILKINSON

Crop Genetics Department, Scottish Crop Research Institute, Invergowrie, Dundee DD2 5DA, UK.

Base Number

Introduction

The cultivated potato (*Solanum tuberosum* L.) is widely regarded as a cytologically difficult species to examine because of its small and relatively numerous chromosomes. This is illustrated by the large number of erroneous chromosome counts published before gametic counts of $n = 24$ and somatic counts of $2n = 48$ (Figure 2.1) were reported by Salaman (1926), Fukuda (1927), Smith (1927) and Stow (1927).

In a study of some tuber-bearing wild relatives of the potato, Smith (1927) discovered gametic numbers of $n = 12$ for *S. jamesii*, *S. chacoense* and *S. fendleri*, and of $n = 36$ for *S. demissum*. This provided the first evidence of a polyploid series in the section and led to the suggested base number of $x = 12$. However, although a general consensus arose in support of this tenet, there was not unanimity. Many authors (Lawrence, 1931; Müntzing, 1933, Ellison, 1936; Schwarz, 1937; Emme, 1938; Choudhuri, 1943; Koopmans, 1951) favoured the theory that $x = 12$ is a derived number, the true base number being $x = 6$. Wanscher (1934) even postulated that the true base number might be $x = 4$, but found no support for this idea. An extensive survey of the evidence suggesting base numbers of $x = 6$ and $x = 12$ was provided by Swaminathan and Howard (1953). At that time, evidence supporting a base number of $x = 6$ rested heavily on evidence of secondary associations and a small degree of autosyndesis (pairing between non-homologous chromosomes of the same genome) during meiosis of triploid and pentaploid hybrids (Lamm, 1945; Bains, 1951; Howard and Swaminathan, 1952; Prakken and Swaminathan, 1952). However, the reviewers produced a large body of evidence supporting a base number of $x = 12$ and suggesting that no recent doubling had occurred during the evolution of the genus. In brief, their argument ran as follows:

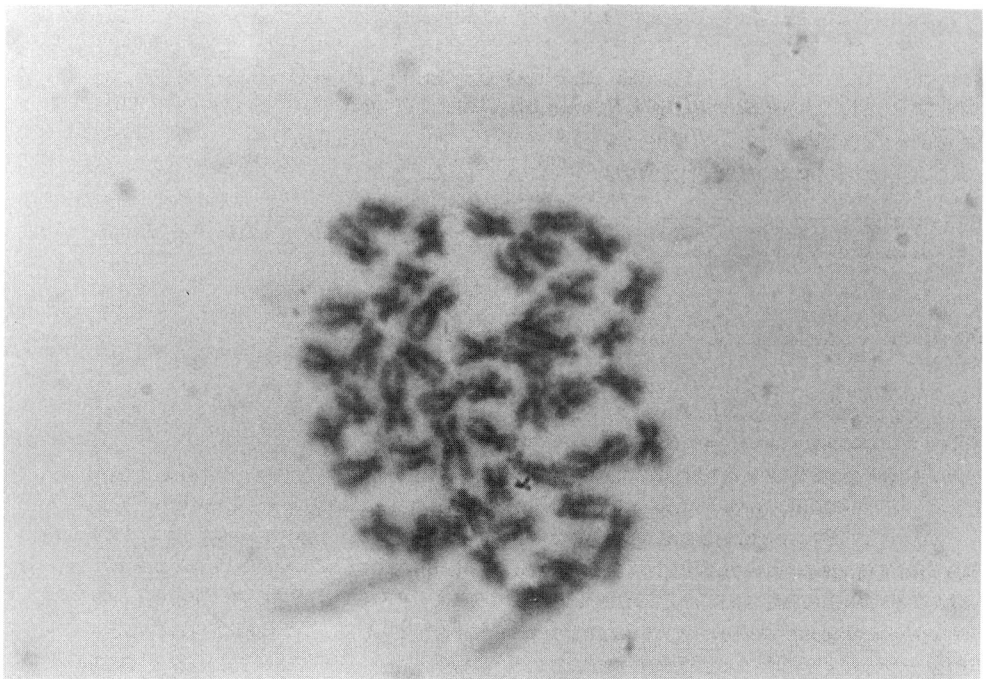

Fig. 2.1. *Solanum tuberosum* cv. Pentland Crown with $2n = 4x = 48$.

1. They agreed with Thomas and Revell (1946), who suggested that secondary associations are merely due to attractions between heterochromatic regions and cannot therefore be taken as evidence of homology within a genome.
2. The autosyndesis reported by some workers in triploid and pentaploid hybrids involved low numbers of chromosomes and so was inconsistent with a recent doubling of the base number from $x = 6$.
3. If the true base number were $x = 6$, then secondary balance (a stable intermediate number) would be expected in hybrids between triploid species and either tetraploids or diploids. In general, this has not been observed.
4. Self-incompatibility is effective in diploid species (i.e. $2n = 24$) but is lost in autotetraploids derived from them (Livermore and Johnstone, 1940; Swaminathan, 1951). 'This suggests that diploids behave genetically as diploids, and not as tetraploids.'
5. No number of less than $2n = 24$ has been reported in *Solanum* ($x = 12$). Most other genera of the Solaneae also have apparent base numbers of $x = 12$ (*Lycium, Atropa, Physalis, Capsicum, Lycopersicon, Cyphomandra, Jaborosa, Salpichroa, Scopolia, Saracha* and *Withania*). The remaining genus of the tribe, *Hyoscyamus*, has basic numbers of 14 and 17. In other tribes base numbers of 7–12 are found but none with $x = 6$. The nearest $x = 6$ is in the related family Scrophulariaceae.

Pairing behaviour of monoploids

Indirect evidence of homoeology within the genome was produced shortly after the review by Swaminathan and Howard (1953) through the genetic analysis of monoploids in the neighbouring genus *Lycopersicon*. Gottschalk (1956) reported autosyndetic pairing in monoploid tomatoes (*Lycopersicon esculentum*) and concluded that the base number of the genus is $x = 6$. Similar results were observed in tomatoes by Ecochard *et al.* (1969), although they were more cautious in drawing the same conclusions. The first extensive meiotic analysis of a potato monoploid was made by Van Breukelen *et al.* (1975). Between none and three bivalents (usually two) were observed during metaphase I of meiosis. Low numbers of trivalents and quadrivalents were also reported. The authors did not suggest that the occurrence of pairing supported a putative ancestral base number of $x = 6$, although, provided the chromosome associations were not an artefact, they did indicate duplicated segments in the genome. The two monoploid potato clones used in the study were produced by successive pollinations with dihaploid and monoploid inducer clones of *S. phureja*. The first meiotic analysis of potato monoploids produced by anther culture techniques (Singh *et al.*, 1988) revealed essentially similar results, with observed bivalent frequencies at metaphase I of 2.07 to 3.0 per PMC, a small number of which were heteromorphic. Cells with three bivalents and six univalents or two bivalents and eight univalents were the most frequently encountered classes. Trivalents were also observed at low frequencies, although no quadrivalents were reported. Meiotic analysis of the parental dihaploid and colchidiploids derived from the monoploids revealed regular pairing, with 85% of the PMCs showing the formation of 12 bivalents while the rest had two to four early disjunct univalents. They concluded that their results suggest the presence of duplicated sequences in the chromosome complement of the monoploids. They also suggested that the trivalents at metaphase I and the presence of chromatin bridges point to exchanges between chromosomes with duplications in common.

Ramanna and Wagenvoort (1976) and Wagenvoort and Ramanna (1979) used pachytene analysis to examine the pairing behaviour of dihaploid addition line material representing 11 of the 12 possible types of primary trisomics. Trivalents involving the three homologous chromosomes were observed in the majority of cells, although a low frequency of non-homologous (autosyndetic) pairing was seen in both studies, again leading to the suggestion that some degree of homoeology may exist within the potato genome.

Evidence from karyotype analysis

Karyotype analysis can reveal valuable information relating to the origin of the base number of a group and to the mechanisms of genome evolution which are operating (for review see Swanson *et al.*, 1981). However, the chromosomes of the cultivated potato and its relatives are characterized by their small size and generally similar appearance. This limits the amount of information which can be derived from such studies.

Swaminathan (1954) attempted to characterize the karyotype of *S. tuberosum* cv. Duke of York. He was unable to identify all chromosomes but divided the genome into seven groups of chromosomes on the basis of length, position of centromere and presence of secondary constrictions. Yeh and Peloquin (1965) used differences in the morphology of pachytene chromosomes when stained with acetocarmine to identify the chromosomes in dihaploids of *S. tuberosum*. Variability in length during pachytene and in the stretching associated with slide preparations prevented the authors from presenting an ideogram, although a dichotomous identification key and extensive descriptions of each of the chromosomes were given. Similarities were noted between chromosomes 1 and 3, between chromosomes 6, 7 and 8, and between chromosomes 11 and 5. Nevertheless, the authors were confident in their ability to distinguish 12 structurally distinct chromosomes in the genome. Mok *et al.* (1974) were the first to attempt Giemsa C-banding of potato chromosomes. Descriptions of each of the 12 chromosomes were accompanied by an ideogram. There were similarities in the descriptions given of chromosomes C, D, E and F, of G and H, and of K and L. The resolution of the banding was insufficiently sharp to detect subtle structural changes to the genome (such as translocations) or to distinguish between chromosomes of *S. phureja*-induced dihaploids of *S. tuberosum* and *S. chacoense*. Lee and Hanneman (1976) attempted to combine the use of Giemsa C-banding with pachytene analysis to identify potato chromosomes in order to overcome the limitations of resolution. However, at this time resolution of the Giemsa C-banding was still relatively poor and pachytene analysis was complex and limited in scope. As a consequence, disagreements over the interpretation of such preparations appeared between some workers. For instance, Lee and Rowe (1975) identified an isochromosome for the short arm of chromosome 9 in a trisomic plant, whereas when Wagenvoort and Ramanna (1979) examined the same plant they concluded that the plant was trisomic for chromosome 4.

In a classical work, Pijnacker and Ferwerda (1984) greatly improved on existing Giemsa C-banding procedures and described a new method of silver-staining nucleolar organizing regions in potatoes. They divided the unbanded karyotype into three groups on the basis of length: chromosomes 1 and 2 were described as long, chromosomes 3–9 as medium length and 10–12 as short. Centromere position was also used to identify the chromosomes: chromosome 11 being described as metacentric, chromosomes 1, 2, 4, 7 and 12 as meta-submetacentric, and chromosomes 3, 5, 6, 8, 9 and 10 as submeta-acrocentric. Chromosome 1 was also described as carrying a non-nucleolar constriction on the proximal part of the short arm and chromosome 2 as containing the nucleolar organizing region on the short arm. Nevertheless, the authors concluded that it is not possible to distinguish chromosomes 5, 6, 8 and 9 on structural grounds. The high-resolution banding enabled the construction of a detailed banded ideogram which contained information from AgNOR staining and Giemsa C-banding (Figure 2.2). Chromosomes 1 and 2 were extremely distinct and clearly were not duplicated elsewhere in the genome. Equally,

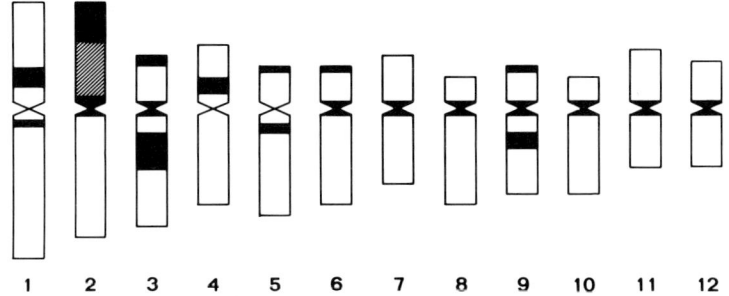

Fig. 2.2. Ideogram of *S. tuberosum* using Giemsa C-banding and silver nitrate staining (according to Pijnacker and Ferwerda, 1984).

there were no chromosomes similar in appearance to chromosomes 4, 5 or 6. However, the authors drew attention to the likeness of chromosomes 3 and 9. The ideogram also suggested close resemblance between chromosomes 8 and 10, and between chromosomes 7, 11 and 12.

These karyological studies suggest that duplication of the base number from an ancestral $x = 6$ is unlikely, at least in the recent history of the genus. Marked similarities between certain chromosomes within the genome instead point to chromosome duplication followed by divergence.

Molecular evidence

Yeh and Peloquin (1965) commented upon the striking similarity which exists between the morphology of potato and tomato chromosomes. Bonierbale *et al.* (1988) used cDNA clones generated in an earlier study of the tomato genome (Bernatzky and Tanksley, 1986) to make a molecular comparison of the two crops. They found that, for nearly every tomato genomic sequence or cDNA tested, it was possible to find a homologous counterpart in the potato, suggesting a high degree of homology between the two genomes. The resultant linkage map of the potato consisted of 135 markers, with most clones having the same copy number as in the tomato. All clones mapped to the same linkage groups. Nine of the 12 chromosomes were identical in linkage order. In the other three, the authors suggested that paracentric inversions may account for the differences. Perhaps the most striking feature of both linkage maps was the lack of duplication between chromosomes. A few multilocus probes were detected but these were generally scattered across the genome and independent of each other. Thus, there was no support provided by these studies for the recent doubling of the base number from $x = 6$, or even for the recent duplication of individual chromosomes. In a more recent map of the potato, Gebhardt *et al.* (1991) mapped 492 fragments to 304 loci on 12 linkage groups. The total map length was 1034 cM, increasing the number of previously mapped loci by 117% and the map length from 690 to 1034 cM.

This corresponds to 80% of the potato genome. Out of the 239 molecular probes used, 189 (78%) mapped to a single locus and 50 (22%) to two or more loci. The multilocus probes were well scattered throughout the genome and did not cluster to suggest pairs or small groups of homoeologous chromosomes within the genome. The authors also drew attention to the extremely close similarity between the potato and tomato genomes. Indeed, of the 64 markers analysed in potato and tomato, only three did not map to homologous positions (see Figure 10.5 for most recent map).

Further evidence indicating the predominance of coding genes with a single copy number or a single chromosomal location in potato and tomato has been provided by other studies. Lapitan *et al.* (1991) made a detailed characterization of the 5S rRNA genes of tomato. They demonstrated by RFLP analysis that the genes were tandomly repeated on chromosome 1 but were present on no other chromosome, and they confirmed their position by *in situ* hybridization. Similarly, Wagenvoort (1982, 1988) found nonduplication of two recessive genes which he mapped to chromosomes 3 and 12 by trisomic analysis. Conversely, in an extensive survey in which 35 accessions of four *Solanum* species were examined using ten isozyme systems, Martinez-Zapater and Oliver (1985) found evidence of gene duplication in the 20 diploid accessions and suggested this provided evidence of an ancestral base number of $x = 6$ for the Solanaceae.

Conclusions

Swaminathan and Howard (1953) pointed out that the base number in the Solanaceae is extremely conserved ($x = 12$). Therefore it would be reasonable to presume that this number is more likely to have evolved early in the evolution of the tribe than to have appeared independently in each of the genera it contains (including *Solanum*). This is supported by evidence from several independent sources:

1. The extreme similarities which exist between the karyotypes and linkage maps of *Solanum* and the neighbouring genus *Lycopersicon* is more likely to have arisen through common ancestry than through convergent evolution.
2. The divergence in the appearance of chromosomes within the potato genome as revealed by Giemsa C-banding would have required a considerable number of large-scale structural rearrangements following any recent doubling in base number.
3. The high number of single-copy cDNA markers and the low number of duplicated sequences in the genome of the potato suggest that considerable divergence has occurred since the most recent change in base number.
4. The cDNA and isozyme duplications in the potato fail to 'cluster' to suggest pairs or small groupings of similar chromosomes within the genome. This is also consistent with considerable divergence having followed the most recent change in base number.

Clearly, then, it is highly improbable that the current base number of $x = 12$ arose through the doubling of an ancestral base number of $x = 6$ since the formation of the genus *Solanum*. On the other hand, the regular appearance of between one and three bivalents during meiosis in potato monoploids may be indicative of a low level of homoeology within the genome. Similarly, non-homologous pairing observed in trisomic plants (Wagenvoort and Ramanna, 1979) and in triploid and pentaploid hybrids (Lamm, 1945; Bains, 1951; Howard and Swaminathan, 1952; Prakken and Swaminathan, 1952) indicates that sufficient homoeology remains between at least six of the chromosomes to cause regular formation of up to three bivalents. The presence of duplicated isozyme genes in a range of diploid taxa (Martinez-Zapater and Oliver, 1985) also provides genetic evidence of at least partial chromosomal homology within the genome. A close examination of the C-banding karyotype produced by Pijnacker and Ferwerda (1984) also reveals three 'pairs' of chromosomes with similar size, centromere position and C-banding pattern.

Swaminathan and Howard (1953) point out that the other tribes in the Solanaceae have various base numbers, from $x = 7$ to $x = 12$. A base number of $x = 12$ may have arisen either from one or more of the lower base numbers through the duplication of individual chromosomes or else through the doubling of an older ancestor with a base number of $x = 6$. The failure of cDNA probes to 'cluster' during linkage analysis is more consistent with an ancient doubling of the whole genome than it is with a more recent duplication of one or more chromosomes. Furthermore, the high level of divergence observed points to an extremely early doubling event and emphasises that the currently accepted base number of $x = 12$ is a fairly accurate reflection of the genetic affinities of the chromosomes contained within the genome.

Evolution Within the Genus

Polyploidy

Early evidence of the presence of a polyploid series extending from diploids ($2n = 24$) to hexaploids ($2n = 72$) in *Solanum* section Petota Dumortier was provided by several detailed cytological studies of the group (Smith, 1927; Longley and Clark, 1930; Rybin, 1930, 1933; Bukasov, 1939; Hawkes, 1944). The vast majority of species are diploid $2n = 2x = 24$. Indeed, of the 183 species in the section for which counts are known, Hawkes (1990) described 136 (74%) as diploid, seven (4%) as triploid, 27 (15%) as tetraploid, three (2%) as pentaploid and ten (5%) as hexaploid. Taxonomically, 15 of the 21 series are exclusively diploid, leaving only six containing some polyploid species and just three (series Acaulia, Longipedicellata and Demissa) lacking any diploid representatives. Thus, polyploidization has played no part in the evolution of the majority of series, although its importance in some others is undeniable. Even within series Tuberosa, to which the modern cultivated potato belongs, the vast majority of species (69 out of 83) are diploids.

Hybridization between diploids has been widely reported in the section (Swaminathan and Howard, 1953). Naturally occurring hybrids between taxa of different species, series and even superseries are relatively frequent (Hawkes, 1958) and the importance of hybridization in speciation is often quoted (e.g. Ugent, 1970; Hawkes, 1990). Most species belonging to the same series cross freely and meiotic pairing in the hybrids is generally as regular as that found in the original parents (Emme, 1937; Swaminathan and Howard, 1953; Magoon *et al.*, 1958). Meiotic observations made by Singh *et al.* (1989) even suggested that chiasma frequencies during male meiosis may be greater sometimes in the hybrids than in either parent. Evidence suggesting otherwise was provided by the molecular linkage study of Gebhardt *et al.* (1991), who noted a reduction in the recombination frequency of interspecific hybrids when compared with intraspecific hybrids. Nevertheless, the readiness of chromosomes of different species to pair with each other has been demonstrated in many studies. For example, Ramanna and Hermsen (1971) examined meiosis in tetraploid 'triple hybrids' between *S. phureja* (P), *S. bulbocastanum* (B) and *S. acaule* (A). The hybrids had the genome constitution AABP and behaved as an autotetraploid, with the formation of quadrivalents involving all 12 groups of four homoeologous chromosomes. The authors concluded that the four parental genomes are not differentiated to the extent of affecting normal pairing.

Complete pairing of genomes in polyploid potatoes, however, is not always a good indicator of the fertility of hybrids at the diploid level. Chavez *et al.* (1988) studied the product of a triparental cross between *S. acaule* (A), *S. etuberosum* (E) and *S. pinnatisectum* (P) and with the genome structure AAAAEP. They observed full pairing between the *S. etuberosum* and *S. pinnatisectum* genomes. These species are widely regarded as being taxonomically well separated and have been placed in different subsections by Hawkes (1990) (Estolonifera and Potatoe, respectively). Hybrids between them are completely male-sterile (Hermsen and Taylor, 1979). This illustrates that there are internal interspecific incompatibility mechanisms operating between fairly remotely related species in *Solanum*, although not generally between species of the same (or even neighbouring) series.

As artificial hybrids form so readily between many diploid species, and since the genomes of even widely separated species pair to such an extent that they result in apparently normal diploid meiosis and sometimes fertile offspring, other isolation mechanisms must be operating to maintain the integrity of the multitude of species described in the section. These must include ecological, geographical, temporal and pollinator preferences of the taxa. However, Den Nijs and Peloquin (1977) formulated a hypothesis, later elaborated by Johnston *et al.* (1980) and by Johnston and Hanneman (1980b), which explained why many closely related species with apparently homoeologous genomes (i.e. genomes that pair fully in polyploid hybrids) failed to produce hybrid offspring when crossed at the diploid level. The authors emphasized the importance of the ratio between the ploidy levels of the embryo and the endosperm in ensuring the viability of the seed. An 'unbalanced' ratio would result in the formation of an inviable seed. The concept of an endosperm balance number (EBN) was introduced to determine whether or not the ploidy of endosperm and embryo are compatible. They described the hypothesis thus:

> Under this hypothesis the genome of each species is assigned a specific value in the endosperm. The value may be different for species of the same ploidy. It is the EBN which determines the effective ploidy in the endosperm and which must be in a 2 maternal : 1 paternal ratio.
> Johnston et al. (1980).

Johnston and Hannemann (1980a, 1981, 1982) produced evidence in support of the theory through an extensive series of crossing experiments. Species with the same EBN cross freely and produce viable offspring, whereas those differing in their EBN fail to produce viable seed. Doubling the ploidy level of a diploid plant or of its gametes also doubles the EBN. Thus, a tetraploid derived from one or two diploid species with EBN = 1 would have EBN = 2. Johnston et al. (1980) identified the importance of EBN as an isolating mechanism. They also drew attention to the fact that, when a diploid with an EBN half the value of a tetraploid species crosses with it, the EBN balance requirement can serve as a strong selection mechanism in favour of unreduced gametes of the diploid. In turn, this would facilitate direct introgression from the diploid into the tetraploid, while maintaining the ploidy integrity of each. An excellent review of the theory is given by Hawkes and Jackson (1992).

The EBN theory may also help to throw light upon why the polyploid species in the section are restricted to only six series. The EBN has been established for species belonging to 15 of the 21 series recognized by Hawkes (1990). All series which are exclusively diploid contain only one EBN. However, series with diploids and polyploids also contain either species with the same ploidy but a different EBN, or else species with the same EBN but a different ploidy level (Table 2.1). Therefore, in series containing a mix of diploids and tetraploids, there is evidence of EBN-mediated incompatibility groupings. Were the series to contain only diploid species, gene flow between the two species groups would not be possible. In time, the EBN isolating mechanism would facilitate divergence between them and might ultimately result in their becoming sufficiently divergent to be recognized as separate series. Alternatively, the EBN isolation mechanism could be overcome through the polyploidization of representatives of the lower EBN grouping. This would result in the formation of diploid and tetraploid plants of different groupings sharing the same EBN, and thereby allowing at least some gene flow to occur between them. The appearance of a more extensive polyploid series within the group would enable higher levels of gene flow to occur.

The extraordinary crossability between diploid species in the section and the fertility of their hybrids would seem to negate some of the forces which normally favour polyploidization (i.e. to facilitate gene flow between otherwise incompatible diploid species). However, the formation of incompatibility groupings through the evolution (possibly through mutation) of different EBNs within a series may restore the need for polyploidy to allow gene flow to occur.

Table 2.1. The distribution of endosperm balance number in *Solanum* section Petota Dumortier.

Series	Ploidy level		
	2x	4x	6x
Etuberosa	EBN = 1		
Bulbocastana	EBN = 1		
Pinnatisecta	EBN = 1		
Commersoniana	EBN = 1		
Circaeifolia	EBN = 1		
Lignicaulia	EBN = 1		
Yungasensa	EBN = 2		
Megistacroloba	EBN = 2		
Cuneoalata	EBN = 2		
Conicibaccata	EBN = 2	EBN = 2	
Piurana	EBN = 2	EBN = 2	
Tuberosa (wild)	EBN = 1 or 2	EBN = 4	EBN = 4
Tuberosa (cultivated)	EBN = 2	EBN = 2 or 4	EBN = 4
Longipedicellata		EBN = 2	
Demissa			EBN = 4

Another factor affecting the ploidy levels of some species has been cultivation by man. Conscious and unconscious selection of the largest and highest-yielding individuals is widely believed to favour the retention of polyploids, which are often larger and more vigorous. In potatoes, all cultivated forms belong to series Tuberosa, which is the largest in terms of numbers of species. It contains 76 wild species, of which 10 (7.6%) are polyploid, and seven cultivated species, four of which (i.e. 57%) are polyploid. The modern cultivated potato (*S. tuberosum*) is generally viewed as having an autopolyploid or segmental allopolyploid origin (Hawkes, 1958; Wagenheim *et al.*, 1960; Howard, 1973; Woodcock and Howard, 1975; Rickeman and Desborough, 1978). In contrast, the triploid cultivated species (*S. chaucha* and *S. juzepczukii*) and the cultivated pentaploid (*S. curtilobum*) are thought to have arisen through hybridization of clearly distinct species (see Hawkes, 1990).

The determination of the most likely origins of any polyploid species of *Solanum* has been complicated by the high degree of meiotic pairing that frequently occurs between the genomes of its possible progenitors. However, recent advances in the molecular cytogenetics of other plant groups may provide a powerful new approach to the problem. Genomic *in situ* hybridization (GISH) utilizes total genomic DNA as probes for *in situ* hybridization rather than the more traditional cloned repeated DNA sequences. This allows the identification of whole parental genomes and has been used in genome allocation of chromosomes in hybrids (Schwarzacher *et al.*, 1989) and in partial hybrids (Schwarzacher *et al.*, 1992), and has been used to determine the genome consitution of an allotetraploid grass (Bennett *et al.*, 1992). The application of the technique to wild and cultivated potatoes will provide valuable new data relating to the origins of the polyploid species in the group.

Aneuploidy

Variation in chromosome number through polyploidy, although scattered in its occurrence, is by no means uncommon in the genus. Aneuploidy, on the other hand, is extremely rare in natural populations and has not played a significant role in speciation or in the diversification of the group. Avanzi (1949) reported the appearance of aneusomatic individuals in a Chilean population of *S. pinnatum* with a chromosome complement of between $2n = 24$ and $2n = 28$. During metaphase, 80% of the cells contained $2n = 26$ chromosomes and the author observed that two of the chromosomes were considerably smaller than the others. It was unclear from the data presented whether the chromosomes were structurally altered autosomes or accessory chromosomes (so-called B chromosomes).

Despite repeated introductions of 'alien' germplasm from both wild and cultivated relatives by modern breeding programmes, the chromosome number of *S. tuberosum* has remained remarkably invariable ($2n = 4x = 48$) (e.g. Smith, 1927; Sree Ramulu *et al.*, 1983; Fish and Karp, 1986; Gill *et al.*, 1986). One notable exception is that of cv. Torridon, which carries two additional chromosomes ($2n = 48 + 2 = 50$). Two related species (*S. phureja* and *S. demissum*) were used repeatedly in the breeding of this cultivar and are thought to be the probable origin of the extra chromosomes (Wilkinson, 1992). Aneusomatic individuals with between 40 and 48 chromosomes have also been reported in the cultivar 'Adretta' (Schreiter, 1988). Clones varied in the proportion of hypotetraploid cells they contained, with the lowest containing 20% and the highest 56.5%. The origin of this variation could be chromosomal chimerism in the shoots giving rise to aneusomatic roots and/or chromosomal elimination during root growth.

Chromosome Instability

In vivo *somatic chromosome instability*

The loss of chromosomes from somatic tissues of living potato plants is a comparatively rare event and has not been widely reported.

An example of the high levels of chromosome stability usually encountered during vegetative growth was provided in a detailed study by Sree Ramulu *et al.* (1984) on a large population of protoclones. Protoplast regenerants contained many aneuploid, mixoploid and/or polyploid individuals. A comparison of the chromosome numbers of these plants with their 116 tuber progenies revealed that, in all but two instances, the original protoclone complement was preserved. The only exceptions were those of mixoploid individuals which segregated into different ploidy types.

Potato dihaploids ($2n = 2x = 24$) can be produced by anther culture (e.g. Powell and Uhrig, 1987) or, more usually, through the pollination of tetraploid *S. tuberosum* using pollen from certain *Solanum phureja* clones

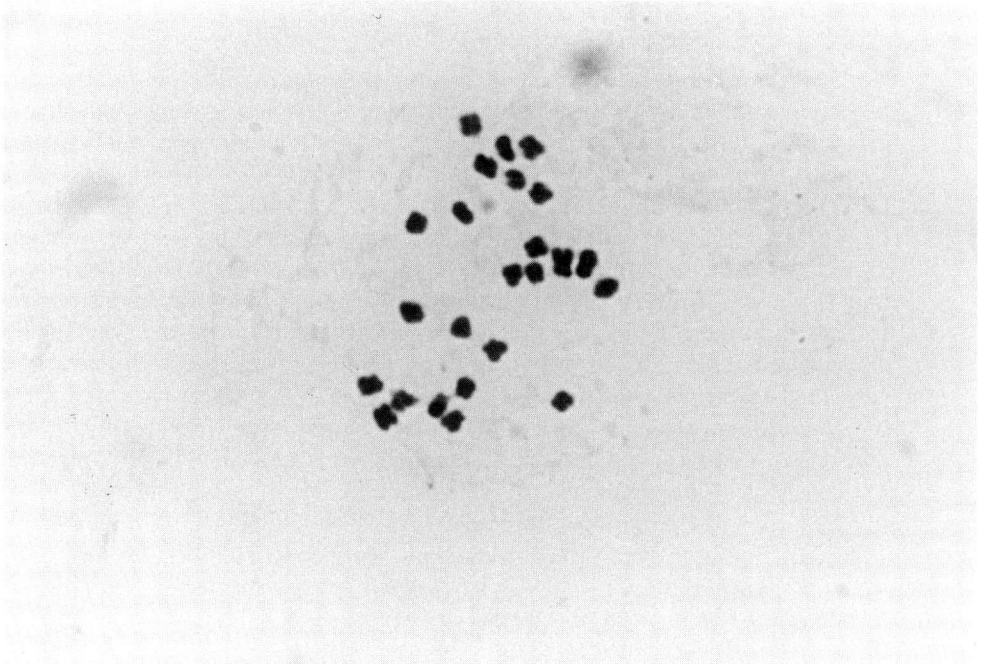

a)

(so-called dihaploid inducers) (e.g. Hougas and Peloquin, 1957). Until recently, it was widely believed that pollen from the dihaploid inducers stimulate unfertilized ovules to develop parthenogenically and that the dihaploid inducer itself does not make any genetic contribution to the dihaploid progeny (Hermsen and Verdenius, 1973; Rowe, 1974; Van Breukelen et al., 1977). Clulow et al. (1991) discovered that many of the dihaploids produced from such crosses were in fact aneusomatic and that, in addition to diploid cells, there were some cells that contained one or two additional chromosomes, and occasional triploid cells (Figure 2.3a–c). The detection of S. phureja-specific cDNA markers (Clulow et al., 1991) and isozyme markers (Clulow et al., 1992) in these plants demonstrated that the extra chromosomes originated from the dihaploid inducer clone and therefore that fertilization must have taken place in these plants. The predominance of cells with a dihaploid chromosome number ($2n = 2x = 24$) probably arose through the systematic loss of the S. phureja chromosomes. The elimination process is probably entirely somatic and occurs after fertilization; otherwise it would be difficult to account for the occasional triploid cells in some individuals. Nearly all modern dihaploid inducer clones are homozygous for a dominant marker gene which produces a purple coloration on the seed of its offspring (embryo spot). The absence of this embryo spot marker from the seeds which gave rise to the dihaploid clones demonstrates that the S. phureja chromosome which carries this gene had been eliminated before seed maturation. Likewise the fact that these plants survived a ploidy screen early in development leads to the suggestion

Genome Evolution in Potatoes

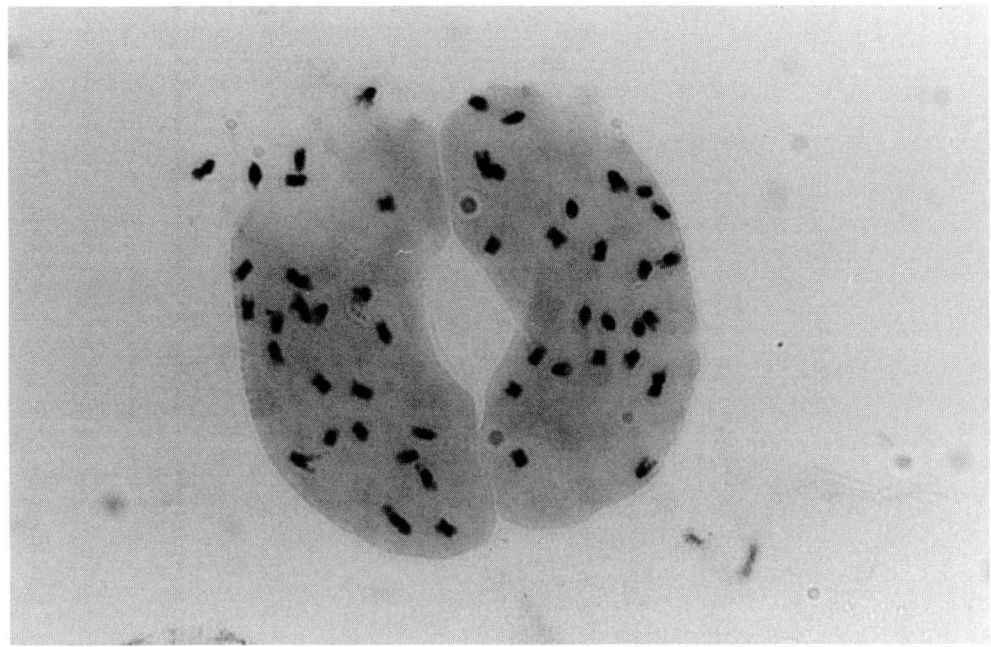

b)

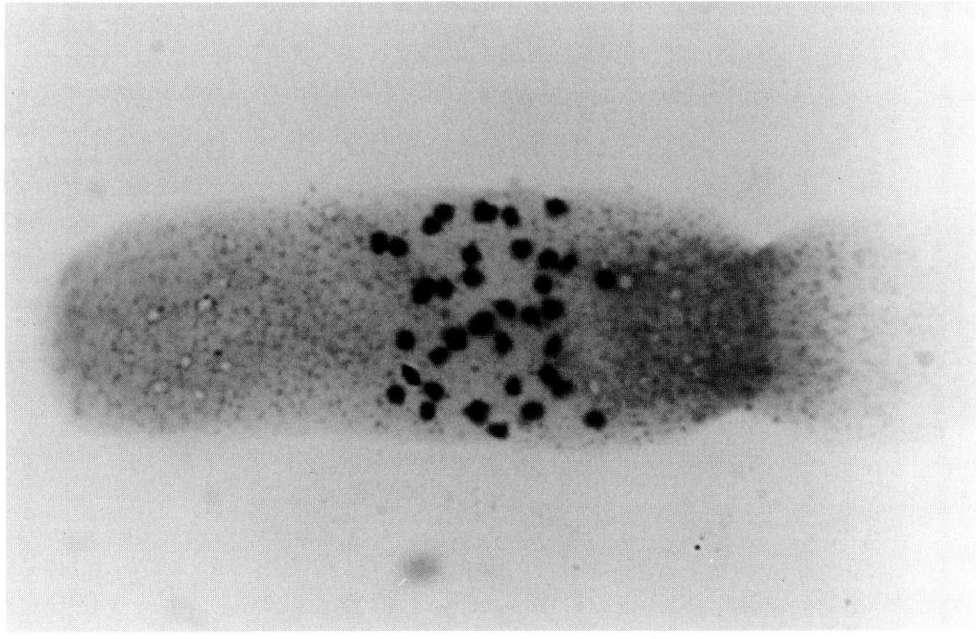

c)

Fig. 2.3. Squash preparations of root cells of: a) PDH 7, 24 chromosomes; b) PDH 7, 25 chromosomes; c) PDH 41, 36 chromosomes.

that most chromosomes are lost at least by the seedling stage and more probably during embryo development. That a poor endosperm can prevent maturation of triploid (and presumably near-triploid) zygotes (Wagenheim et al., 1960; Kasha, 1974) also favours the rapid loss of *S. phureja* chromosomes during seed development.

It seems likely, therefore, that the chromosomes are highly unstable during development of the embryo. However, many of the dihaploids examined have been in existence for many years (up to 20) and yet chromosome profiles of different tuber progenies of the same clone often retain broadly similar aneusomatic profiles (M.J. Wilkinson and S.A. Clulow, unpublished). This suggests either that no further chromosomal loss occurs during vegetative growth or that the elimination rate is extremely slow. Certainly, examination of unpretreated shoot and root meristems reveals no evidence of elimination. This mirrors the situation in *Hordeum vulgare* × *H. bulbosum* hybrids, where, following fertilization, the *H. bulbosum* chromosomes are rapidly eliminated during embryo development (Bennett et al., 1976). However, Thomas and Pickering (1983) observed that, where chromosome elimination does not take place in the embryo, loss of the *H. bulbosum* chromosomes in the somatic tissue is slow and erratic.

The stability of the aneusomatic profiles within and between tuber progenies of the same clone raises the questions: what type of chimerism is generated by chromosome loss and how is it maintained between tuber generations? There are several different types of cytological chimeras which could develop. Cytological differences might arise between different areas, tissues or organs of the plant, between the initial cells which give rise to the differentiated tissues or between groups of initial cells, resulting in differences between tissue layers. It is extremely unlikely that cytological differences between different parts of the plant would survive many tuber generations. In any case, such chimerism would be characterized by cytological differences between roots rather than chimerism within them. Therefore, the most likely alternatives are cytological chimerism between tissue layers (periclinal chimeras) and/or between initial cells within a layer. Howard et al. (1963) recognized three independent layers in the stem apex and labelled them L1, L2 and L3. It was suggested that the growing points of auxiliary buds (and, by inference, tubers) trace back to the L1 and L2 layers. L1 gives rise to the epidermal layers and its ploidy can be assessed by chloroplast counts in stomatal guard cells. L2 gives rise to the male germ line and so its ploidy can be assessed by meiotic analysis or by pollen size. Roots derive from L3 and root squash preparations can be used to provide accurate chromosome numbers for this layer. The application of colchicine treatment to *S.* × *juzepczukii* enabled Howard et al. (1963) to produce periclinal chimeras between these layers. Initial plants obtained from the treatment included individuals with some triploid roots (L3) and hexaploid pollen mother cells (PMCs) (L2), and others with some hexaploid roots and entirely triploid PMCs. The failure to secure stable L1 ($3x$), L2 ($6x$) and L3 ($3x$) or L1 ($3x$), L2 ($3x$) and L3 ($6x$) from tuber progenies of these plants led Howard et al. (1963) to the theory that stolons and tubers were primarily derived from L1 and L2. Over several tuber generations the original L3 component becomes entirely replaced by L2. This hypothesis was strongly supported by later observations of tuber transmission of an L3 chlorophyll mutant (Howard, 1972). The replacement of the L3 layer by the L1 and L2 layers means that any stable interlayer chimerism must be between the L1 and L2 layers. However, since the L1 layer is unlikely to contribute greatly to the L3 replacement (from which root counts are derived), and since there are commonly three different chromosome numbers in the aneusomatic dihaploids, interlayer chimerism is unlikely to account for much, if any, of the observed aneusomatic profiles. By elimination, then, differences between the chromosome numbers of the 'initial' cells of the L2 layer are probably important in determining the nature of the aneusomatic profile. That the aneusomatic profiles are largely conserved between tuber generations implies that there is little reassortment between the initial cell profiles of the mother plant and its tubers, although there is mixing between cell lines of the L2 (Howard *et al.*, 1963). The mechanism by which relative frequencies are preserved therefore requires further investigation.

Differential instability of chromosomes between somatic tissues was also observed in glasshouse- and field-grown *S. tuberosum* cv. Torridon (Wilkinson, 1992). The two additional chromosomes discovered in the cultivar (believed to have originated from *S. phureja* or *S. demissum*) were found to be stable in shoot tissues ($2n = 48 + 2 = 50$), but were progressively eliminated during root growth. In mature plants this led to a discrepancy between the mean chromosome counts of the shoots and those of the roots. Somatic bridges during anaphase were associated with the elimination process and were thought to be implicated with the mechanism of chromosome loss.

The possibility that chromosomes of 'alien' origin (i.e. those from a different species) are more stable in some tissues of the potato than in others is of intrinsic interest and may have implications for work attempting to effect limited chromosome transfer by sexual means or by somatic hybridization. The search for further examples may reveal whether tissue-dependent instability is widespread, and may point towards a mechanism of action.

In vitro *somatic chromosome instability*

In the last 10 years there has been intense interest over the instability of chromosomes in potato plants regenerated from tissue culture. Most of this work has concentrated on somatic hybrids and plants regenerated from protoplast culture. However, there have been several notable works on plants regenerated from leaf or stem callus cultures. Jacobsen (1981) studied changes in chromosome numbers to dihaploid plants regenerated from leaf calli. Evidence of polyploidization and of chromosome elimination leading to aneusomatic individuals was reported amongst the regenerants. Chromosome loss was limited to tetraploid plants, with no elimination being observed at the dihaploid level. Polyploidization was therefore a prerequisite of chromosome elimination from dihaploid material. The proportion of tetraploid and octaploid plants recovered from culture was heavily dependent upon genotype, indicating a possible genetic factor contributing to the polyploidization mechanism. The author also found evidence that the regeneration technique itself contributed towards the tendency for chromosome doubling. A marginally higher frequency of polyploidization was noted in plants regenerated from media containing 2,4-D rather than NAA. An absence of ploidy chimeras in over 300 regenerants suggests either that all shoots derived from a single cell and no doubling occurred after shoot initiation or that shoots were derived from more than one cell of the same chromosome number. Diploid regenerants were all euploid but evidence of chromosome elimination was noticed at the tetraploid level. Mixohypotetraploid plants with between 43 and 48 chromosomes were found to retain the characteristic into the tuber progenies. However, the mean chromosome number per individual fell in at least two of these clones, suggesting that further elimination may have occurred during vegetative growth. It was suggested that somatic bridges observed during the callus phase might be one of the possible causes of chromosome loss.

Chromosome doubling during regeneration is less frequent when tetraploid explants are used, although chromosome loss appears to be more common. In a study of callus-derived regenerants of potato cultivars, Wheeler et al. (1985) found no instances of chromosome doubling but nine cases (out of 62) of aneuploidy. There was little deviation from the euploid complement, with most aneuploids falling into the range $2n = 47-49$. Similar levels of instability were reported by Ooms et al. (1985), who are investigating 42 regenerants of the cultivar Desiree transformed by *Agrobacterium rhizogenes*. Ten of these were aneuploid with 46, 47 or 49 chromosomes. Structural changes to the nucleolus organizer chromosome were observed in two of the aneuploids. In one, the chromosome had an additional segment (the result of translocation or amplification), whereas, in the other, part of the long arm had been deleted (or translocated to another chromosome). Landsmann and Uhrig (1985) used restriction fragment length polymorphism (RFLP) analysis rather than cytological observations to detect changes associated with passage through callus and protoplast culture. They reported no differences detectable between the original clones, their callus-derived regenerants and the tuber progenies of these regenerants.

Studies of regeneration from protoplast culture have generally revealed a higher incidence of chromosome elimination and of chromosome doubling from tetraploid starting material than has been observed from leaf- or stem-derived calli (e.g. Karp et al., 1982; Sree Ramulu et al., 1983; Fish and Karp, 1986; Gill et al., 1987). However, there was considerable variation between these works. For instance, Karp et al. (1982) reported that only one out of 26 regenerated protoclones of cultivar Maris Bard (i.e. 4%) retained its original complement of $2n = 48$, compared with no change in the eight plants regenerated from leaf-disc callus. In a later work (Fish and Karp, 1986), the proportion of tetraploid protoplast regenerants increased to 60%, although they were unable to attribute the cause of this difference to a single factor. Higher frequencies of tetraploids amongst protoplast regenerants were recorded by Sree Ramulu et al. (1983) (64% for cv. Bintje), Wheeler et al.

(1985) (84% for cvs Champion, Myatts, Desiree) and Gill *et al.* (1987) (100% for cv. Russet Burbank). Comparisons between these works reveal little about the factors affecting chromosome instability from protoplast culture because each study used a different methodology and different genotypes.

The regeneration process from protoplast culture is difficult and notoriously dependent upon the genotype of the starting material. In consequence, most studies have concentrated upon observing the performance of only one cultivar or of one somatic hybrid combination. The divergence in the results obtained by these works is consistent with, but not proof of, genotype dependency in the stability of chromosomes during regeneration. In one of few exceptions, Karp *et al.* (1982) made an extensive study of the chromosome instability in cultivars Maris Bard and Fortyfold. Although a degree of aneuploidy and polyploidization was noted in both cultivars, considerable differences were observed between them. Only one of the 26 regenerants of Maris Bard contained 48 chromosomes, with the remainder containing 46–96 chromosomes, all but two of these being hypertetraploid. In comparison, the numbers in Fortyfold varied little: all but one of the 25 examined fell in the range $2n = 46-49$ (one contained $2n = 93$) and eight of these were tetraploid. Minor differences in the regeneration procedures used for the cultivars prevent the observed differences from being attributed categorically to differences in genotype, although this would appear to be the most probable explanation.

Perhaps the most marked difference between plants regenerated from protoplast culture and those from leaf or stem callus culture is the extended period of callus growth in the former. One possible reason for the increased frequency of chromosome elimination during protoplast regeneration, therefore, might be that elimination occurs principally during the callus growth phase. Certainly, a number of authors have indicated that a single callus can give rise to plants with differing chromosome numbers (e.g. Sree Ramulu *et al.*, 1983; Austin *et al.*, 1985; Creissen and Karp, 1985). In some instances regenerants from the same callus differed by only one chromosome or by relatively small numbers of chromosomes. This provides compelling evidence that at least some chromosome elimination occurs during callus growth, since there are no reports of aneusomaty in any cultivar used as an explant source in these studies. However, where 'sibling' callus regenerants differ in their ploidy levels, there is the possibility that the observed difference may have derived from natural endoreduplication in the original explant tissue. D'Amato (1978) stated that endoreduplication occurs in most flowering plants. The appearance of diplochromosomes in material regenerated from tissue culture provides cytological evidence suggesting that endoreduplication also occurs in potatoes (De Vries *et al.*, 1987; Pijnacker *et al.*, 1987). Jacobsen *et al.* (1983) and Sree Ramulu and Dijkhuis (1986) used flow-cytometric techniques to reveal variation in the presence and degree of endoreduplication between different genotypes and ploidy levels, and between different plant tissues. Sree Ramulu and Dijkhuis (1986) found no evidence of endoreduplication in cv. Bintje shoot cultures and used this to follow changes in ploidy level during callus and protoplast culture. Flow-cytometric techniques and mitotic squash preparations demonstrated the gradual appearance of octoploid and aneuploid cells in both forms of *in vitro* culture.

Somatic hybrids differ from other protoplast culture regenerants in that they possess chromosomes from at least two sources (generally from different species). The effect of chromosomes from one genome on the stability of those from another has been examined by a number of workers. De Vries *et al.* (1987) made somatic fusions between a monoploid potato and a nitrate reductase-deficient *Nicotiana plumbaginifolia*. Clear differences were noted in the karyotypes of the two species and the authors were able to determine that the fusion products possessed 40–120 chromosomes from *N. plumbaginifolium* and 9–20 chromosomes from the potato. Preferential elimination of the chromosomes of either species was not detected in over 80 somatic hybrids. Pijnacker *et al.* (1987) used Giemsa C-banding to distinguish between chromosomes of *S. phureja* (P) and *S. tuberosum* (T) in somatic hybrids with the genome constitutions of TTTTPP, TTPPPP or TTTTPPPP. They reported preferential loss of the *S. phureja* satellited chromosomes and suggested that elimination probably occurred during early callus growth. The dosage of the *S. phureja* genomes had a significant effect upon the instability of the two nucleolar chromosomes in that elimination was enhanced when four *S. phureja* genomes were present. The nucleolar chromosomes were also more stable in somatic hybrids containing four *S. tuberosum* genomes than in those containing only two. The authors concluded that the elimination of these chromosomes was under genetic control. In a later work, Pijnacker *et al.* (1989) examined the stability of chromosomes in somatic hybrids between dihaploid *S. tuberosum* (TT) and *S. phureja* (PP). The resultant regenerants included 11 hypotetraploids with between one and seven chromosomes missing. The nucleolar chromosomes of *S. phureja* were again lost preferentially to other chromosomes present and only one hybrid was found to have lost a nucleolar chromosome of *S. tuberosum*. Clearly, then, the stability of the *S. phureja* nucleolar chromosomes is strongly dependent upon the genotype of the somatic hybrid.

Structural instability of potato chromosomes

In addition to changes to chromosome number, a number of workers have described changes to the gross structure of chromosomes associated with passage through tissue culture. Creissen and Karp (1985) provided one of the earliest and clearest examples of karyotype rearrangements in an extensive study of over 200 protoplast regenerants of *S. tuberosum* cv. Majestic. One of these regenerants contained one abnormally large chromosome and another chromosome so small that it approached the dimensions of a centric fragment. The authors concluded that this atypical karyotype is perhaps best explained by an unequal translocation event between the two chromosomes involved. Sree Ramulu *et al.* (1985) reported on the occurrence of dicentric chromosomes in six regenerants of *S. tuberosum* dihaploids and chromosome fragments in a further two regenerants. Similarly, Ooms *et al.* (1985) reported structural changes to the karyotype in two of the ten aneuploids obtained during the *A. rhizogenes*-mediated transformation of cv. Desiree. In the first, part of the long arm of the nucleolar (NOR) chromosome had been deleted and, in the second, the NOR chromosome contained an additional segment, which may have been the result of a translocation or deletion. Pijnacker *et al.* (1987) also found evidence of structural rearrangements in the NOR chromosomes in somatic hybrids between *S. tuberosum* and *S. phureja* and suggested that breakage in the NORs or satellites may precede or be implicated in the process of elimination of the nucleolar chromosomes. Gill *et al.* (1987) studied changes in chromosome structure as revealed through pachytene analysis, and reported that two of the protoplast regenerants of cv. Russet Burbank contained deletion loops in chromosomes 3 and 10 which were not present in the parental cultivar.

Applications of chromosome instability

The clonal nature of the potato crop usually negates the possibility of passage through a seed generation following genetic transformation, because this would result in recombination. Therefore it is not possible to 'correct' heterozygous structural rearrangements such as deletions or translocations by the selection of karyotypically typical individuals in the selfed progeny. In turn, it becomes of high priority to establish regeneration techniques that minimize such structural changes. Loss of whole chromosomes or polyploidization are similarly undesirable traits during the production of genetically transformed potato clones.

The instability of chromosome numbers during regeneration from callus or protoplast culture or during vegetative growth also has several obvious negative implications in the use of wild relatives to improve the genetic base of the modern cultivated potato. In particular, it may be the source of unwanted variation during attempts to genetically modify currently successful cultivars or clones by asymmetric somatic hybridization. It is therefore highly important that a fuller understanding is developed of the factors which determine chromosome elimination, chromosome doubling and structural changes to chromosomes during the passage of potatoes through tissue culture. Evidence of different levels of endoreduplication in different tissues and different genotypes (Jacobsen *et al.*, 1983; Sree Ramulu and Dijkhuis, 1986) means that the choice of the original explant can also affect chromosome number variability amongst the regenerants. The effect of various aspects of the culture technique itself upon the instability of chromosomes also needs careful examination so that methods can be developed to minimize unwanted elimination or doubling. However, not all chromosome instability is necessarily undesired. Indeed, a limited level of deliberately induced chromosome loss could provide a powerful tool in linkage studies. The comparison of clones which differ only in the deliberately induced elimination of one chromosome could reveal molecular markers and dominant genetic traits specific to the missing chromosome. This approach would be particularly useful on monosomic addition lines. Evidence suggesting that elimination can be at least sometimes genome-specific leads to the possibility that, if conditions favouring elimination were identified and adopted, significant numbers of chromosomes from the wild relative could be eliminated. This could be particularly useful where only a relatively small number of traits are sought from the wild species, as it would shorten the backcrossing programme needed to produce material of commercial quality. It is therefore highly important that a greater understanding is developed of the mechanisms controlling chromosome instability in potatoes.

References

Austin, S., Baer, M., Ehlenfeldt, M., Kazmierczak, J.P. and Helgeson, J.P. (1985) Intraspecific fusions of *Solanum tuberosum*. *Theoretical and Applied Genetics* 71, 172-175.

Avanzi, M.G. (1949) Il numero cromosomico di specie e varietà di *Solanum* oriunde del Chile. *Caryologia* 2, 205-222.

Bains, G.S. (1951) Cytogenetical studies in the genus *Solanum*, sect. Tuberarium. Unpublished MSc. dissertation, University of Cambridge.

Bennett, M.D., Finch, R.A. and Barclay, I.R. (1976) The time rate and mechanism of chromosome elimination in *Hordeum* hybrids. *Chromosoma* 54, 175-200.

Bennett, S.T., Kenton, A.Y. and Bennett, M.D. (1992) Genomic *in situ* hybridization reveals the allopolyploid nature of *Milium montianum* (Gramineae). *Chromosoma* 101, 420-424.

Bernatzky, R. and Tanksley, S.D. (1986) Towards a saturated linkage map in tomato based on isozymes and random cDNA sequences. *Genetics* 112, 887-898.

Bonierbale, M.W., Plaisted, R.L. and Tanksley, S.D. (1988) RFLP maps based on a common set of clones reveal modes of chromosomal evolution in potato and tomato. *Genetics* 120, 1095-1103.

Bukasov, S.M. (1939) The origin of potato species. *Physis* (*Buenos Aires*) 18, 41-46.

Chavez, R., Brown, C.R. and Iwanaga, M. (1988) Application of interspecific sesquiploidy to introgression of PLRV resistance from non-tuber-bearing *Solanum etuberosum* to cultivated potato germplasm. *Theoretical and Applied Genetics* 76, 497-500.

Choudhuri, H.C. (1943) Cytological studies in the genus *Solanum*. II. Wild and cultivated diploid potatoes. *Transactions of the Royal Society of Edinburgh* 61, 199-219.

Clulow, S.A., Wilkinson, M.J., Waugh, R., Baird, E., De Maine, M.J. and Powell, W. (1991) Cytological and molecular observations on *Solanum phureja*-induced dihaploid potatoes. *Theoretical and Applied Genetics* 82, 545-551.

Clulow, S.A., Wilkinson, M.J. and De Maine, M.J. (1992) Potato dihaploid production - a new hypothesis. In: Rouselle-Bourgeois, F. and Rousselle, P. (eds) *Proceedings of EAPR/EUCARPIA Potato Sections Conference*. INRA, Landerneau, France, pp. 165-169.

Creissen, G.P. and Karp, A. (1985) Karyotypic changes in potato plants regenerated from protoplasts. *Plant Cell Tissue and Organ Culture* 4, 171-182.

D'Amato, F. (1978) Chromosome number variation in cultured cells and regenerated plants. In: Thorpe T.A. (ed.) *Frontiers of Plant Tissue Culture*. IAPTC, Calgary, pp. 287-295.

Den Nijs, T.P.M. and Peloquin, S.J. (1977) The role of endosperm in hybridization. *American Potato Journal* 54, 488-489 (Abstract).

De Vries, S.E., Ferwerda, M.A., Loonen, A.E.H.M., Pijnacker, L.P. and Feenstra, W.J. (1987) Chromosomes in somatic hybrids between *Nicotiana plumbaginifolia* and a monoploid potato. *Theoretical and Applied Genetics* 75, 170-176.

Ecochard, R., Ramanna, M.S. and De Nettancourt, D. (1969) Detection and cytological analysis of tomato haploids. *Genetica (The Hague)* 40, 181-190.

Ellison, W. (1936) Mciosis and fertility in certain British varieties of the cultivated potato (*S. tuberosum*). *Genetica* 18, 217-254.

Emme, H. (1937) Genetik der Kartoffel II. Bastarde von *Solanum emmeae* Juz. und *S. commersonii* Dun. Arten, welche gegen den Koloradokä fer widerstandsfähig sind. *Biologicheskii Zhurnal* 6, 299-310.

Emme, H. (1938) Studies on interspecific hybridization of tuber bearing potatoes section Tuberarium Bitter, genus *Solanum* L. *Biologicheskii Zhurnal* 7, 1093-1104.

Fish, N. and Karp, A. (1986) Improvements in regeneration from protoplasts of potato and studies on chromosome stability. *Theoretical and Applied Genetics* 72, 405-412.

Fukuda, Y. (1927) Cytological studies on the development of pollen grains in different races of *S. tuberosum* with special reference to sterility. *Botanical Magazine (Tokyo)* 61, 459-476.

Gebhardt, C., Ritter, E., Barone, A., Debener, T., Walkemeier, B., Schachtschabel, U., Kaufmann, H., Thompson, R.D., Bonierbale, M.W., Ganal, M.W., Tanksley, S.D. and Salamini, F. (1991) RFLP maps of potato and their alignment with the homoeologous tomato genome. *Theoretical and Applied Genetics* 83, 49-57.

Gill, B.S., Kam-Morgan, L.N.W. and Shepard, J.F. (1986) Origin of chromosomal and phenotypic variation in potato protoclones. *Journal of Heredity* 77, 13-16.

Gill, B.S., Kam-Morgan, L.N.W. and Shepard, J.F. (1987) Cytogenetic and phenotypic variation in mesophyll cell-derived tetraploid potatoes. *Journal of Heredity* 78, 15-20.

Gottschalk, W. (1956) Die Cytologie der Kulturtomate und ihrer wildwachsenden Verwandten. *Bibliographia Genetica* 17, 1-109.

Hawkes, J.G. (1944) The indigenous American potatoes and their value in plant breeding. *Empire Journal of Experimental Agriculture* 13, 1-40.

Hawkes, J.G. (1958) Potatoes: taxonomy, cytology and crossability. In: Kappert, H. and Rudorf, W. (eds) *Handbuch der Pflanzenzuchtungz*, 2nd edn. Paul Parey, Hamburg, pp. 10-43.

Hawkes, J.G. (1990) *The Potato: Evolution, Biodiversity and Genetic Resources*. Belhaven Press, London.

Hawkes, J.G. and Jackson, M.T. (1992) Taxonomic and evolutionary implications of the endosperm balance number hypothesis in potatoes. *Theoretical and Applied Genetics* 84, 180-185.

Hermsen, J.G.Th. and Taylor, L.M. (1979) Successful hybridization of non-tuberous *Solanum etuberosum* Lind. and tuber-bearing *S. pinnatisectum*. *Euphytica* 28, 1-7.

Hermsen, J.G.Th. and Verdenius, J. (1973) Selection from *Solanum tuberosum* group *phureja* of genotypes combining high frequency haploid induction with homozygosity for embryo spot. *Euphytica* 22, 244-259.

Hougas, R.W. and Peloquin, S.J. (1957) A haploid plant of the potato variety Katahdin. *Nature* 180, 1209-1210.

Howard, H.W. (1972) The stability of an L3 mutant potato chimera. *Potato Research* 15, 374–377.

Howard, H.W. (1973) Calyx forms in dihaploids in relation to the origin of *Solanum tuberosum*. *Potato Research* 16, 43–46.

Howard, H.W. and Swaminathan, M.S. (1952) Species differentiation in the genus *Solanum*, sect. Tuberarium with particular reference to the use of interspecific hybridization in breeding. *Euphytica* 1, 20–28.

Howard, H.W., Wainwright, J. and Fuller, J.M. (1963) The number of independent layers at the stem apex in potatoes. *Genetica* 34, 113–120.

Jacobsen, E. (1981) Polyploidization in leaf callus tissue and in regenerated plants of dihaploid potato. *Plant Cell Tissue and Organ Culture* 1, 77–84.

Jacobsen, E., Tempelaar, M.J. and Bijmolt, E.W. (1983) Ploidy levels in leaf callus and plants of *Solanum tuberosum* determined by cytophotometric measurements of protoplasts. *Theoretical and Applied Genetics* 65, 113–118.

Johnston, S.A. and Hanneman, R.E., Jr (1980a) Support of the endosperm balance number hypothesis utilizing some tuber-bearing *Solanum* species. *American Potato Journal* 57, 7–14.

Johnston, S.A. and Hanneman, R.E., Jr (1980b) The discovery of effective ploidy barriers between diploid solanums. *American Potato Journal* 57, 484–485 (Abstract).

Johnston, S.A. and Hanneman, R.E., Jr (1981) The discovery of one additional endosperm balance number (EBN) in diploid *Solanum* species. *American Potato Journal* 58, 505–506 (Abstract).

Johnston, S.A. and Hanneman, R.E., Jr (1982) Manipulations of endosperm balance number overcome barriers between diploid *Solanum* species. *Science* 217, 446–448.

Johnston, S.A., der Nijs, T.P.M., Peloquin, S.J. and Hanneman, R.E., Jr (1980) The significance of genic balance to endosperm development in interspecific crosses. *Theoretical and Applied Genetics* 57, 5–9.

Karp, A., Nelson, R.S., Thomas, E. and Bright, S.W.J. (1982) Chromosome variation in protoplast-derived potato plants. *Theoretical and Applied Genetics* 63, 265–272.

Kasha, K.J. (1974) Haploids from somatic cells. In: Kasha, K.J. (ed.) *Haploids in Higher Plants: Advances and Potential*. University of Guelph, Guelph, pp. 67–87.

Koopmans, A. (1951) Cytogenetic studies on *Solanum tuberosum* L. and some of its relatives. *Genetica* 25, 193–337.

Lamm, R. (1945) Cytogenetic studies in *Solanum* sect. Tuberarium. *Hereditas (Lund)* 31, 1–128.

Landsmann, J. and Uhrig H. (1985) Somaclonal variation in *Solanum tuberosum* detected at the molecular level. *Theoretical and Applied Genetics* 71, 500–505.

Lapitan, N.L.V., Ganal, M.W. and Tanksley, S.D. (1991) Organization of the 5S ribosomal RNA genes in the genome of tomato. *Genome* 34, 509–514.

Lawrence, W.J.C. (1931) The secondary association of chromosomes. *Cytologia* 2, 352–384.

Lee, H.K. and Hanneman, R.E., Jr (1976) Identification of the extra chromosomes in Giemsa stained somatic cells of pachytene identified trisomics of *Solanum chacoense*. *Canadian Journal of Genetics and Cytology* 18, 297–302.

Lee, K.H. and Rowe, P.R. (1975) Trisomics in *Solanum chacoense*: fertility and cytology. *American Journal of Botany* 62, 593–601.

Livermore, J.R. and Johnstone, F.E. (1940) The effect of chromosome doubling on the crossability of *S. chacoense*, *S. jamesii* and *S. bulbocastanum* with *S. tuberosum*. *American Potato Journal* 17, 169–173.

Longley, A.E. and Clark, C.F. (1930) Chromosome behaviour and pollen production in the potato. *Journal of Agricultural Research* 41, 867–888.

Magoon, M.L., Hougas, R.W. and Cooper, D.C. (1958) Cytogenetic studies of South American diploid *Solanums*, section Tuberarium. *Americal Potato Journal* 35, 375–394.

Martinez-Zapater, J.M. and Oliver, J.L. (1985) Isozyme gene duplication in diploid and tetraploid potatoes. *Theoretical and Applied Genetics* 70, 172–177.

Mok, D.W.S., Heiyoung, K.L. and Peloquin, S.J. (1974) Identification of potato chromosomes with Giemsa. *American Potato Journal* 51, 337–341.

Müntzing, A. (1933) Studies on meiosis in diploid and triploid *S. tuberosum* L. *Hereditas (Lund)* 17, 223–245.

Ooms, G., Karp, A., Burrell, M.M., Turell, D. and Roberts, J. (1985) Genetic modification of potato development using Ri T-DNA. *Theoretical and Applied Genetics* 70, 440–446.

Pijnacker, L.P. and Ferwerda, M.A. (1984) Giemsa C-banding of potato chromosomes. *Canadian Journal of Genetics and Cytology* 26, 415–419.

Pijnacker, L.P., Ferwerda, M.A., Puite, K.J. and Roest, S. (1987) Elimination of *Solanum phureja* nucleolar chromosomes in *S. tuberosum* and *S. phureja* somatic hybrids. *Theoretical and Applied Genetics* 73, 878–882.

Pijnacker, L.P., Ferwerda, M.A., Puite, K.J. and Schaart, J.G. (1989) Chromosome elimination and mutation in tetraploid somatic hybrids of *Solanum tuberosum* and *Solanum phureja*. *Plant Cell Reports* 8, 82–85.

Powell, W. and Uhrig H. (1987) Anther culture of *Solanum* genotypes. *Plant Cell Tissue and Organ Culture* 11, 13–24.

Prakken, R. and Swaminathan, M.S. (1952) Cytological behaviour of some interspecific hybrids in the genus *Solanum*, sect. Tuberarium. *Genetica* 26, 77–101.

Ramanna, M.S. and Hermsen, J.G.Th. (1971) Somatic chromosome elimination and meiotic chromosome pairing in the triple hybrid 6x (*Solanum acaule* × *S. bulbocastanum*) × 2x-*S. phureja*. *Euphytica* 20, 470–481.

Ramanna, M.S. and Wagenvoort, M. (1976) Identification of the trisomic series in diploid *Solanum tuberosum* L. group *tuberosum*. I. Chromosome identification. *Euphytica* 25, 233–240.

Rickeman, V.S. and Desborough, S.L. (1978) Elucidation of the evolution and taxonomy of cultivated potatoes with electrophoresis. I. Groups *tuberosum*, *andigena*, *phureja* and *stenotomum*. *Theoretical and Applied Genetics* 52, 217–220.

Rowe, P.R. (1974) Methods of producing haploids: parthenogenesis following interspecific hybridization. In: Kasha, K.J. (ed.) *Haploids in Higher Plants, Advances and Potential*. University of Guelph, Guelph, pp. 43–52.

Rybin, V.A. (1930) Karyologische Untersuchgen an einigen wilden und einheimischen kultivierten Kartoffeln Amerikas. *Zeitschrift für induktive Abstammungs-Vererbungslehre* 53, 313–354.

Rybin, V.A. (1933) Cytological investigation of the South American cultivated and wild potatoes and its significance for plant breeding. *Bulletin of Applied Botany, Series II* 2, 3–100.

Salaman, R.N. (1926) *Potato Varieties*. Cambridge University Press, Cambridge.

Schreiter, J. (1988) Variability in chromosome number in sprout-borne roots of the potato *Solanum tuberosum* $2n = 4x = 48$. *Archiv für Zuchtungsforschung* 18(3), 169–174.

Schwarz, P.A. (1937) Zytogenetische Untersuchungen der Kartoffel. *Bulletin of the Academy of Sciences of the USSR Natural Series Biology* 1, 59–67.

Schwarzacher, T., Leitch, A.R., Bennett, M.D. and Heslop-Harrison, J.S. (1989) *In situ* localization of parental genomes in a wide hybrid. *Annals of Botany* 64, 315–324.

Schwarzacher, T., Anamthawat-Jonsson, K., Harrison, G.E., Islam, A.K.M.R., Jia, J.Z., King, I.P., Leitch, A.R., Miller, T.E., Reader, S.M., Rogers, W.J., Shi, M. and Heslop-Harrison, J.S. (1992). Genomic *in situ* hybridization to identify alien chromosomes and chromosome segments in wheat. *Theoretical and Applied Genetics* 84, 778–786.

Singh, A.K., Uhrig, H. and Salamini, F. (1988) Implications of chromosome pairing in a monoploid and its colchidiploid of *Solanum tuberosum* ($x = 12$). *Genome* 30, 347–351.

Singh, A.K., Salamini, F. and Uhrig, H. (1989) Chromosome pairing in 14 F_1 hybrids among 11 diploid potato species. *Journal of Genetics and Breeding* 43, 1–5.

Smith, H.B. (1927) Chromosome counts in the varieties of *S. tuberosum* and allied wild species. *Genetics* 12, 84–92.

Sree Ramulu, K.P. and Dijkhuis, P. (1986) Flow cytometric analysis of polysomaty and *in vitro* genetic instability in potato. *Plant Cell Reports* 3, 234–237.

Sree Ramulu, K.P., Dijkhuis, P. and Roest, S. (1983) Phenotypic variation and ploidy level of plants regenerated from protoplasts of tetraploid potato (*Solanum tuberosum* L. cv. Bintje). *Theoretical and Applied Genetics* 65, 329–338.

Sree Ramulu, K.P., Dijkhuis, P. and Roest, S. (1984) Genetic instability in protoclones of potato (*Solanum tuberosum* L. cv. Bintje): new types of variation after vegetative propagation. *Theoretical and Applied Genetics* 68, 515–519.

Sree Ramulu, K.P., Dijkhuis, P., Roest, S., Bokelmann, G.S. and deGroot, B. (1985) Variation in phenotype and chromosome number of plants regenerated from protoplasts of dihaploid and tetraploid potato. *Plant Breeding* 97, 119–128.

Stow, I. (1927) A cytological study on the pollen sterility in *S. tuberosum* L. *Proceedings of the Imperial Academy* 2, 426–430.

Swaminathan, M.S. (1951) Notes on induced polyploids in the tuber-bearing *Solanum* species and their crossability with *S. tuberosum*. *American Potato Journal* 28, 472–489.

Swaminathan, M.S. (1954) Nature of polyploidy in some 48-chromosome species of the genus *Solanum*, section tuberarium. *Genetics* 39, 59–76.

Swaminathan, M.S. and Howard, H.W. (1953) The cytology and genetics of the potato (*Solanum tuberosum*) and related species. *Bibliographia Genetica* 16, 1–192.

Swanson, C.P., Merz, T. and Young, W.J. (1981) *Cytogenetics: the Chromosome in Division, Inheritance and Evolution*. Prentice-Hall, Englewood Cliffs, New Jersey.

Thomas, H.M. and Pickering, R.A. (1983) Chromosome elimination in *Hordeum vulgare* $\times$ *H. bulbosum* hybrids. 1. Comparisons of stable and unstable amphidiploids. *Theoretical and Applied Genetics* 66, 135–140.

Thomas, P.T. and Revell, S.H. (1946) Secondary association and heterochromatic attraction. *Annals of Botany* 9, 159–164.

Ugent, D. (1970) The potato: what is the origin of this important crop plant, and how did it first become domesticated? *Science* 170, 1161–1166.

Van Breukelen, E.W.M., Ramanna, M.S. and Hermsen, J.G.Th. (1975) Monohaploids ($n = x = 12$) from autotetraploid *Solanum tuberosum* ($2n = 4x = 48$) through two successive cycles of female parthenogenesis. *Euphytica* 24, 567–574.

Van Breukelen, E.W.M., Ramanna, M.S. and Hermsen, J.G.Th. (1977) Parthenogenic monohaploid ($2n = x = 12$) from *Solanum tuberosum* L. and *S. verrucosum* Schlechtd., and the production of homozygous potato dihaploids. *Euphytica* 26, 263–271.

Wagenheim, K.H.v., Peloquin, S.J. and Hougas, R.W. (1960) Embryological investigations on the formation of haploids in the potato (*Solanum tuberosum*). *Zeitschrift für Vererbungslehre* 91, 391–399.

Wagenvoort, M. (1982) Location of the recessive gene ym (yellow margin) on chromosome 12 of diploid *Solanum tuberosum* by means of trisomic analysis. *Theoretical and Applied Genetics* 61, 239–243.

Wagenvoort, M. (1988) Chromosomal localization of a recessive gene tp controlling the pleiotropic character topiary in *Solanum*. *Theoretical and Applied Genetics* 75, 712–716.

Wagenvoort, M. and Ramanna, M.S. (1979) Identification of the trisomic series in diploid *Solanum tuberosum* L. group *tuberosum*. II. Trivalent configurations at pachytene stage. *Euphytica* 28, 633–642.

Wanscher, J.H. (1934) The basic chromosome number of higher plants. *New Phytologist* 33, 101–126.

Wheeler, V.A., Evans, N.E., Foulger, D., Webb, K.J., Karp, A., Franklin, J. and Bright, S.W.J. (1985) Shoot formation from explant cultures of fourteen potato cultivars and studies of the cytology and morphology of regenerated plants. *Annals of Botany* 55(3), 309–320.

Wilkinson, M.J. (1992) The partial stability of additional chromosomes in *Solanum tuberosum* cv. Torridon. *Euphytica* 60, 115–122.

Woodcock, K.M. and Howard, H.W. (1975) Calyx type in *Solanum tuberosum* dihaploids, *S. stenotomum*, *S. sparsipilum* and their hybrids. *Potato Research* 18, 460–465.

Yeh, B.P. and Peloquin, S.J. (1965) Pachytene chromosomes of the potato (*Solanum tuberosum* group *andigena*). *American Joural of Botany* 52(10), 1014–1020.

II THEORY AND METHODS OF GENETICAL ANALYSIS

3 Quantitative Genetics Theory for Tetrasomic Inheritance

J.E. BRADSHAW

Crop Genetics Department, Scottish Crop Research Institute, Invergowrie, Dundee DD2 5DA, UK.

Introduction

Quantitative genetics is concerned with the inheritance of those differences between individuals which give rise to continuous variation in populations of living organisms. Such variation is common for economically important traits in domesticated animals and cultivated plants, and for traits affecting survival and fertility in wild species. Quantitative genetics, therefore, has an important role to play in the development of scientific breeding methods for animal and crop improvement, and in theories of how evolution occurs in natural populations.

The foundations of the modern theory of quantitative inheritance were laid by the experimental work of Johannsen (1909), Nilsson-Ehle (1909) and East (1915), and by the theoretical work of Fisher (1918). By progeny-testing individuals, Johannsen showed that heritable and non-heritable factors were jointly responsible for quantitative variation in seed weight of beans (*Phaseolus vulgaris*). Nilsson-Ehle discovered three Mendelian factors for red versus white grain in wheat (*Triticum aestivum*) and showed that the degree of redness was associated with the number of factors present. He and East independently realized that the segregation of a number of Mendelian factors (genes) of similar and cumulative action, together with the effects of non-heritable factors, could account for continuous variation. East used this new multifactorial (polygenic) hypothesis to explain the variation seen in the families and generations derived from a cross between two varieties of tobacco (*Nicotiana longiflora*) that differed in corolla length.

Fisher showed that the correlations observed between human relatives for stature could be deduced from biometrical models on the supposition of Mendelian inheritance, and that due allowance could be made for the Mendelian factors differing in the magnitude of their effects and in their degree of dominance. Genetical information could therefore be inferred from measurements on related individuals, even though individual genes could not be recognized by Mendelian analysis because discrete classes could not be discerned.

The ways in which this biometrical approach has been refined to take account of all the properties of genes known from Mendelian genetics and the ways in which it has been tested experimentally can be found in the following books: Kempthorne (1957), Thompson and Thoday (1979), Bulmer (1980), Mather and Jinks (1982), Becker (1984), Crow (1986), Falconer (1989) and Narain (1990). Applications in plant breeding can be found in Wricke and Weber (1986) and in Mayo (1987). Only Kempthorne (1957) and Wricke and Weber (1986), however, deal with tetraploids and tetrasomic inheritance in any detail.

This chapter reviews theoretical results of relevance to both diploid and tetraploid potatoes. The emphasis, however, is on the latter because the most widely cultivated potato is a tetraploid which displays tetrasomic inheritance and the theory is less well covered in books on quantitative genetics. The theory and uses of $2n$ gametes from diploids are dealt with in a separate chapter by Tai, and the mapping of quantitative trait loci is covered in the next chapter.

Genotype–Environment Interaction

As potatoes are readily propagated by asexual reproduction, many individuals with the same genotype (a clone) can be produced and assessed under different environmental conditions. Observed differences between plants of the same genotype grown in the same growth room, glasshouse or field arise from differences in their microenvironments, 'accidents' of development and errors of measurement. A proper randomization of different genotypes to the units of an experiment, however, ensures that there is no correlation (covariance) between genotype and environment. Hence the observed phenotypic variation, with variance σ_P^2, can be partitioned into two components, a genotypic variance σ_G^2 and a non-genotypic variance σ_e^2:

$$\sigma_P^2 = \sigma_G^2 + \sigma_e^2$$

A good experimental design and careful experimentation can reduce σ_e^2, but replication is usually required to increase the precision of estimates of genotype means, as well as providing an estimate of σ_e^2. The observed variance of genotype means over r replicates is a phenotypic variance $\sigma_{\bar{P}}^2$:

$$\sigma_{\bar{P}}^2 = \sigma_G^2 + \frac{\sigma_e^2}{r}$$

The ratio $\sigma_G^2/\sigma_{\bar{P}}^2$ is called broad-sense heritability.

In designing experiments with potatoes, one must be aware of the possibility of transmission of environmental differences between individuals to all their clonal descendants. In other words, common-origin seed tubers are highly desirable. Furthermore, the non-genetic variance may not be the same for all genotypes because genotypes can vary in their sensitivity to environmental differences. This phenomenon is called genotype–microenvironment interaction.

Genotype–environment interactions can also occur between genotypes and specific macroenvironments such as locations, seasons and fertilizer regimes. The interactions may or may not involve changes in the ranking of genotypes over environments. Provided genotypes (G) are replicated in each environment as well as over environments (E), genotype–environment interactions (GE) can be detected through a two-way analysis of variance, and their relative importance determined from the components of variance:

$$\sigma_P^2 = \sigma_G^2 + \sigma_E^2 + \sigma_{GE}^2 + \sigma_e^2$$

The observed variance of genotype means over n environments with r replicates in each environment is:

$$\sigma_P^2 = \sigma_G^2 + \frac{\sigma_{GE}^2}{n} + \frac{\sigma_e^2}{nr}$$

Hence estimates of the components of variance from a series of variety trials can be used to evaluate the effectiveness of the trials system and to make improvements where necessary (Talbot, 1984). On the other hand, one may wish to know if there are changes in the ranking of varieties such that one variety does consistently better than another in one environment but consistently worse in another environment. Recently, Baker (1988) has shown how a technique developed by Azzalini and Cox (1984) can be used for this purpose.

Sometimes all or most of the genotype–environment interaction variance can be accounted for by the heterogeneity of regressions (β_i) in a joint regression analysis of the interactions on to an environmental index. In the absence of a physical assessment of the environments, the mean value of all genotypes in a given environment (E_j) is used as a biological assessment of that environment (Finlay and Wilkinson, 1963), and is expressed as a deviation from the overall mean (μ):

$$P_{ij} = \mu + G_i + E_j + (GE)_{ij} + \bar{e}_{ij}$$
$$= \mu + G_i + E_j + \beta_i E_j + d_{ij} + \bar{e}_{ij}$$

Where the heterogeneity of regressions accounts for all of the interaction (i.e. the ds all equal zero), the regression coefficients are properties of the genotypes *per se* and measure their responsiveness to environmental change. Where the heterogeneity of regressions accounts for most of the interaction, the regression coefficients still have a predictive value, but they are no longer properties of the genotypes *per se*, nor are the deviations from regression. These points are best seen by considering the response curves and surfaces of genotypes to changes in their physical environment, and translating them into regression lines (Knight, 1970, 1973). A simple example is shown in Figure 3.1, where deviations from regression occur for genotype A (maximum at lower

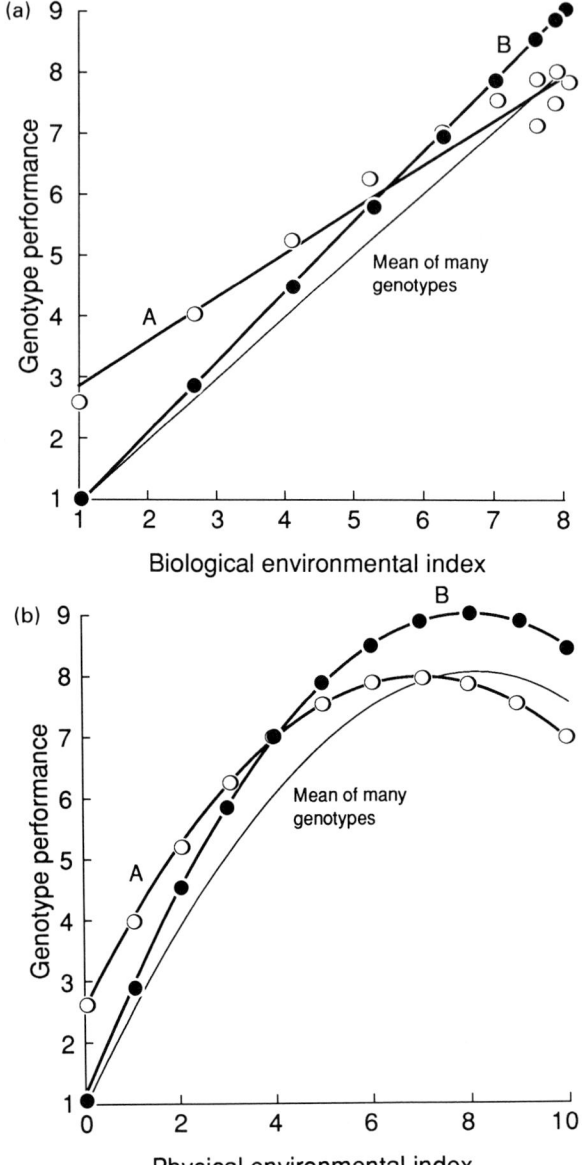

Fig. 3.1. Response curves and their corresponding regression lines.

level of physical index) but not for genotype B (higher maximum at same level of physical index).

The complexity or otherwise of genotype–environment interactions can, in general, be seen from a principal component analysis. If the overall (additive) effects of genotypes and environments are first removed from the data, one is left with multiplicative terms for the interaction (Gauch, 1988):

$$P_{ij} = \mu + G_i + E_j + \sum_{k=1}^{K} \theta_k u_{ki} v_{kj} + \bar{e}_{ij}$$

where θ, u and v are the square roots of the eigenvalues, PCA scores for genotypes and PCA scores for environments, and K is the number of PCA axes retained in the model. The first few principal components may account for most of the interaction and may have a biological meaning, in which case the nature of the interactions can be visualized in biplots (Kempton, 1984), as done by Dale and Phillips (1989) in their assessment of tolerance to potato cyst nematodes.

A consideration of the sensitivity of genotypes to environmental change has led to predictions about the consequences of selecting genotypes in different types of environment. Upward selection in above-average environments and downward selection in below-average environments should result in genotypes with a high sensitivity, whereas upward selection in below-average environments and downward selection in above-average environments should result in ones with a low sensitivity. These predictions have been confirmed in selection experiments with the fungus *Schizophyllum commune* (Jinks and Connolly, 1973, 1975) and with the tobacco *Nicotiana rustica* (Jinks and Pooni, 1982). Similar work could be done with potatoes, as genotypes can be maintained by vegetative propagation for further evaluation.

Falconer (1989) has shown how the concept of genetic correlation (r_G) can be applied to selection problems involving genotype–environment interaction by regarding a trait measured in two environments as two different, but correlated, traits. The improvement of performance in environment Y as a result of selection in environment X is then seen as a correlated response (CR), which depends on the square root of the heritability of performance in each environment and the genetic correlation between the two performances:

$$CR_y = i h_x h_y r_G \sigma_{PY}$$

where i is the intensity of selection, h is the square root of the heritability and σ_{PY} is the square root of the phenotypic variance in Y. The correlated response can be compared with the response (R_y) expected from direct selection in Y:

$$\frac{CR_y}{R_y} = r_G \frac{i_x h_x}{i_y h_y}$$

Further information on genotype–environment interactions can be found in the reviews by Freeman (1973) on statistical methods of analysis, Hill (1975) on the challenge for plant breeding and Lin *et al.* (1986) on stability parameters. References to work on genotype–environment interactions in potatoes can be found in Vermeer (1990). Although genotype–environment interaction is not mentioned in subsequent sections, it is assumed that the experimental designs used will ensure that there is no genotype by environment covariance, and that genotypes are replicated over environments when it is necessary to ensure that the conclusions have general validity.

Tetrasomic Inheritance

In a true autotetraploid the four sets of chromosomes are entirely homologous, meiotic pairing within each group of four homologous chromosomes is completely random and tetrasomic inheritance always occurs. Whilst tetrasomic inheritance of major genes has commonly been reported in tetraploid *S. tuberosum* (see Chapter 12), it may nevertheless be a segmental allotetraploid rather than a true autotetraploid, so that departures from expected Mendelian ratios through non-random pairing of chromosomes cannot be ruled out.

Biometrical models

Biometrical models contain more parameters with tetrasomic than with disomic inheritance. With two alleles at a diploid locus there are three genotypes A_1A_1, A_1A_2 and A_2A_2 which are usually given the genetical values a, d and $-a$ (Fisher, 1918) or d, h and $-d$ (Fisher *et al.*, 1932), which is unfortunate because the different meanings of d can be confusing. The latter notation is extended to digenic interactions between two loci by including parameters i, j and l for homozygote by homozygote, homozygote by heterozygote and heterozygote by heterozygote interactions (Wright, 1979; Mather and Jinks, 1982). With two alleles at a tetraploid locus there are five genotypes, and again different and potentially confusing notations are used for their genetical values:

	(1)	(2)	(3)
$A_1A_1A_1A_1$ quadruplex	$2a$	$2d$	d
$A_1A_1A_1A_2$ triplex	$a + d_3$	$d + 3h + v + w$	h_3
$A_1A_1A_2A_2$ duplex	d_2	$4h$	h_2
$A_1A_2A_2A_2$ simplex	$-a + d_1$	$-d + 3h - v + w$	h_1
$A_2A_2A_2A_2$ nulliplex	$-2a$	$-2d$	$-d$

Notation (1) is used by Wricke and Weber (1986), notation (2) by Easton (1976) and Wright (1979), and notation (3) by Killick (1971) and Mather and Jinks (1982). With four alleles at a locus, there are 35 genotypes and hence a need for simple but sensible models of gene expression with which to explore the consequences of inbreeding and crossbreeding (for example, see Bingham, 1980).

Double reduction

Furthermore, with tetrasomic inheritance, sister chromatids can end up in the same (diploid) gamete as a result of four homologous chromosomes forming a quadrivalent, crossing over occurring between the locus and spindle attachment (centromere), the two chromosomes with the crossover going to the same pole (adjacent disjunction, i.e. genetic non-disjunction) at first anaphase (probability $\frac{1}{3}$ with random disjunction), and the two sister chromatids going to the same pole at second anaphase (probability usually $\frac{1}{2}$). The phenomenon is known as double reduction (Mather, 1936), and the coefficient of double reduction (α) is defined as the probability of two sister chromatids going to the same gamete (Fisher and Mather, 1943). The coefficient has a maximum value of $\frac{1}{6}$, as can be seen in Figure 3.2. With chromosomal segregation, for example when four homologous chromosomes form two bivalents or the locus lies close to the centromere, $\alpha = 0$, whereas when $\alpha = \frac{1}{7}$ it is as though random chromatid segregation has occurred. The gametic output of an autotetraploid follows from the definition of α. Each of the four homologous chromosomes can provide a pair of sister chromatids, so that the probability for any one chromosome is $\frac{1}{4}\alpha$. Then for the $1 - \alpha$ situations where this does not occur, two chromosomes are chosen at random. The outcome is:

	Gametes			
	A_1A_1	A_1A_2	A_2A_2	Divisor
$A_1A_1A_1A_1$	1	—	—	1
$A_1A_1A_1A_2$	$2 + \alpha$	$2(1 - \alpha)$	α	4
$A_1A_1A_2A_2$	$1 + 2\alpha$	$4(1 - \alpha)$	$1 + 2\alpha$	6
$A_1A_2A_2A_2$	α	$2(1 - \alpha)$	$2 + \alpha$	4
$A_2A_2A_2A_2$	—	—	1	1

Unfortunately the symbol α is also used for the average effect of a gene substitution.

Inbreeding

Inbreeding is the intermating of individuals with common ancestors and hence a certain probability of having alleles identical by descent, which they can transmit to their offspring. Alleles are identical by descent when they are copies of a gene occurring previously in the ancestry, or one is a copy of the other (Malécot, 1948). Although many of the formulae derived by Malécot using this concept had already been obtained by Wright (1921) using path coefficients, Malécot's method has greater generality (Kempthorne, 1957). The inbreeding coefficient (F) in a diploid is simply the probability that the two alleles at a locus are identical by descent. It depends on the mating system, with self-pollination the strongest form of inbreeding. Furthermore, for a diploid, the inbreeding coefficient of an individual equals the coancestry of its parents. The coefficient of coancestry (Φ), or parentage, is the probability (P) that a pair of alleles taken at random from each of two individuals, X and Y, are identical by descent:

$$\Phi_{XY} = \tfrac{1}{4}[P(x_1 \equiv y_1) + P(x_1 \equiv y_2) + P(x_2 \equiv y_1) + P(x_2 \equiv y_2)]$$

The probability that two loci are simultaneously identical by descent is F^2 for unlinked loci, but for linked loci no general formula can be given (Wricke and Weber, 1986).

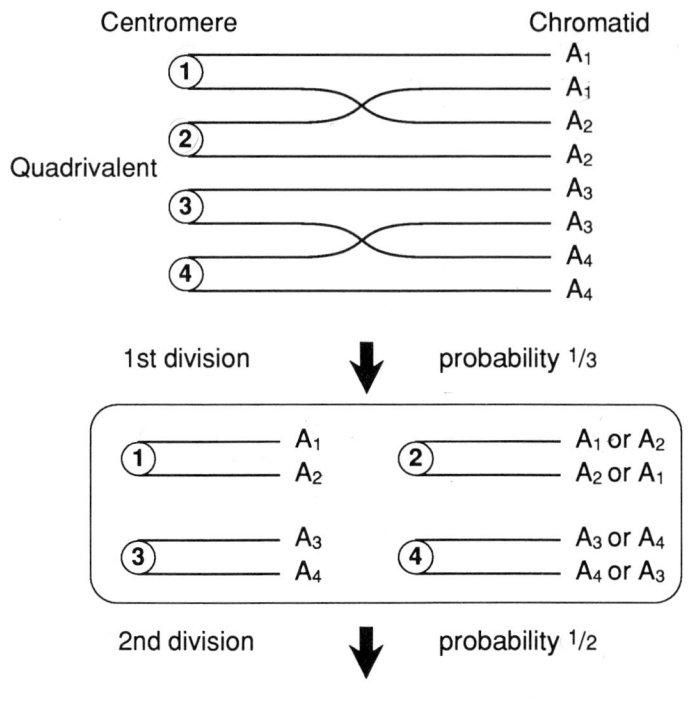

Fig. 3.2. Double reduction.

The inbreeding coefficient (F) in a tetraploid is the probability that a randomly chosen pair of alleles at a locus are identical by descent. Hence, for an individual $X_iX_jY_kY_l$ formed from gametes X_iX_j and Y_kY_l,

$$F = \tfrac{1}{6}P[(x_i \equiv x_j) + (x_i \equiv y_k) + (x_i \equiv y_l) + (x_j \equiv y_k) + (x_j \equiv y_l) + (y_k \equiv y_l)]$$
$$= \tfrac{1}{6}(F_x + 4\Phi_{XY} + F_y)$$

because a gamete can contain two alleles identical by descent, the probability of which is by definition the inbreeding coefficient of the parent from which it came. Mendoza and Haynes (1973, 1974a) used these concepts of coancestry and inbreeding to discuss the effects of inbreeding tetraploid potatoes and their implications for potato breeding, and to determine the genetic relationship among potato cultivars grown in the USA. For a full description of inbreeding, however, further parameters are required. This can most readily be seen when an individual with four non-identical alleles $A_1A_2A_3A_4$ is selfed (Figure 3.3): as well as progeny with only a pair of alleles identical by descent (probability P_3), for example $A_1A_1A_2A_3$, there will also be individuals with two pairs of alleles identical by descent, but different from each other (probability P_2), for example $A_1A_1A_2A_2$. On further selfing, individuals with three identical alleles $A_1A_1A_1A_2$ and ones with four identical alleles $A_1A_1A_1A_1$ are produced, with probabilities P_1 and P_0, respectively. Hence

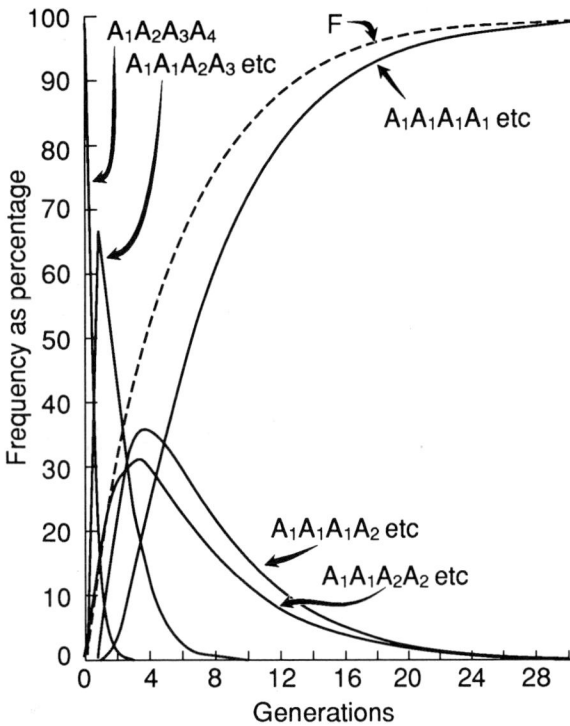

Fig. 3.3. Decline in heterozygosity on selfing.

$$F = P_0 + \tfrac{1}{2}P_1 + \tfrac{1}{3}P_2 + \tfrac{1}{6}P_3$$

With double reduction, the effect of one generation of selfing on F (Kempthorne, 1957) is

$$F_1 = \tfrac{1}{6}(1 + 2\alpha) + \tfrac{1}{6}(5 - 2\alpha)F_0$$

When F_0 and α both equal zero, F_1 equals $\tfrac{1}{6}$ compared with $\tfrac{1}{2}$ for diploids. The results of continuous selfing with chromosome segregation on the four parameters F, $P_0 + \tfrac{1}{4}P_1$, P_0 and P_2 can be found in Gallais (1968) and Wricke and Weber (1986).

The probability method of Malécot (1948) cannot be used to determine the array of genotypes in an arbitrary generation arising from an arbitrary initial population under a particular system of inbreeding. For this, the generation matrix approach of Fisher (1965) is required, which is based on the concept of the genotype and the results of Mendelian segregation. A comparative evaluation of the two methods can be found in Kempthorne (1957). The generation matrix for selfing an autotetraploid is given in Figure 3.4.

In a diploid species, the effects of inbreeding can be removed by one generation of random mating because the offspring of a cross between unrelated parents are not inbred, regardless of the inbreeding coefficients of

(a) $$\begin{bmatrix} u_{t+1} \\ v_{t+1} \\ w_{t+1} \\ x_{t+1} \\ y_{t+1} \end{bmatrix} = \begin{bmatrix} 1 & \dfrac{(2+\alpha)^2}{16} & \dfrac{(1+2\alpha)^2}{36} & \dfrac{\alpha^2}{16} & 0 \\ 0 & \dfrac{(2+\alpha)(1-\alpha)}{4} & \dfrac{2(1+2\alpha)(1-\alpha)}{9} & \dfrac{\alpha(1-\alpha)}{4} & 0 \\ 0 & \dfrac{(2-2\alpha+3\alpha^2)}{8} & \dfrac{(3-4\alpha+4\alpha^2)}{6} & \dfrac{(2-2\alpha+3\alpha^2)}{8} & 0 \\ 0 & \dfrac{\alpha(1-\alpha)}{4} & \dfrac{2(1+2\alpha)(1-\alpha)}{9} & \dfrac{(2+\alpha)(1-\alpha)}{4} & 0 \\ 0 & \dfrac{\alpha^2}{16} & \dfrac{(1+2\alpha)^2}{36} & \dfrac{(2+\alpha)^2}{16} & 1 \end{bmatrix} \begin{bmatrix} u_t \\ v_t \\ w_t \\ x_t \\ y_t \end{bmatrix}$$

(b) $$\begin{bmatrix} u_{t+1} \\ v_{t+1} \\ w_{t+1} \\ x_{t+1} \\ y_{t+1} \end{bmatrix} = \begin{bmatrix} 1 & \tfrac{1}{4} & \tfrac{1}{18} & \tfrac{1}{36} & 0 \\ 0 & \tfrac{1}{2} & \tfrac{4}{9} & \tfrac{2}{9} & 0 \\ 0 & \tfrac{1}{4} & \tfrac{1}{2} & \tfrac{1}{4} & \tfrac{1}{6} \\ 0 & 0 & 0 & \tfrac{1}{2} & \tfrac{2}{3} \\ 0 & 0 & 0 & 0 & \tfrac{1}{6} \end{bmatrix} \begin{bmatrix} u_t \\ v_t \\ w_t \\ x_t \\ y_t \end{bmatrix}$$

Fig. 3.4. a) Generation matrix for selfing an autotetraploid with two alleles and double reduction. u, v, w, x and y are the frequencies of $A_1A_1A_1A_1$, $A_1A_1A_1A_2$, $A_1A_1A_2A_2$, $A_1A_2A_2A_2$ and $A_2A_2A_2A_2$ in generations t and t+1.
b) Generation matrix for selfing an autotetraploid with four alleles but no double reduction. u, v, w, x and y are the frequencies of $A_iA_iA_iA_i$, $A_iA_iA_iA_j$, $A_iA_iA_jA_j$, $A_iA_iA_jA_k$, and $A_iA_jA_kA_l$ in generations t and t+1, where i, j, k and l can each take the values 1 to 4 i.e. there are four different alleles.

the parents. The situation is quite different in a tetraploid, because gametes can contain two alleles identical by descent. The inbreeding coefficient of the offspring depends on the inbreeding coefficients of the parents and the probability that two pairs of alleles are identical by descent is greater than zero if both parents are inbred. The inbreeding coefficients of tetraploids derived from diploids which produce $2n$ gametes have been determined by Haynes (1992).

The theory of inbreeding and crossbreeding in autotetraploids is therefore more complicated than in diploids, and there are differences which have consequences for breeding strategies aimed at exploiting heterosis (hybrid vigour) in hybrid or synthetic varieties. In tetraploids, for example, double-cross hybrids can be superior to single-cross hybrids. Such results are relevant to methods of widening the genetic base of potato breeding and of producing true potato seed. The interested reader is referred to Mendoza and Haynes (1974b), Bingham (1980), Gallais (1981), Sanford and Hanneman (1982) and Wricke and Weber (1986), as well as to chapter 22 by Golmirzaie et al., for further information.

Random Mating Populations in Equilibrium

The genetical structure of a population of interbreeding individuals is of central importance in both the study of evolution in natural populations and the development of long-term methods of crop improvement by recurrent selection, particularly in outbreeding crops. Most wild and cultivated diploid species of potato are outbreeders in which self-fertilization is prevented by a gametophytic self-incompatibility system. Populations consist of highly heterozygous individuals and are intolerant of inbreeding. The cultivated tetraploid potato, although self-compatible, is also regarded as an outbreeder. Again, populations are highly heterozygous and show inbreeding depression.

In studying such populations the concept of a random mating (panmictic) population in equilibrium is extremely important. Only in such a population can the total genetical variation be partitioned into uncorrelated components which are attributable to different causes, and which can be used to determine the covariance of relatives and to predict the response to selection. Furthermore, individuals chosen for inheritance studies based on family relationships must come from such a population for the results to have a genetical rather than a statistical interpretation.

The approach to equilibrium

As neither natural populations under selection nor newly created experimental populations are likely to be in equilibrium, we need to consider how quickly an equilibrium can be achieved.

With disomic inheritance, one generation of random mating establishes the equilibrium between gene and genotype frequencies at a single locus. For multiple alleles A_i at frequencies p_i ($i = 1$ to k), the genotypes and their frequencies can be expressed as $(\Sigma p_i A_i)^2$, the well known Hardy–Weinberg equilibrium. With many loci, the equilibrium frequency of gamete $A_i B_j C_k D_l \ldots$ is $p_i p_j p_k p_l \ldots$, but this gametic-phase equilibrium is not established by one generation of random mating. For two loci A and B, the gametic phase disequilibrium for gamete $A_i B_j$ in generation t is

$$P_t - p_i p_j = (1 - c)^t [P_o - p_i p_j]$$

where c is the recombination frequency (with values from 0 to 0.5) and P_o and P_t are the frequencies of $A_i B_j$ in generations o and t, respectively (Crow and Kimura, 1970).

With tetrasomic inheritance the single-locus equilibrium is not attained in one generation and the genotype frequencies at equilibrium are not the products of the gene frequencies unless $\alpha = 0$. The Hardy–Weinberg principle does, however, hold true in the sense that the zygotic genotype frequencies are given by the products of the appropriate gametic frequencies. For multiple alleles A_i at frequencies p_i, if P_{ii} and $2P_{ij}$ are the frequencies of gametes $A_i A_i$ and $A_i A_j$, respectively, at equilibrium (e)

$$P_{ij,e} = (1-f)p_i p_j$$

and

$$P_{ii,e} = (1-f)p_i^2 + fp_i$$

where

$$f = 3\alpha/(2 + \alpha)$$

When $\alpha = 0$, the gametic and genotypic frequencies at equilibrium are $(\Sigma p_i A_i)^2$ and $(\Sigma p_i A_i)^4$, respectively. With four alleles at equal frequencies, genotypes with three different alleles, such as $A_1 A_1 A_2 A_3$, are the most frequent type and account for 56.25% of the population.

When $\alpha = \frac{1}{7}$, $f = \frac{1}{5}$ and we have the equivalent of 20% inbreeding.

The rate at which equilibrium is approached (Crow and Kimura, 1970) is:

$$P_{ij,t} - P_{ij,e} = \left(\frac{1-\alpha}{3}\right)^t (P_{ij,o} - P_{ij,e})$$

When $\alpha = 0$, three generations of random mating reduce this difference to 3.7% of its initial value. Taken in conjunction with the rate of approach to gametic-phase equilibrium in diploids, at least three generations of random mating would seem desirable before commencing a genetical analysis.

Partitioning the genetical variation

The factorial analysis of variance approach to partitioning the genetical variation is a general method which can deal with multiple alleles at any number of loci and any level of ploidy. It was developed by Kempthorne and others (Kempthorne, 1957) from the work of Fisher (1918). With n loci there are $2n$ factors in diploids and $4n$ factors in tetraploids. The levels of each factor are the different alleles. The main effects of single factors and their successively higher orders of interaction are fitted by least squares. The complexity of the genetical model is determined by the highest order of interaction considered. All possible genotypes and their frequencies are taken into account and effects are measured from the population mean. It is, however, easier to relate the components of variation to genetical phenomena at the level of the gene rather than the population when there are just two alleles at each locus. In these circumstances, Wright (1979) has shown that the components can be derived more easily by the use of the differential calculus.

In a diploid with multiple alleles at two loci A and B, the genotypic value (G) of genotype $A_i A_j B_k B_l$, as a deviation from the population mean, is

$$G = \alpha_i + \alpha_j + \beta_{ij} + \alpha_k + \alpha_l + \beta_{kl} + \alpha\alpha_{ik} + \alpha\alpha_{il} + \alpha\alpha_{jk} + \alpha\alpha_{jl} + \alpha\beta_{ikl} + \alpha\beta_{jkl} + \beta\alpha_{ijk} + \beta\alpha_{ijl} + \beta\beta_{ijkl}$$

where the αs are the main effects of alleles, the βs are the diallelic interactions, and the $\alpha\alpha$s, $\alpha\beta$s + $\beta\alpha$s and $\beta\beta$s are the non-allelic interactions. The corresponding components of variance are:

$$\sigma_G^2 = \sigma_A^2 + \sigma_D^2 + \sigma_{AA}^2 + \sigma_{AD}^2 + \sigma_{DD}^2$$

where A and D stand for additive and dominance, respectively. The contributions of d, h, i, j and l, as defined by Mather and Jinks (1982), to these components in a two-allele model can be found in Wright (1979), but it is worth emphasizing that all five parameters contribute to the additive variance:

$$\sigma_A^2 = 2pq[d_A + (u-v)i + 2uvj_A + (q-p)(h_A + (u-v)j_B + 2uvl)]^2$$
$$+ 2uv[d_B + (p-q)i + 2pqj_B + (v-u)(h_B + (p-q)j_A + 2pql)]^2$$

where p and q are the frequencies of the two alleles at locus A, and u and v are the frequencies at B, and the j suffix refers to the locus which is homozygous.

In a tetraploid with multiple alleles at a single locus, A_i, at frequencies p_i, the genotypic value (G) of genotype $A_iA_jA_kA_l$ is

$$G = \alpha_i + \alpha_j + \alpha_k + \alpha_l + \beta_{ij} + \beta_{ik} + \beta_{il} + \beta_{jk} + \beta_{jl} + \beta_{kl} + \gamma_{ijk} + \gamma_{ijl} + \gamma_{ikl} + \gamma_{jkl} + \delta_{ijkl}$$

where the αs are the main effects of alleles and the βs, γs and δs are diallelic, triallelic and tetra-allelic interactions. With chromosomal segregation, the genotypic variance can be partitioned into four components:

$$\sigma_G^2 = \sigma_A^2 + \sigma_D^2 + \sigma_T^2 + \sigma_Q^2$$

where the components for multiple alleles (Kempthorne, 1957) and two alleles (Wright, 1979) are:

$$\sigma_A^2 = 4\sum_i p_i\alpha_i^2$$
$$= 4pq[d + 3(q-p)h + (6pq-1)v + (q-p)^3w]^2$$
$$\sigma_D^2 = 6\sum_i\sum_j p_ip_j\beta_{ij}^2$$
$$= 24p^2q^2[h - (q-p)v + (q-p)^2w]^2$$
$$\sigma_T^2 = 4\sum_i\sum_j\sum_k p_ip_jp_k\gamma_{ijk}^2$$
$$= 16p^3q^3[v + 2(p-q)w]^2$$
$$\sigma_Q^2 = \sum_i\sum_j\sum_k\sum_l p_ip_jp_kp_l\delta_{ijkl}^2$$
$$= 64p^4q^4w^2$$

In the absence of epistasis the total variance for each component is given simply by summation over loci. The tetraploid model can be extended by analogy with the diploid one to digenic interactions between two loci, but this leads to 14 components, which is far more than can be estimated:

$$\sigma_G^2 = \sigma_A^2 + \sigma_D^2 + \sigma_T^2 + \sigma_Q^2 + \sigma_{AA}^2 + \sigma_{AD}^2 + \sigma_{AT}^2 + \sigma_{AQ}^2 + \sigma_{DD}^2 + \sigma_{DT}^2 + \sigma_{DQ}^2 + \sigma_{TT}^2 + \sigma_{TQ}^2 + \sigma_{QQ}^2$$

The covariances between relatives

In a random mating population in equilibrium, the covariances between relatives are a linear function of the components of genetical variance in the population, provided that the relatives are not inbred (i.e. their parents are not related). In a diploid, the parents can be inbred and related if there is no dominance or just two alleles with equal frequencies at all loci (Wricke and Weber, 1986). In a tetraploid, the parents must not be inbred because the offspring of inbred parents are themselves inbred. Indeed, no simple formulae exist for covariances under inbreeding with tetraploids (Gallais, 1981).

For two unlinked diploid loci

$$\text{cov}(G_x, G_y) = k_A \sigma_A^2 + k_D \sigma_D^2 + k_A^2 \sigma_{AA}^2 + k_A k_D \sigma_{AD}^2 + k_D^2 \sigma_{DD}^2$$

where G_x and G_y are the genotypic values of a random pair of relatives and the ks depend on their coefficient of coancestry. Covariances of particular interest are shown in Table 3.1. In diploids, when the common parent (A) in half-sib families is inbred, k_A is $\frac{1}{4}(1 + F_A)$, and when the parents (A and B) of full-sib families are inbred, k_A is $\frac{1}{4}(2 + F_A + F_B)$ and k_D is $\frac{1}{4}(1 + F_A)(1 + F_B)$. The covariance between parent and offspring is not influenced by linkage (Cockerham, 1956) but the coefficients for epistatic components are increased for half- and full-sib families (Wricke and Weber, 1986).

For a tetraploid locus

$$\text{cov}(G_x, G_y) = k_A \sigma_A^2 + k_D \sigma_D^2 + k_T \sigma_T^2 + k_Q \sigma_Q^2.$$

Gallais (1974) showed that in a non-equilibrium population covariances are induced amongst all the parameters in the genetical model.

Estimating the covariances between relatives and the components of variance

A number of mating designs are available for estimating the covariances between relatives. They are well documented in standard textbooks. Any design not involving inbreeding which can be used for diploids can also be used for tetraploids. All such designs are feasible with potatoes, as a single plant usually has many flowers and more than one plant of any given genotype can be produced by vegetative propagation. However, fertility problems can occur which make some crosses difficult to secure. Clones can be designated female and/or male as necessary, and reciprocal crosses made to check for maternal and/or cytoplasmic effects. Reciprocal differences have, for example, often been reported to occur in crosses between subsp. *tuberosum* and *andigena* parents (Hoopes et al., 1980; Tarn and Miller, 1981; Maris, 1989).

The commonly used designs for estimating the covariances of half and full sibs are the hierarchical (experiment I) and factorial (experiment II) ones of Comstock and Robinson (1952) and various diallels, both complete (Griffing,

Table 3.1. Covariances between relatives (parents unrelated and not inbred).

	σ_A^2	σ_D^2	σ_{AA}^2	σ_{AD}^2	σ_{DD}^2	σ_T^2	σ_Q^2	σ_{AT}^2	σ_{AQ}^2	σ_{DT}^2	σ_{DQ}^2	σ_{TT}^2	σ_{TQ}^2	σ_{QQ}^2
Diploid														
Population	1	1	1	1	1									
Parent-offspring	1/2		1/4											
Half-sib	1/4		1/16											
Full-sib	1/2	1/4	1/4	1/8	1/16									
(FS) − (HS)	1/4	1/4	3/16	1/8	1/16									
(FS) − 2(HS)		1/4	1/8	1/8	1/16									
Tetraploid														
Population	1	1	1	1	1	1	1	1	1	1	1	1	1	1
Parent-offspring	1/2	1/6	1/4	1/12	1/36									
Half-sib	1/4	1/36	1/16	1/144	1/1296									
Full-sib	1/2	2/9	1/4	1/9	4/81	1/12	1/36	1/24	1/72	1/54	1/162	1/144	1/432	1/1296
(FS) − (HS)	1/4	7/36	3/16	15/144	63/1296	1/12	1/36	1/24	1/72	1/54	1/162	1/144	1/432	1/1296
(FS) − 2(HS)	1/4	1/6	1/8	7/72	31/648	1/12	1/36	1/24	1/72	1/54	1/162	1/144	1/432	1/1296

1956) and partial (Kempthorne and Curnow, 1961). All of them produce groups of full-sib families with a common parent, i.e. a group is a half-sib family. An appropriate analysis of variance for each design can be found in standard textbooks (see also chapter 6 by Ortiz and Peloquin). The component of variance for differences between groups of full-sib families is the covariance of half sibs (cov (HS)). In diallels it is often referred to as the general combining ability variance. In factorial and full diallel designs there are both male and female groups, and the variance of the latter can be inflated by maternal effects (i.e. $\text{cov}(HS_f) > \text{cov}(HS_m)$). In a hierarchical design the component of variance for differences between full-sib families within groups is cov (FS) − cov(HS), whereas in factorial and diallel designs the component for variance not accounted for by groups is cov (FS) − cov(HS_m) − cov(HS_f), which equals cov (FS) − 2 cov(HS) in the absence of maternal effects. In diallels, it is often referred to as the specific combining ability variance. In diallel designs which include selfed progeny, it must be remembered that the selfs are not part of the random mating population.

In order to estimate σ_A^2 and σ_D^2 from the covariances of full and half sibs, one has to assume no non-allelic interactions in diploids and also no triallelic or tetra-allelic interactions in tetraploids. The extent to which the presence of interactions will bias the estimates can be determined from Table 3.1. Thompson and Mendoza (1984), for example, used 4 cov (HS) and 6 [cov(FS) − 2 cov(HS)] to estimate $\sigma_A^2 + \frac{1}{9}\sigma_D^2$ and $\sigma_D^2 + \frac{1}{2}\sigma_T^2 + \frac{1}{6}\sigma_Q^2$, respectively, in a heterogeneous potato population propagated from true seed. The genetical variation between clones within full-sib families is the total variation in the population minus the cov (FS).

More elaborate triallel and tetra-allel designs were proposed for diploids by Rawlings and Cockerham (1962a, b) and extended to tetraploids by Levings and Dudley (1963). Although in theory they allow more genetical components of variance to be estimated, Levings and Dudley concluded that for tetraploids the coefficients of the components of genetical variance are too small for them to be of much practical value. The estimation procedure they suggested for alfalfa (a tetraploid) was a partial diallel cross in conjunction with a parent-offspring regression and an analysis of variance among the parental clones, where the parental clones and their progenies are assessed together to avoid problems from genotype–environment interactions. Four genetical variances could in theory be estimated, for example, σ_A^2, σ_D^2, σ_T^2 and σ_Q^2 or σ_A^2, σ_D^2, σ_{AA}^2 and σ_{AD}^2.

Wricke et al. (1982) used a diploid ornamental plant, *Gerbera jamesonii*, to investigate inclusion of clonal parents in a factorial design in order to estimate σ_{AA}^2 as well as σ_A^2 and σ_D^2, and concluded that these three components could not be estimated simultaneously with sufficient accuracy. The value of σ_A^2 was, however, roughly the same when either σ_A^2 and σ_D^2 or σ_A^2 and σ_{AA}^2 were estimated, but there was no way of deciding which non-additive component was more important and should therefore be included. One must therefore have reservations about the proposals of Levings and Dudley (1963) for tetraploids.

Mather and Jinks (1982) showed that the precision of estimates of σ_A^2

in hierarchical and factorial designs is improved by including the parent-offspring regression. They also showed that even in a factorial design with diploids, where estimates of σ_A^2 and σ_D^2 are uncorrelated in large samples, the precision of σ_D^2 is considerably less than that of σ_A^2. Wricke and Weber (1986) reached similar conclusions about precision in their book.

My own tentative conclusion from these considerations is that with potatoes, whether diploid or tetraploid, the most promising design is a factorial one which includes the parental clones so that the parent–offspring regression can be calculated, and that weighted least-squares estimates of σ_A^2 and σ_D^2, which may be biased, are better than imprecise estimates of four components.

In diploids, Kearsey (1970) has shown that a comparison of outcrossed and inbred family means provides a sensitive test for non-additive heritable variation. Furthermore, in a complete diallel set of crosses including selfs, Mather and Jinks (1982) think that a Hayman analysis of variance is probably the most sensitive means available of detecting non-additive variation, and maternal sources of reciprocal differences and dominance if the model is adequate. Again, in diploids, if inbred lines can be selected from the population for high and low manifestations of the trait under investigation, the most efficient analysis of the population is provided by the triple test cross of Kearsey and Jinks (1968). A further advantage of this mating design is that it provides a test for non-allelic interactions that is valid for the genes for which the inbred testers differ. However, it is not clear to me that any of these advantages extend to autotetraploids, so perhaps this is an area that requires further attention. Dessureaux (1959), for example, was unable to give a generalized analysis of the autotetraploid diallel, but did calculate a number of genetical parameters for the special case of random mating with equal allele frequency.

General and Specific Combining Ability in Non-equilibrium Populations

Two types of factorial design are commonly used by plant breeders, including potato breeders, to determine the combining abilities of clones of interest to them, but which are not drawn from random mating populations in equilibrium. In the first, one set of clones is crossed in all combinations with another set of clones which complements it for desirable traits. In the second, a diallel set of crosses is made amongst clones showing a range of values for some trait of interest. In these situations, the general combining abilities (GCA) of clones are an assessment of their gametic output, as judged by the mean performance of their progenies, and are estimated using the following model:

$$\text{progeny}_{ij} = \mu + \text{GCA}_i + \text{GCA}_j + \text{SCA}_{ij}$$

where SCA_{ij}, the specific combining ability, is the departure of the progeny mean from that expected on the basis of the general combining abilities of its parents.

The total genetical variation between families for a factorial design is

$$\sigma_G^2 = \sigma_{GCA}^2(\text{set } 1) + \sigma_{GCA}^2(\text{set } 2) + \sigma_{SCA}^2$$

and for a diallel design is

$$\sigma_G^2 = 2\sigma_{GCA}^2 + \sigma_{SCA}^2$$

However, these components cannot be used to estimate σ_A^2 and σ_D^2 because of the covariances between A and D in a non-equilibrium population.

Nevertheless, general combining abilities can be of use to plant breeders in choosing parents for crossing. Phenotypically similar clones may differ in GCA or there may be a high correlation between GCA and parental phenotype so that parents can be chosen on the basis of their own phenotype. If, however, all of the variation between progenies is due to SCA, no useful predictions can be made about the performance of progenies. Examples are given in Chapter 21 on breeding strategies for clonally propagated potatoes.

Response to Selection

When individual members of a random mating population in equilibrium are selected and used as parents to form a new random mating population which is also in equilibrium, the mean of the new population will differ from that of the old population if changes in gene frequencies have occurred. The expected change in population mean can be predicted from the regression of gene frequency on phenotype (Empig *et al.*, 1972; Hill and Haag, 1974).

For a single locus with two alleles A_1 and A_2, at frequencies p and q, we have in diploids:

	A_1A_1	A_1A_2	A_2A_2
	p^2	$2pq$	q^2
frequency of A_1	1	$\frac{1}{2}$	0
genotypic value	G_{11}	G_{12}	G_{22}

and in tetraploids with chromosomal segregation:

	$A_1A_1A_1A_1$	$A_1A_1A_1A_2$	$A_1A_1A_2A_2$	$A_1A_2A_2A_2$	$A_2A_2A_2A_2$
	p^4	$4p^3q$	$6p^2q^2$	$4pq^3$	q^4
frequency of A_1	1	$\frac{3}{4}$	$\frac{1}{2}$	$\frac{1}{4}$	0
genotypic value	G_{1111}	G_{1112}	G_{1122}	G_{1222}	G_{2222}

The covariance of gene frequency and phenotypic value equals the covariance of gene frequency and genotypic value provided that there is no correlation between genotypic and non-genotypic effects. In diploids and tetraploids with chromosomal inheritance, this covariance equals $pq\alpha$, where α is the average effect of a gene substitution in diploids (α_d) and tetraploids (α_t) respectively (i.e. α is $\alpha_1 - \alpha_2$ is where α_1 and α_2 are the main effects of A_1 and A_2, respectively). Hence the regression (β) of gene frequency on phenotypic value is

$$\beta = \frac{pq\alpha}{\sigma_P^2}$$

where σ_P^2 is the phenotypic variance.

If the phenotypic difference between the selected group and the population mean is S, called the selection differential, then the expected change in gene frequency (Δp) is βS.

For a small change in gene frequency the expected change (R) in population mean (M) is

$$R = \Delta p \frac{\mathrm{d}M}{\mathrm{d}p}$$

where $\mathrm{d}M/\mathrm{d}p$ equals $2\alpha_d$ for diploids and $4\alpha_t$ for tetraploids. Hence, for diploids

$$R = \frac{2pq\alpha_d^2 S}{\sigma_P^2} = \frac{\sigma_A^2 S}{\sigma_P^2}$$

and for tetraploids

$$R = \frac{4pq\alpha_t^2 S}{\sigma_P^2} = \frac{\sigma_A^2}{\sigma_P^2} S$$

where σ_A^2 is the additive genetic variance. The ratio σ_A^2/σ_P^2 is called narrow-sense heritability.

The selection method just considered was individual mass selection before pollination. The same principles can be used to predict the response to selection based on any criterion or unit of selection, e.g. family means, combined between and within family selection, selection using an index and indirect selection. The methods and prediction formulae can be found in standard textbooks and a more extensive treatment of index selection in Baker (1986).

A few important points require further consideration.

One generation of random mating does not establish an equilibrium population in tetraploids, or in diploids when more than one locus is considered. Hence the immediate response to selection may not be the final equilibrium response. Wright (1975) showed that for any ploidy level the response measured at equilibrium depends only on the additive variance, whereas the immediate response is proportional to the covariance of offspring and parent. Thus the immediate response to individual selection before pollination in diploids is

$$R = \frac{(\sigma_A^2 + \frac{1}{2}\sigma_{AA}^2)}{\sigma_P^2} S,$$

but the part due to additive × additive epistasis is lost during subsequent generations of random mating, as first described by Griffing (1960). The immediate response in tetraploids is

$$R = \frac{S}{\sigma_P^2} (\sigma_A^2 + \tfrac{1}{3}\sigma_D^2 + \tfrac{1}{2}\sigma_{AA}^2 + \tfrac{1}{6}\sigma_{AD}^2 + \tfrac{1}{18}\sigma_{DD}^2),$$

and the variance still contributing to the response after m generations of random mating, for unlinked loci, is

$$\sigma_G^2(m) = \sigma_A^2 + (\tfrac{1}{3})^m \sigma_D^2 + (\tfrac{1}{2})^m \sigma_{AA}^2 + (\tfrac{1}{6})^m \sigma_{AD}^2 + (\tfrac{1}{18})^m \sigma_{DD}^2.$$

Selection induces a gametic-phase disequilibrium which reduces the observed additive variance. Bulmer (1980) showed that a constant variance is reached under continuous truncation selection in a diploid when the reduction of variance due to selection is balanced by the increase of variance from random mating following selection:

$$\sigma_A^2(t+1) = \tfrac{1}{2}[1 - h_n^2(t)i(i - x_s)]\sigma_A^2(t) + \tfrac{1}{2}\sigma_A^2$$

where $h_n^2(t)$ is the narrow-sense heritability in generation t, i is the intensity of selection and x_s is the truncation point of the selected fraction. Gallais (1975a) showed that the response due to σ_D^2 in tetraploids changes from cycle to cycle of selection. Two-thirds of the dominance variance from earlier cycles is removed by random mating but a new component from the last cycle is added. For mass selection before pollination, the total after n cycles of selection is $\tfrac{1}{2}[1 - (\tfrac{1}{3})^n]\sigma_D^2$.

An important difference between diploids and tetraploids occurs in progeny testing schemes where the parents or their selfed progenies can be used to form the new population. The response to selection is the same for both methods in diploids, but not in tetraploids, where recombining selfed progenies for n generations leads to an inbreeding coefficient of

$$F_n = \frac{1 - \left(\frac{5}{18}\right)^n}{13}$$

Although the maximum value of F is only $\tfrac{1}{13}$, Gallais (1981) thought that the effect of inbreeding on heterozygosity could be much greater and recommended avoiding the problem by recombining the parents rather than the selfed progenies. This can easily be done in potatoes, as the parents can be maintained by vegetative propagation.

In both diploids and tetraploids the assessment of selfed families is theoretically attractive (Empig et al., 1972; Hill and Haag, 1974). The response to selfed family selection, however, cannot be expressed in terms of σ_A^2. In diploids the formula is relatively simple: the covariance required is $2pq\alpha\alpha_k$ where α_k is $d + 2^{-k}(q-p)h$ and k is the number of generations of selfing from non-inbred plants (Wricke and Weber, 1986). In tetraploids very complex expressions occur (Gallais, 1975b). When a mixture of natural selfing and random mating is considered, the formulae for diploids also become very complex (Wright and Cockerham, 1985, 1986). The immediate response can be estimated from the parent–offspring covariance and the phenotypic variance, but this can differ from the equilibrium response. The equivalent formulae for tetraploids have not to my knowledge been derived, although they would be relevant to selection following natural pollination in potatoes, where 80% selfing has been reported (Glendinning, 1989).

The final important conclusion from this section on selection theory is that after selection a population may not be in equilibrium, but in a crop like potatoes, which can be clonally propagated, disequilibrium can be maintained indefinitely.

Inbreeding Depression

In a diploid population with two alleles at each of two loci, A and B, in gametic-phase equilibrium, the mean of the population at inbreeding level F is

$$M_F = M - 2[pqh_A + uvh_B + pq(u-v)j_{BA} + uv(p-q)j_{AB} + 4pquvl]\,F + [4pquvl]F^2$$

where M is the mean when $F=0$, and p and q, and u and v, are the allele frequencies at A and B, respectively. Hence when there is no interaction between the dominance effects at the two loci ($l=0$), the change in M on inbreeding is a linear function of F (Crow and Kimura, 1970).

In tetraploids the situation is considerably more complicated. The effect on the population mean of n generations of selfing (Gallais, 1981) is:

$$M_n = M_o + 6F_{(n)} \cdot E(\beta_{ii}) + P_{2(n)} \cdot E(\delta_{iijj}) + 4(P_{o(n)} + \tfrac{1}{4}P_{1(n)}) \cdot E(\gamma_{iii}) + P_{o(n)} \cdot E(\delta_{iiii})$$

where F, P_o, P_1 and P_2 are the probabilities of the identities by descent given in the earlier section on inbreeding and $E(\beta_{ii})$, $E(\gamma_{iii})$, $E(\delta_{iijj})$ and $E(\delta_{iiii})$ are expected values. For example, with two alleles A_1 and A_2, at frequencies p and q, and chromosomal segregation:

$$6F_{(n)} \cdot E(\beta_{ii}) = 6[1 - (\tfrac{5}{6})^n] \cdot (p\beta_{11} + q\beta_{22})$$

The effect of selfing is a linear function of F in the absence of all interactions except those between two alleles. The equation has been used to estimate the importance of the different interallelic interactions in *Viola wittrockiana* (Trang, 1979), but not to my knowledge in potatoes. Mendoza and Haynes (1973) did, however, estimate that 92.7% of the variance of the percentage reduction in yield (in a previously published study) on selfing potatoes for six generations was due to its linear regression on the inbreeding coefficient of the successive generations. In contrast, the results of work with $2n$ gametes (see chapter by Tai) support the importance of tri- and tetra-allelic interactions in tetraploid potatoes.

It is perhaps easier to see what is happening on selfing tetraploids by considering chromosomal segregation with just two alleles, A_1 and A_2, at frequencies p and q. Following Bennett (1976) and using the notation of Wright (1979):

$$M_n = M_o - 2[1 - (\tfrac{5}{6})^n]pq(6h + w) - 4[1 - (\tfrac{1}{2})^n]pq(p-q)\,v - 2[1 - (\tfrac{1}{6})^n]pq(p-q)^2 w$$

Bennett points out that, when p is much less than q, the last two terms could contribute to a very rapid inbreeding depression on selfing, and also makes the following interesting point. The equilibrium frequency of a deleterious recessive gene arising by mutation is higher with tetrasomic than with disomic inheritance, and could explain the greater inbreeding depression which is sometimes seen in established autotetraploids when compared with newly synthesized ones.

In an inbred diploid population, the total genetic variance can be partitioned into an additive and a dominance component, where the inbred additive variance ($\sigma^2_{A(F)}$) is:

$$\sigma^2_{A(F)} = (1+F)2pq\alpha^2_F$$

where

$$\alpha_F = d + (q-p)\frac{(1-F)}{(1+F)}h,$$

but it cannot be partitioned in terms of F and the non-inbred additive variance, i.e. $(1+F)2pq\alpha^2$, unless we allow a covariance term between α and h (Wricke and Weber, 1986). The number of such variances and covariances in an inbred tetraploid population is very large (Gallais, 1981).

Crosses Between Two Pure Lines

With continued inbreeding, true-breeding homozygous lines are eventually produced. Seven generations of selfing are required to make an initially heterozygous diploid 99% homozygous, whereas in tetraploids 27 generations are required with chromosomal segregation and 20 with chromatid segregation (Figure 3.5). If a gametophytic self-incompatibility system prevents selfing in a diploid, 21 generations of sib mating are required. In potatoes, however, monoploids can now be produced by parthenogenesis or androgenesis and then doubled to give homozygous diploids, which in turn can be doubled to give homozygous tetraploids. Although Jacobsen and Ramanna are optimistic in Chapter 7, it remains to be seen how easily vigorous and fertile lines can be produced for genetical research, and whether or not they are self-incompatible.

Disomic inheritance

In the generations that can be derived from a cross between two true-breeding (homozygous) lines, there are two alleles of known frequency at all segregating loci, whatever the ploidy level. This is a big advantage for studying quantitative traits and with disomic inheritance the biometrical approach has reached a high level of sophistication. It can accommodate linkage and non-allelic interaction (epistasis), as well as genotype–environment interaction, although

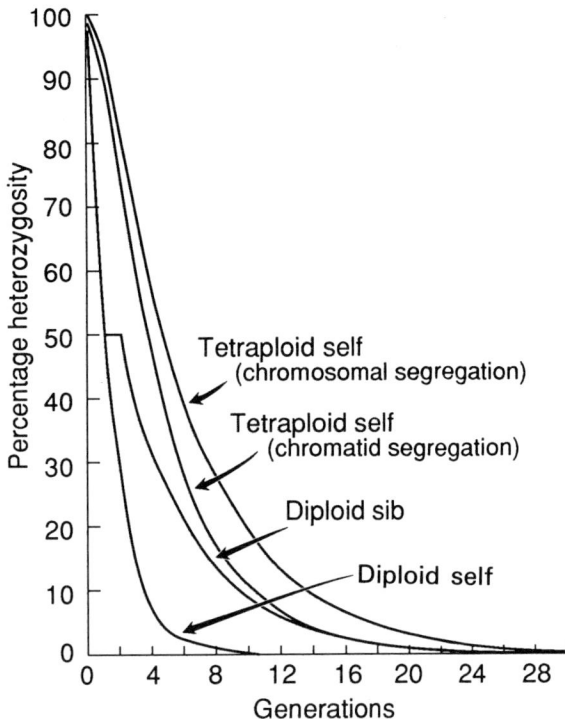

Fig. 3.5. Decline in heterozygosity on inbreeding.

one may wish to seek a simpler model with better predictive power by transforming the data if there is no obvious biological scale of measurement. The methods of analysis have been explained and illustrated in detail by Mather and Jinks (1982).

The design they recommend for initiating a biometrical genetical analysis in a new situation combines the parental (P_1 and P_2), F_1, F_2 and first backcross (B_1 and B_2) generations of a cross with an F_2 triple-test cross, in which a number of F_2 individuals are each crossed to P_1, P_2 and F_1 to provide sets of three families, L_1, L_2 and L_3, respectively. Before fitting a digenic interaction model to the generation means and an additive-dominance model to the variances, simple tests can be used to check for the presence of more complex effects. The B_1 and $F_2 \times P_1$ and B_2 and $F_2 \times P_2$ generations have the same mean in the presence of trigenic interactions but not in the presence of linked digenic interactions. In the F_2 triple-test cross the comparison ($L_1 + L_2 - 2L_3$) for each set of three families provides an unambiguous test for non-allelic interaction which is independent of linkage. The relationship between the variances of F_2 and L_3, B_1 and L_1, and B_2 and L_2 provide three independent variance ratio tests for linkage; and heterogeneity of the three variances V_{P1}, V_{P2} and V_{F1} unambiguously detects the presence of genotype–microenvironmental interaction. In the absence of these complex effects, good estimates of [d], [h], [i], [j] and [l] can be obtained from the generation means, where [ ] indicates summation over all k segregating loci and pairs of loci, i.e. they are net effects. In the absence of non-allelic interactions, good estimates of $D (=\Sigma d_i^2)$ and $H (=\Sigma h_i^2)$ can be obtained from the variation in $L_1 + L_2$ and $L_1 - L_2$, respectively. One can then determine the genetic basis of heterosis because $(H/D)^{\frac{1}{2}}$ is a weighted average dominance ratio; predict the proportion of recombinant inbred lines expected to exceed the better parent and the F_1 from $[d]/\sqrt{D}$ and $[h]/\sqrt{D}$, respectively; and infer the effect of past selection from features of the genetical architecture of the trait, e.g. directional dominance, $|[h]| > 0$, is evidence of directional selection (Jinks, 1979).

Tetrasomic inheritance

The high degree of sophistication achieved with disomic inheritance has not been repeated with tetrasomic inheritance. The information that can be obtained from the generations derived from a cross between two autotetraploid homozygous lines was considered by Killick (1971), but assuming no genotype–environment interaction, no linkage and no epistasis. He derived the expected means of the 21 generations produced by all possible matings within and between the six generations P_1, P_2, F_1, F_2, B_1 and B_2. With double reduction, there were 21 parameters in the genetical model, but these reduced to nine if parameters with α raised to a power greater than 1 were neglected. In every generation, double reduction caused the proportions of quadruplex and/or nulliplex genotypes to increase, and consequently the overall proportion of heterozygous genotypes (simplex, duplex and triplex) to decline. Killick found two combinations of generation means involving second backcrosses (e.g. $B_{11} = B_1 \times P_1$) that are expected to be zero (scaling tests) if a '$[d]$, $[h_3]$, $[h_2]$ and $[h_1]$' model is adequate, and which are unaffected by the degree of double reduction. They are given in Table 3.2, together with the expected means (reduced model) of 12 of the 21 generations. The diploid digenic interaction model for pairs of unlinked genes is included for comparison. Here the scaling tests fail for trigenic and higher-order interactions but not for linked digenics. Standard methods for testing the adequacy of these models and for estimating the parameters are given by Mather and Jinks (1982).

With second-degree statistics, Killick (1971) thought there was little to be gained from taking double reduction into account, although his results were extended by Tan (1976) to include double reduction. Ten parameters (d^2, h_3^2, h_2^2, h_1^2, h_3h_2, h_3h_1, h_2h_1, dh_3, dh_2 and dh_1) are required for the full single-locus model, although the six h terms and the three dh terms can be summed to give a reduced model with three parameters, d^2, $\Sigma h_i h_j$ and Σdh_i. Even so, considerable difficulty is likely to be encountered in their estimation because the small but variable coefficients involved are expected to lead to estimates with large sampling errors. Killick (1971) therefore

Table 3.2. Reduced tetraploid model and diploid digenic interaction model for pairs of unlinked genes.

	Diploid						Tetraploid								
	m	$[d]$	$[h]$	$[i]$	$[j]$	$[l]$	m	$[d]$	$[\alpha d]$	$[h_3]$	$[\alpha h_3]$	$[h_2]$	$[\alpha h_2]$	$[h_1]$	$[\alpha h_1]$
P_1	1	1	0	1	0	0	1	1	0	0	0	0	0	0	0
P_2	1	−1	0	1	0	0	1	−1	0	0	0	0	0	0	0
F_1	1	0	1	0	0	1	1	0	0	0	0	1	0	0	0
F_2	1	0	1/2	1/4	0	1/4	1	0	0	2/9	2/9	1/2	−2/3	2/9	2/9
B_1	1	1/2	1/2	1/4	1/4	1/4	1	1/6	1/3	2/3	−2/3	1/6	1/3	0	0
B_2	1	−1/2	1/2	1/4	−1/4	1/4	1	−1/6	−1/3	0	0	1/6	1/3	2/3	−2/3
B_{11}	1	3/4	1/4	9/16	3/16	1/16	1	19/36	5/18	4/9	−5/9	1/36	5/18	0	0
B_{12}	1	−1/4	3/4	1/16	−3/16	9/16	1	−1/36	−5/18	0	0	19/36	5/18	4/9	−5/9
B_{21}	1	1/4	3/4	1/16	3/16	9/16	1	1/36	5/18	4/9	−5/9	19/36	5/18	0	0
B_{22}	1	−3/4	1/4	9/16	−3/16	1/16	1	−19/36	−5/18	0	0	1/36	5/18	4/9	−5/9
B_{1s}	1	1/2	1/4	1/4	1/8	1/16	1	1/3	1/3	10/27	−7/18	1/4	−5/18	1/27	5/18
B_{2s}	1	−1/2	1/4	1/4	−1/8	1/16	1	−1/3	−1/3	1/27	5/18	1/4	−5/18	10/27	−7/18

$$P_1 - F_1 + 2B_{21} - 2B_{11} = 0 \qquad\qquad P_1 - F_1 + 2B_{21} - 2B_{11} = 0$$
$$P_2 - F_1 + 2B_{12} - 2B_{22} = 0 \qquad\qquad P_2 - F_1 + 2B_{12} - 2B_{22} = 0$$

concluded that generation mean analysis is likely to be a more fruitful source of information in tetraploids than second-degree statistics. Such an analysis may give an indication of the overall importance of double reduction, which of course varies from locus to locus, and of whether or not h_1, h_2 and h_3 differ much. But, even with true-breeding lines, there are likely to be severe limitations on any genetical analysis at the tetraploid level, where powerful designs such as the diploid F_2 triple test cross are not available.

Conclusion

It is unlikely that the high degree of sophistication achieved in the biometrical genetical analysis of crosses between true-breeding lines that display disomic inheritance can ever be matched where tetrasomic inheritance occurs. Indeed, detailed information on the genetical architecture of quantitative traits in potatoes and the location of quantitative trait loci (next chapter) is much more likely to come from diploid species and dihaploids derived from tetraploids, although such analyses will not detect interactions between more than two alleles which can occur in tetraploids. Furthermore, some genetical information of value to plant breeders can be obtained at the tetraploid level, for example: the extent and nature of genotype–environment interactions; estimates of the additive genetic variance in equilibrium populations, which can be used to predict and compare the outcome of various recurrent selection schemes; and estimates of the general combining abilities of clones with desirable phenotypes which are not from equilibrium populations, but which may be used in crossing schedules.

References

Azzalini, A. and Cox, D.R. (1984) Two new tests associated with analysis of variance. *Journal of the Royal Statistical Society B* 46, 335–343.

Baker, R.J. (1986) *Selection Indices in Plant Breeding*. CRC Press, Boca Raton, Florida.

Baker, R.J. (1988) Tests for crossover genotype–environmental interactions. *Canadian Journal of Plant Science* 68, 405–410.

Becker, W.A. (1984) *Manual of Quantitative Genetics*, 4th edn. Academic Enterprises, Washington.

Bennett, J.H. (1976) Expectations for inbreeding depression on self-fertilization of tetraploids. *Biometrics* 32, 449–452.

Bingham, E.T. (1980) Maximizing heterozygosity in autopolyploids. In: Lewis, W.H. (ed.) *Polypoidy: Biological Relevance*. Plenum Press, New York and London, pp. 471–489.

Bulmer, M.G. (1980) *The Mathematical Theory of Quantitative Genetics*. Clarendon Press, Oxford.

Cockerham, C.C. (1956) Effects of linkage on the covariances between relatives. *Genetics* 41, 138–141.

Comstock, R.E. and Robinson, H.F. (1952) Estimation of average dominance of genes. In: Gowen, J.W. (ed.) *Heterosis.* Iowa State College Press, Ames, pp. 494-516.

Crow, J.F. (1986) *Basic Concepts in Population, Quantitative, and Evolutionary Genetics.* Freeman, New York.

Crow, J.F. and Kimura, M. (1970) *An Introduction to Population Genetics Theory.* Harper and Row, New York, Evanston and London.

Dale, M.F.B. and Phillips, M.S. (1989) Genotype by environment interaction in the assessment of tolerance of partially resistant potato clones to potato cyst nematodes. *Annals of Applied Biology* 119, 69-75.

Dessureaux, L. (1959) Introduction to the autotetraploid diallel. *Canadian Journal of Genetics and Cytology* 1, 94-101.

East, E.M. (1915) Studies on size inheritance in *Nicotiana. Genetics* 1, 164-176.

Easton, H.S.(1976) Étude comparative d'effects génétiques chez des plantes diploïdes et tetraploïdes isogéniques de *Festuca pratensis* Huds. Thèse de Doctorat d'État des Sciences Naturelles, Université de Paris-Sud.

Empig, L.T., Gardner, C.O. and Compton, W.A. (1972) *Theoretical Gains for Different Population Improvement Procedures.* MP26, University of Nebraska College of Agriculture, Nebraska.

Falconer, D.S. (1989) *Introduction to Quantitative Genetics*, 3rd edn. Longman Scientific and Technical, Harlow.

Finlay, K.W. and Wilkinson, G.N. (1963) The analysis of adaptation in a plant-breeding programme. *Australian Journal of Agricultural Research* 14, 742-754.

Fisher, R.A. (1918) The correlation between relatives on the supposition of Mendelian inheritance. *Transactions of the Royal Society of Edinburgh* 52, 399-433.

Fisher, R.A. (1965) *The Theory of Inbreeding*, 2nd edn. Oliver and Boyd, London.

Fisher, R.A. and Mather, K. (1943) The inheritance of style length in *Lythrum salicaria. Annals of Eugenics* 12, 1-23.

Fisher, R.A., Immer, F.R. and Tedin, O. (1932) The genetical interpretation of statistics of the third degree in the study of quantitative inheritance. *Genetics* 17, 107-124.

Freeman, G.H. (1973) Statistical methods for the analysis of genotype-environment interactions. *Heredity* 31, 339-354.

Gallais, A. (1968) Interactions between alleles and their variability in autotetraploid cross-fertilized plants: consequences for selection. *Genetica Agraria* 23, 312-323.

Gallais, A. (1974) Covariance between arbitrary relatives in autotetraploids with panmictic disequilibrium. *Genetics* 76, 587-600.

Gallais, A. (1975a) Prévision de la vigeur et sélection des parents d'une variété synthétique. *Annales de l'Amélioration des Plantes* 25, 233-264.

Gallais, A. (1975b) Selection with truncation in autotetraploids - comparison with diploids. *Theoretical and Applied Genetics* 46, 387-394.

Gallais, A. (1981) Quantitative genetics and breeding theory of autopolyploids. In: Gallais, A. (ed.) *Quantitative Genetics and Breeding Methods.* INRA, Lusignan, pp. 189-216.

Gauch, H.G. (1988) Model selection and validation for yield trials with interaction. *Biometrics* 44, 705-715.

Glendinning, D.R. (1989) Some aspects of autotetraploid population dynamics. *Theoretical and Applied Genetics* 78, 233-242.

Griffing, B. (1956) Concept of general and specific combining ability in relation to diallel crossing systems. *Australian Journal of Biological Sciences* 9, 463-493.

Griffing, B. (1960) Theoretical consequences of truncation selection based on the individual phenotype. *Australian Journal of Biological Sciences* 13, 307–343.

Haynes, K.G. (1992) Some aspects of inbreeding in derived tetraploids of potatoes. *Journal of Heredity* 83, 67–70.

Hill, J. (1975) Genotype-environment interactions – a challenge for plant breeding. *Journal of Agricultural Science* 85, 477–493.

Hill, R.R. and Haag, W.L. (1974) Comparison of selection methods for autotetraploids. 1. Theoretical. *Crop Science* 14, 587–590.

Hoopes, R.W., Plaisted, R.L. and Cubillos, A.E. (1980) Yield and fertility of reciprocal cross Tuberosum–Andigena hybrids. *American Potato Journal* 57, 275–284.

Jinks, J.L. (1979) The biometrical approach to quantitative variation. In: Thompson, J.N. and Thoday, J.M. (eds) *Quantitative Genetic Variation*. Academic Press, New York, San Francisco and London, pp. 81–109.

Jinks, J.L. and Connolly, V. (1973) Selection for specific and general response to environmental differences. *Heredity* 30, 33–40.

Jinks, J.L. and Connolly, V. (1975) Determination of the environmental sensitivity of selection lines by the selection environment. *Heredity* 34, 401–406.

Jinks, J.L. and Pooni, H.S. (1982) Determination of the environmental sensitivity of selection lines of *Nicotiana rustica* by the selection environment. *Heredity* 49, 291–294.

Johannsen, W. (1909) *Elemente der exakten Erblichkeitslehre*. Fischer, Jena.

Kearsey, M.J. (1970) Experimental sizes for detecting dominance variation. *Heredity* 25, 529–542.

Kearsey, M.J. and Jinks, J.L. (1968) A general method of detecting additive, dominance and epistatic variation for metrical traits: 1. Theory. *Heredity* 23, 403–409.

Kempthorne, O. (1957) *An Introduction to Genetic Statistics*. Wiley, New York.

Kempthorne, O. and Curnow, R.N. (1961) The partial diallel cross. *Biometrics* 17, 229–250.

Kempton, R.A. (1984) The use of biplots in interpreting variety by environment interactions. *Journal of Agricultural Science* 103, 123–135.

Killick, R.J. (1971) The biometrical genetics of autotetraploids 1. Generations derived from a cross between two pure lines. *Heredity* 27, 331–346.

Knight, R. (1970) The measurement and interpretation of genotype-environment interactions. *Euphytica* 19, 225–235.

Knight, R. (1973) The relation between hybrid vigour and genotype-environment interactions. *Theoretical and Applied Genetics* 43, 311–318.

Levings, C.S. and Dudley, J.W. (1963) Evaluation of certain mating designs for estimation of genetic variance in autotetraploid alfalfa. *Crop Science* 3, 532–535.

Lin, C.S., Binns, M.R. and Lefkovitch, L.P. (1986) Stability analysis: where do we stand? *Crop Science* 26, 894–900.

Malécot, G. (1948) *Les Mathématiques de l'hérédité*. Masson, Paris.

Maris, B. (1989) Analysis of an incomplete diallel cross among three ssp. *tuberosum* varieties and seven long-day adapted ssp. *andigena* clones of the potato (*Solanum tuberosum* L.). *Euphytica* 41, 163–182.

Mather, K. (1936) Segregation and linkage in autotetraploids. *Journal of Genetics* 32, 287–314.

Mather, K. and Jinks, J.L. (1982) *Biometrical Genetics*, 3rd edn. Chapman and Hall, London and New York.

Mayo, O. (1987) *The Theory of Plant Breeding*, 2nd edn. Clarendon Press, Oxford.

Mendoza, H.A. and Haynes, F.L. (1973) Some aspects of breeding and inbreeding in potatoes. *American Potato Journal* 50, 216-222.

Mendoza, H.A. and Haynes, F.L. (1974a) Genetic relationship among potato cultivars grown in the United States. *HortScience* 9, 328-330.

Mendoza, H.A. and Haynes, F.L. (1974b) Genetic basis of heterosis for yield in the autotetraploid potato. *Theoretical and Applied Genetics* 45, 21-25.

Narain, P. (1990) *Statistical Genetics*. Wiley Eastern Limited, New Delhi.

Nilsson-Ehle, H. (1909) Kreuzunguntersuchungen an Hafer und Weizen. *Acta Univ. Lund*. Ser 2, 5, no 2, 1-122.

Rawlings, J.O. and Cockerham, C.C (1962a) Analysis of double cross hybrid populations. *Biometrics* 18, 229-244.

Rawlings, J.O. and Cockerham, C.C. (1962b) Triallel analysis. *Crop Science* 2, 228-231.

Sanford, J.C. and Hanneman, R.E., Jr (1982) A possible heterotic threshold in the potato and its implications for breeding. *Theoretical and Applied Genetics* 61, 151-159.

Talbot, M. (1984) Yield variability of crop varieties in the UK. *Journal of Agricultural Science* 102, 315-321.

Tan, W.Y. (1976) On the biometrical genetics of autotetraploid from a cross between two pure lines. *Botanical Bulletin of Academia Sinica* 17, 1-17.

Tarn, T.R. and Miller, S.M. (1981) Flower abnormalities and male sterilities in *Solanum tuberosum* groups Tuberosum and Andigena, and in inter-group hybrids. *American Potato Journal* 58, 521 (Abstract).

Thompson, J.N. and Thoday, J.M. (eds) (1979) *Quantitative Genetic Variation*. Academic Press, New York, San Francisco and London.

Thompson, P.G. and Mendoza, H.A. (1984) Genetic variance estimetes in a heterogenous potato population propagated from true seed. *American Potato Journal* 61, 697-702.

Trang, Q.S. (1979) Study of the inheritance of quantitative characters in autopolyploid *Viola wittrockiana* Gams. *Zeitschrift für Pflanzenzuchtung* 83, 97-113.

Vermeer, H. (1990) Optimising potato breeding. 1. The genotypic, environmental and genotype-environment coefficients of variation for tuber yield and other traits in potato (*Solanum tuberosum* L.) under different experimental conditions. *Euphytica* 49, 229-236.

Wricke, G. and Weber, W.E. (1986) *Quantitative Genetics and Selection in Plant Breeding*. Walter de Gruyter, Berlin and New York.

Wricke, G., Weber, W.E. and Ottleben, R. (1982) Die Analyse der genetischen Varianz bei quantitativ Vererbten Merkmalen von *Gerbera jamesonii*. *Zeitschrift für Pflanzenzuchtung* 89, 329-336.

Wright, A.J. (1975) Phenotypic selection in autotetraploids. *Heredity* 35, 282-286.

Wright, A.J. (1979) The use of differential coefficients in the development and interpretation of quantitative genetic models. *Heredity* 43, 1-8.

Wright, A.J. and Cockerham, C.C. (1985) Selection with partial selfing. 1. Mass selection. *Genetics* 109, 585-597.

Wright, A.J. and Cockerham, C.C. (1986) Selection with partial selfing. II. Family selection. *Crop Science* 26, 261-268.

Wright, S. (1921) Systems of mating. *Genetics* 6, 111-178.

4 Theory for Locating Quantitative Trait Loci

J.E. BRADSHAW

Crop Genetics Department, Scottish Crop Research Institute, Invergowrie, Dundee DD2 5DA, UK.

Introduction

The biometrical approach has not provided any accurate methods of estimating the number of segregating loci involved in quantitative variation. Various methods have been proposed, but all of them usually give underestimates (Mather and Jinks, 1982; Zeng, 1992). However, as the genetics of the fruit fly (*Drosophila melanogaster*) progressed from the work of Morgan (1910), a wealth of easily recognizable marker genes and special stocks became available, and these enabled the contributions of particular chromosomes to quantitative variation to be determined. Breese and Mather (1957), for example, showed that one or more genes for differences in the number of abdominal chaetae were located in each of six portions of chromosome III.

Once a quantitative difference has been traced to a section of chromosome between two markers, Thoday (1961, 1979) and his colleagues showed that it is possible to determine if more than one locus is segregating by progeny-testing sufficient marker recombinants for discrete classes to appear. The method does, however, break down when there are more than two loci segregating with mixed coupling and repulsion linkages between them (McMillan and Robertson, 1974; Thoday, 1979). Using these techniques in *D. melanogaster*, Thoday and his colleagues often found that segregation at rather few loci accounted for most of the genetical variation under investigation, and that different loci sometimes influenced the same trait in developmentally quite different ways (Thoday and Thompson, 1976).

Hence separable genetic loci whose segregation affects quantitative traits could be isolated and studied, but it was not known how often differences in structural genes, control genes, repeated sequences, etc., or, indeed, in blocks of functionally related genes, contributed to the polygenic systems. These loci are now called quantitative trait loci, or QTLs for short (Gelderman, 1975), a term that Thoday (1979) considered etymologically superior to his polygenic loci (Thompson and Thoday, 1974). With the advent of molecular markers, particularly restriction fragment length polymorphisms (RFLPs), high-density linkage maps have become available in many organisms, including the potato (Gebhardt *et al.*, 1991; Chapter 10 by Watanabe). It should therefore be possible to map and study QTLs for any quantitative trait of interest. In *Arabidopsis thaliana* there is even talk of cloning quantitative genes using artificial chromosome libraries from yeast (Griffing and Scholl, 1991). The potato, however, is not an ideal organism for such genetical research.

Contributions of Particular Chromosomes to Quantitative Variation

Apart from the work with the fruit fly, *D. melanogaster*, mentioned in the introduction, probably the best-known chromosome assays are those of Law and his colleagues in wheat, *Triticum aestivum* (Law and Gale, 1979). They used intervarietal chromosome substitution lines to study the quantitative genetic differences between two varieties of wheat, chromosome by chromosome. Then each chromosome difference was analysed further, using single-chromosome recombinant inbred lines derived from single-chromosome heterozygotes. These techniques were possible because wheat, unlike the cultivated potato, is an allopolyploid which displays disomic inheritance, tolerates aneuploidy, has chromosomes with useful cytological markers, and is an inbreeder so that true-breeding lines can be readily produced.

The best prospect for determining the contribution of particular chromosomes to quantitative variation in potatoes appears to be the manipulation of the aneusomatic clones that arise from somatic hybrids (Pijnacker *et al.*, 1989) and from tetraploids pollinated by dihaploid inducers (Clulow *et al.*, 1991), and which are then stable during plant propagation.

Quantitative Trait Loci

Now that high-density linkage maps based on molecular markers are available in many organisms, rapid progress is expected in mapping and studying the effects of quantitative trait loci (QTLs). In the tomato, for example, a start has been made to comparing QTLs across species, generations and environments (Paterson *et al.*, 1991).

Similar work with potatoes is less advanced because true-breeding lines have not been available for analysis. Recently, Ritter *et al.* (1990) have shown how RFLP linkage analysis can be done from any type of F_1 population, including those from heterozygous individuals, and encompassing F_2 and backcross populations from homozygous inbred lines as special cases. Care is required, however, with computer programs like MAPMAKER (Lander *et al.*, 1987), which use multiple-point estimates to determine gene order, because they cannot be applied to a whole data set where fragments segregating with 3:1 and 1:1 ratios are intermixed. Using these techniques RFLP maps of potato have been obtained from diploids and aligned with the homoeologous tomato genome (Gebhardt et al., 1991). QTL mapping in potatoes will no doubt also initially proceed in diploids, to avoid the complexities of tetrasomic inheritance, will take advantage of vegetative clones which can be reproduced indefinitely (Soller and Beckmann, 1990), and will use extensions of the methods developed for crosses between true-breeding diploid lines, which ideally should be extremes for the quantitative trait. These methods will now be briefly reviewed. For more detailed reviews the reader is referred to Beckmann and Osborn (1992).

Evidence of linkage between a segregating marker locus with codominant alleles M_1 and M_2, and a segregating QTL with alleles Q_1 and Q_2, is provided by differences in the means of the marker genotype classes with respect to the quantitative trait. Such differences, however, cannot be used to estimate QTL effects or the recombination frequency (r) between marker locus and QTL because the two are confounded. In the F_2 from the cross $M_1M_1Q_1Q_1 \times M_2M_2Q_2Q_2$, for example, the difference between the means of the marker classes M_1M_1 and M_2M_2 equals $(1 - 2r)(\mu_{Q11} - \mu_{Q22})$, where $(\mu_{Q11} - \mu_{Q22})$ is the genotypic difference between Q_1Q_1 and Q_2Q_2.

The means (μ) and variances (σ^2) of Q_1Q_1, Q_1Q_2 and Q_2Q_2, and r, can, however, be estimated from the means ($\bar{X}$) and variances (S^2) of the three marker classes, although the system of six equations does not have a unique solution because there are seven unknowns. Weller (1986) argued that deviations of the marker genotype statistics from their expectations will cause greater deviations in the estimates of the heterozygote parameters than in the estimates of the homozygote parameters. He therefore used the method of maximum likelihood to derive estimates of μ_{Q12}, and σ^2_{Q12} and r, and then used the method of moments to obtain estimates of the other four parameters from $\bar{X}_{M11}$, $\bar{X}_{M22}$, S^2_{M11} and S^2_{M22}. Luo and Kearsey (1989, 1991) went a stage further and described the use of likelihood functions for F_2, backcross and doubled haploid populations in which r was the only unknown parameter. They also used the method of moments to estimate the other parameters. However, Darvasi and Weller (1992) have pointed out that these procedures are only approximate maximum likelihood methods because they do not simultaneously maximize the likelihood for all seven parameters, a task which the reader may have realized is rather difficult. Further work is in progress to determine those situations in which the approximate estimates are satisfactory, and to obtain more rapid numerical methods for solving the seven 'likelihood equations' (e.g. Luo and Woolliams, 1993).

More efficient methods of locating QTLs and estimating their effects involve flanking marker genetical models, in which a QTL (Q_1,Q_2) is hypothesized to lie between linked marker loci, with codominant alleles (A_1,A_2) and (B_1,B_2), and with recombination frequencies r_1 and r_2 between A and Q, and Q and B, respectively. Hence these models are functions of the means of quantitative trait locus genotypes and recombination frequencies between marker and quantitative trait loci. For a simple example, consider the F_2 from the cross $A_1A_1Q_1Q_1B_1B_1 \times A_2A_2Q_2Q_2B_2B_2$. In the absence of double crossovers, Weller (1987) showed that $\mu_{Q11} - \mu_{Q22}$, r_1 and r_2 can be estimated as follows:

$$\mu_{Q11} - \mu_{Q22} = \bar{X}_{A11B11} - \bar{X}_{A22B22}$$
$$r_1 = \tfrac{1}{2} - (\bar{X}_{A11} - \bar{X}_{A22})/[2(\bar{X}_{A11B11} - \bar{X}_{A22B22})]$$
$$r_2 = \tfrac{1}{2} - (\bar{X}_{B11} - \bar{X}_{B22})/[2(\bar{X}_{A11B11} - \bar{X}_{A22B22})]$$

In the interval mapping method of Lander and Botstein (1989), the probability of a putative QTL being located in a given interval is estimated using the method of log-odds. For any map position between two markers, the likelihood of the data is calculated assuming that a QTL is present. This likelihood is then compared with the likelihood of the data assuming no QTL (the odds ratio). The $\log_{10}$ of this likelihood ratio (LOD) is then plotted along the genetic map, and a support interval shown for the position of the assumed QTL, based on a given fall-off value from the maximum LOD. This allows efficient detection of QTLs whilst limiting the overall occurrence of false positives. QTL effects can then be estimated. The details for backcross populations can be found in Lander and Botstein (1989) and for F_2 populations in Paterson et al. (1991) and Luo and Kearsey (1992). The marker genotypes are shown in Table 4.1. Flanking marker genetical models for these and other common types of populations can also be found in Knapp et al. (1990), together with maximum-likelihood methods for estimating their parameters.

Knapp (1991) extended the backcross, doubled haploid, recombinant inbred, and test-cross progeny models to allow simultaneous estimation of the parameters required for multiple quantitative trait loci, both unlinked and linked but in adjacent flanking marker segments. He found that the means of certain QTL genotypes were not estimated efficiently with linked loci and this made it difficult to resolve ambiguities, such as whether differences found for a given segment were in fact due to the effects of a QTL in an adjacent segment. He did not address the problem of multiple QTLs within a particular segment.

Stam (1991) showed that, if two linked QTLs have opposite effects, the result is a downward bias of LOD values in the Lander and Botstein (1989) model, such that both QTLs may go unnoticed. If on the other hand they have unidirectional effects, the result is an upward bias in the map region between the actual QTL positions, and one may incorrectly infer the presence of a single QTL. Stam (1991) also showed that an alternative multiple regression approach to QTL location is worthy of consideration, because the effects of all QTLs are simultaneously absorbed in the regression coefficients, and the method benefits from additional markers.

Haley and Knott (1992) have also shown that it is possible to fit flanking marker models by multiple regression and that the results are very similar to the maximum-likelihood method. Their regression method provided good estimates for the positions and the effects of QTLs. Furthermore, its relative simplicity and computational rapidity made it easier to fit models for two or more linked and/or interacting QTLs, and these were shown to give good estimates of QTL effects. All of the steps in the analysis can be performed with

Table 4.1a. Frequencies of the eight possible gametes from $A_1 A_2 Q_1 Q_2 B_1 B_2$, where r_1 and r_2 are the recombination frequencies between A and Q, and Q and B, respectively, and δ is the coefficient of coincidence, which equals the observed number of double crossovers divided by the expected number, assuming no interference.

$A_1 Q_1 B_1$	$\frac{1}{2}(1 - r_1 - r_2 + \delta r_1 r_2)$
$A_2 Q_2 B_2$	$\frac{1}{2}(1 - r_1 - r_2 + \delta r_1 r_2)$
$A_1 Q_2 B_2$	$\frac{1}{2}(r_1 - \delta r_1 r_2)$
$A_2 Q_1 B_1$	$\frac{1}{2}(r_1 - \delta r_1 r_2)$
$A_1 Q_1 B_2$	$\frac{1}{2}(r_2 - \delta r_1 r_2)$
$A_2 Q_2 B_1$	$\frac{1}{2}(r_2 - \delta r_1 r_2)$
$A_1 Q_2 B_1$	$\frac{1}{2}(\delta r_1 r_2)$
$A_2 Q_1 B_2$	$\frac{1}{2}(\delta r_1 r_2)$

Table 4.1b. Genotypes in F_2 (all) and backcrosses of $A_1 A_2 Q_1 Q_2 B_1 B_2$ to $A_1 A_1 Q_1 Q_1 B_1 B_1$ (1) and $A_2 A_2 Q_2 Q_2 B_2 B_2$ (2); frequencies can be determined from Table 4.1a.

Marker genotypes	QTL genotypes		
	$Q_1 Q_1$	$Q_1 Q_2$	$Q_2 Q_2$
$A_1 A_1 B_2 B_2$			
$A_1 A_1 B_1 B_2$	1	1	
$A_1 A_1 B_1 B_1$	1	1	
$A_1 A_2 B_1 B_1$	1	1	
$A_1 A_2 B_1 B_2$	1	1/2	2
$A_1 A_2 B_2 B_2$		2	2
$A_2 A_2 B_2 B_2$		2	2
$A_2 A_2 B_1 B_2$		2	2
$A_2 A_2 B_1 B_1$			

any general computer software package for multiple regression. It could therefore be worth while extending the models to tetrasomic inheritance.

Martinez and Curnow (1992) used such a regression approach to show that when two linked QTLs are present between the first and second and between the third and fourth of four markers, the residual sum-of-squares function for flanking markers can have a global minimum between markers two and three, and hence a 'ghost' effect can be wrongly interpreted as a true QTL. (The residual sum-of-squares function behaves in a way approximately proportional to the inverse of the LOD score.) The solution is to simultaneously use all of the information provided by the four markers, but this can only be done if the sample size is large enough to have sufficient representation in each of the different marker classes.

Finally, the reader is referred to Edwards (1992) for a good review of breeding strategies for the use of molecular markers in the evaluation and introgression of genetic diversity for quantitative traits from diverse germplasm sources.

References

Beckmann, J.S. and Osborn, T.C. (eds) (1992) *Plant Genomes: Methods for Genetic and Physical Mapping*. Kluwer Academic Publishers, Dordrecht, Boston and London.

Breese, E.L. and Mather, K. (1957) The organisation of polygenic activity within a chromosome in *Drosophila* 1. Hair characters. *Heredity* 11, 373-395.

Clulow, S.A., Wilkinson, M.J., Waugh, R., Baird, E., De Maine, M.J. and Powell, W. (1991) Cytological and molecular observations on *Solanum phureja*-induced dihaploid potatoes. *Theoretical and Applied Genetics* 82, 545-551.

Darvasi, A. and Weller, J.I. (1992) On the use of the moments method of estimation to obtain approximate maximum likelihood estimates of linkage between a genetic marker and a quantitative locus. *Heredity* 68, 43-46.

Edwards, M. (1992) Use of molecular markers in the evaluation and introgression of genetic diversity for quantitative traits. *Field Crops Research* 29, 241-260.

Gebhardt, C., Ritter, E., Barone, A., Debener, T., Walkemeier, B., Schachtschabel, U., Kaufmann, H., Thompson, R.D., Bonierbale, M.W., Ganal, M.W., Tanksley, S.D. and Salamini, F. (1991) RFLP maps of potato and their alignment with the homoeologous tomato genome. *Theoretical and Applied Genetics* 83, 49-57.

Gelderman, H. (1975) Investigations on inheritance of quantitative characters in animals by gene markers 1. Methods. *Theoretical and Applied Genetics* 46, 319-330.

Griffing, B. and Scholl, R.L. (1991) Qualitative and quantitative genetic studies of *Arabidopsis thaliana*. *Genetics* 129, 605-609.

Haley, C.S. and Knott, S.A. (1992) A simple regression method for mapping quantitative trait loci in line crosses using flanking markers. *Heredity* 69, 315-324.

Knapp, S.J. (1991) Using molecular markers to map multiple quantitative trait loci: models for backcross, recombinant inbred, and doubled haploid progeny. *Theoretical and Applied Genetics* 81, 333-338.

Knapp, S.J., Bridges, W.C. and Birkes, D. (1990) Mapping quantitative trait loci using molecular marker linkage maps. *Theoretical and Applied Genetics* 79, 583-592.

Lander, E.S. and Botstein, D. (1989) Mapping Mendelian factors underlying quantitative traits using RFLP linkage maps. *Genetics* 121, 185-199.

Lander, E.S., Green, P., Abrahamson, J., Barlow, A., Daly, M.J., Lincoln, S.E. and Newburg, L. (1987) MAPMAKER: an interactive computer package for constructing primary genetic linkage maps of experimental and natural populations. *Genomics* 1, 174-181.

Law, C.N. and Gale, M.D. (1979) Cytological markers and quantitative variation in wheat. In: Thompson, J.N. and Thoday, J.M. (eds) *Quantitative Genetic Variation*. Academic Press, New York, San Francisco and London, pp. 275-293.

Luo, Z.W. and Kearsey, M.J. (1989) Maximum likelihood estimation of linkage between a marker gene and a quantitative locus. *Heredity* 63, 401-408.

Luo, Z.W. and Kearsey, M.J. (1991) Maximum likelihood estimation of linkage between a marker gene and a quantitative trait locus. II. Application to backcross and doubled haploid populations. *Heredity* 66, 117-124.

Luo, Z.W. and Kearsey, M.J. (1992) Interval mapping of quantitative trait loci in an F_2 population. *Heredity* 69, 236-242.

Luo, Z.W. and Woolliams, J.A. (1993) Estimation of genetic parameters using linkage between a marker gene and a locus underlying a quantitative character in F_2 populations. *Heredity* 70, 245-253.

McMillan, I. and Robertson, A. (1974) The power of methods for the detection of major genes affecting quantitative characters. *Heredity* 32, 349-356.

Martinez, O. and Curnow, R.N. (1992) Estimating the locations and the sizes of the effects of quantitative trait loci using flanking markers. *Theoretical and Applied Genetics* 85, 480-488.

Mather, K. and Jinks, J.L. (1982) *Biometrical Genetics*, 3rd edn. Chapman and Hall, London and New York.

Morgan, T.H. (1910) Sex limited inheritance in *Drosophila*. *Science* 32, 120-122.

Paterson, A.H., Damon, S., Hewitt, J.D., Zamir, D., Rabinowitch, H.D., Lincoln, S.E., Lander, E.S. and Tanksley, S.D. (1991) Mendelian factors underlying quantitative traits in tomato: comparison across species, generations, and environments. *Genetics* 127, 181-197.

Pijnacker, L.P., Ferwerda, M.A., Puite, K.J. and Schaart, J.G. (1989) Chromosome elimination and mutation in tetraploid somatic hybrids of *Solanum tuberosum* and *S. phureja*. *Plant Cell Reports* 8, 82-85.

Ritter, E., Gebhardt, C. and Salamini, F. (1990) Estimation of recombination frequencies and construction of RFLP linkage maps in plants from crosses between heterozygous parents. *Genetics* 125, 645-654.

Soller, M. and Beckmann, J.S. (1990) Marker-based mapping of quantitative trait loci using replicated progenies. *Theoretical and Applied Genetics* 80, 205-208.

Stam, P. (1991) Some aspects of QTL analysis. In: Pesek, J. and Hartmann, J. (eds) *Biometrics in Plant Breeding*. Proceedings of the 8th Meeting of the Eucarpia Section Biometrics in Plant Breeding, July 1-6 1991, Brno, Czechoslovakia, pp. 23-32.

Thoday, J.M. (1961) Location of polygenes. *Nature* 191, 368-370.

Thoday, J.M. (1979) Polygene mapping: uses and limitations. In: Thompson, J.N. and Thoday, J.M. (eds) *Quantitative Genetic Variation*. Academic Press, New York, San Francisco and London, pp. 219-233.

Thoday, J.M. and Thompson, J.N. (1976) The number of segregating genes implied by continuous variation. *Genetica* 46, 335-344.

Thompson, J.N. and Thoday, J.M. (1974) A definition and standard nomenclature for 'polygenic loci'. *Heredity* 33, 430-437.

Weller, J.I. (1986) Maximum likelihood techniques for the mapping and analysis of quantitative trait loci with the aid of genetic markers. *Biometrics* 42, 627-640.

Weller, J.I. (1987) Mapping and analysis of quantitative trait loci in *Lycopersicon*. *Heredity* 59, 413-421.

Zeng, Z.B. (1992) Correcting the bias of Wright's estimates of the number of genes affecting a quantitative character: a further improved method. *Genetics* 131, 987-1001.

5 Use of 2*n* Gametes

G.C.C. TAI

Agriculture Canada Research Station, PO Box 20280, Fredericton, New Brunswick, Canada, E3B 4Z7.

Introduction

The cultivated potato, *Solanum tuberosum* L., is a tetraploid ($2n = 4x = 48$) which displays tetrasomic inheritance. Dihaploids ($2n = 2x = 24$) were first induced from tetraploid varieties through a parthenogenic process (Hougas and Peloquin, 1958; Hougas *et al.*, 1964). Dihaploid induction, using either parthenogenesis or anther culture, is now a routine procedure in many breeding programmes. Many wild and cultivated potato species are diploid ($2n = 2x = 24$). Hybrids ($2n = 2x = 24$) can be obtained between the *S. tuberosum* dihaploids and diploid species. More importantly, many diploid species and dihaploid–species hybrids are found to produce $2n$ unreduced gametes. Mendiburu and Peloquin (1976) gave the term diplandroid and diplogynoid for $2n$ pollen and egg, respectively. The unusual meiotic mechanisms that result in $2n$ gametes provide new opportunities for genetic analysis with tetrasomic inheritance and for breeding highly heterotic tetraploid varieties.

Modes of 2*n* Gamete Formation

A number of mechanisms during the premeiotic, meiotic and postmeiotic stages of gamete formation may induce $2n$ gametes (Veilleux, 1985). Peloquin *et al.* (1989a) listed six distinct possible modes of $2n$ gamete formation: (i) premeiotic doubling; (ii) first division restitution (FDR); (iii) chromosome replication during meiotic interphase; (iv) second division restitution (SDR); (v) postmeiotic doubling; and (vi) apospory. In potatoes, FDR and SDR are the two important modes of $2n$ gamete formation.

For both FDR and SDR, the abnormal event that leads to the formation

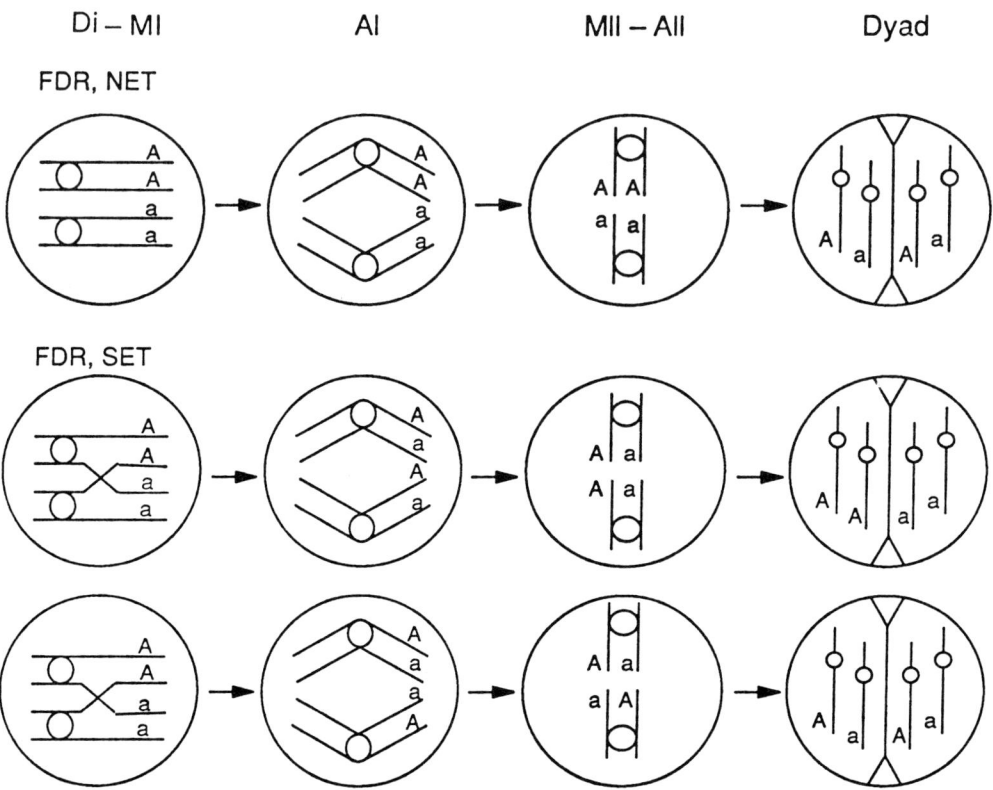

Fig. 5.1. Formation of 2n gametes by FDR. Di = diakinesis; M = metaphase; A = anaphase.

of 2n gametes occurs during anaphase II of meiosis. Figure 5.1 shows the cytological events during FDR for a pair of chromosomes of a diploid individual with the genotype Aa. An abnormal parallel orientation of the spindles during anaphase II prevents cell division and consequently two 2n microspores are formed following cytokinesis. Mok and Peloquin (1975a) reported that 'parallel spindles' are inherited as a simple recessive (ps). No-exchange tetrads (NET) are formed when there is no crossing over between chromatids carrying A and a. All 2n gametes are of the genotype Aa. Single-exchange tetrads (SET) are formed when there is crossing over between chromatids carrying A and a. As shown in Figure 5.1, two types of chromosomal orientation are possible with SET. The first orientation results in 2n gametes with genotypes AA and aa in a 1:1 ratio, whereas in the second one all gametes have the Aa genotype. Assuming that the frequency of SET is β, the ratio of AA, Aa and aa gametes can be calculated as $\beta/4 : (1 - \beta/2) : \beta/4$. Ramanna (1979) and Veilleux et al. (1982) attributed FDR to fused spindles during the second meiotic division. This mechanism, however, still gives 2n gametes with the same genetic composition.

SDR results from a different meiotic abnormality. The cytological events of SDR are shown in Figure 5.2. The first division is followed by a premature

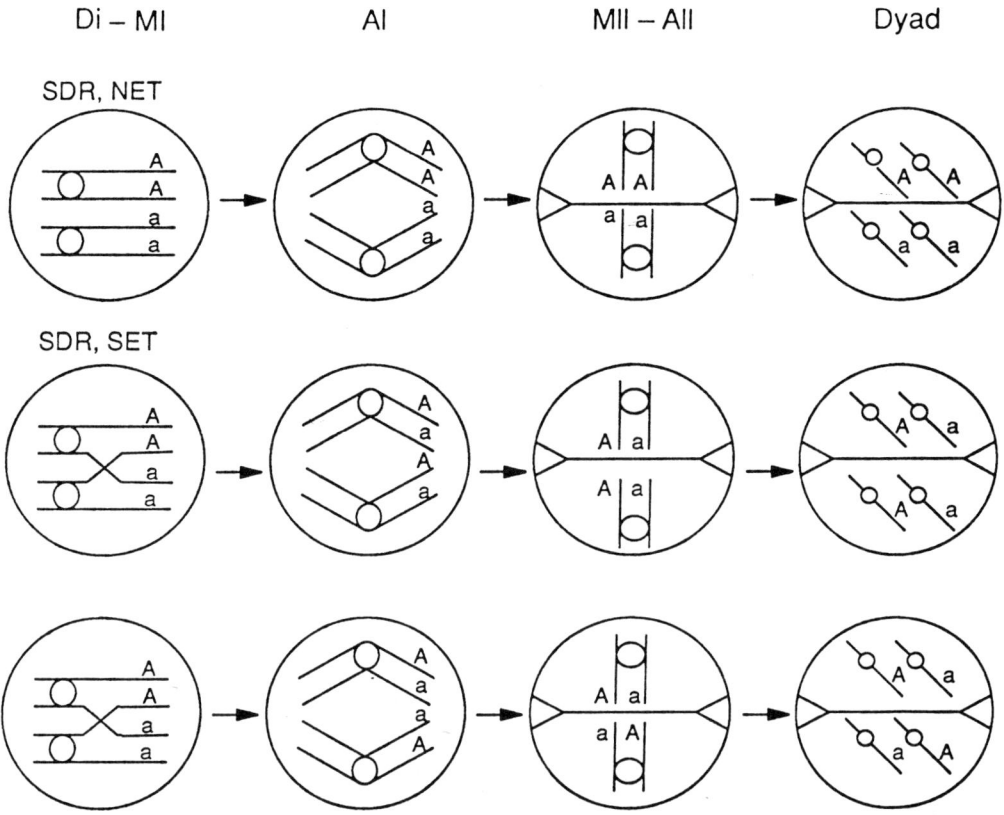

Fig. 5.2. Formation of 2n gametes by SDR. Di = diakinesis; M = metaphase; A = anaphase.

cytokinesis. The second division does not take place, but the chromatids fall apart. Consequently a dyad of two 2n microspores is formed. Mok and Peloquin (1975a) called this type of SDR mechanism premature cytokinesis. It is also simply inherited (*pc*-1 and *pc*-2). With NET, the 2n gametes have the genotypes *AA* and *aa* with equal frequency. Two types of chromosomal orientation are possible in the case of SET. As shown in Figure 5.2, both types produce gametes with the *Aa* genotype. Again, assuming the frequency of SET is β, the ratio of *AA*, *Aa* and *aa* gametes is $(1 - \beta)/2 : \beta : (1 - \beta)/2$.

The range of the coefficient β is from 0 to 1 for both FDR and SDR. Chromosomal segregation occurs when $\beta = 0$. Random chromatid segregation takes place when $\beta = \frac{2}{3}$ and results in a 1:4:1 ratio of *AA:Aa:aa* with both FDR and SDR. In contrast, the *AA*, *Aa* and *aa* gametes produced by the equivalent tetraploid genotype *AAaa* have the ratio $(1 + 2\alpha)/6 : (2 - 2\alpha)/3 : (1 + 2\alpha)/6$, where α is the rate of double reduction (Fisher and Mather, 1943). The range of α is from 0 to $\frac{1}{6}$. Chromosomal and chromatid segregation occur when $\alpha = 0$ and $\frac{1}{7}$, respectively. There are thus three known corresponding cytological events in a diploid and tetraploid during meiosis, i.e. $\alpha = 0$

and $\beta = 0$ for chromosomal segregation, $\alpha = \frac{1}{7}$ and $\beta = \frac{2}{3}$ for chromatid segregation and $\alpha = \frac{1}{6}$ and $\beta = 1$ for maximum value. The three pairs of values give an empirical relationship between α and β as $\beta = 0.0822 \, [\exp(15.4635\alpha) - 1]$.

Synaptic mutants have been found which cause poor pairing and/or reduced chiasma frequencies in microsporogenesis (Jongedijk and Ramanna, 1988; Peloquin et al., 1989a). The synaptic mutant, sy-3, when combined with 'parallel spindles', can produce FDR 2n pollen with no crossing over (NCO), i.e. for any gene, $\beta = 0$. This mechanism of 2n gamete formation has been referred to as FDR-NCO (Hermundstad and Peloquin, 1987). The mutant sy-3 was found in S. phureja–dihaploid hybrids (Okwuagwu and Peloquin, 1981).

Parallel spindles, premature cytokinesis and synaptic mutants have all been identified in microsporogenesis and thus are all mechanisms for the formation of diplandroids. The cytological mechanisms of diplogynoid formation in megasporogenesis have also been studied (Iwanaga and Peloquin, 1979; Jongedijk, 1985; Stelly and Peloquin, 1986; Werner and Peloquin, 1987; Parrot and Hanneman, 1988). Werner and Peloquin (1990) indicated that omission of the second meiotic division is the predominant mechanism of 2n egg formation. It is controlled by a single recessive gene (os). Another mechanism of 2n egg formation is the failure of cytokinesis after the second meiotic division, which is controlled by a single recessive gene (fc). The genetical structure of 2n eggs formed by either mechanism follows that of SDR. Recently, Jongedijk et al. (1991b) identified diploid clones which produce FDR 2n eggs through a direct equational division of univalent chromosomes at anaphase I (pseudohomotypic division). These diploids possess the desynaptic gene ds-1.

Diploid species and dihaploid–species hybrids in general produce both 2n and n gametes (Mendiburu and Peloquin, 1977). Matings between 4x and 2x parents (4x-2x and 2x-4x crosses) give rise to almost entirely 4x progeny due to a 'triploid block' mechanism (Marks, 1966). Matings between 2x parents (2x-2x crosses) produce both 2x and 4x progeny. Frequencies of 2x and 4x progeny vary between interdiploid crosses (Mendiburu and Peloquin, 1977). The 4x progenies from 2x-2x crosses have to be distinguished by chromosome counts or some other means such as establishing the mean number of chloroplasts in stomata (Wagenvoort and Zimnoch-Guzowska., 1992).

Gene–Centromere Mapping

Mendiburu and Peloquin (1979) developed a method of gene–centromere mapping using the 4x-2x cross $aaaa \times Aa$, where A is dominant and a recessive. The map distance is determined by measuring the frequency of the nulliplex genotype (aaaa) in the tetraploid progenies. It is based on the fact that the production of aa gametes by FDR from an Aa diploid is due to the formation of SET, whose frequency is strictly a function of the distance between the locus and centromere. Assuming that double or higher-order crossovers do not occur, as considered reasonable for the potato by Hermsen

(1984a), the expected gametic output of an FDR diploid (Aa) is $AA : Aa : aa = \beta/4 : 1 - \beta/2 : \beta/4$, where β is the frequency of SET. Let d be the map distance between the locus Aa and the centromere so that $d = (\beta/2) \times 100$ cM. The coefficient β is estimated from the number of nulliplex progenies in an experiment. Let a total of N progenies be scored in an experiment, and let n of them be nulliplex. Then $\beta = 4n/N$ and $d = (2n/N) \times 100$ cM. With isozyme or RFLP loci, duplex, simplex and nulliplex genotypes can sometimes be identified when the Aa alleles are codominant. Douches and Quiros (1987) proposed a maximum-likelihood procedure to estimate map distance, using this type of segregation data. Jongedijk et al. (1991a) suggested that segregation data from similar tetraploid families obtained from two or more crosses can be pooled together for mapping when they are homogeneous, based on a χ^2 test. A confidence interval for the map distance can then be obtained based on the binomial distribution (Jongedijk et al., 1991a).

The precision of an estimated map distance can be inferred from the size of the confidence interval. A confidence interval estimated from segregation data of a single $4x$–$2x$ cross or pooled data from several crosses takes into account the effect of sample size. The variation due to the instability of the meiotic behaviour *per se* is not included. This can only be achieved by taking into consideration the variation of estimates of map distance obtained from a series of crosses. A more appropriate method of mapping based on segregation data from multiple crosses is the weighted least-squares procedure (Grizzle et al., 1969; Tai, 1989). A simple example is presented here.

Consider a locus with alleles A (dominant) and a (recessive). Let T1, T2 and D represent two tetraploid and a diploid parent, respectively. The tetraploid parents can be any genotype except $AAAA$. The diploid must be a heterozygote (Aa) for the monogenic characteristic under study and produce abundant diplandroids by FDR. T1, T2 and D produce aa gametes with expected frequencies $t1$, $t2$ and f, respectively. The parents are used to make three crosses: T1 × T2, T1 × D and T2 × D. A total of $N1$, $N2$ and $N3$ progenies are used in an experiment and $n1$, $n2$ and $n3$ nulliplexes ($aaaa$) are recorded from the three crosses, respectively. The observed frequencies of nulliplex in the three crosses are $w1 = n1/N1$, $w2 = n2/N2$ and $w3 = n3/N3$. Their expectations are $E(w1) = t1 \times t2$, $E(w2) = t1 \times f$ and $E(w3) = t2 \times f$. Taking the logarithm on the observed frequencies, we have

$$\mathbf{w} = \begin{bmatrix} \log w1 \\ \log w2 \\ \log w3 \end{bmatrix} \quad E(\mathbf{w}) = \begin{bmatrix} \log t1 + \log t2 \\ \log t1 + \log f \\ \log t2 + \log f \end{bmatrix}$$

Corresponding to the mating design, a 'design' matrix ($\mathbf{x}$) and a 'gamete' matrix ($\mathbf{g}$) can be formed as follows

$$\mathbf{x} = \begin{bmatrix} 1 & 1 & 0 \\ 1 & 0 & 1 \\ 0 & 1 & 1 \end{bmatrix} \quad \mathbf{g} = \begin{bmatrix} \log t1 \\ \log t2 \\ \log f \end{bmatrix}$$

Obviously, $E(\mathbf{w}) = \mathbf{xg}$. To estimate the frequencies in $\mathbf{g}$ by means of weighted least squares, a variance–covariance matrix of elements in $\mathbf{w}$ is needed. This can be estimated by

$$\mathbf{V} = \begin{bmatrix} (1-w1)/N1w1 & 0 & 0 \\ 0 & (1-w2)/N2w2 & 0 \\ 0 & 0 & (1-w3)/N3w3 \end{bmatrix}$$

$\mathbf{g}$ can be estimated by $\mathbf{g} = (\mathbf{x}'\mathbf{V}^{-1}\mathbf{x})^{-1}\mathbf{x}'\mathbf{V}^{-1}\mathbf{w}$. Only f is required for calculating the map distance. f is estimated by $f = \exp(\mathbf{cg})$ where $\mathbf{c} = [0\ \ 0\ \ 1]$, and $\exp(\mathbf{cg})$ means e to the power of $\mathbf{cg}$. The map distance of the locus A/a is estimated by

$$d = 2f \times 100\,\text{cM} = 2[\exp(\mathbf{cg})] \times 100\,\text{cM}$$

A confidence interval for d can be derived using the χ^2 distribution. The statistic $(\log f - \mathbf{cg})^2/u$, where $u = \mathbf{c}(\mathbf{x}'\mathbf{V}^{-1}\mathbf{x})^{-1}\mathbf{c}'$, is asymptotically distributed as χ^2 with one degree of freedom. Thus, a $(1-\alpha)$ confidence interval for d can be established as

$$d \times \exp[-(\chi_\alpha^2 u)^{\frac{1}{2}}] < d < d \times \exp[(\chi_\alpha^2 u)^{\frac{1}{2}}]$$

where χ_α^2 is the tabulated χ^2 value at probability level α with one degree of freedom.

The map distance is estimated by the following formula when an SDR diploid hybrid is used in the mating scheme

$$d = (0.5 - f) \times 100\,\text{cM} = [0.5 - \exp(\mathbf{cg})] \times 100\,\text{cM}$$

The confidence interval for d is

$$50 - (50-d) \times \exp[(\chi_\alpha^2 u)^{\frac{1}{2}}] < d < 50 - (50-d) \times \exp[-(\chi_\alpha^2 u)^{\frac{1}{2}}]$$

The above mapping procedure is based on the frequency of $2n$ gametes of the recessive genotype produced by the $2x$ parent. The procedure can also be used for mapping codominant monogenic markers. In this case, the frequencies of both AA and aa gametes from the Aa parent should be jointly considered and the data from $AAAA$ and $aaaa$ offspring in all crosses should be pooled together for mapping.

The $2x$ parent used for mapping should produce $2x$ gametes by either FDR or SDR, but not a mixture of both. Wagenvoort and Zimnoch-Guzowska (1992) suggest that the $2x$–$4x$ cross is advantageous in gene–centromere mapping studies. This is because in certain diploids SDR $2n$ eggs occur exclusively (Jongedijk, 1985; Stelly and Peloquin, 1986; Douches and Quiros, 1987). Wagenvoort and Zimnoch-Guzowska (1992) also point out that the mapping result for a gene from $2x$–$4x$ crosses can be used to discriminate between FDR and SDR pollen in $4x$–$2x$ crosses when the gene is located close to the centromere.

Table 5.1. Frequencies of 2n gametes of different genotypes produced by tetraploid and diploid parents.

Parental genotype		2n gamete		
		AA	Aa	aa
Tetraploid				
Quadruplex	(AAAA)	1	0	0
Triplex	(AAAa)	$(2+\alpha)/4$	$(1-\alpha)/2$	$\alpha/4$
Duplex	(AAaa)	$(1+2\alpha)/6$	$(2-2\alpha)/3$	$(1+2\alpha)/6$
Simplex	(Aaaa)	$\alpha/4$	$(1-\alpha)/2$	$(2+\alpha)/4$
Nulliplex	(aaaa)	0	0	1
Diploid				
FDR	(AA)	1	0	0
FDR	(Aa)	$\beta/4$	$(2-\beta)/2$	$\beta/4$
FDR	(aa)	0	0	1
SDR	(AA)	1	0	0
SDR	(Aa)	$(1-\beta)/2$	β	$(1-\beta)/2$
SDR	(aa)	0	0	1

Biometrical Genetical Analysis with Tetrasomic Inheritance

The tetrasomic inheritance of quantitative traits in autotetraploids is a complex subject. Consider a single locus with two alleles A/a. There are five possible genotypes, i.e. quadruplex (*AAAA*), triplex (*AAAa*), duplex (*AAaa*), simplex (*Aaaa*) and nulliplex (*aaaa*). The interaction pattern between four genes is undoubtedly more complicated than that between two genes in a diploid. The ratio of *AA*, *Aa* and *aa* gametes produced by each of these genotypes is shown in Table 5.1. The coefficient of double reduction, α, complicates the expression of the ratio when compared with that of *A* and *a* gametes produced by a diploid. Both factors contribute to the difficulty of quantitative analysis of tetrasomic inheritance, using any type of mating design. Killick (1971) derived results for the biometrical genetics of autotetraploids (see chapter 3, by Bradshaw). Using an additive–dominance model, first- and second-degree statistics are obtained from the generations derived from a cross between two autotetraploid homozygous lines. The first-degree statistics for the generation means involve higher powers of α. The second-degree statistics for the variances of various generations contain cross-products of genetic parameters, even where $\alpha = 0$ (i.e. segregation is chromosomal). There is thus a severe problem with the estimation of genetic parameters based on generations derived from tetraploid parents.

The simpler structure of the ratio of 2n gametes produced by FDR and SDR diploids, however, makes the analysis a lot easier. Tai (1982a, b, 1986, 1989) proposed several mating designs using $2x$ (FDR and SDR) and $4x$

parents for the biometrical genetical analysis of tetrasomic inheritance based on family means. The detailed analysis of one mating design, which involves crosses of the type $2x$(SDR) $\times$ $2x$(FDR), is given here.

Assume that two homozygous diploid parents are available and that their diploid F_1 hybrid is capable of producing $2n$ pollen by FDR and $2n$ eggs by SDR. Let the genotypes of the parents be AA and aa. By backcrossing the F_1 to both parents, we have

$$AA \times Aa \rightarrow \beta/4 \; [AAAA] : (1-\beta/2) \; [AAAa] : \beta/4 \; [AAaa]$$
$$Aa \times AA \rightarrow (1-\beta)/2 \; [AAAA] : \beta \; [AAAa] : (1-\beta)/2 \; [AAaa]$$
$$aa \times Aa \rightarrow \beta/4 \; [AAaa] : (1-\beta/2) \; [Aaaa] : \beta/4 \; [aaaa]$$
$$Aa \times aa \rightarrow (1-\beta)/2 \; [AAaa] : \beta \; [Aaaa] : (1-\beta)/2 \; [aaaa]$$

The effects of the five tetraploid genotypes are defined by the additive-dominance model of Killick (1971) and Mather and Jinks (1982). Let m be the mid-homozygote value, $[AAAA] = m + d$, $[AAAa] = m + h_3$, $[AAaa] = m + h_2$, $[Aaaa] = m + h_1$ and $[aaaa] = m - d$. The genetic structures of the means of the four families listed above are

$$[AA \times Aa] = m + h_3 + \beta/4(d + h_2 - 2h_3)$$
$$[Aa \times AA] = m + \tfrac{1}{2}(d + h_2) - \beta/2(d + h_2 - 2h_3)$$
$$[aa \times Aa] = m + h_1 - \beta/4(d - h_2 + 2h_1)$$
$$[Aa \times aa] = m + \tfrac{1}{2}(h_2 - d) + \beta/2(d - h_2 + 2h_1)$$

The $2x$ progenies of the above four crosses can be backcrossed to both parents to generate eight more $4x$ hybrid families. Together with the tetraploids of the parents and their F_1, the following mating scheme is achieved

Mating scheme	Symbol	Mean
1. AA doubled	P1 ($AAAA$)	$m + d$
2. aa doubled	P2 ($aaaa$)	$m - d$
3. Aa doubled	F1 ($AAaa$)	$m + h_2$
4. $AA \times Aa$	B1	$m + h_3 + f_1$
5. $Aa \times AA$	rB1	$m + \tfrac{1}{2}(d + h_2) - 2f_1$
6. $aa \times Aa$	B2	$m + h_1 - f_2$
7. $Aa \times aa$	rB2	$m + \tfrac{1}{2}(h_2 - d) + 2f_2$
8. $AA \times (AA \times Aa)$	B11	$m + \tfrac{1}{2}(d + h_3) + \tfrac{1}{2}f_1$
9. $aa \times (AA \times Aa)$	B12	$m + \tfrac{1}{2}(h_2 + h_1) - \tfrac{1}{2}f_2$
10. $AA \times (aa \times Aa)$	rB11	$m + \tfrac{1}{2}(h_2 + h_3) + \tfrac{1}{2}f_1$
11. $aa \times (aa \times Aa)$	rB12	$m + \tfrac{1}{2}(h_1 - d) - \tfrac{1}{2}f_2$
12. $(AA \times Aa) \times AA$	B21	$m + \tfrac{1}{4}(3d + h_2) - f_1$
13. $(AA \times Aa) \times aa$	B22	$m + \tfrac{1}{4}(3h_2 - d) + f_2$
14. $(aa \times Aa) \times AA$	rB21	$m + \tfrac{1}{4}(d + 3h_2) - f_1$
15. $(aa \times Aa) \times aa$	rB22	$m + \tfrac{1}{4}(h_2 - 3d) + f_2$

in which $f_1 = \beta/4(d + h_2 - 2h_3)$ and $f_2 = \beta/4(d - h_2 + 2h_1)$. Let **g** and **O** be the vectors for the genetic parameters and the observed generation means, respectively

$$\mathbf{g}' = [m \ d \ h_3 \ h_2 \ h_1 \ f_1 \ f_2]$$

$$\mathbf{O}' = [\bar{P}1 \ \bar{P}2 \ \bar{F}1 \ \bar{B}1 \ r\bar{B}1 \ \bar{B}2 \ r\bar{B}2 \ \bar{B}11 \ \bar{B}12 \ r\bar{B}11 \ r\bar{B}12 \ \bar{B}21 \ \bar{B}22 \ r\bar{B}21 \ r\bar{B}22]$$

Also let **C** and **V** be the matrices of coefficients for the various parameters in the means, and estimates of the variances for the means, respectively

$$\mathbf{C} = \begin{bmatrix} 1 & 1 & 0 & 0 & 0 & 0 & 0 \\ 1 & -1 & 0 & 0 & 0 & 0 & 0 \\ 1 & 0 & 0 & 1 & 0 & 0 & 0 \\ 1 & 0 & 1 & 0 & 0 & 1 & 0 \\ 1 & \frac{1}{2} & 0 & \frac{1}{2} & 0 & -2 & 0 \\ 1 & 0 & 0 & 0 & 1 & 0 & -1 \\ 1 & -\frac{1}{2} & 0 & \frac{1}{2} & 0 & 0 & 2 \\ 1 & \frac{1}{2} & \frac{1}{2} & 0 & 0 & \frac{1}{2} & 0 \\ 1 & 0 & 0 & \frac{1}{2} & \frac{1}{2} & 0 & -\frac{1}{2} \\ 1 & 0 & \frac{1}{2} & \frac{1}{2} & 0 & \frac{1}{2} & 0 \\ 1 & -\frac{1}{2} & 0 & 0 & \frac{1}{2} & 0 & -\frac{1}{2} \\ 1 & \frac{3}{4} & 0 & -\frac{1}{4} & 0 & -1 & 0 \\ 1 & -\frac{1}{4} & 0 & \frac{3}{4} & 0 & 0 & 1 \\ 1 & \frac{1}{4} & 0 & \frac{3}{4} & 0 & -1 & 0 \\ 1 & -\frac{3}{4} & 0 & \frac{1}{4} & 0 & 0 & 1 \end{bmatrix}$$

$$\mathbf{V} = \text{diag}[\hat{V}(\text{P1}) \ \hat{V}(\text{P2}) \ \ldots \ \hat{V}(\text{rB21}) \ \hat{V}(\text{rB22})]$$

g can be estimated by $(\mathbf{C}'\mathbf{V}^{-1}\mathbf{C})^{-1}\mathbf{C}'\mathbf{V}^{-1}\mathbf{O}$. The validity of the additive-dominance model can be verified from a joint scaling test. For example, $\frac{1}{2}$ (P1 + P2) + F1 = B11 + B12 + rB11 + rB12 − B1 − B2.

The mating design uses chosen homozygous diploid genotypes as parents and allows two alleles per locus. The estimation of genetic parameters involved in tetrasomic inheritance is based on means of $4x$ progenies of the families in the mating scheme. Homozygous lines of dihaploids and dihaploid–species hybrids are not available at present. Tai (1982a) did, however, propose a mating design which uses a heterozygous diploid and its doubled tetraploid as parents. It should also be noted that both $2x$ and $4x$ progenies are produced within a family. They have to be distinguished so that experimental work for estimating genetic parameters can be carried out on $4x$ progenies.

The above work has not been extended to second-degree statistics because α/β and cross-products of genetic parameters complicate the estimation procedure. Boudec *et al.* (1989a, b) carried out a theoretical study of the genetical structure of first and second degree statistics for various types of $4x$–$2x$ and $2x$–$2x$ crosses based on a generalized model and the concept of identity by descent. Special values of α and β have to be assumed in order to estimate the various genetic parameters.

Haynes (1990) focused on the genetical structure of the covariances between $2x$ parents and $4x$ offspring in $4x$–$2x$ crosses. Removing the limitation of two alleles per locus but assuming $\alpha = 0$ and no epistasis, covariances of $2x$ parent–$4x$ offspring are derived by computing the coefficients of coancestry and double coancestry. The covariance between $2x$ (FDR) parent–$4x$ offspring is $2\sigma_{aa'} + (1 - \beta/2)\sigma_{dd'} + (\beta/2)\sigma_{a'd''}$ where σ represents the covariance, a' and d' the additive and digenic effects of the $2x$ parent, and a, d and d'' the additive, digenic and homozygous digenic (i.e. the digenic effect of AA in $AAaa$) effects of $4x$ offspring, respectively. The covariance between $2x$ (SDR) parent–$4x$ offspring is $2\sigma_{aa'} + \beta\sigma_{dd'} + (1 - \beta)\sigma_{a'd''}$.

All of the theoretical work discussed above is based on $2n$ gametes produced by FDR or SDR diploids. Conicella *et al.* (1991) reported a diploid *S. tuberosum* × *S. chacoense* hybrid that produced $2n$ eggs by a mixture of SDR–FDR. Caution must therefore be practised in choosing diploid parents which produce FDR or SDR $2n$ gametes but not both for experimental work. Haynes *et al.* (1991) conducted a theoretical study on estimating the amount of preferential pairing of homologous chromosomes in the $4x$–$2x$ tetraploid hybrid. They cautioned that preferential pairing may be fairly large in interspecific $4x$–$2x$ tetraploid hybrids.

Marker-based Analysis of Tetrasomic Inheritance of Quantitative Traits

Linkage maps constructed by means of molecular markers such as isozyme and restriction fragment length polymorphism (RFLP) loci have proved useful for locating quantitative trait loci (QTL) (Lander and Botstein, 1989). The unique genetic behaviour of $2n$ gametes provides an ideal tool to locate QTLs based on the linkage map of chromosomes and to determine their role in the tetrasomic inheritance of the quantitative trait under investigation. Consider the situation for a QTL Qq located between two flanking markers M_1m_1 and M_2m_2. M_1m_1 is located closer to the centromere than M_2m_2. Let the SET frequencies for the two markers be β_1 and β_2, and that of the QTL be β_q, respectively. The two alleles of each of the two markers are assumed to be codominant. Two diploid parents with genotypes $M_1M_1QQM_2M_2$ and $m_1m_1qqm_2m_2$ are hybridized to produce a diploid F_1 of the genotype $M_1m_1QqM_2m_2$. The F_1 is backcrossed to both parents. Assume that all three genotypes also produce $2n$ gametes. Tetraploid progenies are produced and identified from the backcross families. These families are sorted according to marker genotypes and scored for the quantitative trait. The data are then used to see if a QTL is present between the marker loci and, if so, to estimate various genetic parameters in tetrasomic inheritance. As an example, the results for $M_1m_1QqM_2m_2 \times M_1M_1QQM_2M_2$ in the form of FDR × FDR are given in detail.

Four separate events should be considered with regard to chromosomal behaviour during meiosis of the genotype $M_1m_1QqM_2m_2$. They are: (i) no crossover between M_2m_2 and the centromere; (ii) a crossover in the region between Qq and M_2m_2; (iii) a crossover in the region between M_1m_1 and

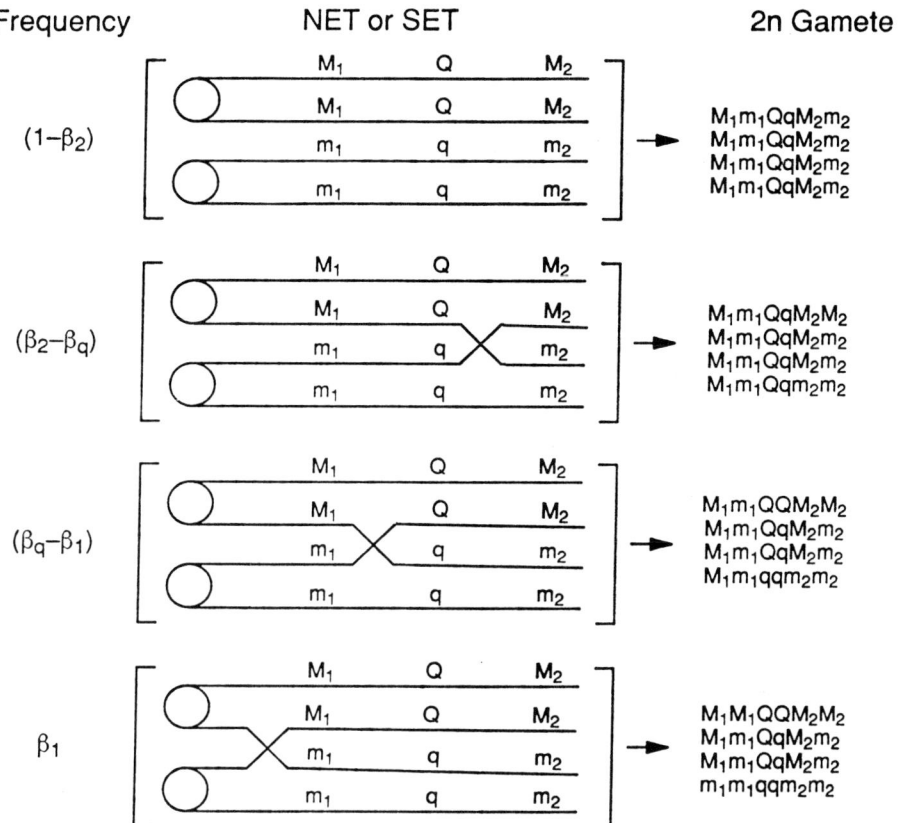

Fig. 5.3. Genotypes of 2n gametes produced by the parent $M_1m_1QqM_2m_2$.

Qq; and (iv) a crossover in the region between the centromere and M_1m_1. The relative frequencies of these four events are $(1-\beta_2)$, $(\beta_2-\beta_q)$, $(\beta_q-\beta_1)$ and β_1, respectively. The genotypes of the 2n gametes produced by each of these events are shown in Figure 5.3. The genetic structure of progenies obtained from backcrossing the F_1 to the parent $M_1M_1QQM_2M_2$ are shown in Table 5.2. The following steps are taken for detecting QTL effects between the flanking markers and estimating the effects of various genetic parameters involved in the tetrasomic inheritance of the quantitative trait:

1. The segregating $4x$ progenies in the backcross generation are identified and sorted according to the five marker genotypes (see Table 5.2). The five marker genotypes can be distinguished because both markers M_1m_1 and M_2m_2 are codominant.

2. Means of the quantitative trait for the five marker classes are compared to detect the possible existence of a QTL effect.

3. Progenies are sorted according to genotypes of the M_1m_1 locus. Their relative frequencies are used to obtain an estimate of β_1.

4. Progenies are then sorted according to genotypes of the M_2m_2 locus. Their relative frequencies are used to obtain an estimate of β_2.

Table 5.2. QTL effects of marker genotypes for the situation that a QTL lies between two flanking markers M_1/m_1 and M_2/m_2 in the backcross generation $M_1M_1QQM_2M_2 \times M_1m_1QqM_2m_2$.

Marker	QTL	Frequency	Expected	Observed
$M_1M_1M_1M_1M_2M_2M_2M_2$	$QQQQ$	$\beta_1/4$	$\mu_1 + d$	x_1
$M_1M_1M_1m_1M_2M_2M_2M_2$	$QQQQ$, $QQQq$	$(\beta_2-\beta_1)/4$	$\mu_1 + k_1d + k_2h_3$	x_2
$M_1M_1M_1m_1M_2M_2M_2m_2$	$QQQq$	$1-\beta_2/2$	$\mu_1 + h_3$	x_3
$M_1M_1M_1m_1M_2M_2m_2m_2$	$QQQq$, $QQqq$	$(\beta_2-\beta_1)/4$	$\mu_1 + k_1h_2 + k_2h_3$	x_4
$M_1M_1m_1m_1M_2M_2m_2m_2$	$QQqq$	$\beta_1/4$	$\mu_1 + h_2$	x_5

$k_1 = [(\beta_q - \beta_1)/(\beta_2 - \beta_1)]$; $k_2 = [(\beta_2 - \beta_q)/(\beta_2 - \beta_1)]$.

5. Let x_1, x_2, x_3, x_4 and x_5 be the means of the quantitative trait for the five marker classes listed in Table 5.2. The SET coefficient of the QTL is estimated by

$$\hat{\beta}_q = \hat{\beta}_1 + (\hat{\beta}_2 - \hat{\beta}_1)(x_2 + x_4 - 2x_3)/(x_1 + x_5 - 2x_3)$$

6. The genetic parameters of the QTL Q/q involved in the means of the marker classes are estimated by a weighted least-squares analysis: $\mathbf{g} = (\mathbf{C'V^{-1}C})^{-1}\mathbf{C'V^{-1}x}$, where $\mathbf{g'} = [\mu_1\ d\ h_2\ h_3]$, in which μ_1 is the 'background' effect of all QTLs not located between the markers,

$$\mathbf{C} = \begin{bmatrix} 1 & 1 & 0 & 0 & 0 \\ 1 & k_1 & 0 & 0 & k_2 \\ 1 & 0 & 0 & 0 & 1 \\ 1 & 0 & 0 & k_1 & k_2 \\ 1 & 0 & 0 & 1 & 0 \end{bmatrix}$$

in which $k_1 = (\hat{\beta}_q - \hat{\beta}_1)/(\hat{\beta}_2 - \hat{\beta}_1)$ and $k_2 = (\hat{\beta}_2 - \hat{\beta}_q)/(\hat{\beta}_2 - \hat{\beta}_1)$,

$$\mathbf{x'} = [x_1\ x_2\ x_3\ x_4\ x_5]$$

and

$$\mathbf{V} = \text{diag}[V_1\ V_2\ V_3\ V_4\ V_5]$$

in which the Vs are the estimates of variances of the five means.

Genetic Consequences of $2n$ Gametes in Unilateral and Bilateral Sexual Polyploidization

The genetic consequence of $2n$ gametes in their role of sexual polyploidization has been discussed in many reports (Mendiburu et al., 1974; Mendiburu and Peloquin, 1977; Hermsen, 1984a; Veilleux, 1985; Tai, 1989; Watanabe et al., 1991; Werner and Peloquin, 1991). Hermsen (1984a) calculated that the average percentage of the parental heterozygosity present in FDR and SDR gametes for four types of chromosome with known sites of centromere and crossovers are 80.2% and 39.6%, respectively. Tai (1989) compared means and variances of six types of crosses (FDR × FDR, FDR × SDR, SDR × SDR, $4x$ × FDR, $4x$ × SDR, $4x$ × $4x$). These crosses were assumed to be produced by parents of the diallelic genotypes Aa and $AAaa$ under different types of hypothetical models (additive, dominance and overdominance). The mean and variance of FDR progenies are expected to be larger and smaller, respectively, than those of SDR ones when a locus is close to the centromere. The situation is gradually reversed when the location of a locus is near the end of the long arm of a chromosome. The 'position effect' of a gene in SDR gametes appears to be stronger than that in FDR gametes. The $2n$ gametes exert a stronger influence in $2x$–$2x$ than in $4x$–$2x$ crosses. The performance of the FDR × SDR type of $2x$–$2x$ cross appears to fall between those of FDR × FDR and SDR × SDR crosses. Computer simulation was used by Watanabe *et al.* (1991) to compare the genetic consequences for different populations of tetraploid progenies. These populations were assumed to be produced by a population of heterozygous diploids through asexual (somatic doubling) and sexual ($2n$ gametes) polyploidization. Multiallelic situations were considered in their study. The $2n$ gametes transmit about 80% (FDR) and 40% (SDR) of the parental heterozygosity to the $4x$ offspring whereas somatic doubling transmits all of the parental heterozygosity. Tetra-allelic and triallelic genotypes are possible in $4x$ progeny produced by $2n$ gametes. All of the loci are duplex or balanced diallelic genotypes in $4x$ progeny produced by somatic doubling. This results in more inbreeding and fewer allelic interactions at a locus than in the sexually produced $4x$ progenies. Among the three mechanisms involved in sexual polyploidization, FDR × FDR has less inbreeding and more genotypic diversity than SDR × SDR in $4x$ progeny when a locus is close to the centromere. SDR × SDR has a lower degree of inbreeding than FDR × FDR when the locus is distal to the centromere. SDR × FDR, on the other hand, is the least affected by the change of the probability of a SET between the centromere and a locus. It is the most stable mechanism against extreme recombination events and produces a high proportion of tri- and tetra-allelic genotypes. These results, however, are not realistic in the sense that the real goal is to compare the genetic consequences of $2n$ gametes produced by chosen $2x$ (FDR and SDR) and $4x$ parents on their tetraploid progenies *per se*, rather than the breeding behaviour of various tetraploid populations produced by a diploid population of infinite size. Werner and Peloquin (1991) conducted a theoretical and experimental study on the importance of allelic diversity and modes of $2n$ egg formation for recovering heterotic $4x$ progeny from $2x$–$4x$ crosses. Closely related parents produced progeny with drastic yield decline because of increased homozygosity and loss of intra- and interlocus interactions. Tri- and tetra-allelic interactions appear important in generating a heterotic effect in polyploids with tetrasomic inheritance. The difference between FDR and SDR $2n$ gametes increases when the genetic diversity decreases. Thus, the overall advantage of FDR in transmitting heterozygosity from $2x$ parents to $4x$ progeny depends on the allelic diversity present in the parents.

Table 5.3. Frequencies of homozygous and heterozygous 2n gametes produced by 2x and 4x parents and number of gametic genotypes in them.

		2n gametes			
		Homozygotes		Heterozygotes	
Parent	Genotypic example	Frequency	No.	Frequency	No.
Tetraploid					
Monoallelic	$A_1A_1A_1A_1$	1	1	0	0
Unbalanced diallelic	$A_1A_1A_1A_2$	$(1+\alpha)/2$	2	$(1-\alpha)/2$	1
Balanced diallelic	$A_1A_1A_2A_2$	$(1+2\alpha)/3$	2	$2(1-\alpha)/3$	1
Triallelic	$A_1A_1A_2A_3$	$(1+5\alpha)/6$	3	$5(1-\alpha)/6$	3
Tetra-allelic	$A_1A_2A_3A_4$	α	4	$1-\alpha$	6
Diploid					
Monoallelic	A_1A_1	1	1	0	0
Diallelic FDR	A_1A_2	$\beta/2$	2	$(2-\beta)/2$	1
Diallelic SDR	A_1A_2	$1-\beta$	2	β	1

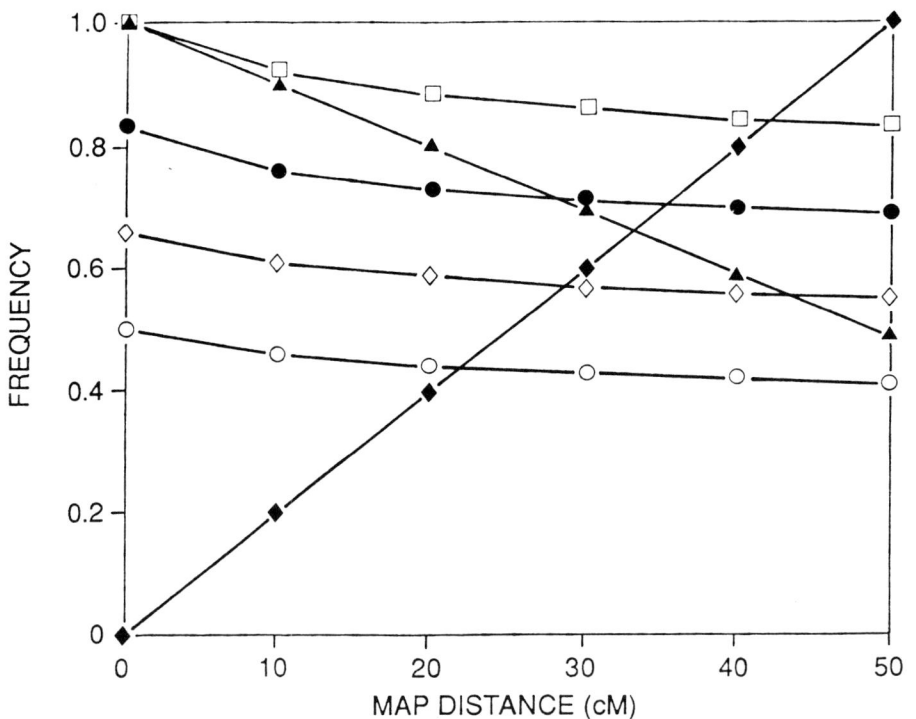

Fig. 5.4. Frequencies of heterozygous 2n gametes produced by tetraploid and diploid parents. (○) Un. diallelic; (◇) Ba. diallelic; (●) triallelic; (□) tetra-allelic; (▲) FDR; (◆) SDR.

As we are interested in the degree of heterosis that a parent can pass on to its progenies through 2n gametes, the simplest way to compare the genetic differences between 2x (FDR and SDR) and 4x parents in their role of unilateral sexual polyploidization is to investigate the frequency and type of heterozygous 2n gametes produced by a heterozygous parent. For a diploid parent there is one type of heterozygous genotype A_1A_2, and the heterozygous 2n gametes are all A_1A_2. In contrast, there are four types of heterozygous tetraploid parents when we consider four alleles per locus, i.e. balanced diallelic, unbalanced diallelic, triallelic and tetra-allelic genotypes. They each produce heterozygous 2n gametes of one or more gametic genotypes. Table 5.3 gives the frequencies of homozygous and heterozygous 2n gametes produced by various 2x and 4x parents, as well as the number of gametic genotypes in each of the two categories of gametes.

Assuming one crossover per pair of chromosomes (Hermsen, 1984a) and using the empirical relationship between α and β discussed earlier, the frequencies of the heterozygous 2n gametes for the six heterozygous parents listed in Table 5.3 are shown in Figure 5.4 for the entire map distance (0–50 cM). The frequencies of heterozygous gametes of 4x parents follow the order tetra-allelic > triallelic > balanced allelic > unbalanced diallelic. Their trends with respect to map distance are roughly parallel to one another and show a slow decline in frequency of heterozygous gametic genotypes as a locus moves away from the centromere. The decline of frequency varies from 0.17 for a tetra-allelic locus to 0.09 for an unbalanced diallelic locus. It should be noted that the gametes of a triallelic locus consist of three gametic genotypes, whereas those of a tetra-allelic locus have six genotypes. The heterozygous gametes of a diallelic FDR parent, on the other hand, have only one gametic genotype and its frequency is equal to or higher than the heterozygous ones of a triallelic parent between 0 and 30 cM. A frequency of 0.49 is maintained for the heterozygous gametic genotype when the map distance is at the maximum. The most drastic change of frequency is with the lone heterozygous gametic genotype produced by the diallelic SDR parent which varies from 0 at 0 cM to 1 at 50 cM. The frequency of the heterozygous gametic genotype produced by the FDR parent is higher than that of the SDR parent for a locus located between 0 and 33 cM. The superiority of a heterozygous FDR parent over a heterozygous 4x parent in terms of producing highly heterotic and uniform hybrid families is obvious from the present analysis: the FDR parent produces a higher frequency of heterozygous 2n gametes than that of three of the four possible heterozygous 4x parents for map distances between 0 and 30 cM (i.e. higher heterotic effect) and keeps the number of gametic genotypes at a minimum (i.e. better uniformity of progenies).

Another issue to be considered here is the genetic consequences of different types of bilateral sexual polyploidization, i.e. FDR × FDR, FDR × SDR and SDR × SDR hybridizations. Consider two diploid parents of genotypes A_1A_2 and A_3A_4, respectively. Hybridization between them leads to the production of the following three groups of hybrids

Table 5.4. Frequencies of diallelic (I), triallelic (II) and tetra-allelic (III) genotypes in the progenies of three types of 2x-2x cross.

	I	II	III
FDR × FDR	$\beta^2/4$	$\beta(1-\beta/2)$	$(1-\beta/2)^2$
FDR × SDR	$\beta(1-\beta)/2$	$(1-\beta)(1-\beta/2)+\beta^2/2$	$\beta(1-\beta/2)$
SDR × SDR	$(1-\beta)^2$	$2\beta(1-\beta)$	β^2

Group I–Diallelic $A_1A_1A_3A_3$ $A_1A_1A_4A_4$ $A_2A_2A_3A_3$ $A_2A_2A_4A_4$

Group II–Triallelic $A_1A_1A_3A_4$ $A_2A_2A_3A_4$ $A_1A_2A_3A_3$ $A_1A_2A_4A_4$

Group III–Tetra-allelic $A_1A_2A_3A_4$

The expected frequencies of the above three groups of genotypes in the progenies of FDR × FDR, FDR × SDR and SDR × SDR crosses are shown in Table 5.4.

The trends of the frequencies of the three groups of progeny genotypes in the three types of 2x–2x crosses with respect to the position of a locus on the chromosome are shown in Figure 5.5. The FDR × FDR cross produces more tetra-allelic progenies than the other two crosses for a locus located within 30 cM of the centromere. The same situation is true for triallelic progenies in the FDR × SDR cross. The frequency of triallelic progenies in the FDR × FDR cross and that of tetra-allelic progenies in the FDR × SDR cross is 0 at 0 cM and continuously increases to 0.5 at 50 cM. There is a greater drop in frequency of tetra-allelic progenies in the FDR × FDR cross than of triallelic ones in the FDR × SDR cross when a locus is located beyond 30 cM. The frequency of diallelic progenies in the FDR × FDR cross continuously increases from 0 at 0 cM to 0.25 at 50 cM, whereas that in the FDR × SDR cross is at a maximum of 0.125 at 25 cM and 0 at 0 and 50 cM. The degree of heterosis generated by tri- and tetra-allelic loci determines the relative size of hybrid vigour for a quantitative trait between FDR × FDR and FDR × SDR progenies. The FDR × SDR cross, however, has one clear advantage over the FDR × FDR cross in that it has a higher combined proportion of tetra-allelic and triallelic genotypes at a locus located beyond 33 cM.

For the progenies of the SDR × SDR cross, the frequency of diallelic genotypes at a locus located within 25 cM and of tetra-allelic genotypes at a locus located beyond 35 cM is higher than that of the remaining two groups of genotypes. The frequency of both has a range of 0 to 1 but in opposite directions (Figure 5.5). The frequency of triallelic progenies is 0 at 0 and 50 cM and reaches a maximum value of 0.455 at 25 cM. It should be noted that the frequency of diallelic progenies reaches a maximum value of 0.25 at 50 cM for the FDR × FDR cross and 0.15 at 25 cM for the FDR × SDR cross. Assuming a regular distribution along the chromosome of parental heterozygous loci affecting a quantitative trait, the SDR × SDR progenies have an excess of diallelic loci at the expense of tri- and tetra-allelic loci when compared with the other two types of crosses. The SDR × SDR cross is thus expected to be inferior to the FDR × FDR and FDR × SDR crosses for

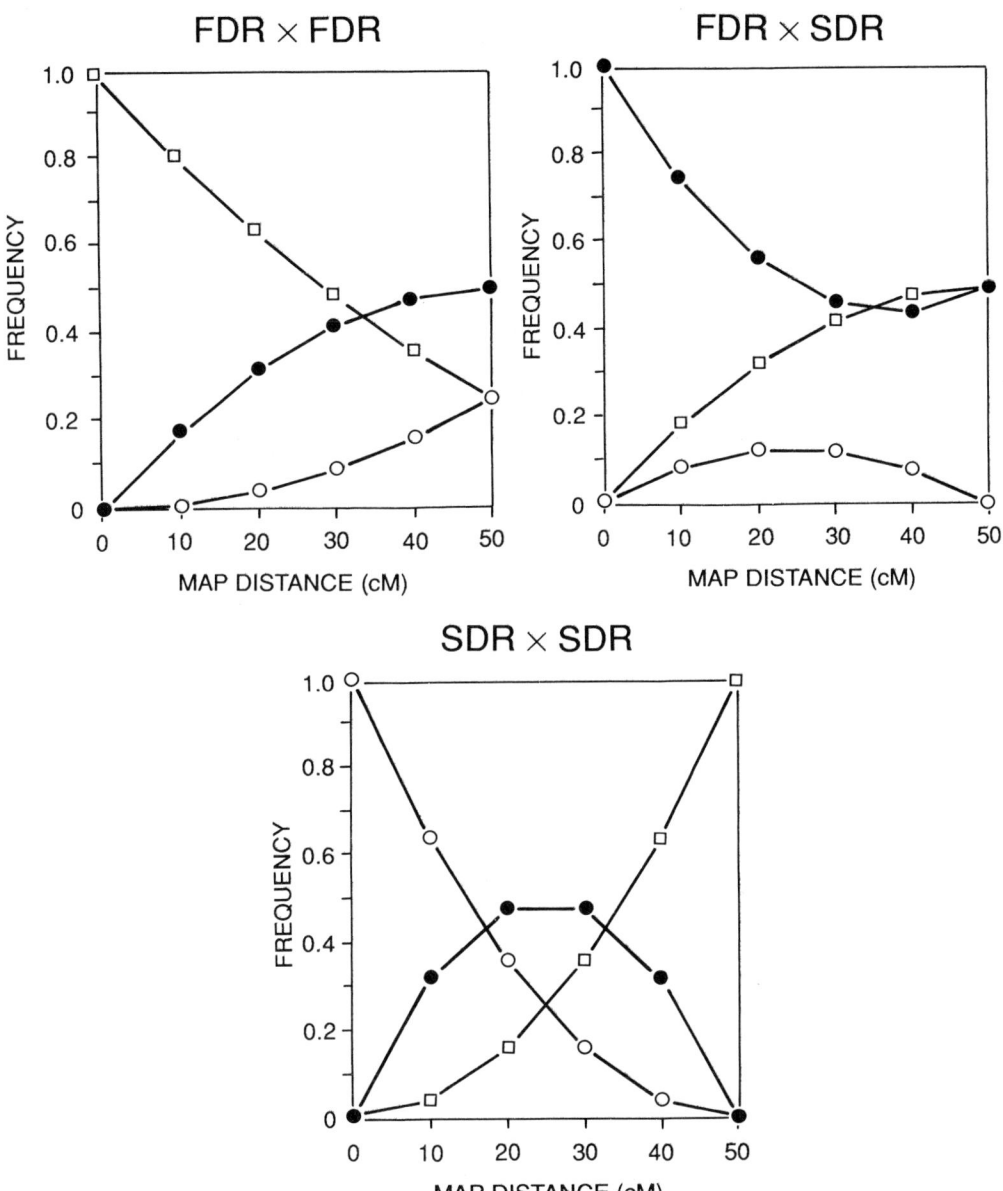

Fig. 5.5. Frequencies of diallelic (○), triallelic (●) and tetra-allelic (□) progenies in FDR × FDR, FDR × SDR, and SDR × SDR.

maintaining hybrid vigour in the progenies unless the diallelic genotypes have the same degree of heterosis as the tri- and tetra-allelic ones.

As indicated above, the number of possible genotypes in the di-, tri- and tetra-allelic groups are four, four and one, respectively, for the 2x–2x cross of the type $A_1A_2 \times A_3A_4$. Thus, the genetic variability within a 2x–2x cross can be kept at a low level when the progenies carry an excess amount of tetra-allelic loci. This would occur for an FDR × FDR cross when all quantitative gene loci affecting a trait are located close to the centromere, or an SDR × SDR cross when they are all located far away from the centromere, say 40 cM or more. This kind of extreme situation would never be achieved in an FDR × SDR cross (see Figure 5.5). The FDR × FDR cross is therefore favoured for the possibility of creating a genetically more uniform hybrid population than the other two types of crosses, since it has a higher frequency of tetra-allelic genotypes for loci located up to 30 cM, and a low frequency of diallelic genotypes for loci located anywhere in a chromosome.

Breeding New Varieties

The value of unilateral and bilateral sexual polyploidization as a breeding tool has been confirmed by many research workers (see reviews by Hermsen, 1984b; Ross, 1986; Hermundstad and Peloquin, 1987; Birhman and Garg, 1989; Carroll and De Maine, 1989; Iwanaga et al., 1989). Hermundstad and Peloquin (1987) have developed high-quality $2n$ gamete-producing parents from dihaploid S. tuberosum-diploid species hybrids. Tetraploid progenies obtained from them in $4x$-$2x$ crosses not only show good yielding ability but, unlike those obtained in earlier experiments, acceptable tuber traits, such as long dormancy, high specific gravity, uniform size, medium set and attractive appearance. The diploid parents also serve as a bridge for the introduction of resistances to diseases, pests and environmental stresses from other diploid species into breeding material (Iwanaga, 1985; Carroll, 1987; Iwanaga et al., 1989). There are, however, a number of issues in relation to the use of $2n$ gametes in breeding $4x$ varieties which require further discussion.

The most straightforward use of $2n$ gametes in breeding would be the direct use of those produced by the dihaploids of $4x$ S. tuberosum varieties. Maris (1990) concluded that such an approach serves no purpose in breeding because, without the introduction of new germplasm, dihaploidization followed by repolyploidization can only be expected to result in tetraploids with heterozygosity equal to or below that of their parents. Earlier studies using $2x$ FDR parents derived from dihaploid-S. phureja hybrids in $4x$-$2x$ crosses had found various degrees of heterosis for total yield based on family means or the segregation of high-yielding progenies (Hanneman and Peloquin, 1967; Mendiburu and Peloquin, 1971; Quinn and Peloquin, 1973; Kidane-Mariam and Peloquin 1975; Mok and Peloquin, 1975b; De Jong and Tai, 1977). McHale and Lauer (1981a) and Veilleux and Lauer (1981) employed selected clones of S. phureja as pollen (FDR) parents in $4x$-$2x$ crosses. Mean total yield of these hybrids exceeded that using dihaploid-S. phureja diploids as $2x$ parents, but did not differ from the yield of the $4x$ parents. The frequency distribution of total yield revealed the existence of high-yielding segregants. Both types of hybrids, however, showed a number of problems: low marketable yield accompanied by high set and small size, late maturity and vigorous haulm, and poor tuber appearance with deep eyes. Multivariate analysis (Tai and De Jong, 1980) indicated that the *S. phureja*-based 4x–2x hybrids represented a distinct group of genetical material in comparison with the *S. tuberosum* varieties. The *S. phureja*-based 2x parents should therefore be improved further for horticultural performance before their use in 4x–2x crosses. Quinn and Peloquin (1973) suggested intercrossing the 4x–2x families to improve their horticultural performance. McHale and Lauer (1981b), on the other hand, advocated the use of phenotypic recurrent selection in the *S. phureja* population as an effective way to increase the frequency of alleles conditioning desirable tuber traits. Nevertheless, high-yielding and stable advanced breeding lines were identified (De Jong *et al.*, 1981), and *S. phureja*-based 4x clones have been released as new varieties (Johnston and Rowberry, 1981; International Potato Center, 1984).

Experimental results from 4x–2x crosses using improved *S. phureja*-based 2x parents and others, such as dihaploid *S. tuberosum–S. chacoense* hybrids continue to appear in the literature (Chujoy, 1985; Buso, 1986; Concilio, 1987; Masson and Peloquin, 1987; De Jong and Tai, 1991; Tai and De Jong, 1991). The widespread occurrence of the *ps* gene among wild diploid species (Quinn *et al.*, 1974; Camadro and Peloquin, 1980) stimulated the use of dihaploid *S. tuberosum*-wild species hybrids as potential 2x parents in 4x–2x and 2x–2x crosses (Hermundstad and Peloquin, 1987; Peloquin *et al.*, 1989b; Jansky *et al.*, 1990). The following wild species are able to produce male fertile hybrids when crossed with *S. tuberosum* dihaploids: *S. chacoense*, *S. berthaultii*, *S. boliviense*, *S. canasense*, *S. microdontum*, *S. raphanofolium*, *S. sanctae-rosae*, *S. kurtzianum*, *S. bukasovii*, *S. spegazzinii*, *S. sparsipilum* and *S. tarijense* (Hermundstad and Peloquin, 1987). Selected tetraploid clones obtained from 4x–2x (dihaploid–*S. tarijense*) have already demonstrated their breeding value for yield, tuber appearance, specific gravity, chip colour and maturity in comparison with 4x varieties (Peloquin *et al.*, 1989b). Jansky *et al.* (1990) indicated the need to improve the dihaploid–wild species hybrids before they are used in 4x–2x and 2x–2x crosses. The wild species parents must be selected so that their progenies with the dihaploids tuberize, possess good tuber traits, tolerance to environmental stresses and resistances to diseases and pests, have a reasonable level of fertility and, specifically, produce many 2n gametes.

It is crucial to assess the merit of 2x and 4x parents before they are widely used in a breeding programme based on 2n gametes. The easiest way is to directly evaluate the performance of parents. This type of evaluation is useful if there is a strong parent–offspring correlation. Schroeder and Peloquin (1983) reported that vine maturity of 4x families was correlated to both 2x parental and midparental ratings, but not with 4x parental ratings. Keijzer-van der Stoel *et al.* (1991) reported a high parent–offspring correlation for underwater weight of tubers. De Jong and Tai (1991) found a low degree of determination of 4x and 2x parents on the performance of their hybrid families for total and marketable yield, tuber number, mean tuber weight, chipping score and specific gravity. A low parent–offspring correlation for total yield has, in fact, been found in all studies (McHale and Lauer, 1981a; Veilleux and

Lauer, 1981; Schroeder, 1983; Keijzer-van der Stoel *et al.*, 1991). It appears that progeny test experiments are required to assess the merit of parents and cross combinations in a 4x-2x and 2x-2x breeding programme.

Combining-ability analysis based on data obtained from a factorial experiment provides the most convenient method of evaluating 4x and 2x parents and their hybrid families. This has indeed been carried out in all experimental work on 4x-2x crosses. A general survey of all studies (De Jong and Tai, 1977; Mendiburu and Peloquin, 1977; McHale and Lauer, 1981a, b; Veilleux and Lauer, 1981; Schroeder, 1983; Ortiz *et al.*, 1988; Tai and De Jong, 1991) indicates the importance of general combining ability (GCA) for both 2x and 4x parents on yield and yield components, whereas specific combining ability (SCA) between 4x and 2x parents has not been detected often. Significant GCA effects have also been reported for such traits as maturity, hollow heart, underwater weight and tuber protein content in the afore-cited works and Veilleux *et al.* (1981).

All results of combining-ability analysis appear to support the value of conducting progeny test experiments to assess the merit of parents in a 4x-2x breeding programme. A problem of using the factorial mating design to evaluate hybrid progenies is that some of the crossing combinations may not be obtainable due to fertility problems. Evaluating the parental value of a large number of 2x parents by crossing them with a number of 4x parents also requires a large amount of work. Schroeder (1983) suggested that a 2x clone must be crossed with at least two 4x varieties of different maturity to verify its parental value. Ortiz *et al.* (1988) recommended the use of a single 4x parent with a low GCA as tester. Tai (1989) and Tai and De Jong (1991) developed a procedure which can be used to evaluate parental values of 2x parents based on data obtained from an incomplete series of matings between 4x and 2x parents.

The above discussions concentrate on the potential of using 4x-2x crosses to breed new varieties. The hybrid seed from 4x-2x crosses also has the potential to be used as true seed (TPS) varieties (Macaso and Peloquin, 1983; Kidane-Mariam and Peloquin, 1985). Other issues not fully explored in this chapter are related to the use of 2x-2x crosses in a 2n gamete-based breeding programme – i.e. the feasibility and relative merit of FDR × FDR, FDR × SDR and SDR × SDR crosses, and the value of 2n gametes produced by FDR-NCO 2x clones. Experimental work in these areas is still in a preliminary stage.

References

Birhman, R.K. and Garg, K.C. (1989) Yield and yield component of meiotic tetraploids of potato. *Potato Research* 32, 457-462.

Boudec, P., Masson, M. and Dattee, Y. (1989a) A quantitative genetics model for the estimation of variances, covariances between relatives in crosses using 2n gametes, in potato. In: Louwes, K.M., Toussaint, H.A.J.M. and Dellaert, L.M.W. (eds) *Parental Line Breeding and Selection in Potato Breeding.* Centre for Agricultural Publishing and Documentation, PUDOC, Wageningen, The Netherlands, pp. 43-48.

Boudec, P., Masson, M. and Dattee, Y. (1989b) A comparison between different mating systems on means in potato. In: Louwes, K.M., Toussaint, H.A.J.M. and Dellaert, L.M.W. (eds) *Parental Line Breeding and Selection in Potato Breeding*. Centre for Agricultural Publishing and Documentation, PUDOC, Wageningen, The Netherlands, pp. 158–162.

Buso, J.A. (1986) Evaluation of families and clones from the $4x \times 2x$ breeding scheme in potato. Unpublished PhD thesis, University of Wisconsin-Madison, Madison, Wisconsin.

Camadro, E.L. and Peloquin, S.J. (1980) The occurrence and frequency of $2n$ pollen in three diploid *Solanums* from northwest Argentina. *Theoretical and Applied Genetics* 56, 11–15.

Carroll, C.P. (1987) The use of diploid *Solanum phureja* germplasm. In: Jellis, G.J. and Richardson, D.E. (eds), *The Production of New Potato Varieties: Technological Advances*. Cambridge University Press, Cambridge, pp. 231–234.

Carroll, C.P. and De Maine, M.J. (1989) The agronomic value of tetraploid F_1 hybrids between potatoes of Group Tuberosum and Group Phureja/Stenotomum. *Potato Research* 32, 447–456.

Chujoy, J.E. (1985) Tuber yields of $2x$ and $4x$ progeny from $2x \times 2x$ crosses in potato: barriers to interspecific hybridization between *Solanum chacoense* Bitt. and *S. commersonii* Dun. Unpublished PhD thesis, University of Wisconsin-Madison, Madison, Wisconsin.

Concilio, L. (1987) Evaluation of the tetraploid × diploid ($4x \times 2x$) mating scheme and its efficiency in a new breeding program. Unpublished PhD thesis, University of Wisconsin-Madison, Madison, Wisconsin.

Conicella, C., Barone, A., Del Giudice, A., Frusciante, L. and Monti, L.M. (1991) Cytological evidences of SDR-FDR mixture in the formation of $2n$ eggs in a potato diploid clone. *Theoretical and Applied Genetics* 81, 59–63.

De Jong, H. and Tai, G.C.C. (1977) Analysis of tetraploid–diploid hybrids in cultivated potatoes. *Potato Research* 20, 111–121.

De Jong, H. and Tai, G.C.C. (1991) Evaluation of potato hybrids obtained from tetraploid–diploid crosses in an incomplete mating design. 1. Parent-offspring relationships. *Plant Breeding* 107, 177–182.

De Jong, H., Tai, G.C.C., Russell, W.A., Johnston, G.R. and Proudfoot, K.G. (1981) Yield potential and genotype–environment interactions of tetraploid–diploid ($4x$–$2x$) potato hybrids. *American Potato Journal* 58, 191–199.

Douches, D.S. and Quiros, C.F. (1987) Use of $4x$–$2x$ crosses to determine gene–centromere map distances of isozyme loci in *Solanum* species. *Genome* 29, 519–527.

Fisher, R.A. and Mather, K. (1943) The inheritance of style length in *Lythrum salicaria*. *Annals of Eugenics* 12, 1–23.

Grizzle, J.E., Starmer, C.F. and Koch, G.G. (1969) Analysis of categorical data by linear models. *Biometrics* 25, 489–504.

Hanneman, R.E. and Peloquin, S.J. (1967) Crossability of 24-chromosome potato hybrids with 48-chromosome cultivars. *European Potato Journal* 10, 62–72.

Haynes, K.G. (1990) Covariances between diploid parent and tetraploid offspring in tetraploid × diploid crosses of *Solanum tuberosum* L. *Journal of Heredity* 81, 208–210.

Haynes, K.G., Potts, W.E. and Camp, M.J. (1991) Estimation of preferential pairing in tetraploid × diploid hybridizations. *Theoretical and Applied Genetics* 81, 504–508.

Hermsen, J.G.Th. (1984a) Mechanisms and genetic implications of $2n$-gamete formation. *Iowa State Journal of Research* 58(4), 421–434.

Hermsen, J.G.Th. (1984b) The potential of meiotic polyploidization in breeding allogamous crops. *Iowa State Journal of Research* 58(4), 435–448.

Hermundstad, S.A. and Peloquin, S.J. (1987) Breeding at the 2x level and sexual polyploidization. In: Jellis, G.J. and Richardson, D.E. (eds) *The Production of New Potato Varieties: Technological Advances.* Cambridge University Press, Cambridge, pp. 197–210.

Hougas, R.W. and Peloquin, S.J. (1958) The potential of potato haploids in breeding and genetic research. *American Potato Journal* 35, 701–707.

Hougas, R.W., Peloquin, S.J. and Gabert, A.C. (1964) Effect of seed-parent and pollinator on frequency of haploids in *Solanum tuberosum. Crop Science* 4, 593–595.

International Potato Center (1984) *Potatoes for the Developing World.* International Potato Center, Lima, Peru.

Iwanaga, M. (1985) Haploids, ploidy manipulation, and meiotic mutants in potato breeding. In: *Biotechnology in International Agricultural Research.* IRRI, Manila, Philippines, pp. 139–148.

Iwanaga, M. and Peloquin, S.J. (1979) Synaptic mutant affecting only megasporogenesis in potatoes. *Journal of Heredity* 70, 385–389.

Iwanaga, M., Jatala, P., Oritz, R. and Guevara, E. (1989) Use of FDR 2n pollen to transfer resistance to root-knot nematodes into cultivated 4x potatoes. *Journal of American Society of Horticultural Science* 114, 1008–1013.

Jansky, S.H., Yerk, G.L. and Peloquin, S.J. (1990) The use of potato hybrids to put 2x wild species germplasm into a usable form. *Plant Breeding* 104, 290–294.

Johnston, G.R. and Rowberry, R.G. (1981) Yukon gold: a new yellow-fleshed, medium early, highly quality table and french fry cultivar. *American Potato Journal* 58, 241–244.

Jongedijk, E. (1985) The pattern of megasporogenesis and megagametogenesis in diploid *Solanum* species hybrids: its relevance to the origin of 2n eggs and the induction of apomixis. *Euphytica* 34, 599–611.

Jongedijk, E. and Ramanna, M.S. (1988) Synaptic mutants in potato, *Solanum tuberosum* L. I. Expression and identity of genes for desynapsis. *Genome* 30, 664–670.

Jongedijk E., Hutten, R.C.B., van der Wolk, J.M.A.S.A and Schuurmans Stekhoven, S.I.J. (1991a) Synaptic mutants in potato, *Solanum tuberosum* L. III. Effect of the *Ds-1/ds-1* locus (desynapsis) on genetic recombination in male and female meiosis. *Genome* 34, 121–130.

Jongedijk E., Ramanna, M.S., Sawor, Z. and Hermsen, J.G.Th. (1991b) Formation of first division restitution (FDR) 2n-megaspores through pseudohomotypic division in *ds-1* (desynapsis) mutants of diploid potato: routine production of tetraploid progeny from 2x FDR × 2x FDR crosses. *Theoretical and Applied Genetics* 82, 645–656.

Keijzer-van der Stoel, M.C., Pegels-van Deelen, M.W. and Neele, A.E.F. (1991) An analysis of the breeding value of diploid potato clones comparing 2x-2x and 4x-2x crosses. *Euphytica* 52, 131–136.

Kidane-Mariam, H.-M. and Peloquin, S.J. (1985) Method of diplandroid formation and yield of progeny from reciprocal (4x-2x) crosses. *Journal of American Society of Horticultural Science* 100, 602–603.

Killick, R.J. (1971) The biometrical genetics of autotetraploids. I. Generations derived from a cross between two pure lines. *Heredity* 27, 331–346.

Lander, E.S. and Botstein, D. (1989) Mapping Mendelian factors underlying quantitative traits using RFLP linkage maps. *Genetics* 121, 185–199.

Macaso, A.C. and Peloquin, S.J. (1983) Tuber yields of families from open pollinated and hybrid true potato seed. *American Potato Journal* 60, 645-651.

McHale, N.A. and Lauer, F.I. (1981a) Inheritance of tuber traits from *phureja-tuberosum* hybrids. *American Potato Journal* 58, 93-102.

McHale, N.A. and Lauer, F.L. (1981b) Breeding value of 2n pollen from diploid hybrids and *phureja* in 4x-2x crosses in potatoes. *American Potato Journal* 58, 365-374.

Maris, B. (1990) Comparison of diploid and tetraploid potato families derived from *Solanum phureja* × dihaploid *S. tuberosum* hybrids and their vegetatively doubled counterparts. *Euphytica* 46, 15-34.

Marks, G.E. (1966) The origin and significance of intraspecific polyploidy: experimental evidence from *Solanum chacoense*. *Evolution* 20, 552-557.

Masson, M.F. and Peloquin, S.J. (1987) Heterosis for tuber yields and total solids content in 4x × 4x FDR-CO crosses in potato. In: Jellis, G.J. and Richardson, D.E. (eds) *The Production of New Potato Varieties: Technological Advances*. Cambridge University Press, Cambridge, pp. 213-217.

Mather, K. and Jinks, J.L. (1982) *Biometrical Genetics*, 3rd. edn. Chapman Hall, London.

Mendiburu, A.O. and Peloquin S.J. (1971) High yielding tetraploids from 4x-2x and 2x-2x matings. *American Potato Journal* 48, 300-301.

Mendiburu A.O. and Peloquin, S.J. (1976) Sexual polyploidization and depolyploidization: some terminology and definitions. *Theoretical and Applied Genetics* 48, 137-143.

Mendiburu, A.O. and Peloquin, S.J. (1977) The significance of 2n gametes in potato breeding. *Theoretical and Applied Genetics* 49, 53-61.

Mendiburu, A.O. and Peloquin, S.J. (1979) Gene-centromere mapping by 4x-2x matings in potatoes. *Theoretical and Applied Genetics* 54, 177-180.

Mendiburu, A.O., Peloquin, S.J. and Mok, D.W.S. (1974) Potato breeding with haploids and 2n gametes. In: Kasha, K.J. (ed.) *Haploids in Higher Plants*. Guelph University Press, Guelph, pp. 249-258.

Mok, D.W.S. and Peloquin, S.J. (1975a) Three mechanisms of 2n pollen formation in diploid potatoes. *Canadian Journal of Genetics and Cytology* 17, 217-225.

Mok, D.W.S. and Peloquin, S.J. (1975b) Breeding value of 2n pollen (diplandroids) in tetraploid × diploid crosses in potatoes. *Theoretical and Applied Genetics* 46, 307-314.

Okwuagwu, C.O. and Peloquin, S.J. (1981) A method of transferring the intact parental genotype to the offspring via meiotic mutants. *American Potato Journal* 58, 512-513.

Ortiz, R., Iwanaga, M. and Mendoza, H.A. (1988) Combining ability and parental effects in 4x-2x crosses for potato breeding. *Potato Research* 31, 643-650.

Parrot, W.A. and Hanneman R.E., Jr (1988) Megasporogenesis in normal and a synaptic mutant (*sy-2*) of *Solanum commersonii* Dun. *Genome* 30, 536-539.

Peloquin, S.J., Yerk, G.L. and Werner, J.E. (1989a) Ploidy manipulations in the potato. In: Adolph, K.W. (ed.) *Chromosomes: Eukaryotic, Prokaryotic, and Viral*. CRC Press, Boca Raton, Florida, Vol. II, pp. 167-178.

Peloquin, S.J., Yerk, G.L., Werner, J.E. and Darmo, E. (1989b) Potato breeding with haploids and 2n gametes. *Genome* 31, 1000-1004.

Quinn, A.A. and Peloquin, S.J. (1973) Use of experimental tetraploids in potato breeding. *American Potato Journal* 50, 415-420.

Quinn, A.A., Mok, D.W.S. and Peloquin, S.J. (1974) Distribution and significance of diplandroids among the diploid *Solanums*. *American Potato Journal* 51, 16-21.

Ramanna, M.S. (1979) A re-examination of the mechanisms of $2n$ gamete formation in potato and its implications for breeding. *Euphytica* 28, 537-561.

Ross, H. (1986) *Potato Breeding - Problems and Perspectives*. Verlag Paul Parey, Berlin.

Schroeder, S.H. (1983) Parental value of $2x$, $2n$ pollen clones and $4x$ cultivars in $4x \times 2x$ crosses in potato. Unpublished PhD thesis, University of Wisconsin.

Schroeder, S.H. and Peloquin, S.J. (1983) Parental effects for yield and tuber appearance on $4x$ families from $4x \times 2x$ crosses. *American Potato Journal* 60, 819.

Stelly, D.M. and Peloquin, S.J. (1986) Formation of $2n$ megagametophytes in diploid tuber-bearing *Solanums*. *American Journal of Botany* 73, 1351-1363.

Tai, G.C.C. (1982a) Estimation of double reduction and genetic parameters in autotetraploids. *Heredity* 49, 63-70.

Tai, G.C.C. (1982b) Estimation of double reduction and genetic parameters in autotetraploids based on $4x-2x$ and $4x-4x$ matings. *Heredity* 49, 331-335.

Tai, G.C.C. (1986) Biometrical genetical analysis of tetrasomic inheritance based on matings of diploid parents which produce $2n$ gametes. *Heredity* 57, 315-317.

Tai, G.C.C. (1989) Biometrical methods in investigating $2n$ gametes in tetraploid-diploid and diploid-diploid crosses. In: Louwes, K.M., Toussaint, H.A.J.M. and Dellaert, L.M.W. (eds) *Parental Line Breeding and Selection in Potato Breeding*. Centre for Agricultural Publishing and Documentation, PUDOC, Wageningen, The Netherlands, pp. 15-21.

Tai, G.C.C. and De Jong, H. (1980) Multivariate analysis of potato hybrids. I. Discrimination between tetraploid-diploid hybrid families and their relationship to cultivars. *Canadian Journal of Genetics and Cytology* 22, 227-235.

Tai, G.C.C. and De Jong, H. (1991) Evaluation of potato hybrids obtained from tetraploid-diploid crosses in an incomplete mating design. II. Progeny analysis. *Plant Breeding* 107, 183-189.

Veilleux, R.E. (1985) Diploid and polyploid gametes in crop plants: mechanisms of formation and utilization in plant breeding. *Plant Breeding Review* 3, 253-288.

Veilleux, R.E. and Lauer, F.L. (1981) Breeding behavior of yield components and hollow heart in tetraploid-diploid vs conventionally derived potato hybrids. *Euphytica* 30, 547-561.

Veilleux, F.I., Lauer, F.L and Desborough, S.L. (1981) Breeding behavior for tuber protein in *Solanum tuberosum* and *tuberosum-phureja* hybrids. *Euphytica* 30, 563-578.

Veilleux, R.E., McHale, N.A. and Lauer, F.L. (1982) $2n$ gametes in diploid *Solanum*: frequency and types of spindle abnormalities. *Canadian Journal of Genetics and Cytology* 24, 301-314.

Wagenvoort, M. and Zimnoch-Guzowska, E. (1992) Gene-centromere mapping in potato by half-tetrad analysis: map distances of H_1, Rx, and Ry and their possible use for ascertaining the mode of $2n$-pollen formation. *Genome* 35, 1-7.

Watanabe, K., Peloquin, S.J. and Endo, M. (1991) Genetic significance of mode of polyploidization: somatic doubling or $2n$ gametes? *Genome* 34, 28-34.

Werner, J.E. and Peloquin, S.J. (1987) Frequency and mechanisms of $2n$ egg formation in haploid *tuberosum* wild species F_1 hybrids. *American Potato Journal* 64, 641-654.

Werner, J.E. and Peloquin, S.J. (1990) Inheritance and two mechanisms of $2n$ egg formation in $2x$ potatoes. *Journal of Heredity* 81, 371-374.

Werner, J.E. and Peloquin, S.J. (1991) Significance of allelic diversity and $2n$ gametes for approaching maximum heterozygosity in $4x$ potatoes. *Euphytica* 58, 21-29.

6 Use of 24-Chromosome Potatoes (Diploids and Dihaploids) for Genetical Analysis

R. ORTIZ AND S.J. PELOQUIN

Departments of Genetics and Horticulture, University of Wisconsin-Madison, 1575 Linden Drive, Madison, Wisconsin 53706, USA.

Advantages of Genetical Analysis at Diploid Level

The presence of four sets of chromosomes and tetrasomic inheritance patterns makes the genetics of the tetraploid potato ($2n = 4x = 48$) more complicated than that of diploid potatoes ($2n = 2x = 24$), which have only two sets of chromosomes. The utilization of 24-chromosome potatoes for inheritance studies should therefore represent an advantage because it simplifies the genetical analysis. For example, after selfing a diploid individual which is heterozygous (*Aa*) for a specific locus, three genotypes (*AA*, *Aa*, *aa*) are expected. In a comparable tetrasomic polyploid (*AAaa*), five different genotypes are possible in its selfed progeny, *AAAA* (quadruplex), *AAAa* (triplex), *AAaa* (duplex), *Aaaa* (simplex) and *aaaa* (nulliplex), and further segregation results after selfing any of the three genotypes: *AAAa*, *AAaa* and *Aaaa*.

A detailed comparison of disomic and tetrasomic inheritance is presented in Table 6.1. After selfing a heterozygous diploid (*Aa*), the chance of recovering a homozygous recessive individual is 1/4. In the comparable tetraploid (*AAaa*), the chance of obtaining a nulliplex individual after selfing is subject to the extent of double reduction and is 1/36 with chromosome segregation and 9/196 with chromatid segregation.

Double reduction is a phenomenon associated with tetrasomic inheritance. It occurs if the two chromosomes in a gamete are derived from two sister chromatids, or, in other words, if sister chromatids end up in the same gamete. Double reduction requires quadrivalent formation and that a single crossover occurs between the centromere and the locus so that the sister chromatids are attached to two different centromeres; that centromeres with sister chromatids go to the same pole in anaphase I; and that sister chromatids go to the same pole in anaphase II.

Table 6.1. Comparison of tetrasomic versus disomic inheritance.

Generation	Diploid level	Tetraploid Level
Parents	AA × aa	AAAA × aaaa
F_1	Aa	AAaa
Gametes	A, a	AA, Aa, aa
		1/6 4/6 1/6 (chromosome segregation)
		3/14 8/14 3/14 (chromatid segregation)

F_2 (selfing F_1)				Chromosome	Chromatid
	AA	25%	AAAA	3%	5%
			AAAa	22%	24%
	Aa	50%	AAaa	50%	42%
			Aaaa	22%	24%
	aa	25%	aaaa	3%	5%

Table 6.2. Expected frequencies of gametes formed in a tetrasomic polyploid as a function of the coefficient of double reduction, alpha (α).

Tetraploid parent	Genotype		Gametes			Divisor
			AA	Aa	aa	
Quadruplex	AAAA	(A_4)	1	0	0	1
Triplex	AAAa	(A_3a)	$2+\alpha$	$2(1-\alpha)$	α	4
Duplex	AAaa	(A_2a_2)	$1+2\alpha$	$4(1-\alpha)$	$1+2\alpha$	6
Simplex	Aaaa	(Aa_3)	α	$2(1-\alpha)$	$2+\alpha$	4
Nulliplex	aaaa	(a_4)	0	0	1	1

The frequency of double reduction is expressed mathematically by the parameter α (alpha) $= qea/2$, where q is the quadrivalent frequency, e is the frequency of equational separation, which depends on the gene–centromere map distance, and a is the frequency of genetic non-disjunction (normally equal to 1/3). A value of α not significantly different from zero indicates that the gene of interest lies close to the centromere and chromosome segregation must be expected. If α is equal to 1/7, chromatid segregation for that locus should occur. A special case, called maximal equational segregation, occurs when α is equal to 1/6. The expected frequencies of gametes formed in tetrasomic polyploids are indicated in Table 6.2. (See also chapter 3 by Bradshaw and chapter 5 by Tai.)

The use of $2x$ material is not only valuable for genetic analysis but also more convenient than $4x$ populations for breeding purposes. The advantages of breeding at the $2x$ level are that it may shorten the time required to produce a new variety, more rapidly eliminate deleterious recessive alleles, and enable a more efficient introduction of desired characters from $2x$ species. Gametes with the sporophytic chromosome number ($2n$ gametes) can be used to restore the $4x$ ploidy level, using only those selected and superior $2x$ clones with specific attributes as parents. (See also chapter 5 by Tai.)

Use of Dihaploids

The authors of this chapter normally use the term haploid for a sporophyte with the gametic chromosome number. However, to avoid confusion between this and other chapters, haploids of tetraploid ($2n = 4x = 48$) potatoes are referred to as dihaploids, and haploids of diploid ($2n = 2x = 24$) potatoes are called monohaploids. The term dihaploid is not used for doubled haploids.

Ivanovskajya (1939) accidentaly found potato dihaploid plants among the progeny of a cross between $4x$ *S. tuberosum* and *S. phureja*. Hougas *et al.* (1958) initiated potato dihaploid production on a large scale by a systematic utilization of $4x \times 2x$ crosses between $4x$ Group Tuberosum cultivars and Group Phureja 'pollinators'.

Extraction

Sporophytes with the gametic chromosome number of polysomic polyploids have been used for genetic research and breeding applications (Jansky *et al.*, 1990; Peloquin *et al.*, 1990). Parthenogenesis has been the method most commonly followed to extract them from polyploid plants. Maternal dihaploids ($2n = 2x = 24$) are easily obtained in potato from $4x$ parents ($2n = 48$) via $4x \times 2x$ crosses (Peloquin *et al.*, 1966; Hermsen and Verdenius, 1973). When a $4x$ potato is pollinated with $2x$ clones of Group Phureja (e.g. 1.22 or IvP-35), dihaploids develop parthenogenetically ('pseudogamy') as a result of both male gametes fertilizing the polar nuclei. Embryo sacs with a $2x$ embryo and $6x$ endosperm are produced which will develop normally into seeds (Hougas *et al.*, 1964). The $4x \times 2x$ method of obtaining dihaploids was improved through the utilization of the decapitation technique (Peloquin and Hougas, 1959), and the use of élite 'pollinators' (pollen sources) in $4x \times 2x$ crosses (Hougas *et al.*, 1964).

In vitro anther culture has also been used with success to reduce the ploidy level of the tetraploid potato. However, the application of this approach is limited due to the lack of responsiveness of many genotypes to anther culture (Bartels *et al.*, 1988). Two causes for the failure to produce viable embryos through anther culture have been proposed: (i) recessive lethal genes in the $4x$ parent which are unmasked at the $2x$ level (Wenzel, 1980); (ii) genotype sensitivity to culture medium components (Wenzel *et al.*, 1980). Furthermore $4x$ individuals originating from $2n$ pollen or sporophytic tissue are sometimes recovered after anther culture, in which case anther culture may require more screening than interploidy crosses for dihaploid production. (See also chapter 7 by Jacobsen and Ramanna and chapter 8 by Wenzel.)

Utilization

Hougas and Peloquin (1958) have pointed out two advantages of using dihaploids: the direct gene transfer from the wild and cultivated $2x$ tuber-bearing *Solanum* species and the benefits of disomic rather than tetrasomic inheritance. These advantages facilitate the breeding of potatoes at the $2x$ level which combine multiple pest and disease resistance along with acceptable agronomic traits, plus the opportunity to conduct genetic studies based on the more simple disomic inheritance pattern, which requires smaller populations to detect recessive genes than $4x$ intercrossing (Peloquin *et al.*, 1990).

As random gametic samples of the $4x$ seed parent, dihaploids can be used to determine the breeding value of those $4x$ parents (Peloquin and Hougas, 1960; Kotch, 1987). Thus, dihaploids can be tools for the determination of the number of genes for a specific trait (Cipar and Lawrence, 1972; Simon and Peloquin, 1980; Landeo and Hanneman, 1982; De Maine, 1984; Quiros and McHale, 1985) and the genotype of the tetraploid parent (Cipar *et al.*, 1964; Desborough and Peloquin, 1967; Iwanaga, 1984a; Douches *et al.*, 1990; Ortiz *et al.*, 1991).

Dihaploids have also been used in cytogenetic research to study the nature of polyploidy and to conduct genome analyses which relate to the origin of the common cultivated tetraploid potato (Yeh *et al.*, 1964; Matsubayashi, 1979). They have also been used to produce aneuploid series, particularly trisomics through parthenogenetic aneuploid offspring following $4x \times 2x$ crosses (Hermsen *et al.*, 1970) or $3x \times 2x$ (dihaploid) crosses (Wagenwoort and Lange, 1975). Most of the progeny in both cases were primary trisomics. These were used to locate the locus for yellow margin (ym) and a chlorophyll deficiency on chromosome 12 (Hermsen *et al.*, 1973; Wagenwoort, 1982).

The nature of polyploidy in potato was examined by cytological analysis of potato dihaploids (Yeh *et al.*, 1964). Some of the dihaploids had regular chromosome pairing and others low or high frequencies of univalents. However, the F_1 progeny from crossing Group Tuberosum dihaploids with irregular meiosis and Group Andigena dihaploids with normal chromosome pairing had normal meiosis. This evidence indicates that the potato is a tetrasomic polyploid. Conversely, Matsubayashi (1979) concluded that potato was a segmental allotetraploid based on segregation ratios of dihaploids of the cultivar Chippewa. He observed that segregation ratios of several traits fit either disomic or tetrasomic ratios (chromosome segregation). It may be argued, however, that those loci with 'disomic segregation' were located far away from the centromere and thus were the product of chromatid segregation. A re-examination of his results suggests that this was indeed the case (Table 6.3).

Dihaploids have been used in quantitative genetic research to calculate the heritability of several traits (Landeo and Hanneman, 1982; Gaur *et al.*, 1983; Kotch, 1987), the genetic load of the tetraploid parents (Kotch *et al.*, 1991) and the relative importance of both inbreeding coefficient and loss of intra- and interlocus interactions for yield and vigour of potatoes (Mendiburu *et al.*,

Table 6.3. Chromatid segregation for those traits reported as having disomic instead of tetrasomic segregation. Expected chromatid segregation ratio assuming duplex genotype for cv. Chippewa.

Trait		Observed ratio[1]	Expected chromatid segregation	
Stem pubescens				
Pubescent	$(P\_)$	48	51	
Glabrous	(pp)	17	14	$\chi^2 = 0.82$ n.s.
Stolon length				
Short	$(S\_)$	46	51	
Long	(ss)	19	14	$\chi^2 = 2.23$ n.s.
Meiotic behaviour				
Normal	$(N\_)$	28	30	
Desynaptic	(nn)	10	8	$\chi^2 = 0.63$ n.s.
Corolla shape				
Stellate	(SS)	16	12	
Pentagonal	(SR)	28	32	
Subrotate	(RR)	12	12	$\chi^2 = 1.80$ n.s.
Photoperiod response for tuberization				
Short day	(SS)	18	14	
Day-neutral	(SL)	33	37	
Long day	(LL)	14	14	$\chi^2 = 1.57$ n.s.

[1] From Matsubayashi, 1979.

1974; Kotch, 1987). Dihaploids are also now being used for molecular analysis in the potato, based on isozymes (Ortiz et al., 1991) and restriction fragment length polymorphisms (Gebhart et al., 1989).

Dihaploids have been used along with wild species for the evaluation of wild species for tuber traits (Leue and Peloquin, 1980; Hermundstad and Peloquin, 1985; Yerk and Peloquin, 1988; Jansky et al., 1990), for breaking undesirable linkage blocks within wild species (Jansky et al., 1990), for maintenance of species germplasm in clonal form and for population improvement at the $2x$ level (Iwanaga, 1984b; Zimmoch-Guzowska, 1986).

Determination of coefficient of inbreeding and coancestry in dihaploids

Mendiburu (1971) developed a procedure to estimate the coefficient of inbreeding in dihaploids from tetrasomic polyploids which takes account of double reduction in producing the $4x$ parent as well as in producing the dihaploids. The value of the coefficient of inbreeding of dihaploids from a non-inbred $4x$ clone ($abcd$) at any locus is a function of the coefficient of double reduction, α. The coefficient of inbreeding, F, can also be estimated

Table 6.4. 4x genotype, probability of occurrence and proportion of homozygous dihaploids by descent derived from common 4x parent. (Modified from Mendiburu, 1971.)

4x Genotype	Probability of occurrence (1)	Proportion of homozygous dihaploids by descent (2)	1 × 2
aabb	α^2	$\alpha + (1-\alpha)/3$	$(2\alpha^3 + \alpha^2)/3$
aabc	$2\alpha(1-\alpha)$	$(5\alpha + 1)/6$	$(\alpha + 4\alpha^2 - 5\alpha^3)/3$
abcd	$(1-\alpha)^2$	α	$\alpha - 2\alpha^2 + \alpha^3$
Total F			$\alpha(4-\alpha)/3$

α = alpha = coefficient of double reduction.

Table 6.5. F values for dihaploid in relationship to locus segregation. (Modified from Mendiburu, 1971.)

Segregation	α	F	$(F-\alpha) \times 100$
Chromosome	0	0	
Chromatid	$1/7 = 0.143$	$9/49 = 0.184$	4.1%
Maximal equational division	$1/6 = 0.167$	$23/108 = 0.213$	4.6%

for 4xs with the genotypes *aabb* and *aabc*, in which both or one of the gametes that gave rise to the 4x were the product of double reduction at that locus. Table 6.4 indicates the probability of each *abcd*, *aabc* and *aabb* genotype from parents *a--d* and *-bc-*, the probability of obtaining a homozygous haploid by descent and the coefficient of inbreeding of the haploid. The F values (Table 6.5) are a function of the coefficient of double reduction and range from 0% (for those loci close to the centromere) to 21.3% (for loci far removed from the centromere).

Mendiburu (1971) indicated that the F value of a potato dihaploid for a character depending on a large number of loci would be zero if double reduction for those loci was not frequent. Mendiburu *et al.* (1974) considered that the decrease in yield following dihaploidization was a consequence of the loss of favourable intra- and interlocus interactions, rather than inbreeding.

Chujoy (1985) derived the coefficient of coancestry between 'full-sib' dihaploids (dihaploids extracted from the same 4x parent). There are six different dihaploid genotypes possible if double reduction does not occur at a given locus. Therefore, 36 different pairwise combinations of dihaploids are possible. The frequencies and the respective coefficient of coancestry for dihaploids sharing two, one and no alleles in common are indicated in Table 6.6.

The coefficient of coancestry between 'full-sib' dihaploids (ϕ_{FSH}) is obtained by summing the products of frequency × coefficient of coancestry for the 36 combinations of pairs of dihaploids. Thus

Table 6.6. Frequencies and coefficients of coancestry of three possible 'full-sib' dihaploid genotype pairs. (From Chujoy, 1985.)

Pair of dihaploids	Frequency	Coefficient of coancestry
x_1x_2, x_1x_2	1/36	$2/4 + 2/4\ F_a$
x_1x_2, x_2x_4	1/36	$1/4 + 2/4\ \phi_{ab} + 1/4\ F_a$
x_1x_2, x_3x_4	1/36	$4/4\ \phi_{ab}$

Note The genotype of the 4x clone was $x_1x_2x_3x_4$, which had parents *a* and *b* with coefficients of inbreeding F_a and F_b and coefficient of coancestry ϕ_{ab}.

$$\phi_{FSH} = \tfrac{1}{4} + \tfrac{1}{2}\phi_{ab} + \tfrac{1}{8}F_a + \tfrac{1}{8}F_b,$$

where F_a and F_b are the coefficients of inbreeding of the 4x progenitors of the 4x clone from which the dihaploids were extracted, and ϕ_{ab} is the coefficient of coancestry between the tetraploid parents. ϕ_{FSH} can be expressed as a function of the coefficient of inbreeding of the 4x clone (F_c) from which the dihaploids were extracted, thus

$$\phi_{FSH} = \tfrac{1}{4} + \tfrac{3}{4}F_c.$$

Hence $\phi_{FSH} = \tfrac{1}{4}$ if they were derived from a non-inbred 4x clone.

Chujoy (1985) indicated that the coefficient of coancestry between a 4x individual and its dihaploid is

$$\tfrac{1}{4} + \tfrac{1}{8}F_a + \tfrac{1}{8}F_b + \tfrac{1}{2}\phi_{ab}, \quad \text{or} \quad \tfrac{1}{4} + \tfrac{3}{4}F_c$$

in terms of the coefficient of inbreeding of the 4x parent from which the dihaploids were extracted, which is the same as ϕ_{FSH}.

Estimation of the coefficient of double reduction using dihaploids

The best unbiased estimates of α using information from dihaploid populations with complete (no dominance) and incomplete (dominance) classifications can be obtained through the maximum-likelihood method (MLM).

The expected frequencies of dihaploid formation in tetrasomic polyploids as a function of the parameter of double reduction, α, are indicated in Table 6.7 for different 4x genotypes. The MLM provides estimates of the parameter α that maximize the probability of the observed results.

Unbalanced diallelic ($a_ia_ja_ja_j$)

The probability of the observed numbers of dihaploids, A (a_ia_i), B (a_ia_j), and C (a_ja_j), as a function of the unknown parameter α for an unbalanced diallelic 4x, is the likelihood of α

Table 6.7. Expected frequencies of dihaploid types in each marker class from different tetrasomic polyploids.

	Genotype of the 4x parent		Number of dihaploids per marker class
Genotype of Dihaploid	Simplex or unbalanced diallelic	Duplex or balanced diallelic	
AA or $a_i a_i$	$\alpha/4$	$(1+2\alpha)/6$	A
Aa or $a_i a_j$	$[2-2(\alpha)]/4$	$4(1-\alpha)/6$	B
aa or $a_j a_j$	$(2+\alpha)/4$	$(1+2\alpha)/6$	C
Total			**N**

$$L(\alpha) = (N!/[A!B!C!])(\{\alpha/4\}^A)(\{2(1-\alpha)/4\}^B)(\{[2+\alpha]/4\}^C)$$

where $N = A + B + C$.

The value of α that maximizes the probability will also maximize the natural logarithm of L

$$\ln L(\alpha) = A \ln(\alpha/4) + B \ln[(1-\alpha)/2] + C \ln[(2+\alpha)/4] + K$$

where K is a constant.

The derivative of $\ln L$ with respect to α is equated to 0 to find the value of α which maximizes L

$$d \ln L/d\alpha = A/\alpha - B/(1-\alpha) + C/(2+\alpha) = 0$$

Using algebra

$$0 = A(1-\alpha)(2+\alpha) - B(\alpha)(2+\alpha) + C(\alpha)(1-\alpha)$$
$$0 = -\alpha^2(A+B+C) - \alpha(A+2B-C) + 2A$$
$$0 = N\alpha^2 + (A+2B-C)\alpha + (-2A)$$

and taking the positive root of this quadratic equation

$$\alpha = \{-(A+2B-C) + [(A+2B-C)^2 - 4N(-2A)]^{\frac{1}{2}}\}/2N \qquad (1)$$

Balanced diallelic ($a_i a_i a_j a_j$)

In a similar way, the likelihood of the observed numbers of dihaploids, A ($a_i a_i$), B ($a_i a_j$) and C ($a_j a_j$) as a function of the unknown parameter α for a balanced diallelic, is

$$L(\alpha) = (N!/[A!B!C!])(\{[1+2\alpha]/6\}^A)(\{4[1-\alpha]/6\}^B)(\{[1+2\alpha]/6\}^C)$$

and the maximum-likelihood estimate (MLE) of α is

$$\alpha = [2(A+C) - B]/2N \qquad (2)$$

Triallelic loci ($a_i a_j a_k a_k$)

$$\alpha = \{-(A' + 2B' - C') + [(A' + 2B' - C')^2 + 8NA']^{\frac{1}{2}}\}/2N \qquad (3)$$

where A' = dihaploids with $a_i a_i$ or $a_j a_j$ genotypes, B' = dihaploids with $a_i a_k$ or $a_j a_k$ genotypes and C' = dihaploids with $a_i a_j$ or $a_k a_k$ genotypes.

Using dihaploids from simplex ($Aaaa$)

$$\alpha' = 2(D - E)/N \qquad (4)$$

where D = dihaploids with recessive 'a' phenotype and E = dihaploids with dominant 'A' phenotype.

Using dihaploids from duplex ($AAaa$)

$$\alpha' = (5D - E)/2N \qquad (5)$$

The variance for the MLE of α

The theoretical variance for an MLE of α (V_α) is given by the expected value of the negative reciprocal of the second derivative of $\ln L(\alpha)$ if the sample size is large and is normally distributed. Thus

$$1/V_\alpha = -E[d^2 \ln L(\alpha)/d\alpha^2]$$

where E means 'replace the observed quantities with their maximum-likelihood estimates'. For multinomial data, the expected value is found by replacing each count wherever it occurs by its expected value. The standard error for the MLE of α is $V_\alpha^{\frac{1}{2}}$.

Examples

The MLM provided unbiased estimates of the parameter α with the smallest possible variances. In effect, the estimators were satisfactorily tested for chromosome segregation ($\alpha = 0$), chromatid segregation ($\alpha = 1/7$), and maximal equational segregation ($\alpha = 1/6$), assuming the expected gametic output of the $4x$ to be the frequency of each genotypic (no dominance, complete classification) or phenotypic (dominance, incomplete classification) class in the dihaploid segregating population.

COMPLETE DOMINANCE

A total of 47 dihaploids from the cultivar Atlantic were scored for $2n$ pollen production by Kotch (1987). The most common mechanism of $2n$ pollen formation, parallel spindles in the second meiotic division, is controlled by a recessive meiotic mutant *ps*. Atlantic dihaploids were grouped as non-$2n$ pollen producers ($Ps/-$) or $2n$ pollen producers (ps/ps). The number of

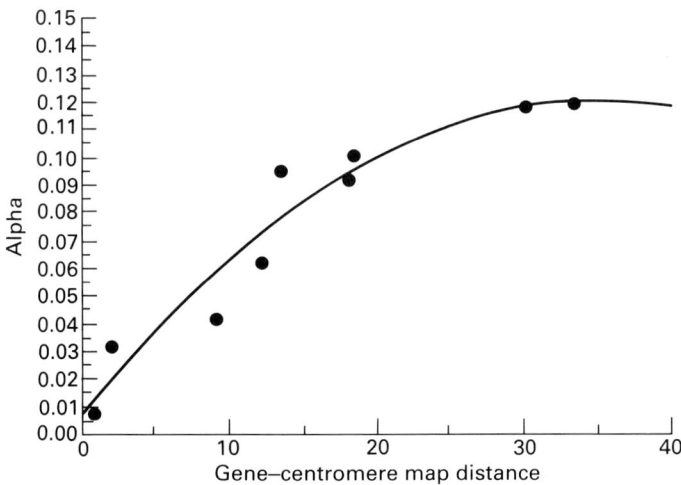

Fig. 6.1. Coefficient of double reduction (α) as a function of gene-centromere map distance (GCMD). $\alpha = 0.00736 + 0.00649 \, \text{GCMD} - 0.00009 \, \text{GCMD}^2$; $r^2 = 0.93076$.

dihaploids with $2n$ pollen production was 23, the remainder being non-$2n$ pollen producers. Alpha was estimated as 0.042 by using eqn (4).

INCOMPLETE DOMINANCE

Tuber and corolla shape data were taken from Matsubayashi (1979), who observed a trimodal distribution (codominance) in the dihaploid population of cv. Chippewa for both traits. The α estimates were obtained through eqn (2). They were 0.062 and 0.125 for tuber and corolla shape, respectively.

Alpha and gene-centromere map distances

Using data from segregating dihaploid populations evaluated for nine different morphological and isozyme loci (R. Ortiz, unpublished), the following significant quadratic relationship was derived between α and gene-centromere map distances (GCMD) (Figure 6.1):

$$[\alpha = 0.00736 + 0.00649 \, \text{GCMD} - 0.00009 \, \text{GCMD}^2]$$

with a coefficient of determination, r^2, equal to 93%. Burnham (1984) indicated that α values may be of use only in determining relative distances between loci and centromeres. In effect, low values of α indicate that the locus is proximal to the centromere. Conversely, high values of α are expected for loci distal to the centromere. Burnham (1984) pointed out that map distances could not be determined directly without considering the other factors affecting the occurrence of double reduction. However, the quadratic relationship found between the MLE of α and the GCMD calculated through half-tetrad analysis (Mendiburu and Peloquin, 1979) appears to remedy this problem.

The theoretical maximum for α is equal to $\frac{1}{6}$ for loci which are located far from their centromere and which have maximal equational segregation. Maximal equational segregation is, however, rarely observed due to the requirements necessary for its occurrence (Burnham, 1984). In this respect, the highest MLE of α was 0.144 for the *Mdh*-2 locus (R. Ortiz, unpublished) and none of the 14 loci examined had maximal equational segregation.

Douches and Quiros (1987), using theory developed by Mather (1935), established that the gene–centromere map distance limit should be 33.3 cM because, with an infinite number of chiasmata between a locus and centromere, the proportion of equational separation is $\frac{2}{3}$ and the maximum map distance is $50 \times \frac{2}{3}$. But Mather (1935) points out that the situation is complicated by the fact that changes of partner, resulting in crossings over with more than one other chromosome, may occur between the locus and the centromere. The quadratic relationship between α and GMCD supports this argument because values for α close to $\frac{1}{7}$ (chromatid segregation) should be expected for those loci positioned 33.3 cM from their centromere. They must be considered as segregating independently of their respective centromeres.

In practical breeding terms, the estimation of α is very important, because an increase in the proportion of homozygous gametes is expected for loci which have high α values.

Analysis of frequency distribution of dihaploid populations, type of gene action and genetic load

The evaluation of dihaploids from different $4x$ parents and the comparison of the frequency distributions of the dihaploid populations from each $4x$ parent permits detection of genotypic differences among the $4x$ parents. Moreover, the use of third- and fourth-degree statistics, such as skewness (the degree of departure of a distribution from symmetry) and kurtosis (the peakedness of a distribution), allows differences within haploid populations to be detected and genetic control of the trait to be inferred (Kotch et al., 1991).

Narrow-sense heritability can be estimated by $2x$ offspring–$4x$ parent regression, where $h^2 = b_{OP}$ because each progeny has only one parent. The mean of each dihaploid family is regressed on the mean of its respective $4x$ parent.

Deleterious genes increase the frequency of low-yielding plants and reduce the frequency of high-yielding plants. The degree of deviation is proportional to the load of deleterious genes fixed and may be measured by the coefficient of variation (CV = standard deviation/mean). The greater the load of deleterious genes, the more asymmetrical the frequency distribution and the larger the CV. A disproportionate reduction in yielding capacity between classes will increase the CV because the reduction in the mean is not accompanied by a proportional reduction in the standard deviation (Fasoulas, 1988).

Use of Diploid Species

Genetic system

Dodds and Paxman (1962) established that the cultivated $2x$ potatoes were obligate crossbreeders due to self-incompatibility.

Wang (1971) studied the F_2 progeny from intra- and interspecific crosses of four Argentine $2x$ (2 EBN) tuber-bearing *Solanum species*: *S. chacoense*, *S. infundibuliforme*, *S. kurtzianum* and *S. microdontum* subsp. *gigantophyllum*. She investigated species relationships among these species using morphological, cytogenetic and isozyme marker approaches. Particular interest centred on the occurrence of F_2 breakdown, e.g. appearance of weak or sterile plants among F_2 of both inter- and intraspecific crosses. She found that F_1 hybrids were intermediate in phenotype and more vigorous than the parental species. The interspecific F_2 progenies were very variable in their phenotypes. Parental-like phenotypes were recovered in small F_2 populations. F_2 breakdown occurred in progeny from both intra- and interspecific crosses with similar frequency. She concluded that 'F_2 breakdown' was due to the effect of imposed inbreeding on natural outcrossers rather than a result of 'cryptic structural differentiation'. Her conclusion was supported by the chromosome behaviour of both intra- and interspecific hybrids. Pachytene analysis revealed no structural differences between the chromosomes of the four species.

Unusual meiotic behaviour (poor pairing at pachytene, ring bivalents at diakinesis and univalents in metaphase I) was observed in progeny from inter- and intraspecific crosses (Wang, 1971). However, these meiotic abnormalities were found to be under genetic control, each inherited as a simple recessive. Her results indicate that meiotic behaviour must be interpreted with caution as evidence of genome differentiation between species.

Production of aneuploids for genetic analysis

Aneuploids, mainly trisomics, have been produced from wild species (Vogt and Rowe, 1968; Lee, 1970; Kessel, 1972; Kessel and Rowe, 1974; Kessel *et al.*, 1975). They are characterized by good vigour and fertility but poor tuberization under long-day environments, which complicates their utilization and maintenance. The loci for topiary (*tp*) and glucosylation of rutin were assigned to chromosome 3 (Wagenwoort, 1988) and the long arm of chromosome 9 (Lee and Rowe, 1975), respectively, by means of trisomic analysis (Hermsen, 1970). In a similar way, the locus for albinism (*a*) was placed on chromosome 9 of *S. chacoense* (Lam and Erickson, 1971).

Utilization of cultivated diploids in potato breeding and genetic analysis

In 1966 Haynes (1980) initiated a breeding project using $2x$ cultivated species Group Phureja and Group Stenotomum at North Carolina State University. The main emphasis of this project was adaptation to long days, resistance to early blight and soft rot, high dry-matter content and tuber dormancy. This project made several contributions to the genetics of the cultivated $2x$ potato in areas such as response to photoperiod (Mendoza and Haynes, 1976, 1977), heat tolerance (Haynes, 1980), specific gravity (Ruttencutter et al., 1979), tuber dormancy (Thompson et al., 1980) and early blight resistance (Herriot et al., 1986). Parent-offspring regression analysis and North Carolina designs I (nested or hierarchical) and II (factorial) were used for the genetic analysis of the data.

Generation mean analysis was used by Mendoza and Haynes (1977) to study response to photoperiod of different $2x$ populations. They developed eight genetic sets in which each set included the parental clones (P_1 and P_2), selfed progeny of both parents (F_1s and F_2s) and backcrosses to P_1 and P_2. One-way analysis of variance was carried out for each generation within each set to estimate the within-plot variance and to provide variance components for the estimation of broad-sense heritability (h^2) for tuber initiation as influenced by photoperiod:

$$h^2 = 100(V_{F_2} - V_P)/V_{F_2}$$

in which V_{F_2} was the variance of the F_2 generation and V_P was the pooled variance among plants within clones of all parental clones used in their research. V_P provided an estimate of the environmental variance.

Extensive evaluation of $2x$ wild relatives has been done by the Inter-Regional Introduction Project at Sturgeon Bay, Wisconsin (Hanneman and Bamberg, 1986). The quantitative genetic variation among different diploid taxa was estimated by Rowe (1969), using nested and factorial mating designs.

Inbreeding depression in cultivated diploid potatoes, using self-compatible diploids, was evaluated by De Jong and Rowe (1971). They found that the average inbreeding depression coincided with that expected on the basis of loss of heterozygosity in a diploid organism.

Pineda-Colorado (1990) used generation mean analysis to analyse progeny derived from inbred lines of S. chacoense. He developed five sets of families from intermating by pairs five inbred lines. He included six generations in his analysis; both parents, F_1, F_2, and backcrosses to each parent. Then he tested the additive-dominance model for the different traits evaluated: number of inflorescences, flowers per inflorescence, tuber yield and components, and specific gravity.

The generation means for each character and the standard error of the mean were calculated based on individual data. The means were weighted to allow for unequal precision in their estimation by the reciprocal of the squared standard error. The weighted least-squares estimates for m (mean), $[d]$

Table 6.8. Generation mean analysis based on an additive-dominance model. Genetic parameters based on the use of inbred lines as parents (P_1 and P_2).

Generation	Phenotypic mean	Genetic parameters[1]
P_1	M_1	$m + [d]$
P_2	M_2	$m - [d]$
F_1	M_3	$m + [h]$
F_2	M_4	$m + \frac{1}{2}[h]$
BC_{P_1}	M_5	$m + \frac{1}{2}[d] + \frac{1}{2}[h]$
BC_{P_2}	M_6	$m - \frac{1}{2}[d] + \frac{1}{2}[h]$

[1] m = mean; $[d]$ = additivity; $[h]$ = dominance; and [ ] indicates net effects over loci.

(additivity) and $[h]$ (dominance) were calculated using matrix algebra from the model in Table 6.8. Then, χ^2 tests of goodness of fit were used to test the adequacy of the additive–dominance models for each trait within each set.

Pineda-Colorado (1990) indicated that his results for number of inflorescences and flowers per inflorescence did not fit the additive–dominance model and that the genotype–environment interaction was significant in the expression of both traits. Directional dominance was the more important type of gene action controlling the number of tubers per plant, while the alleles for tuber size had additive to partially dominant gene action.

Diploid cultivated and wild species have been an important source of naturally occurring mutants, which can be used as genetic markers for further research (Simmonds, 1965). Extensive reviews of the use of both $2x$ cultivated and wild species have been published elsewhere (Swaminathan and Howard, 1953; Howard, 1960, 1970).

Use of Dihaploid–Wild Species Hybrids

Genome analysis and F_2 breakdown

Groups Tuberosum and Andigena dihaploids can be easily crossed with most of the $2x$ species to produce fertile hybrids. The dihaploid–species hybrids generally have normal chromosome pairing and crossing over (Peloquin, 1983). Furthermore, recovery of parental types was found in F_2 progenies from dihaploid–species hybrids (Flewelling, 1987). The percentages of recovery of parental types were used to estimate the number of major genes responsible for specific traits among $2x$ *Solanum* species. Flewelling (1987) concluded that the differences between cultivated and wild tuber-bearing *Solanum* species for leaf type, haulm type, tuberization and haulm maturity were controlled by approximately three to five major genes.

Molecular genetics of the potato

Dihaploid–species hybrids have been used to generate a molecular-genetic linkage map of potato, using genomic and cDNA probes from tomato (Bonierbale et al., 1988). Many isozyme and restriction fragment length polymorphisms were placed on the map. Twelve linkage groups were identified. Each corresponded to a specific chromosome of the potato (see Figure 10.5 for latest map).

Recently, Freyre and Douches (1991) found linkages between several of these molecular markers and tuber traits such as specific gravity and tuber dormancy.

Utilization of dihaploid–species hybrids in potato breeding and genetics

The strategy for utilization of $2x$ wild *Solanum* species in potato breeding for cultivar development (Ross, 1986; Hermundstad and Peloquin, 1987) or in breeding populations (Peloquin and Ortiz, 1992) combines the utilization of dihaploids, species and $2n$ gametes. This approach is based on: (i) simple disomic inheritance and population improvement in the $2x$ population; (ii) the contribution of the dihaploid parent to the improvement of agronomical traits; (iii) the use of wild species as source of genes for resistance to different pests/diseases and abiotic stresses as well as those for processing quality. The wild species also contribute allelic diversity to the cultivated gene pool. The final goal of breeding at the $2x$ level is to develop a $2x$ clone with specific attributes, good breeding value for agronomic traits and $2n$ gametes. Thus, the favourable alleles of the $2x$ parent can be transferred to the $4x$ level via $2n$ gametes.

Genetic studies in dihaploid–species hybrids have been carried out extensively by researchers at the University of Wisconsin-Madison. They were able to: (i) determine the importance of non-additive variance for tuber yield (Mendiburu and Peloquin, 1977; Ortiz, 1991); (ii) demonstrate a two-locus system with dominant–recessive epistasis in the inheritance of scab resistance (Alam, 1972); (iii) determine the contribution of epistatic gene action in the inheritance of tuber greening (Parfitt and Peloquin, 1981); (iv) show that additivity was the most important type of gene action for tuber dormancy (Flewelling, 1987); (v) determine the types of gene action and number of genes controlling chipping ability after reversion and reconditioning (Accatino, 1973); (vi) determine the genetics of tuber yield and other tuber traits (Landeo, 1979); (vii) determine the inheritance of shoot inhibition (Davies, 1981) and tuber initiation under long days (Leue, 1983). Similarly, researchers of the International Potato Center (CIP) were able to establish the genetical basis of resistance for root-knot nematode (Iwanaga et al., 1989), early blight (Ortiz, 1991) and potato tuber moth (Ortiz et al., 1990).

The researchers used different types of matings, such as diallel crosses, nested and factorial designs and parent–offspring regression analysis, as well

Table 6.9. Variance components for 2x genotypes (g), with a half- (HS) and nested full-sib (FS) family genetic structure, evaluated in randomized complete-block designs with r replications. k_1, k_2 and k_3 are weighting factors due to unequal HS and FS family sizes.

Source of variation	Mean square	Expected mean square
Replications	M_1	$V_e + gV_r$
Genotypes	M_2	$V_e + rV_g$
Half-sibs	M_{21}	$V_e + rV_w + rk_2 V_{FS} + rk_3 V_{HS}$
Full-sibs within half-sibs	M_{22}	$V_e + rV_w + rk_1 V_{FS}$
Clones within full-sib families	M_{23}	$V_e + rV_w$
Error	M_3	V_e

Note Genetic parameters:
Additive variance

$$V_A = 4V_{HS} = 4\left([k_1 \times M_{21}] - [k_2 \times M_{22}] + [\{k_2 - k_1\} \times M_{23}]\right)/(rk_1 k_3)$$

Dominance variance

$$V_D = 4(V_{FS} - V_{HS}) = -4\left([k_1 \times M_{21}] - [\{k_2 + k_3\} \times M_{22}] + [\{k_3 + k_2 - k_1\} \times M_{23}]\right)/(rk_1 k_3)$$

as gene substitution analysis (Davies, 1981) for the genetic interpretation of their data. The analysis of variance and variance components for mating designs using half- and full-sib family genetic structures is indicated in Tables 6.9 and 6.10. Different methods of analysis for determining heritability are summarized and compared in Table 6.11.

The evaluation of F_2 progeny from F_1 within family intercrosses provided an estimate of the number of loci, N, (effective factors) controlling the trait of interest:

$$N = D^2/(8[V_{F_2} - V_E]^2)$$

where D is the difference between the largest and the smallest value, V_{F_2} is the total variance in the F_2 progeny and V_E is the environmental variance. Such estimates, however, are not all that reliable.

Genetic–cytoplasmic interactions

Dihaploids and species as parents in reciprocal crosses were used to study genetic–cytoplasmic interactions between different tuber-bearing *Solanum* taxa for photoperiod response (Sanford and Hanneman, 1979) and male sterility (Grun *et al.*, 1962). The differences observed in the phenotypic expression of progeny from reciprocal crosses were considered an indication of cytoplasmic inheritance of the traits.

Table 6.10. Genetical interpretations of components of variance which can be derived from mean squares in a nested design. (From Parfitt and Peloquin, 1981.)

Source	Mean square	Components of variance and genetic interpretation
Locations L	M_1	V_{LOC}
Blocks B/L	M_2	V_{BLOCKS}
Males M	M_3	$V_M = (1/4)V_A + (1/16)V_{AA} + (1/64)V_{AAA}$
Females F/M	M_4	$V_{F/M} = (1/4)V_A + (1/4)V_D + (3/16)V_{AA} + (1/8)V_{AD} + (1/16)V_{DD} + (7/64)V_{AAA} + V_{MATERNAL}$
Clones/(M × F)	M_5	$V_{C/(MF)} = (1/2)V_A + (3/4)V_D + (3/4)V_{AA} + (7/8)V_{AD} + (15/16)V_{DD} + (7/8)V_{AAA}$
Plants/clones	M_6	V_{PLANTS}
M × L	M_7	V_{ML}
F × L	M_8	V_{FL}
M × B/L	M_9	$V_{MB/L}$
F × B/L	M_{10}	$V_{FB/L}$

V_{GE} = variance due to genotype by environment interaction, i.e. M × L, F × L, M × B/L and F × B/L interactions
    = $V_{ML} + V_{FL} + V_{MB/L} + V_{FB/L}$.
V_A = additive variance.
V_D = dominance variance.
V_{AA}, V_{AD}, V_{DD}, V_{AAA} = epistatic interactions.

Table 6.11. Methods of determining narrow-sense heritability (h^2) at the 2x level.

		Variances included in h^2 estimate		
Analysis	h^2	Additive	Epistasis	Maternal
Male-offspring regression	$2b$	+	+	
Female-offspring regression	$2b$	+	+	+
Midparent-offspring regression	b	+	+	some
Nested design	$4V_M/V_{TOTAL}$	+	+	
Factorial design	$4V_M/V_{TOTAL}$	+	+	
Factorial design	$\dfrac{2(V_M + V_F)}{V_{TOTAL}}$	+	+	+

b = coefficient of regression.
V_M = male variance.
V_F = female variance.

Production of aneuploids and trisomic analysis

The main limitations of trisomic series derived from either dihaploids or wild species have been the poor vigour and sterility of the former and the lack of tuberization of the latter, which limit their maintenance and utilization.

A new method of production of aneuploids worth considering is the production of trisomics by $3x \times 2x$ crosses. The $3x$ parent would be derived from crosses between $4x$ Group Tuberosum cultivars and $2x$ Group Phureja; the $2x$ parent could be either a dihaploid-species hybrid or a male-fertile Group Tuberosum dihaploid. The trisomics derived using this approach should have improved tuberization under long days, due to an increase of adapted Group Tuberosum germplasm. They would also have good vigour due to their broad genetic base. Initial attempts to obtain trisomics using this method have been carried out with limited success (Adiwilaga, 1986).

Acknowledgements

This is a paper from the laboratory of genetics. Research was supported by the College of Agricultural and Life Sciences, the International Potato Center, USDA-CRGO-88-37234 3619 and Frito-Lay, Inc.

References

Accatino, P.L. (1973) Inheritance of the potato chip color at the diploid and tetraploid levels of ploidy. Unpublished PhD thesis, University of Wisconsin-Madison.

Adiwilaga, K.D. (1986) Production of potato trisomics and identification of potato chromosomes using Giemsa C-banding technique. Unpublished MS thesis, University of Wisconsin-Madison.

Alam, Z. (1972) Inheritance of scab resistance in 24-chromosome potatoes. Unpublished PhD thesis, University of Wisconsin-Madison.

Bartels, D., Gebhardt, C., Knapp, S., Rohde W,, Thompson, R., Uhrig, H. and Salamini, F. (1988) Combining conventional plant breeding procedures with molecular based approaches. *Genome* 31, 1014–1026.

Bonierbale, M.W., Plaisted, R.L. and Tanksley S.D. (1988) RFLP maps based on a common set of clones reveal modes of chromosomal evolution in potato and tomato. *Genetics* 120, 1095–1103.

Burnham, C.R. (1984) *Discussion in Cytogenetics*. Burgess Publishing Co., Minneapolis.

Chujoy, J.E. (1985) Tuber yield of $2x$ and $4x$ progeny from $2x \times 2x$ crosses in potato; barriers to interspecific hybridization between *Solanum chacoense* Bitt. and *S. commersonii* Dun. Unpublished PhD thesis, University of Wisconsin-Madison.

Cipar, M.S. and Lawrence, C.H. (1972) Scab resistance of haploids from two *Solanum tuberosum* cultivars. *American Potato Journal* 49, 117–119.

Cipar, M.S., Peloquin, S.J. and Hougas, R.W. (1964) Haploidy and the identification of self-incompatibility alleles in cultivated diploid species. *Euphytica* 13, 163–172.

Davies, C.S. (1981) I. Genetics and physiology of shoot inhibition and II. Variation

in tuber ascorbic acid among diploids and tetraploids in *Solanum*. Unpublished PhD thesis, University of Wisconsin-Madison.
De Jong, H. and Rowe, P.R. (1971) Inbreeding in cultivated diploid potatoes. *Potato Research* 14, 74-83.
De Maine, M.J. (1984) Patterns of variation in potato dihaploid families. *Potato Research* 27, 1-11.
Desborough, S. and Peloquin, S.J. (1967) Esterase isozymes from *Solanum* tubers. *Phytochemistry* 6, 989-994.
Dodds, K.S. and Paxman, G.J. (1962) The genetic system of cultivated diploid potatoes. *Evolution* 16, 154-167.
Douches, D.S. and Quiros, C.F. (1987) Use of $4x$-$2x$ crosses to determine gene-centromere map distances of isozyme loci in *Solanum* species. *Genome* 29, 519-527.
Douches, D.S., Jansky, S., Liu, C.A. and Thompson, D. (1990) Segregation of isozyme loci in haploid progeny derived from cultivated tetraploid potato of *Solanum tuberosum* subsp. *tuberosum*. *American Potato Journal* 67, 546-547.
Fasoulas, A.C. (1988) *The Honeycomb Methodology of Plant Breeding*. Thessaloniki, Greece.
Flewelling, H.S. (1987) Use of haploid Tuberosum- wild *Solanum* species F_1 hybrids to study the relationship between the cultivated and wild potatoes, and to analyze the genetic control of tuber dormancy. Unpublished MS thesis, University of Wisconsin-Madison.
Freyre, R. and Douches, D.S. (1991) Use of molecular markers for the study and analysis of quantitative tuber traits in diploid potatoes (*Solanum* spp.). In: *Abstracts of the Symposium on Plant Breeding in the 1990s*. Department of Crop Science, Research Report No. 130, North Carolina State University, Raleigh, p. 61.
Gaur, P.C., Gopal, L. and Rana, M.S. (1983) Combining ability for yield and tuber dry matter in potato. *Indian Journal of Agricultural Science* 53, 876-879.
Gebhardt, C., Ritter, E., Debener, T., Schachtschabel, U., Walkemeier, B., Uhrig H. and Salamini, F. (1989) Restriction fragment length polymorphism analysis and linkage mapping in *Solanum tuberosum*. *Theoretical and Applied Genetics* 78, 65-75.
Grun, P., Aubertin, M. and Radlow, A. (1962) Multiple differentiation of plasmons of diploid species of *Solanum*. *Genetics* 47, 1321-1333.
Hanneman R.E., Jr, and Bamberg, J. (1986) *Inventory of Tuber-bearing Solanum species*. Bulletin 533, Research Division, College of Agricultural and Life Sciences, University of Wisconsin, Madison, Wisconsin.
Haynes, F.L. (1980) Progress and future plans for the use of Phureja-Stenotomum populations. In: *Report of CIP Planning Conference on Utilization of the Genetic Resources of the Potato III*. International Potato Center, Lima, Peru, pp. 80-88.
Hermsen, J.G.Th. (1970) Basic information for the use of primary trisomics in genetics and breeding research. *Euphytica* 19, 125-140.
Hermsen, J.G.Th. and Verdenius, J. (1973) Selection from *Solanum tuberosum* group Phureja of genotypes combining high frequency haploid induction with homozygosity for embryo spot. *Euphytica* 22, 244-259.
Hermsen, J.G.Th., Wagenvoort, M. and Ramanna, M.S. (1970) Aneuploids from natural and colchicine induced autotetraploids of *Solanum*. *Canadian Journal of Genetics and Cytology* 12, 601-613.
Hermsen, J.G.Th., Ramanna, M.S. and Vogel, J. (1973) The location of a recessive gene for chlorophyll deficiency in diploid *Solanum tuberosum* by means of trisomic analysis. *Canadian Journal of Genetics and Cytology* 15, 807-813.

Hermundstad, S.A. and Peloquin, S.J. (1985) Germplasm enhancement with potato haploids. *Journal of Heredity* 76, 463–467.

Hermundstad, S.A. and Peloquin, S.J. (1987) Breeding at the 2x level and sexual polyploidization. In: Jellis, G.J. and Richardson, D.E. (eds) *The Production of New Potato Varieties: Technological Advances.* Cambridge University Press, Cambridge, pp. 197–210.

Herriot, A.B., Haynes, F.L. and Shoemaker, P.B. (1986) The heritability of resistance to early blight in diploid potatoes (*S. tuberosum* subspp. *phureja* and *stenotomum*). *American Potato Journal* 63, 229–232.

Hougas, R.W. and Peloquin, S.J. (1958) The potential of potato haploids in breeding and genetic research. *American Potato Journal* 35, 701–707.

Hougas, R.W., Peloquin, S.J. and Ross, R.W. (1958) Haploids of the common potato. *Journal of Heredity* 49, 103–107.

Hougas, R.W., Peloquin, S.J. and Gabert, A.C. (1964) Effect of seed-parent and pollinator on the frequency of haploids in *Solanum tuberosum*. *Crop Science* 4, 593–595.

Howard, H.W. (1960) Potato cytology and genetics, 1952–1959. *Bibliographia Genetica* 19, 87–216.

Howard, H.W. (1970) *Genetics of Potato.* Springer Verlag, New York.

Ivanovskajya, E.V. (1939) A haploid plant of *Solanum tuberosum*. *Comptes Rendus de L'Académie de Science, URSS* 24, 517–520.

Iwanaga, M. (1984a) Discovery of a synaptic mutant in potato haploids and its usefulness for potato breeding. *Theoretical and Applied Genetics* 68, 87–93.

Iwanaga, M. (1984b) Haploids, ploidy manipulation, and meiotic mutants in potato. In: *Biotechnology in International Agricultural Research.* IRRI, Manila, Philippines, pp. 139–148.

Iwanaga, M., Jatala, P., Ortiz, R. and Guevara, E. (1989) Use of FDR 2n pollen to transfer resistance to root-knot nematodes into cultivated 4x potatoes. *Journal of the American Society for Horticultural Science* 114, 1008–1013.

Jansky, S.H., Peloquin, S.J. and Yerk, G.L. (1990) Use of potato haploids to put 2x wild species germplasm in a usable form. *Plant Breeding* 104, 290–294.

Kessel, R. (1972) Production and use of inter- and intraspecific aneuploids in the genus *Solanum*. Unpublished PhD thesis, University of Wisconsin-Madison.

Kessel, R. and Rowe, P.R. (1974) Interspecific aneuploids in the genus *Solanum*. *Canadian Journal of Genetics and Cytology* 16, 515–528.

Kessel, R., Lee, H.K. and Rowe, P.R. (1975) Production of intraspecific aneuploids in the genus *Solanum*. *Euphytica* 24, 585–595.

Kotch, G.P. (1987) The production of haploids and their use in genetic studies in potatoes. Unpublished PhD thesis, University of Wisconsin-Madison.

Kotch, G.P., Ortiz, R. and Peloquin, S.J. (1991) Genetic analysis by use of potato haploid populations. *Genome* 36, 103–108.

Lam, S.L. and Erickson, H.T. (1971) Location of a mutant gene causing albinism in a diploid potato. *Journal of Heredity* 59, 369–373.

Landeo, J.A. (1979) Breeding potential of Group Andigena haploid potatoes. Unpublished PhD thesis, University of Wisconsin-Madison.

Landeo, J.A. and Hanneman R.E., Jr (1982) Genetic variation in *Solanum tuberosum* Gp. Andigena haploids. *Theoretical and Applied Genetics* 62, 311–351.

Lee, H.K. (1970) Production and cytogenetic studies of triploids and trisomics in tuber bearing *Solanum* species. Unpublished PhD thesis, Purdue University.

Lee, H.K. and Rowe, P.R. (1975) Genetic segregation of the deformed flower in trisomics of *Solanum chacoense*. *Euphytica* 25, 313–320.

Leue, E.F. (1983) The use of haploids, 2n gametes, and the topiary mutant in the adaptation of wild *Solanum* germplasm and its incorporation into Tuberosum. Unpublished PhD thesis, University of Wisconsin-Madison.

Leue, E.F. and Peloquin, S.J. (1980) Selection for 2n gametes and tuberization in *S. chacoense*. *American Potato Journal* 57, 189-195.

Mather, K. (1935) Reductional and equational separation of the chromosomes in bivalents and multivalents. *Journal of Genetics* 30, 53-78.

Matsubayashi, M. (1979) Genetic variation in dihaploid potato clones, with special reference to phenotypic segregation in some characters. *Science Report of Faculty of Agriculture Kobe University* 16, 1-9.

Mendiburu, A.O. (1971) The significance of 2n gametes in potato breeding and genetics. Unpublished PhD thesis, University of Wisconsin-Madison.

Mendiburu, A.O. and Peloquin, S.J. (1977). The significance of 2n gametes in potato breeding. *Theoretical and Applied Genetics* 49, 53-61.

Mendiburu, A.O. and Peloquin, S.J. (1979) Gene-centromere mapping by $4x \times 2x$ matings in potato. *Theoretical and Applied Genetics* 54, 177-180.

Mendiburu, A.O., Peloquin, S.J. and Mok, D.W.S. (1974) Potato breeding with haploids and 2n gametes. In: Kasha, K.J. (ed.) *Haploids in Higher Plants*. University of Guelph, Guelph, pp. 249-258.

Mendoza, H.A. and Haynes, F.L. (1976) Variability for photoperiod reaction among diploid and tetraploid potato clones from three taxonomic groups. *American Potato Journal* 53, 319-332.

Mendoza, H.A. and Haynes, F.L. (1977) Inheritance of tuber initiation in tuber bearing *Solanum* as influenced by photoperiod. *American Potato Journal* 54, 243-252.

Ortiz, R. (1991) Efficiency of potato breeding using 2n gametes; male sterility and 2n pollen in 4x potato. Unpublished PhD thesis, University of Wisconsin-Madison.

Ortiz, R., Iwanaga, M., Raman, K.V. and Palacios, M. (1990) Breeding for resistance to potato tuber moth, *Phthorimaea opercullela* (Zeller), in diploid potatoes. *Euphytica* 50, 119-126.

Ortiz, R., Douches, D.S., Kotch, G.P. and Peloquin, S.J. (1991) Genetic analysis in polysomic polyploids through the utilization of haploids and molecular markers. *Agronomy Abstracts*. ASA, Madison, Wisconsin, p. 109.

Parfitt, D.E. and Peloquin, S.J. (1981) The genetic basis for tuber greening in 24-chromosome potatoes. *American Potato Journal* 58, 299-304.

Peloquin, S.J. (1983) Conservation and utilization of exotic germplasm to improve varieties. In: *Report of the 1983 Plant Breeding Research Forum*. Pioneer Hi-Bred International, pp. 147-158.

Peloquin, S.J. and Hougas, R.W. (1959) Decapitation and genetic markers as related to haploidy in *Solanum tuberosum*. *European Potato Journal* 2, 176-183.

Peloquin, S.J. and Hougas, R.W. (1960) Genetic variation among haploids of the common potato. *American Potato Journal* 37, 176-183.

Peloquin, S.J. and Ortiz, R. (1992) Techniques for introgressing unadapted germplasm to breeding populations. In: Stalker, H.T. and Murphy, J.P. (eds) *Plant Breeding in the 1990s*. CAB International, Wallingford, pp. 485-507.

Peloquin, S.J., Hougas, R.W. and Gabert, A.C. (1966) Haploidy as a new approach to the cytogenetics and breeding of *Solanum tuberosum*. In: Riley, R. and Lewis, K.R. (eds) *Chromosome Manipulations and Plant Genetics*. Oliver and Boyd, Edinburgh, pp. 21-28.

Peloquin, S.J., Werner, J.E. and Yerk, G.L. (1990) The use of potato haploids in genetics and breeding. In: Gupta, P.K. and Tsuchiya, T. (eds) *Chromosome Engineering in Plants*. Elsevier, Barking, Essex, England, Part B, pp. 79-92.

Pineda-Colorado, R. (1990) Quantitative and genetic analysis in *Solanum chacoense* Bitt. using inbred lines. Unpublished PhD thesis, University of Wisconsin-Madison.

Quiros, C.F. and McHale, N. (1985) Genetic analysis of isozyme variants in diploid and tetraploid potatoes. *Genetics* 111, 131–145.

Ross, H. (1986) *Potato Breeding: Problems and Perspectives*. Paul Parey, Berlin and Hamburg.

Rowe, P.R. (1969) Quantitative variation in diploid potatoes. *American Potato Journal* 46, 14–17.

Ruttencutter, G.E., Haynes, F.L. and Moll, R.H. (1979) Estimation of narrow-sense heritability for specific gravity in diploid potatoes (*S. tuberosum* subsp. *phureja* and *stenotomum*). *American Potato Journal* 56, 447–453.

Sanford, J.C. and Hanneman R.E., Jr. (1979) Reciprocal differences in the photoperiod reaction of hybrid populations in *Solanum tuberosum*. *American Potato Journal* 56, 531–540.

Simmonds, N.W. (1965) Mutant expression in diploid potatoes. *Heredity* 20, 65–72.

Simon, P.W. and Peloquin, S.J. (1980) Inheritance of electrophoretic variants of tuber proteins in *Solanum tuberosum* haploids. *Biochemical Genetics* 18, 1055–1063.

Swaminathan, M.S. and Howard, H.W. (1953) The cytology and genetics of the potato (*Solanum tuberosum* L.) and related species. *Bibliographia Genetica* 16, 1–192.

Thompson, P.G., Haynes, F.L. and Moll, R.H. (1980) Estimation of genetic variance components and heritability for tuber dormancy in diploid potato. *American Potato Journal* 57, 39–46.

Vogt, G.E. and Rowe, P.R. (1968) Aneuploids from triploid–diploid crosses in the series Tuberosa of the genus *Solanum*. *Canadian Journal of Genetics and Cytology* 10, 479–486.

Wagenwoort, M. (1982) Location of the recessive gene *ym* (yellow margin) on chromosome 12 of diploid *Solanum tuberosum* by means of trisomic analysis. *Theoretical and Applied Genetics* 61, 239–243.

Wagenwoort, M. (1988) Chromosomal localisation of a recessive gene *tp* controlling the pleiotropic character topiary in *Solanum*. *Theoretical and Applied Genetics* 75, 712–716.

Wagenwoort, M. and Lange, W. (1975) The production of aneudihaploids in *Solanum tuberosum* L. Group Tuberosum (the common potato). *Euphytica* 24, 731–741.

Wang, H. (1971) Species relations among four Argentine diploid tuber-bearing Solanums. Unpublished PhD thesis, University of Wisconsin-Madison.

Wenzel, G. (1980) Recent progress in microspore culture of crop plants. In: Davies D.R. and Hopwood, D.A. (eds) *The Plant Genome*. The John Innes Institute, Chanty, Norwich, pp. 185–196.

Wenzel, G., Meyer, C., Uhrig, H. and Schieder, O. (1980) Current status of exploitation of monohaploids and protoplast fusion and potential in potato breeding. In: *Report of CIP Planning Conference on Utilization of the Genetic Resources of the Potato III*. International Potato Center, Lima, Peru, pp. 169–183.

Yeh, B.P., Peloquin, S.J. and Hougas, R.W. (1964) Meiosis in *Solanum tuberosum* haploids and haploid-species F_1 hybrids. *Canadian Journal of Genetics and Cytology* 6, 393–402.

Yerk, G.L. and Peloquin, S.J. (1988) 2n pollen in eleven 2x, 2 EBN wild species and their haploid × wild species hybrids. *Potato Research* 31, 581–589.

Zimmoch-Guzowska, E. (1986) Breeding of diploid potatoes and associated research in the Institute for Potato Research in Poland. In: Beekman G.B., Louwes, K.M., Dellaert, L.M.W. and Neele, A.E.F. (eds) *Potato Research of Tomorrow*. PUDOC, Wageningen, The Netherlands, pp. 115–119.

7 Production of Monohaploids of *Solanum tuberosum* L. and Their Use in Genetics, Molecular Biology and Breeding

E. JACOBSEN AND M.S. RAMANNA

The Graduate School of Experimental Plant Sciences and Department of Plant Breeding (IVP), Agricultural University, PO Box 386, 6700 AJ Wageningen, The Netherlands.

Introduction

The cultivated potato, *Solanum tuberosum*, is a tetraploid ($2n = 4x = 48$) with tetrasomic inheritance for many, if not all, of its characters. According to Hawkes (1956), *S. tuberosum* is derived from the doubling of the chromosome number of the F_1 hybrid between two closely related diploid species ($2n = 2x = 24$), *S. stenotomum* and *S. sparsipilum*, of which the former is cultivated and the latter is a wild weedy species. Hawkes (1990 and chapter 1, this volume) does, however, acknowledge that some authorities believe that *S. tuberosum* is a straight autotetraploid of *S. stenotomum*. The situation has been further complicated by modern-day breeders, who have introgressed variation from a wide range of other species. Besides being a polyploid of probable interspecific origin, the potato is effectively an outbreeder, which has accumulated a large number of deleterious recessive genes as well as genes for male sterility. Because of these drawbacks the genetics of the potato is rather complicated and relatively less well advanced than that of some of the other major economic crops.

In order to facilitate a critical genetic analysis as well as to simplify breeding procedures, it is theoretically attractive to use dihaploids ($2n = 2x = 24$) derived from tetraploid cultivars. Dihaploids are advantageous because of their expected disomic inheritance, which makes Mendelian genetic analysis more straightforward. Although such dihaploids can be produced on a large scale from potato cultivars (Hermsen and Verdenius, 1973), there are also problems in using dihaploids for genetic analyses. Except in very rare instances, almost all dihaploids are male-sterile, and those rare cases of male-fertile ones are mostly self-incompatible. In addition, dihaploids generally suffer inbreeding depression and segregate for deleterious recessive genes (Hermsen *et al.*, 1978), problems which obviously hamper genetic analysis using dihaploids as such.

One method of producing dihaploids, or diploid *S. tuberosum*, suitable for critical genetic analysis is through the creation of diploid potatoes which are completely homozygous, vigorous, fertile (both male and female), self-compatible and devoid of deleterious recessive genes. For this purpose, the use of monohaploids of potato ($2n = x = 12$) could offer a solution for the following reasons. Doubling of chromosome numbers of monohaploids will give rise to diploids which are 100% homozygous. Since monohaploids are gametic samples, a rigid selection for vigour can produce highly vigorous diploids. Because of the hemizygous condition of the monohaploids, any lethal and semi-lethal genes are selected out. And, finally, fertile and self-compatible doubled monohaploids have been shown to occur in *S. chacoense* (Cappadocia *et al.*, 1986), a self-incompatible species, and can be anticipated in *S. tuberosum*.

In view of the attractiveness of the monohaploids of potato for genetic studies as well as for breeding, some aspects of the production and utilization of potato monohaploids are briefly reviewed in this chapter.

Production of Monohaploids

Monohaploids of angiosperms, as a rule, originate from true diploids. Unlike these, however, the monohaploids of potato derive from a natural tetraploid, in which two successive cycles of chromosome number reductions are involved. In the first place, a tetraploid is reduced to a dihaploid and, in the following step, a dihaploid is reduced to a monohaploid. Both of these steps of reducing chromosome numbers can be achieved either through haploid parthenogenesis or through androgenesis. Since success has been achieved through both methods, they will be considered separately below.

Parthenogenetic method: dihaploids and monohaploids from tetraploids

The early work of Hougas and Peloquin (1957) demonstrated the feasibility of obtaining potato dihaploids, and an efficient method of obtaining large numbers of dihaploids was reported by Hermsen and Verdenius (1973). In this method, advantage is taken of a dominant genetic marker 'embryo spot' from the diploid species *S. phureja*, in which clones with homozygosity for 'embryo spot' genes have been selected. When such clones are used as pollinators, the F_1 seeds are expected to possess embryo spots. The rare 'spotless' seeds are mostly the products of haploid parthenogenesis so that they can be easily selected for obtaining dihaploids from $4x$ cultivars. Through this method a large number of dihaploids have been obtained in numerous cultivars, several polyploid wild species and interspecific hybrids (Irikura, 1975a, b, 1976). It is clear from these investigations that the frequency of dihaploids is determined by the genotype of the pistillate parent as well as the *S. phureja* pollinator.

Table 7.1. Production of monohaploids from diploid S. *tuberosum* and diploid *Solanum* species through the parthenogenetic method.

Species, hybrid code	No. of monohaploids per 100 berries	No. of monohaploids per 1000 seeds	Total no. of monohaploids	Reference
S. *tuberosum*, G609	16.0	0.49	64	Breukelen *et al.*, 1977
S. *tuberosum*, M9 (1982)	18.3	0.66	76	Uijtewaal *et al.*, 1987
S. *tuberosum* (pooled results)[1]	10.5	0.43	376	Uijtewaal *et al.*, 1987
S. *tuberosum*, 880004-9	50.0	3.15	5	Jacobsen *et al.*, 1991
S. *verrucosum*, A3	0.4	0.14	1	Breukelen *et al.*, 1977
S. *verrucosum* WAC3338	1.4	1.16	2	Breukelen *et al.*, 1977

[1] From 3 years (1982-1985), using six different genotypes.

The production of monohaploids through the parthenogenetic method from potato dihaploids was first reported by Breukelen *et al.* (1975). Essentially, this method is identical to that described previously for the production of dihaploids, except that a dihaploid is used as a pistillate parent for obtaining monohaploids. Through this method, monohaploids have been successfully produced from dihaploids of different genotypes, as well as from wild diploid species (Table 7.1). From the table it is evident that the frequencies of monohaploids obtained per 100 berries vary considerably. As in the case of dihaploid production, the frequencies of monohaploids depend on the genotype of the pistillate parent as well as the S. *phureja* pollinator (Uijtewaal *et al.*, 1987).

Anther culture for the production of di- and monohaploids

Attempts have been made to obtain dihaploids from cultivars through anther culture (Dunwell and Sunderland, 1973; Irikura, 1975b, 1976). Irikura (1975b) defined media for culturing anthers from 117 clones, including 41 tuberous *Solanum* species and several interspecific hybrids. Out of these, he successfully obtained plantlets from anthers of 19 species and four interspecific hybrids.

The monohaploid frequencies in some of the diploid S. *tuberosum*, wild species and interspecific hybrids reported in the literature are summarized in Table 7.2. It is evident that, although the frequencies of monohaploids vary between genotypes, they can nevertheless be produced in large numbers. Jacobsen and Sopory (1978) showed that genotypes yielding high frequencies of monohaploids could be selected in the offspring of crosses between responding parents. This indicated that genetic factors were involved in determining the frequencies of anther-derived monohaploids. Similar results have also been reported by Uhrig (1983, 1985).

Because di- and monohaploids can be produced both by parthenogenetic and androgenetic methods, the question arises as to which of these methods is more desirable. From the point of view of recovering a large number, and

Table 7.2. Some reported cases of successful anther culture in diploid *S. tuberosum*, diploid primitive cultivars and interspecific hybrids (all $2n = 2x = 24$).

Species, hybrids and code	No. of anthers plated (response)	No. of plants regenerated (per 100 anthers)	Percentage of monohaploids in the regenerated plants	Reference
S. phureja	100 (15)	8 (8.00)	50.00	Irikura, 1975a,b
S. stenotomum	150 (21)	4 (2.67)	25.00	Irikura, 1975a,b
S. verrucosum	360 (34)	121 (33.6)	68.59	Irikura, 1975a,b
S. bulbocastanum	120 (33)	8 (6.66)	100.00	Irikura, 1975a,b
S. tuberosum				
(H7801/10)	2452 (1620)	517 (21.08)	12.19	Uhrig and Wenzel, 1981
(H7801/27)	3765 (2880)	2564 (68.10)	15.63	Uhrig and Wenzel, 1981
S. tuberosum x S. chacoense				
(IP354 x IP 33)	18,258 (921)	303 (1.66)	16.37	Cappadocia et al., 1984
Complex hybrid Ip 56	4351 (106)	4 (0.09)	25.00	Cappadocia et al., 1984
S chacoense (IP33)	2645 (288)	197 (7.45)	74.61	Cappadocia et al., 1984
S. phureja	1416 (363)	125 (8.83)	23.20	Veilleux, 1990 Veilleux et al., 1985

thus numerous genotypes of monohaploids, anther or microspore culture is better, as the number of microspores far exceeds the number of ovules in an ovary.

Uses of Monohaploids in Potato Genetics

The question of basic chromosome number

Although *S. tuberosum* has been considered to be a tetraploid, there are divergent opinions regarding its basic chromosome number. The generally accepted basic number is $x = 12$, as is the case for most of the solanaceous species, but for various reasons $x = 6$ has also been considered (Grun, 1990; see chapter by Wilkinson). Meiotic pairing behaviour in monohaploids can unequivocally settle this question. One critical study of meiotic pairing behaviour of chromosomes in monohaploids (Breukelen *et al.*, 1975) revealed a predominant formation of univalents (Figure 7.1), which was similar to that of a monohaploid of a true diploid such as *Lycopersicon esculentum*.

Apart from the criterion of chromosome pairing, a recent comparison of the RFLP maps of the genomes of tomato and potato (Bonierbale *et al.*, 1988)

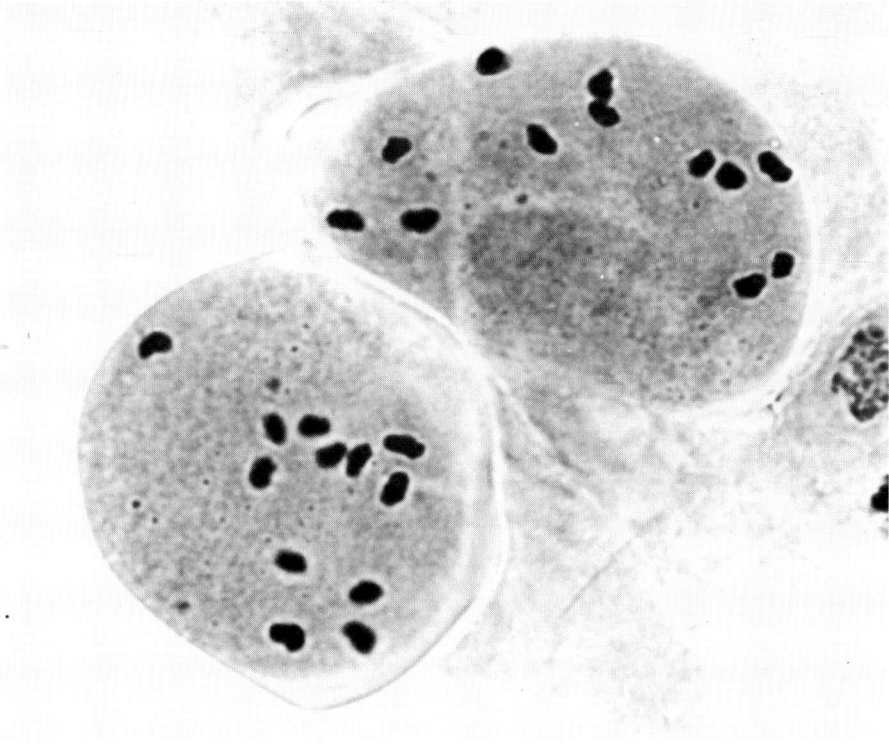

Fig. 7.1. Microsporocytes of a monohaploid of S. tuberosum showing 12 univalents at metaphase I stage. This is evidence for $x = 12$ as the basic number of chromosomes in potato.

has indicated an almost identical order of DNA markers on the linkage maps of these two genera. As each of the 12 different chromosomes has a unique order of markers, it can be concluded that the basic number conforms to the earlier suggestion of $x = 12$.

Mutation studies using monohaploids

For two reasons monohaploids are attractive for mutation induction and selection. Firstly, there have been suggestions that monohaploids are expected to be more mutable (Davies, 1980). This has been related, among other things, to the differences in DNA repair processes between mono- and dihaploids. Secondly, because of the hemizygous condition of the genome, recessive mutations readily express their phenotypes in monohaploids. This facilitates easy and accurate selection of mutations, induced either through mutagens or through transposons, if a detection system is available. Recent successful research on altering the starch composition of potato represents an ideal example of mutation breeding using monohaploids.

Strategies for altering potato starch through mutation and selection

Amylose-free starch

Monohaploids are especially useful for obtaining recessive (or deletion) mutations that were hitherto undetected or undescribed in this plant species. A good example in this context is the amylose-free (*amf*) mutant of potato, which was induced, selected and characterized through the use of a monohaploid. This *amf* mutation is comparable to the so-called *waxy* mutations in other plants such as maize, rice, barley and millets (Kanzaki and Noda, 1988). In *waxy* mutants, the endosperm and pollen grains lack amylose in their starch because of the deficiency or absence of an enzyme, granule-bound starch synthase (GBSS). Such mutations can be detected simply by staining starch, either in the endosperm or in pollen grains, with iodine–potassium iodide solution. The starch stains red when only amylopectin is present, whereas it stains blue when amylose or a mixture of amylose and amylopectin are present. Through this simple test, a procedure was developed for the detection and isolation of an amylose-free potato mutant (Hovenkamp-Hermelink *et al.*, 1987). The salient features of the mutation induction were as follows. A monohaploid (AM79.7322) that could flower and set seed in a low frequency after chromosome doubling (Uijtewaal *et al.*, 1987) was used. It also had a high ability for adventitious shoot induction and microtuber formation on leaf and stem explants, respectively (Hovenkamp-Hermelink *et al.*, 1988). An indirect selection marker (chlorophyll deficiency) was used for the optimalization of mutation induction through X-ray irradiation. The cut surface of the microtubers was highly convenient for the detection of mutants through staining with iodine vapour. After irradiating leaf explants of the wild type (AM79.7322), 5000 adventitious shoots were regenerated, from which 12,000 microtubers were induced. On testing these, one microtuber gave rise to the mutant AM86.040 which was still a monohaploid ($2n = x = 12$). The chromosome number of this mutant was doubled mitotically, it was sexually hybridized (1400 pollinations) with wild-type diploids and the embryos obtained were rescued from the too-early-dropping berries. This gave rise to several fertile F_1 hybrids which were *Amf/amf* (Jacobsen *et al.*, 1989). A remarkable feature of the potato *amf* mutant is that, unlike the *waxy* mutant of maize, amylose-free starch can be detected in cells of all types of tissues containing starch – for example, tuber cells, microspores, leaf and guard cells and collumella cells of root tips.

The possibilities for detecting the starch phenotype in various tissues have obvious advantages for breeding amylose-free starch varieties of potato, because selection for mutant phenotypes can be made from the seedling stages to the adult plants. At present, fertile amylose-free diploids and tetraploids are being used in breeding programmes. The various steps required in the manipulation of ploidy levels are indicated in Figure 7.2, and the relevance of $2n$ gametes in such a procedure has been discussed by Jacobsen *et al.* (1991). About 30% of $2n$ microspores of *Amf/amf* diploids were amylose-

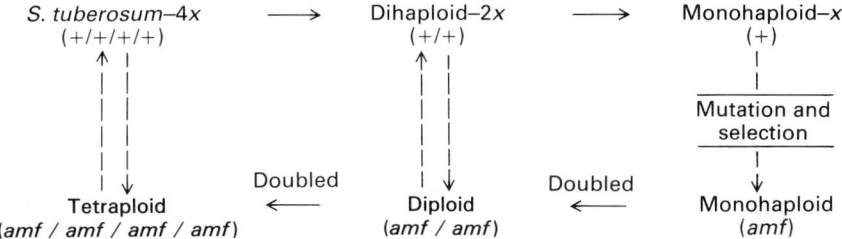

Fig. 7.2. Steps involved in producing a nulliplex amylose-free tetraploid (*amf/amf/amf/amf*) starting from a wildtype *S. tuberosum* (+/+/+/+). Broken arrows indicate that breeding and selection can be carried out either at the diploid or tetraploid level.

free, enabling the direct use of *amf/amf* and *Amf/amf* diploids in $4x$ (amf^4) × $2x$ crosses for *amf* offspring at the tetraploid level. The *Amf/amf* genotype is also helpful for detecting FDR and SDR.

Debranched amylopectin

Another method of altering potato starch composition is through the modification of the branching enzyme that participates in the synthesis of amylopectin. This branching enzyme can be modified either through classical mutation induction methods or through transposon mutagenesis.

CLASSICAL MUTATION INDUCTION

In maize, for example, mutations are known in which one of the branching enzymes is altered. The *amylose extender* (*ae*) is one such mutant, having a deficiency of the branching enzyme IIb and giving rise to decreased branching of amylopectin. In the recombinant *aewx* (*amylose extender* and *waxy* are both recessive), only loosely branched amylopectin is synthesized. Because of the loose branching, amylopectin stains blue instead of red with iodine (Shannon and Garwood, 1984).

If the above principles were to operate in potato, it should also be possible to detect *aewx*-like mutants by using monohaploid mutants such as AM86.040 as a source material. A reversion to blue-staining starch in an *amf* mutant is a potential source of *aewx*-like mutations. After irradiating axillary buds, microtubers were induced on the stem explants of cultured shoots. On the cut surface of these microtubers, blue-staining sectors were observed after treatment with iodine vapour. When such blue sectors were subcultured, several solid blue-staining variants were obtained (Jacobsen *et al.*, 1990). These variants, however, were not *aewx*-type mutants, because they contained amylose and GBSS activity which was similar to that of the wild-type monohaploid AM79.7322. Molecular comparison of the GBSS gene of these revertants with that of the original wild-type monohaploid AM79.7322 showed no differences (Leij *et al.*, 1991b). These observations suggest that the originally selected mutant AM86.040 was probably chimeric for the mutated *amf* character and therefore was not suitable for the induction of *aewx*-like mutants. Recently we have produced monohaploids by the parthenogenetic method from *Amf/amf* genotypes (Jacobsen et al., 1991) and some of them are expected to be solid *amf* mutants (i.e. non-chimeric). These monohaploids, however, are unstable *in vitro* with respect to their ploidy level. Therefore attempts are under way to obtain solid *amf* monohaploids whose ploidy level remains stable *in vitro*.

TRANSPOSON MUTAGENESIS

Besides physical and chemical mutagens, there is another novel method of inducing mutations in plants. This is the so-called transposon mutagenesis. Transposons, or the 'controlling elements' of McClintock (1950), are movable genetic elements, which have been shown to induce mutations in maize through their physical insertion into genes, e.g. the *waxy* gene of maize. Such elements not only induce mutations through insertion but also cause reversions to wild-type through their excisions from mutant genes. Some of these elements can insert (transpose) and excise autonomously, for example, *Activator* or *Ac* of maize, whilst others are non-autonomous and can transpose or excise only in the presence of another active element in the same cell, for example, *Dissociation* or *Ds* of maize, which is active only in the presence of *Ac*. Such elements have been detected in numerous plant species and they have also been found to be mutagenic, as in maize. During the last decade some of the plant transposons have been molecularly isolated (cloned) and characterized (Shure et al., 1983; Nelson, 1988). Remarkably, on transferring such elements from one species to a totally alien species (from maize to potato, for example) through genetic transformation (Knapp et al., 1988), they have been found to transpose in the same way.

In potato, maize transposons can be active when introduced through genetic transformation, using the *Agrobacterium* system (Pereira et al., 1991, 1992). In view of this, it should be possible to develop an alternative approach for mutating the branching enzyme through transposon targeting, using monohaploids of potato. Detection of *aewx*-like mutations should be easy in an *amf* monohaploid, if a large number of monohaploid transformants could be obtained, or in pollen of an *amf* diploid, because blue-staining starch granules can be readily recognized in otherwise red-staining monohaploid cells. A scheme for the induction and selection of branching-enzyme-deficient mutants through transposon targeting, using an *amf* diploid, is presented in Figure 7.3.

This scheme consists of: (i) creating a library of diploid transformants with *Ac*-containing DNA; (ii) isolating and amplifying DNA probes flanking the tDNA (transfer DNA) with IPCR (inverse polymerase chain reaction); (iii) localizing these probes through RFLP mapping and selecting for clones with linkage to the branching-enzyme gene; (iv) screening of the transformants with close linkage between tDNA and branching-enzyme gene for instability, as evidenced by the presence of blue-staining starch in microspores stained with iodine.

A close linkage between *Ac* and the branching-enzyme gene is essential, because it is evident in maize that *Ac* transposes preferentially within its close

(i) Creation of a library of transformants of *amf-* diploid with *Ac* insertions
↓
(ii) Selection of transformants showing blue-staining microspores (potential insertions into BE gene)
↓
(iii) Induction of monohaploids through androgenesis or parthenogenesis
↓
(iv) Selection of BE-deficient mutant

either directly as a mutated gamete   or indirectly from a mutated somatic cell or a sector

Fig. 7.3. A scheme for the induction and selection of a branching enzyme (BE) deficient mutant through transposon mutagenesis in a diploid *amf*-potato clone.

vicinity on the same chromosome. The frequency of insertion of *Ac* into a closely linked target gene in maize has been found to be 10^{-3} to 10^{-4} (Döring, 1989), but much less when such a linkage is absent. It is important to note that, in an *amf* mutant, the microspores spontaneously revert to blue-staining ones at a rate of $\leq 10^{-5}$ (Jacobsen *et al.*, 1989). Therefore, it is the diploid transformants, which show an increased level of instability for this trait, as observed from microspores, that have to be used for the isolation of monohaploids (see Figure 7.3). The *aewx*-like mutant can be obtained directly, out of a mutant gamete (i.e. stable monohaploid), or indirectly, out of a monohaploid which is unstable for this trait in regenerated adventitious shoots. The strategy of transposon-induced mutation in monohaploids is useful only when recessive (deletion) mutations are required, the mutant phenotype is easily recognizable and no alternative method exists. In Figure 7.4 it is shown that the *amf* mutant can be used in combination with the isolated genomic GBSS gene for excision analysis of *Ac* or *Ds*.

The Use of Monohaploids in Molecular Biology

Analysis of multigene families

The fact that the cultivated potato is a tetraploid is a complicating factor, not only for its genetics and breeding, but also for its molecular analysis. This is well illustrated in the case of analyses of some of the 'multigene families'. Some of the proteins in potato are encoded by several genes instead of the usual one or two. For example, patatin (a tuber protein), heat-shock proteins and proteinase inhibitor are encoded by multigene families. About 10–15 genes are involved in the synthesis of patatin (Willmitzer *et al.*, 1990). In all patatin genes that have been sequenced so far, the protein and promoter coding regions up to position −87 are highly homologous. However, sequence divergence has been found upstream of position −87 and forms the basis for

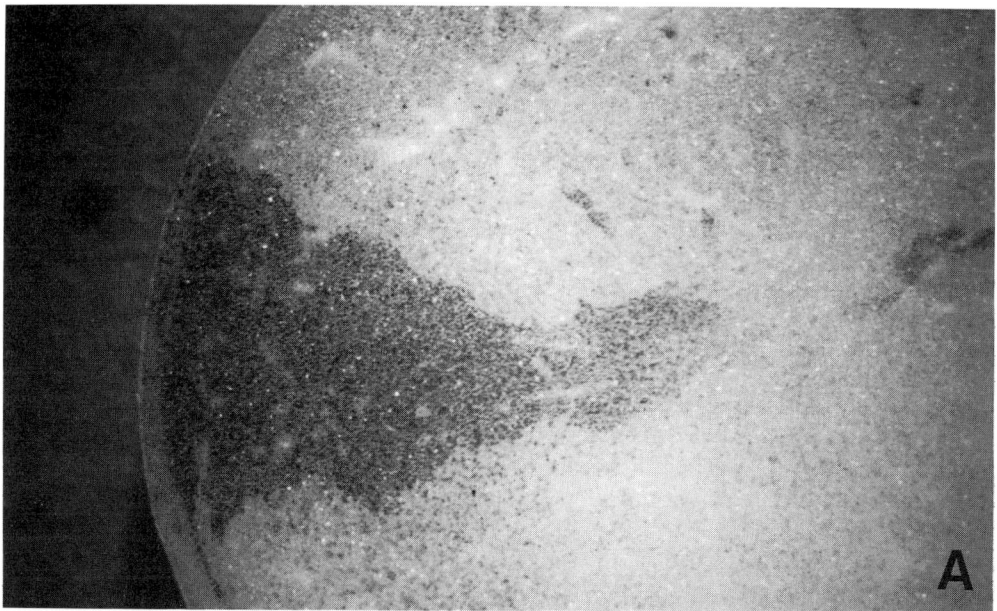

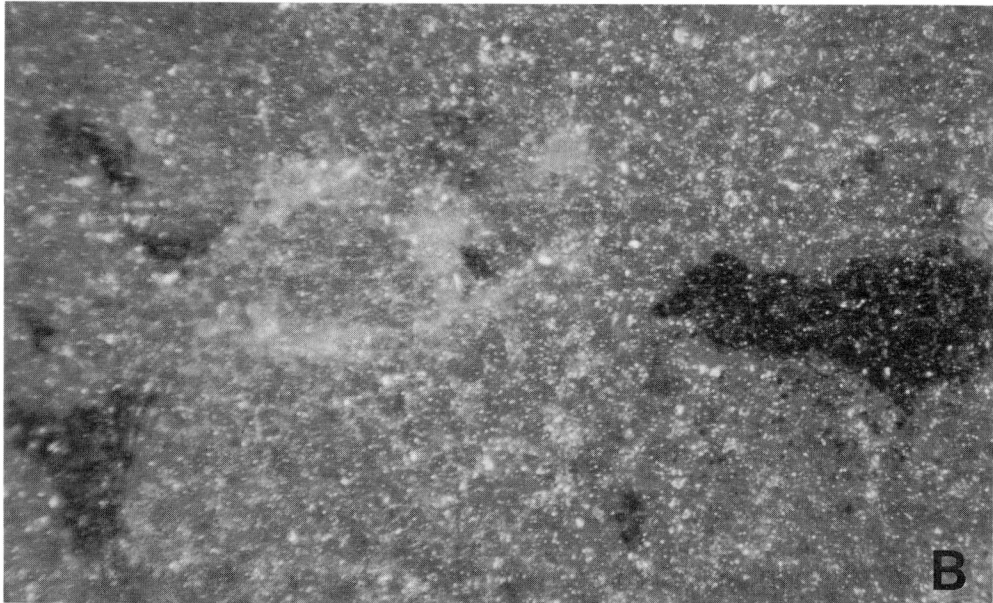

Fig. 7.4. A-B Tuber slices of *amf* potato, stained with iodine vapour, showing dark sectors. In a molecularly isolated genomic clone of GBSS gene an *Ac* transposon was introduced, interrupting the expression of the former, and was transferred through *A. tumefaciens* into *amf* potato. Excision of *Ac* restores the function of GBSS and causes complementation of *amf*. A. Early excision event in Tm15-26 causing complementation, represented by a large dark sector. B. Early and late excision events in Tm15-24 causing large and small dark sectors.

the subdivision of patatin into two classes, viz. 1 and 2, which differ in their pattern of expression. If the genomic DNA of only one allele per locus of these two classes has to be isolated, monohaploids are advantageous because only a single allele per locus will be present. A detailed molecular analysis of such isolated genomic sequences facilitates the selection and characterization of various active and silent member genes. Genomic patatin genes have been molecularly isolated by Mignery et al. (1984), Twell and Ooms (1987) and Bevan et al. (1986), using cvs Superior, Desiree and Maris Piper, respectively. Only Rosahl et al. (1986) used a genomic library of an anther-derived monohaploid, AM80.5793, for cloning the sequences of patatin genes. This material, however, was not selected with the purpose of isolating all possible members of the patatin gene families.

Comparison of a mutated allele with its wild-type version

A mutation induced and selected at the monohaploid level is advantageous when the molecular analysis of both the mutated as well as the original parental wild-type allele of a locus is to be carried out. This has been successfully demonstrated by investigating the genomic sequences of the GBSS gene of the *amf* mutant AM86.040 and its original wild-type AM79.7322, mentioned earlier (Hovenkamp-Hermelink et al., 1987). The isolated wild-type allele complemented the *amf* mutant after transformation (Leij et al., 1991a). The mutation event was approximately localized within the GBSS gene by a transformation test for complementation of the mutant with different chimeric/mutant gene constructs. On DNA sequencing of the relevant part of the GBSS gene, it was shown that a single base-pair deletion in the part coding for the transit peptide of the preprotein was involved in the mutation (Leij et al., 1991b). The strategy adopted in cloning and comparing the sequences of the mutated and the original wild-type alleles prevented potential problems arising from any small intra-allelic variation that might have naturally existed between different wild-type alleles.

Use of monohaploids in molecular mapping

Because of the limitations of using diploid *S. tuberosum* for genetic analysis, a classical linkage map of the potato genome is not available. Nevertheless, molecular maps of all the 12 possible linkage groups have been constructed recently through the use of restriction fragment length polymorphisms (RFLPs) (Bonierbale et al., 1988; Gebhardt et al., 1989, 1991). One difficulty of using diploid potatoes for this purpose is that they are highly heterozygous and heterogeneous, because of the presence of gametophytic self-incompatibility. This makes the classification of the possible polymorphisms in potato more complex than those of F_2 populations in a self-fertilizer like the tomato, as shown in Figure 7.5. In the tomato, for example, starting from the progeny

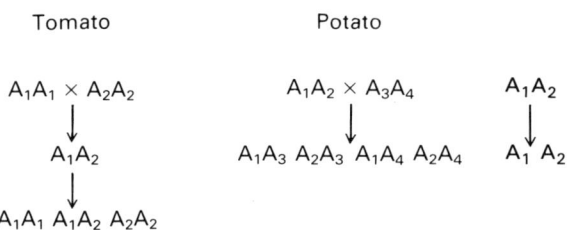

Fig. 7.5. Comparison of allelic segregations in tomato, starting with 2 homozygous parental lines, and in potato starting with 2 different heterozygous parental genotypes or with 1 heterozygous genotype used for monohaploid production.

of a cross between two homozygous parental genotypes, one should expect segregation for two alleles per locus in an F_2 population. In the diploid potato, on the other hand, two to four different alleles can segregate per locus and thus complicate the classification of the progeny. In order to overcome such difficulties, monohaploids (doubled) are attractive.

A start has been made with monohaploids of *S. chacoense* derived from anther culture (Rivard *et al.*, 1989). However, for molecular mapping, recombinant inbred (RI) lines have been found to be highly advantageous in several organisms (Burr and Burr, 1991), but are not yet available in potato. One method of producing such RI lines would be to produce monohaploids from the hybrids derived from crossing two homozygous diploid genotypes. On doubling the chromosome numbers of a fairly large number of such monohaploids (about 75–100), a population suitable for molecular mapping can be set up. In this connection, fertile and vigorous doubled monohaploids will be of great value as the initial diploid parents from which to produce F_1 hybrids.

The Use of Monohaploids in Breeding

The direct use of monohaploids, or their doubled forms, in practical breeding may prove difficult. By producing the monohaploids – or gametic samples – of a highly selected tetraploid cultivar, the latter can potentially be splintered into millions of genotypes, and it will be a formidable task to select suitable parents from such a wide array of genotypes and to combine them into a superior cultivar. But this is true of any breeding programme, and the use of doubled monohaploids could increase the efficiency of long-term recurrent selection programmes.

Nevertheless, monohaploids can be indirectly useful in breeding, as shown by the amylose-free mutant. It has been demonstrated that breeding lines can be produced through anther culture, with increased levels of resistances to diseases such as *Phytophthora infestans* (Behnke, 1979, 1980), *Globodera pallida*, X, Y and leaf roll viruses (Uhrig and Wenzel, 1981; Wenzel and Uhrig, 1981). Suggestions have also been made for combining several types of resistances through protoplast fusion of selected monohaploids (Uijtewaal, 1987). Also, knowledge from investigating quantitative inheritance could be of value in breeding. Finally, when the so-called quantitative trait loci (QTLs) are mapped in relation to RFLPs, as was pointed out earlier, marker-dependent selection for quantitative characters may become a reality in potato.

References

Behnke, M. (1979) Selection of potato callus for resistance to culture filtrates of *Phytophthora infestans* and regeneration of resistant plants. *Theoretical and Applied Genetics* 55, 69-71.

Behnke, M. (1980) General resistance to late blight of *Solanum tuberosum* plants generated from callus resistant to culture filtrates of *Phytophthora infestans*. *Theoretical and Applied Genetics* 56, 151-152.

Bevan, M., Baker, R., Goldsbrough, A., Jarvis, M., Kavanagh, T. and Iturriaga, G. (1986) The structure and transcription start site of a major potato tuber protein gene. *Nucleic Acid Research* 14, 4625-4638.

Bonierbale, M.W., Plaisted, R.L. and Tanksley, S.D. (1988) RFLP maps based on a common set of clones reveal modes of chromosomal evolution in potato and tomato. *Genetics* 120, 1095-1103.

Breukelen, E.W.M. van, Ramanna, M.S. and Hermsen, J.G.Th. (1975) Monohaploids ($n = x = 12$) from autotetraploid *Solanum tuberosum*, ($2n = 4x = 48$) through two successive cycles of female parthenogenesis. *Euphytica* 24, 567-574.

Breukelen, E.W.M. van, Ramanna, M.S. and Hermsen, J.G.Th. (1977) Partheno-genetic monohaploids ($2n = x = 12$) from *Solanum tuberosum* L. and *S. verrucosum* Sclechtd and the production of homozygous potato diploids. *Euphytica* 26, 263-271.

Burr, B. and Burr, F. (1991) Recombinant inbreds for molecular mapping in maize: theoretical and practical considerations. *Trends in Genetics* 7, 55-60.

Cappadocia, M., Cheng, D.S.K. and Ludlum, R. (1984) Plant regeneration from *in vitro* cultures of anthers of *Solanum chacoense* Bitt. and interspecific diploid hybrids *S. tubersoum* × *S. chacoense* Bitt. *Theoretical and Applied Genetics* 69, 139-143.

Cappadocia, M., Cheng, D.S.K. and Ludlum-Simonette, R. (1986) Self-compatibility in doubled haploids and their F_1 hybrids, regenerated via anther culture in self-incompatible *Solanum chacoense* Bitt. *Theoretical and Applied Genetics* 72, 66-69.

Davies, D.R. (1980) Some of the implications of haploidy for the angiosperms. In: Davies, D.R. and Hopwood, D.A. (eds) *The Plant Genomes: Fourth John Innes Symposium and the Second International Haploid Conference, Norwich, UK*, pp. 161-169.

Döring, H.P. (1989) Tagging genes with maize transposable elements: an overview. *Maydica* 34, 73-88.

Dunwell, J.M. and Sunderland, N. (1973) Anther culture of *Solanum tuberosum* L. *Euphytica* 22, 317-323.

Gebhardt, C., Ritter, E., Debener, T., Schachtschabel, U., Walkemier, B., Uhrig, H. and Salamini, F. (1989) RFLP analysis and linkage mapping in *Solanum tuberosum*. *Theoretical and Applied Genetics* 78, 65-75.

Gebhardt, C., Ritter, E., Barone, A., Debener, T., Walkemeier, B., Schachtschabel, U., Kaufmann, H., Thompson, R.D., Bonierbale, M.W., Ganal, M.W., Tanksley, S.D. and Salamini, F. (1991) RFLP maps of potato and their alignment with the homoeologous tomato genome: *Theoretical and Applied Genetics* 83, 49–57.

Grun, P. (1990) The evolution of cultivated potatoes. *Economic Botany* 44, (Suppl. 3), 39–55.

Hawkes, J.G. (1956) Taxonomic studies on tuber-bearing Solanums. I. *Solanum tuberosum* and the tetraploid species complex. *Proceedings of the Linnean Society* 166, 97–144.

Hawkes, J.G. (1990) *The Potato: Evolution, Biodiversity and Genetic Resources*: Belhaven Press, London.

Hermsen, J.G.Th. and Verdenius, J. (1973) Selection from *Solanum tuberosum* group Phureja of genotypes combining high-frequency haploid induction with homozygosity for embryo-spot. *Euphytica* 22, 244–259.

Hermsen, J.G.Th., Taylor, L.M., Breukelen, E.W.M. van and Lipsky, A. (1978) Inheritance of genetic markers from two potato dihaploids and their respective parent cultivars. *Euphytica* 27, 681–687.

Hougas, R.W. and Peloquin, S.J. (1957) A haploid plant of the potato variety Katahdin. *Nature* 180, 1209–1210.

Hovenkamp-Hermelink, J.H.M., Jacobsen, E., Ponstein, A.S., Visser, R.G.F., Vos-Scheperkeuter, G.H., Bijmolt, E.W., Vries, J.N. de, Witholt, B. and Feenstra, W.J. (1987) Isolation of an amylose-free starch mutant of potato (*Solanum tuberosum* L.). *Theoretical and Applied Genetics* 75, 217–227.

Hovenkamp-Hermelink, J.H.M., Jacobsen, E., Pijnacker, L.P., Vries, J.N. de, Witholt, B. and Feenstra, W.J. (1988) Cytological studies on adventitious shoots and microtubers of a monohaploid potato clone. *Euphytica* 39, 213–219.

Irikura, Y. (1975a) Induction of haploid plants by anther culture in tuber-bearing species interspecific hybrids of *Solanum*. *Potato Research* 18, 133–140.

Irikura, Y. (1975b) Cytogenetic studies on haploid plants of tuber-bearing *Solanum* species. I. Induction of haploid plants of tuber-bearing Solanums. *Research Bulletin of Hokkaido National Agricultural Experimental Station* 112, 1–67.

Irikura, Y. (1976) Cytogenetic studies on the haploid plants of tuber-bearing *Solanum* species. II. Cytogenetical investigations on haploid plants and interspecific hybrids by utilizing haploidy. *Research Bulletin of Hokkaido Agricultural Experimental Station* 115, 1–80.

Jacobsen, E. and Sopory, S.K. (1978) The influence and possible recombination of genotypes on the production of microspore embryoids in anther culture of *Solanum tuberosum* and dihaploid hybrids. *Theoretical and Applied Genetics* 52, 119–123.

Jacobsen, E., Hovenkamp-Hermelink, J.H.M., Krijgsheld, H.T., Nijdam, H., Pijnacker, L.P., Witholt, B. and Feenstra, W.J. (1989) Phenotypic and genotypic characterization of an amylose-free starch mutant of potato. *Euphytica* 44, 43–48.

Jacobsen, E., Krijgsheld, H.T., Hovenkamp-Hermelink, J.H.M., Ponstein, A.S., Witholt, B. and Feenstra, W.J. (1990) Revertants of the amylose-free potato clone 86.040 ($2n = 1x = 12$). *Plant Science* 67, 177–182.

Jacobsen, E., Ramanna, M.S., Huigen, D.J. and Sawor, Z. (1991) Introduction of an amylose-free (*amf*) mutant into breeding of cultivated potato, *Solanum tuberosum* L. *Euphytica* 53, 247–253.

Kanzaki, K. and Noda, K. (1988) Glutinous (*Waxy*) endosperm starch mutant of *Triticum monococcum* L. *Japanese Journal of Breeding* 38, 423–427.

Knapp, S., Coupland, G., Uhrig, H., Salamini, F. and Starlinger, P. (1988) Transposition of the maize transposable element *Ac* in *Solanum tuberosum*. *Molecular and General Genetics* 213, 285–290.

Leij, F.R. van der, Visser, R.G.F., Oosterhaven, V., Kop, D.A.M. van der, Jacobsen, E. and Feenstra, W.J. (1991a) Complementation of the amylose-free starch mutant of potato (*Solanum tuberosum* L.) by the gene encoding granule bound starch synthase. *Theoretical and Applied Genetics* 82, 289–295.

Leij, F.R. van der, Visser, R.G.F., Ponstein, A.S., Jacobsen, E. and Feenstra, W.J. (1991b) Sequence of the structural gene for granule bound starch synthase of potato (*Solanum tuberosum* L.). *Molecular and General Genetics* 228, 240–248.

McClintock, B. (1950) The origin and behaviour of mutable loci in maize. *Proceedings of the National Academy of Sciences of the USA* 36, 344–355.

Mignery, G.A., Pikaard, C.S., Hannapel, D.J. and Park, W.D. (1984) Isolation and sequence analysis of cDNAs for the major potato tuber protein, patatin. *Nucleic Acid Research* 12, 7987–8001.

Nelson, O. (1988) *Plant Transposable Elements*. Plenum Press, New York.

Pereira, A., Aarts, M., Agtmaal, S. van, Stiekema, W.J. and Jacobsen, E. (1991) *Waxy* variegation in transgenic potato. *Maydica* 36, 323–327.

Pereira, A., Jacobs, J.M.E., Lintel-Hekkert, W., Rutgers, E., Jacobsen, E. and Stiekema, W.J. (1992) Towards the isolation of resistance genes by transposon tagging in potato. *Netherlands Journal of Plant Pathology* 98 (Suppl. 2), 215–221.

Rivard, S.R., Cappadocia, M., Vincent, G., Brisson, N. and Landry, B.S. (1989) Restriction fragment length polymorphism (RFLP) analyses of plants produced by *in vitro* anther culture of *Solanum chacoense* Bitt. *Theoretical and Applied Genetics* 78, 49–56.

Rosahl, S., Schmidt, R., Schell, J. and Willmitzer, L. (1986) Isolation and characterization of a gene from *Solanum tuberosum* encoding patatin, the major storage protein of potato tubers. *Molecular and General Genetics* 203, 214–220.

Shannon, J.C. and Garwood, A.L. (1984) Genetics and physiology of starch development. In: Whistler, R.L., BeMiller, J.N. and Passhall, E.F. (eds) *Starch: Chemistry Technology* 2nd edn. Academic Press, Orlando, pp. 25–86.

Shure, M., Wessler, S. and Fedoroff, F. (1983) Molecular identification and isolation of the *waxy* locus in maize. *Cell* 25, 225–233.

Twell, D. and Ooms, G. (1987) The 5' flanking DNA of a patatin gene directs tuber specific expression of a chimeric gene in potato. *Plant Molecular Biology* 9, 365–375.

Uhrig, H. (1983) Breeding for *Globodera pallida* resistance in potatoes. I. Improvement of androgenic capacity in some resistant dihaploid clones. *Zeitschrift für Pflanzenzüchtung* 91, 211–218.

Uhrig, H. (1985) Genetic selection and liquid medium conditions improve the yield of androgenic plants from diploid potatoes. *Theoretical and Applied Genetics* 71, 455–460.

Uhrig, H. and Wenzel, G. (1981) *Solanum gourlayi* Hawkes as a source of resistance against white cyst nematode *Globodera pallida* Stone. *Zeitschrift für Pflanzenzüchtung* 86, 148–157.

Uijtewaal, B.A. (1987) The production and evaluation of monohaploid potatoes ($2n = x = 12$) for breeding research on cell and plant level. Unpublished PhD thesis, Agricultural University of Wageningen, The Netherlands.

Uijtewaal, B.A., Huigen, D.J. and Hermsen, J.G.Th. (1987) Production of potato monohaploids ($2n = x = 12$) through prickle pollination. *Theoretical and Applied Genetics* 73, 751–758.

Veilleux, R.E. (1990) Anther culture and induction of haploids in a cultivated diploid potato species. In: Bajaj, Y.P.S. (ed.) *Biotechnology in Agriculture and Forestry*, Vol. 12, *Haploids in Crop Improvement. I.* Springer-Verlag, Berlin, Heidelberg.

Veilleux, R.E., Booze-Daniels, J. and Pehu, E. (1985) Anther culture of a $2n$ pollen producing clone of *Solanum phureja* Juz. & Buk. *Canadian Journal of Genetics and Cytology* 27, 559–564.

Wenzel, G. and Uhrig, H. (1981) Breeding for nematode and virus resistance in potato via anther culture. *Theoretical and Applied Genetics* 59, 333–340.

Willmitzer, L., Basner, A., Frommer, W., Höfgen, R., Liu, X.J., Köster, M., Prat, S., Rocha-Sosa, M., Sonnewald, U. and Vancanneyt, G. (1990) Tuber-specific gene expression in transgenic potato plants. In: Lycett, G.W. and Grierson, D. (eds) *Genetic Engineering in Crop Plants*. Butterworths, London, pp. 105–114.

III CELLULAR AND MOLECULAR GENETICS

8 Tissue Culture

G. WENZEL

Federal Centre for Breeding Research on Cultivated Plants, Institute for Resistance Genetics, Graf-Seinsheimstrasse 23, D-85461 Grünbach, Germany.

Introduction

For classical genetic studies the potato presents certain difficulties because of its tetraploid nature and high frequency of self-sterility. Nevertheless, for breeding purposes, its reproduction by clonal propagation provides some advantages. For research into the physiological aspects of cell culture the potato is in an outstanding position as a solanaceous crop, as it can also easily be cloned in cell culture, and, in terms of gene technology, even cloning of its genes is no longer a serious problem. A number of biotechnological methods developed on model plants like *Petunia* or tobacco can be successfully transferred to potato, thus demonstrating the applicability of novel approaches on an important crop plant. However, it is important to be aware that the potato is a major object of biotechnology, more by chance than due to its breeding needs or its economic importance, and these research activities will move to even more important crops as soon as these other crop species can be handled like potato in tissue culture and gene technology.

Shortly after the clonal propagation of ornamentals *in vitro* became economic (see review of Vasil and Vasil, 1980), it was used for potatoes. The use of *in vitro* techniques for clonal mass propagation is the most advanced application in potato tissue culture. German potato breeders, for example, have used tens of thousands of rapidly propagated plants per year since the middle 1980s and will continue to do so. Other tissue culture techniques, such as *in vitro* selection, microspore and protoplast regeneration and somatic fusion, are now on the verge of application, although most have been available since the late 1970s. It has been demonstrated that on a few genotypes these techniques work well in principle. Thus, the basic researcher has moved on and the applied researcher has the problem of making the procedures reproducible, universal and cheap so that practical potato breeders will be prepared to take up the unconventional methods at their own economic risk.

Potato tissue culture may also be used as a source for secondary product formation but this topic will not be treated here since little genetics is involved, and the reader is referred to the review of Heinstein and Emery (1988).

In this chapter an attempt is made to demonstrate where tissue culture can contribute to targeted hybridization programmes. Breeding mainly deals with complex characteristics, the inheritance of which follows the laws of population genetics. Tissue culture techniques offer the opportunity to combine such quantitatively inherited traits more efficiently than by classical methods, which require enormous populations combined with complex selection procedures. Consequently, tissue culture techniques will only be successful if they can help to increase the efficiency of combining several complex traits in a new cultivar.

The production of cultivars is only one part of breeding; of similar importance are the maintenance, propagation and distribution of a particular genotype. Today these rely on tissue culture techniques, which allow rapid propagation and maintenance of living collections of valuable genotypes and cultivars. Both aspects will be treated here.

Use of Tissue Culture in Germplasm Maintenance and Propagation

The oldest tissue culture technique, rapid *in vitro* propagation, has already reached the applied level and proved its economic advantages (Bajaj, 1991). Asexual *in vitro* multiplication can be achieved by enhancing axillary bud development, by the production of adventitious buds, and by organogenesis or somatic embryogenesis, initiated, for example, from protoplast-derived calli. The process of bud growth may either start from stem segments with axillary buds, when it is called rapid propagation, or from tiny apical meristems, when it is called meristem culture. It is important to stress that, during rapid propagation, the genotype of the plant should stay stable; this cannot be guaranteed as soon as callus formation is observed (see chapter 9 by Kumar).

Rapid propagation

In rapid propagation there is an interplay between the genotype, the tissue source, the culture medium and the culture conditions. The easiest sources are shoots, where 'shoots' describes rootless sprouts growing *in vitro* on agar-solidified media. They are initiated from decontaminated botanical seeds or meristems and grown under axenic conditions. Seeds, sprouts or other tissue sources are surface-sterilized for 15–20 min in an aqueous solution of 25% NaOCl, resulting in 2.5% active chlorine, and a few drops of Tween. Sometimes additional surface sterilization with 70% aqueous ethanol precedes or follows. Once an axenic culture is established, it serves as source for further multiplication cycles, making antiseptic treatments no longer necessary. From such aseptic cultures protoplasts can also be isolated.

Stem explants

The easiest procedure of rapid propagation starts from axenic shoot cultures. Shoots, including an axillary bud, are cut into 1 cm pieces. Each bud grows within 3–5 weeks to a size that allows a multiplication step to be repeated. Neither the culture medium (normally the Murashige and Skoog (1962) (MS) medium) nor the culture conditions (about 22–26°C, 3000–4000 lux for 16 h) are very critical. Today most potato breeders use such rapidly propagated plants for basic maintenance, and then transfer them from *in vitro* culture to the greenhouse in early spring. Tubers from such plants are harvested before aphids begin to fly and transfer virus diseases.

As an alternative to stem cuttings, De Block (1988) reported on leaf discs. They are of particular importance in gene transformation experiments but the technique needs a means to prevent unorganized growth. As long as this is guaranteed, rapid propagation will not alter the genotype and always results in uniform clones. At present it is better to accept additional transfer work instead of waiting too long and finding undesired mutations due to slow growth or callus formation.

Protoplasts

A much more rapid propagation of a given genotype is offered by the maceration enzymatically of potato leaves or shootlets into single cells and protoplasts. Such protoplasts are cultured in hypertonic liquid media – for potato preferably in the VK medium (Binding *et al.*, 1978) or in media given by Shahin (1984). In most cases culture starts from shoot cultures that have been rapidly propagated *in vitro*, which provide starting material already preconditioned for *in vitro* handling. A current standard isolation procedure (Haberlach *et al.*, 1985) can be summarized as follows. Axenic shoot cultures (1 g) are cut in 0.44 M mannitol in 1 mm pieces. Mannitol is replaced by 15 ml enzyme solution (2 g l^{-1} Macerocyme, 8 g l^{-1} cellulase suspended in phytohormone-free MS medium with 110 g l^{-1} sucrose, 1 g l^{-1} potassium dextran sulphate, 10 g l^{-1} PVP 15 and 1.25 g l^{-1} bovine serum albumin; pH 5.6, 550–600 mosm kg^{-1}). The incubation takes 14–17 h at room temperature. The protoplasts are collected by sieving and gentle centrifugation and transferred after repeated washings to the VK culture medium.

As in most other crops, the genotype exerts a strong influence on the regeneration frequency of potato protoplasts. For breeding programmes, a much more genotype-independent response is required. This can be achieved in a number of genotypes by addition of substances to the culture medium that inhibit ethylene biosynthesis. Perl *et al.* (1988) reported that the addition of silver thiosulphate increased the protoplast yield. Möllers *et al.* (1992)

found that 1.5 mg l^{-1} silver thiosulphate led to an increase of leaf material harvested per culture vessel. Depending on genotype, the increase ranged from 1.2 to 2.3 times the yield obtained from normal medium. The addition of silver thiosulphate also promoted the shoot regeneration from protoplast-derived calli.

Although numerous protocols exist today for an efficient protoplast regeneration of potato, and from one leaf 500 to 2000 plants can easily be regenerated, this technique is not used routinely for rapid propagation in maintenance breeding. One reason is its rather laborious and sophisticated nature, and the other is that protoclonal variation often appears. Landsmann and Uhrig (1985) showed that two out of 12 protoplast-derived dihaploid potato clones were variant when probed with one of a set of random potato DNA clones. As the polymorphic DNA clone turned out to represent 25S rDNA, the two variants can be regarded as having mutated in ribosomal RNA genes. Thus it is unlikely that a genotype will stay stable when it is propagated via protoplasts, and a crucial prerequisite of rapid propagation is not fulfilled. By shortening the callus phase, this variability may be reduced but not prevented absolutely.

Meristem culture

Rapid propagation techniques can also suffer because the total offspring from a virus-diseased plant will be diseased as well. Meristem culture provides a means of overcoming this problem. Normally, up to 40% of regenerants from meristems are virus-free. By culturing the starting material at 30–35°C, the percentage of healthy plants can be increased. However, the optimal size for the meristem should always be tested first. The larger they are, the easier the procedure, but the greater the likelihood that viruses may be transferred. In a comparative study with different potato viruses, it was demonstrated that for PVX the size is very critical, while for the elimination of PVS even 2 mm tips gave virus-free regenerants (Meyer and Foroughi-Wehr, 1987). These large tips regenerate much more easily and never produce callus, whilst smaller ones tend to proliferate in an unorganized manner. As meristem culture remains a sophisticated technique, applied programmes try to avoid it. Thus, for rapid propagation programmes, normally a healthy mother culture is initially established, from which microcuttings are taken.

Bacteria may also infect such cultures. All available diagnostic techniques require pathogen numbers for certain detection which are higher than the titre of slightly infected cultures. Thus, for monitoring the axenity of *in vitro* cultures, such techniques are only of limited use. Leifert and Waites (1990) identified 293 bacterial strains from two commercial micropropagation laboratories. Under such conditions it is not practical to work with very specific methods such as DNA hybridization, but one can use a rich bacterial growth medium and monitor turbidity.

Mini- and microtubers

As storage and transport of *in vitro* cultures is not always easy, two alternatives are being explored: the use of mini- and microtubers. Minitubers are produced from *in vitro*-propagated plantlets immediately after transfer to soil by slightly pulling the plant out of the soil, whereas microtubers are produced *in vitro* by increased sucrose concentration (8–10%), short day (6–8 h) and no phytohormones (Garner and Blake, 1989). A problem for using microtubers is insufficient dormancy and the relatively low number of microtubers per single-node microcutting. Marinus (1990) concludes that we know too little about tuberization *in vitro*. Apparently these techniques are a special way of potato growing with its own rules, partly deviating from those for plants grown in soil. Probably the problem of planting such small tubers mechanically will be solved earlier than these physiological ones. The system is much more under physiological than under genetical control.

Living collections

Maintenance of genetic resources offers the best guarantee for solving demands of potato genetics in the future. For this purpose, rapid propagation is started from a healthy mother clone, stopped after some time, transferred to slightly modified media and then stored at temperatures between 4°C and 10°C. On MS medium with only 5 g l^{-1} sucrose and 40 g l^{-1} mannitol, a transfer between 6 months and 2 years, depending on the genotype, is sufficient. Bajaj and Sopory (1986) summarized procedures for cryopreservation, allowing much longer maintenance at a temperature of $-196°C$. At present, the technique still has a number of problems, particularly the *in vitro* culture stress, which increases the genetic instability often observed in *in vitro* cultures (Sree Ramulu and Dijkhius, 1986). Thus, cryopreservation is not yet ready for application in potato.

Use of Tissue Culture in Breeding New Cultivars

Breeding of a new cultivar requires access to a gene pool from which to select specific clones, and then hybridization and subsequent selection cycles are basic requirements. As long as it is possible to detect new genetic resources and to collect and evaluate them, conventional potato breeding is not limited. Conventional breeding is, however, very time-consuming. In the heterozygous tetraploid potato the probability on crossing heterozygotes ($AAaa \times AAaa$) of obtaining a clone with n homozygous loci ($AAAA$) is one in 6^{2n}. In combining quantitatively inherited traits, the population size and the time needed for selection are critical. As a result, the conventional approach responds slowly to new demands of the market. Tissue culture offers the following possibilities of increasing the efficiency of the conventional approach:

1. *In vitro* culture is less influenced by the environment, which can make it easier to measure small quantitative differences in polygenically inherited characters.
2. Large numbers of individual genotypes can be handled as protoplasts, embryoids or calli in a very small space.
3. The use of microspores and haploids exposes recessive traits.
4. *In vitro* cells may allow transfer of genetic information either via symmetrical or asymmetrical fusion, via vectors or by direct DNA transfer techniques, and bridge the gap between genetic engineering and plant breeding.

It should be stressed however that, regardless of whether selection or recombination is involved, tissue culture must be accompanied by and combined with classical breeding procedures, and any *in vitro*-derived clone has to prove its quality in the field.

Haploids

Solanum tuberosum is an autotetraploid with $2n = 4x = 48$ chromosomes, but reasonable percentages of dihaploid potatoes ($2n = 2x = 24$) can be produced parthenogenetically (Hougas and Peloquin, 1957) by pollinating it with particular $2x$ *Solanum phureja* clones. This technique also works for the reduction of the ploidy level from $2x$ to $1x$ monohaploids (van Breukelen *et al.*, 1975). These reductions of the ploidy level increase the probability of finding n homozygous loci ($AAAA$, AA or A) from one in 6^{2n} for tetraploids ($AAaa \times AAaa$) via 2^{2n} for dihaploids ($Aa \times Aa$) to 2^n for monohaploids ($A \times a$).

Production of haploids

Production of dihaploids via parthenogenesis can be efficient, with dihaploids produced from up to 46% of fertilized flowers (Wenzel and Foroughi-Wehr, 1984). The production of monohaploids from dihaploids has reached nearly 10% (van Breukelen *et al.*, 1975) but parthenogenetic haploid production is rather laborious, because the triploid-block lethality of the normal hybrids in $4x \times 2x$ crosses does not work. Since the 1970s, androgenetic procedures have been developed (for review see Foroughi-Wehr and Wenzel, 1993). The main emphasis was on monohaploid induction in early work, because here parthenogenesis needed good markers and very specific genotypes to produce limited numbers of monohaploids. When androgenesis worked reproducibly from $2x$ to $1x$, the technique was transferred to the $4x-2x$ reduction (Uhrig and Salamini, 1987; Zitzlsperger and Wenzel, 1990). Since Clulow *et al.* (1991) have recently demonstrated that *S. phureja*-induced dihaploids are not always completely parthenogenetic in origin, the androgenetic procedure might have additional advantages.

ANDROGENETIC MONOHAPLOIDS

The production of androgenetic monohaploids started by culture of microspores within the anther (anther culture) on agar-solidified media has been described by Foroughi-Wehr et al. (1977) and Sopory et al. (1978). In potatoes, culture is performed in liquid media, as described by Uhrig (1985). This is the most efficient procedure and the central steps can be summarized as follows. Anthers from dihaploid clones grafted on tomato rootstocks are kept for 3 days at 6°C before excision; 50 anthers from surface-sterilized buds are placed in 25 ml Erlenmeyer flasks and shaken at 26°C with 120 strokes min^{-1} in half-strength Linsmaier and Skoog (1965) (LS) medium, supplemented with $0.1 \, mg \, l^{-1}$ indolylacetic acid, 0.05% activated charcoal and 6% sucrose. For different genotypes and in other laboratories, some additions to the medium have proved useful, e.g. $2.5 \, mg \, l^{-1}$ benzylaminopurine (BAP; Uhrig, 1985), or $1 \, mg \, l^{-1}$ BAP, $400 \, mg \, l^{-1} \, NH_4NO_3$, $750 \, mg \, l^{-1}$ L-glutamine and $60 \, g \, l^{-1}$ maltose (Möllers et al., 1993). After 2–3 weeks, growth of embryoids outside the anthers can be observed. One week later the embryoids can be harvested and plated on 0.8% agar-solidified regeneration medium (phytohormone-free half-strength LS with 2% sucrose). Table 8.1 gives the frequency of embryoids which can be induced under these culture conditions. Unfortunately, only an average of 5% of these phenotypically perfect embryoids finally form plantlets. The majority of the regenerated plants are dihaploids, with rare monohaploid exceptions. The dihaploids either are a result of spontaneous doubling of monohaploids or arise from unreduced gametes. In the latter case, they are more homozygous if originating from second-division restitution nuclei than from first-division restitution processes, which are under genetical control (Jongedijk et al., 1991). Both origins are possible, as demonstrated by the segregation after selfing or by RFLP analysis (Wenzel and Foroughi-Wehr, 1984; Meyer, 1991). The proportion of completely or partially heterozygous regenerants originating from unreduced microspores in preselected cultures was up to 90% (Meyer, 1991). This rather low rate of pure homozygotes necessitates careful checks for homozygosity in androgenetic clones.

ANDROGENETIC DIHAPLOIDS

Parthenogenetic dihaploid production, though very reliable, is rather labour-intensive, and experiments to achieve the reduction from $4x$ to $2x$ via microspore androgenesis have been reported. Irikura reached this goal in 1975, but the general application of androgenesis remains difficult. Other laboratories have tried to transfer the technique to tetraploids (Wenzel et al., 1982; Mix, 1983; Johansson, 1986), but tetraploids with interesting agronomic performance have shown a disappointingly poor response in anther culture. J. Zitzlsperger (unpublished) regenerated microspores from 29 out of 58 dihaploid clones, but only 7 out of 40 tetraploids.

GENETICS OF REGENERATION CAPACITY

Crucial to increasing regeneration success in microspore culture was the selection of clones with a high capacity to respond in anther culture and to

Table 8.1 Embryoid production and plantlet formation from dihaploid anther donor clones in optimized liquid culture medium (Meyer, 1991).

Genotype	Embryoids per 100 anthers	Plants per 100 embryoids
H 87.2001/4	874	2
H 87.2002/3	656	5
H 87.2002/4	269	10
H 87.2002/5	1264	3
H 87.2002/6	358	23
H 87.2003/5	1459	2
H 87.2004/4	261	17
H 87.2004/5	545	5
H 87.2002/6	1601	1
H 87.2004/7	702	5
H 87.2002/8	623	2
H 87.2004/9	278	11
H 87.2005/2	383	2
H 87.2005/3	424	2
H 87.2005/4	314	6

transfer this ability to progenies by sexual crosses (Jacobsen and Sopory, 1978; Wenzel, 1980a). Wenzel and Uhrig (1981) demonstrated that regeneration capacity is heritable and can be transferred by conventional breeding to clones with low response. Hybridization of two heterozygous clones (probably also heterozygous for the character regeneration capacity) resulted in a rather quantitative response (Figure 8.1A); however, hybridization of a probable homozygous responsive clone with a probable homozygous wild type without any tissue culture responsiveness resulted in a qualitative segregation (Figure 8.1B). These results were confirmed by Sonnino et al. (1989), who combined a dihaploid clone with regeneration capacity with a root-knot nematode-resistant clone. Among 19 F_1 progenies tested, a wide range of androgenetic capability was found. The authors concluded that regeneration capacity is controlled by more than one major gene. This is in contrast to the conclusion of Singsit and Veilleux (1989), who deduced from data from a backcross between a highly responsive genotype and its unresponsive parent that anther culture capacity may be under control of a single dominant gene. Meyer (1991) analysed the genetics of regeneration capacity even further and claimed that only a few genes were involved, perhaps two major genes, one of which controls the induction of embryoid formation and the other its frequency, as well as minor genes. Using this procedure for anther culture improvement, Uhrig and Salamini (1987) obtained more than 100 embryoids per flower after a prebreeding phase for regeneration capacity. Although these genetical analyses were performed with dihaploid anther donor clones, the genetic basis is also true for $4x$ anther donor clones. Using material selected for androgenesis, the $4x$–$2x$ reduction also becomes possible, using exactly the same culture procedures as described for microspore culture from dihaploids.

Tissue Culture

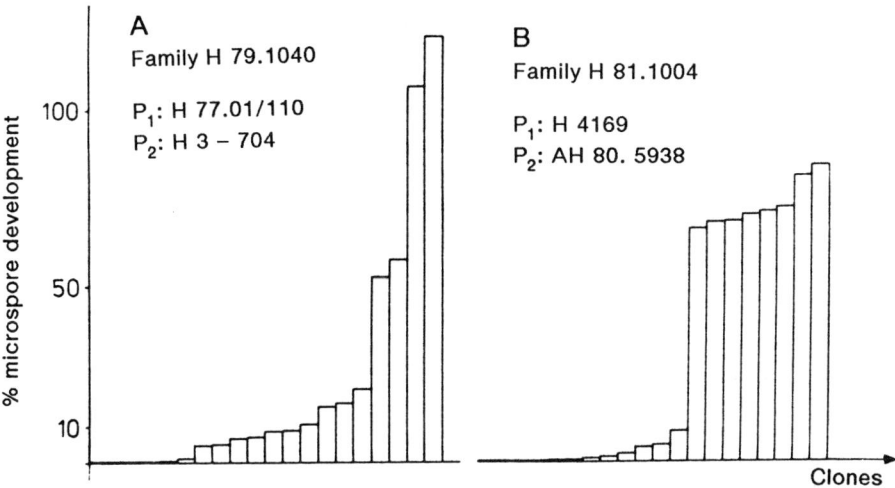

Fig. 8.1. Embryoid and callus formation per 100 anthers. A, from 20 different clones of an F_1 hybrid family the parents of which were heterozygous (probably also for the character regeneration capacity); B, from 20 clones of an F_1 hybrid family, where the parent AH 80.5938 was of androgenetic origin and possessed the regeneration capacity in the homozygous condition.

Use of androgenetic haploids

While the theory of using dihaploids and monohaploids is described in chapters in Part II, it should be stressed that in principle there is no difference whether the ploidy level reduction is performed by partheno- or androgenesis, unless recent research suggesting that the former are not entirely parthenogenetic in origin is confirmed (Clulow et al., 1991). Either way, the higher number of microspores increase the chance of finding a desired genotype in androgenesis. However, de Vincente and Tanksley (1991) found in tomato that the frequency of meiotic recombination differs between male and female gametes. A single F_1 plant was backcrossed to each of *Lycopersicon esculentum* and *L. pennelli*, as the male and the female parent. With 85 RFLP markers, they constructed a map of two backcross populations. Since both recurrent parents were homozygous, recombination measured in each population reflected crossing-over rates leading to male and female gametes. Significantly less recombination was observed in microsporogenesis. Since the numbers required in the recombination process are extremely large, and the probability of finding a specific type is very low, the significantly lower recombination rate of the microspores in most cases should not be too critical. Wenzel and Uhrig (1981) produced androgenetic doubled monohaploid clones possessing PVX, PVY, PLRV and *Globodera pallida* resistance in the homozygous condition. These experiments included monogenically inherited traits (PVX/Y resistance) and polygenic ones (PLRV, *G. pallida* resistance), demonstrating that recombination frequency is sufficient. Combination of characters at the dihaploid level is only a first step in the production of a new cultivar and at present no superior tetraploid cultivars owe their origin to doubled monohaploids. The technique, however, is used in experimental programmes, particularly for the establishment of strong field resistance to PLRV, *G. pallida* and *Phytophthora*. In diploid crops, the use of doubled haploids is possible to achieve varietal status more quickly, and consequently lines derived from microspores of barley are already in official yield trials. Foroughi-Wehr and Wenzel (1990) developed an example of a barley breeding scheme, using recurrent selection with repeated haploid steps as a rapid method for combining quantitatively inherited traits. This scheme can also be applied in programmes of androgenetic doubled monohaploid production in potato.

In vitro *selection on single cells*

Selection in the field is strongly influenced by environmental conditions. Consequently there was a feeling of euphoria when it became possible to grow plants *in vitro* under perfect artificial conditions. Additionally, systems which start with single cells can help to circumvent problems of cross-feeding and chimerism. Protoplasts and microspores offer such single-cell systems from the sporophyte or the gametophyte.

Selection of protoplasts

Shepard (1981) postulated the usefulness of protoplasts for creating new genotypes. He found tremendous variability in potato 'protoclones' from the tetraploid cultivar Russet Burbank. Up to 30% of the regenerated protoclones were claimed to be useful variants; although many were grossly aberrant, some showed improved resistance to *Alternaria solani* (Matern et al., 1978) or *Phytophthora infestans*. However, not all these variants were genetically stable, even during vegetative propagation. These may have originated by some changes other than genetic mutation. Protoplast culture plantlets from the potato cv. Judith could be regenerated with increased resistance to *Phytophthora infestans* (Meulemanns and Fouarge, 1986). Clones with significantly enhanced tuber resistance to *Erwinia* have also been found in non-selected populations of protoplast-derived regenerants (Taylor et al., 1988). In experiments with dihaploid potatoes, Wenzel (1980b) regenerated more than 3000 clones from protoplasts. He found only a few grossly aberrant types, most of which turned out to be aneuploids with no practical value. This difference in the behaviour of tetraploid and diploid potatoes shows the strong influence of the ploidy level on the survival of variants. So it can be concluded that the degree of variants depends on the ploidy level and on the genotype. Somaclonal variation is considered further in chapter 9 by Kumar.

The single-cell system can be used more precisely for resistance selection when the toxic compounds responsible for the virulence are used to detect sensitive protoplasts. Potato protoplasts of dihaploid clones were regenerated in the presence of selective concentrations of extracts from *Fusarium sulphurium, F. coeruleum* or *Phytophthora infestans*. A significantly higher number of protoclones proved not to be affected by the toxin in secondary callus or in a leaf test, compared with regenerated protoclones that had developed without selection pressure. The correlation to subsequent field tests, however, was unreliable (Wenzel and Foroughi-Wehr, 1990). This may have been due to the rather artificial *in vitro* screening and the unreliable field tests, or due to insignificant genetic changes.

For quantitative comparisons of the effect of a toxin on protoplast development, the plating efficiency should be reproducibly high and constant over a range of plating densities. With a number of genotypes used by Möllers and Wenzel (1992) this was not possible, despite the fact that genotype-specific reactions occurred when a purified toxin from late blight was applied, confirming the results of Foroughi-Wehr and Stolle (1985). Also Berggren and Sjödin (1987) observed no effect on the protoplast development of the potato cultivars Bintje and Matilda during the first 5 days after toxin application. It is therefore concluded that for *Fusarium* and *Phytophthora in vitro* selection on protoplasts among several millions of protoplasts was ineffective.

Selection of microspores

While new variability in sporophytic cell populations is presumed to arise from spontaneous mutations, which appear at random, gametophytic cell populations express a variability as segregating genotypes of the parents. The gametophytic population contains all parental traits in new combinations after meiotic recombination. Thus, the chances of finding valuable combinations of agronomically important characters are higher.

The potential advantages of using microspores for selection purposes are:

1. A huge population of single haploid cells is available.
2. A lower level of somaclonal variation will result in very few plants with novel and mostly undesired characteristics.
3. The selection can be carried out at the earliest developmental or breeding stage.
4. Recessive traits are expressed and selected individuals will be homozygous after doubling.

The genes one is interested in come from the parents. This means that, even if screening is insufficient to detect the complex reaction resulting in a phenotypic behaviour as expressed in the field, the technique may be sufficient to identify one step of this complex. As long as this single step is part of the whole process, and as most probably the whole process is often transferred, such a system, using, for example, a toxin as biochemical marker, may work in segregating homozygous material.

For potato such a selection system was applied using the crude toxin from *Phytophthora infestans* as biochemical marker (Zitzlsperger and Wenzel, 1990; Möllers *et al.*, 1992). The microspore population from seven dihaploid donor plants was screened with a relative *Phytophthora* toxin concentration of 0.5, 1.5 and 2.5, where 1.0 is the natural concentration of toxins in the culture filtrate of the fungi prior to purification. After the addition of toxin, three genotypes reacted with a general increase of the embryoid regeneration rate per anther. Among the concentrations tested, optimal regeneration was obtained at a relative toxin concentration of 0.5, with a maximum of 160 embryoids from the microspores isolated from 100 anthers. This is indicative of single anthers, the microspores of which regenerated 30–50 embryoids. Other genotypes reacted with a decrease in the regeneration rate at the lowest toxin concentration applied. However, neither was there a correlation to the *Phytophthora* resistance of the donor plants, nor were surviving and regenerated androgenetic plants more resistant than the starting material, so this selection system cannot be used for *Phytophthora* screening.

Protoplast fusion

With respect to practical applications in breeding programmes, the regeneration of protoplasts has only limited importance. However, not only can protoplasts be used for rapid propagation, but they can also be fused to result in somatic hybrids (for review see Glimelius et al., 1991). The most attractive application of fusion is the possibility of combining asexually – without meiotic segregation – two complex nuclear genomes. This is most important where a need exists to combine quantitatively inherited traits. As polyploid hybrids in the potato are of limited interest, it will usually be necessary to reduce the ploidy level to dihaploidy before constructing a new tetraploid clone by fusion.

The technique

Fusion can be induced chemically by the polycation polyethyleneglycol (PEG) or via electrofusion. After initial emphasis on PEG fusion, the electrofusion technique is now applied in most laboratories since it is less harmful to the cells. The electrofusion equipment used has a lamellar fusion chamber at a voltage of 100–200 V cm^{-1} and a frequency of 1 MHz. After 30–40 s the majority of the protoplasts are aligned, and fusion is obtained by applying a pulse (2000 V cm^{-1} for 20–40µs). Although efficient methods exist (Austin et al., 1985; Schilde-Rentschler et al., 1988; Masson et al., 1989; Waara et al., 1989; Chaput et al., 1990), for applied purposes some bottlenecks must be overcome. Most important is the widening of the fusion procedure to many more genotypes. This could be achieved to some extent by optimizing the culture conditions (Deimling et al., 1988; Möllers and Wenzel, 1992). Culture and regeneration media are identical to those used for normal protoplast regeneration, described earlier in the section on rapid propagation.

Identification of somatic hybrids

One limitation in the application of somatic fusion is the difficulty in differentiation of heterocaryotic fusion products from unfused and homocaryotically fused protoplasts. It is not practical to regenerate all plants and to carry out the selection at the plant level. As long as interspecific fusions are made, phenotypic markers are useful and even intraspecific fusion phenotypes can normally be distinguished. Since the most rapid progress from combining genotypes will result for *S. tuberosum* (+) *S. tuberosum* hybridizations, such phenotypic markers are rare. A number of selection systems use complementation or partial lethality (Schieder and Vasil, 1980), but hybrid vigour of fusion products may prove a more universal selection system. Such a system was used in potato by Debnath and Wenzel (1987) and Waara *et al.* (1989). After electrofusion, 50% of the callus clones selected after 21 days were heterocaryotic fusions, but with increasing developmental period the number of real hybrids selected was reduced (Wenzel *et al.*, 1991). Plants of clones from very vigorous calli also showed hybrid vigour. The procedure, however, is rather empirical, as only combinations expressing a positive heterosis are found; any combining abilities in somatic hybridization leading to negative heterosis will be overlooked. Isozyme analysis of both parents and the presumptive hybrids is a recommended method for a final proof (Austin *et al.*, 1985; Hein and Schieder, 1986; Debnath and Wenzel, 1987; Uijtewaal *et al.*, 1987). On the other hand, if the frequency of heterocaryotic fusion products reaches on average the 40% level, early identification will no longer be a crucial prerequisite for the application of somatic hybridization.

Direct DNA diagnoses allow the immediate detection of hybrids in cases where isoenzymes do not. They further allow the use of undifferentiated calli, so that waiting for morphogenesis is not necessary. For DNA probing, repetitive as well as single-copy probes can be used (Schweizer *et al.*, 1990; Thach *et al.*, 1993). In such tests, using repetitive DNA probes, Schweizer *et al.* (1990) found a possible correlation between a high amount of repetitive DNA and a good agronomic performance of a clone; on the other hand, clones with a lower level of repetitive sequences expressed good regeneration capacity. An increase in the amount of repetitive sequences might therefore promote successful somatic hybridization.

The use of DNA probes usually demands radioactive labelling and a substantial quantity of tissue from which sufficient amounts of DNA can be extracted. In 1985 the polymerase chain reaction (PCR) was developed (Saiki *et al.*, 1985), which allows hybridization in very small portions of plant material, down to 1 mg fresh weight (Dehmer *et al.*, 1991). Thus, the presence of a defined nucleotide sequence can be determined in a very early stage of development and free of radioactivity. In combination with PCR, the whole spectrum of techniques making use of molecular markers can now be applied for *in vitro* selection, not only of fusion products, but also of microspores or other segregating material. However, high costs of DNA extraction and PCR procedures remain an economic problem if such techniques are to be applied to huge numbers of clones.

Hybridization via somatic fusion

The most attractive application of somatic fusion is to combine asexually two heterozygous nuclear genomes without meiotic recombination and to pool polygenic characters. The genes of the mitochondrial genome might be recombined as well, but the chloroplasts from the two parents are normally not mixed in one cell and thus cpDNA will not recombine.

POLYGENIC TRAITS

During selection, polygenic traits are already used as undefined markers, e.g. for hybrid vigour. Thus it is common to use leaf size, plant height and thickness of the stem as hybrid criteria. In several experiments, the tuber yield of the hybrids was also much higher compared with the dihaploid parents and the doubled-up protoclones, indicating a different hybrid vigour or a good combining ability of the parents (Möllers et al., 1991). Quite often, quantitatively inherited traits are intended to be transferred from rather distant somatic hybridizations, in particular from hybridization between *S. tuberosum* and *S. brevidens*. Austin et al. (1985) and Gibson et al. (1988) combined resistance to PLRV and late blight in *S. brevidens* × *S. tuberosum* somatic hybrids. They have also demonstrated *Erwinia* resistance in such hybrids (Austin et al., 1988). Since the relationship between the fusion parents in these combinations is rather wide, applied use of the hybrids still requires further conventional breeding work. More rapid success, akin to conventional breeding, is expected from fusion within the *S. tuberosum* genome. In such experiments, Möllers et al. (1991) demonstrated high flower production, pollen fertility of 60 to 90% and a substantial yield increase of the hybrids in comparison with the mean tuber yield of the parents. There were, however, large differences between different combinations. Differences in combining ability is a possible reason for this. In general, however, it could be demonstrated in these experiments that somatic hybrids could reach the yield of cultivars grown as controls. Besides the yield differences of different fusion combinations, yield differences were observed between clones from the same fusion parents.

In a more detailed analysis of the level of resistance to *Phytophthora infestans*, it was found that, although most of the hybrids showed an intermediate resistance, transgressive variation in both directions existed (Figure 8.2). The same was true for field resistance to PVX and PVY, indicating that fusion experiment hybrids originating from the same two parents showed variability in quantitatively inherited traits. This variability cannot solely be explained as due to somaclonal variation as a consequence of callus culture. Quite probably, the novel 'hybrid' cytoplasm and nuclear–cytoplasmic interactions are responsible as well. It has been demonstrated, by using chloroplast-specific DNA probes, that the chloroplasts segregate soon after fusion. In the material described, the hybrids contain the chloroplasts of either parent in a 1:1 ratio (Möllers et al., 1991).

Furthermore, it should be stressed that there exist clear differences in the suitability of clones as fusion parents. Specific and general combining ability effects for this trait have been observed.

As it is difficult to explain the variability of quantitatively inherited traits,

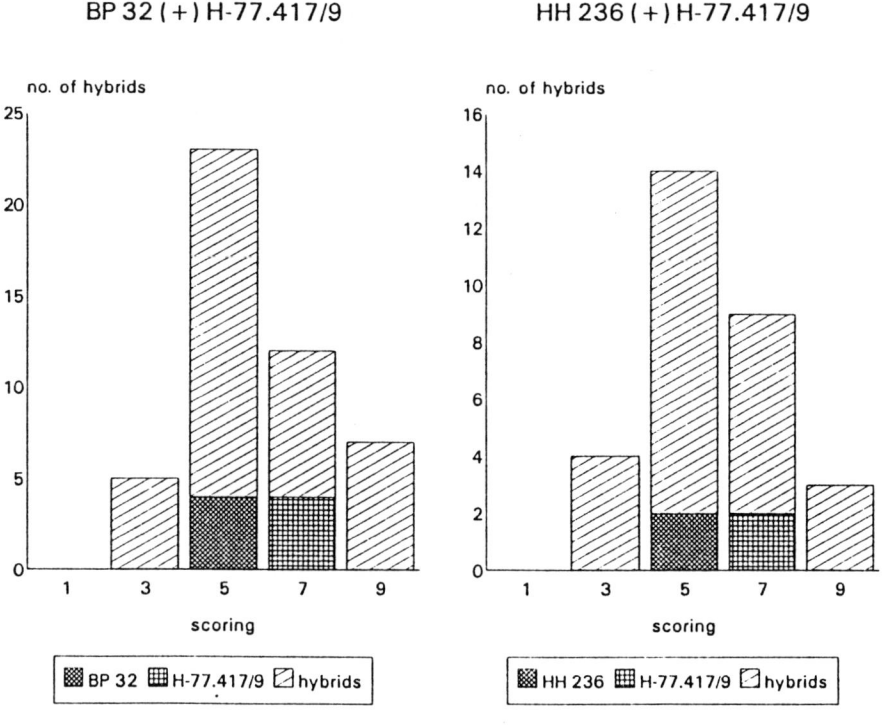
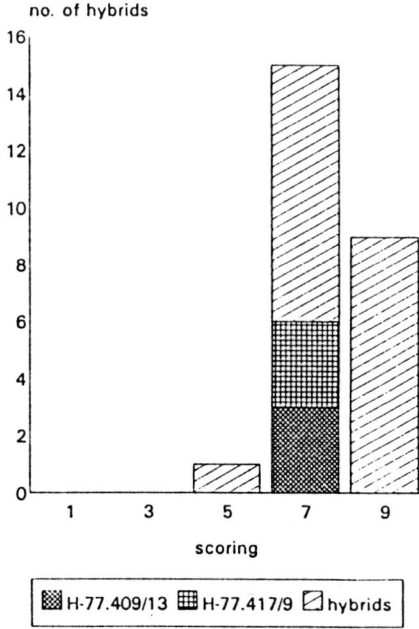

Fig. 8.2. Scoring of *Phytophthora infestans* infections of somatic hybrids from three fusion events, one parent of which possessed a medium field resistance; the other one (H 77.417/9) was highly susceptible (1 = resistant; 9 = susceptible).

Table 8.2 Combination and transfer of resistances after somatic hybridization of dihaploid potato clones (data from Möllers et al., 1991 and Thach et al., 1993).

Fusion		Somatic hybrid clones				
P₁	P₂	Total	Tested	Resistances		
PVX (+)	PVY	54	27	PVX/Y 25	PVX 1	none 1
PVY (+)	PVX	23	5	PVX/Y 5		
Ro 1 (+)	-	61	8	nematode-resistant 8		
met+ (+)	-	4	4	metribuzin-resistant 4		
		45	10	10		

somatic genetics should first be examined via the segregation of monogenic traits.

MONOGENIC TRAITS

Gibson et al. (1988) transferred resistance to PVX from *S. brevidens* into *S. tuberosum*. Möllers and Wenzel (1992) fused metribuzin-tolerant and metribuzin-sensitive clones. De Jong (1983) had previously shown that the sensitivity of potato clones to this herbicide is inherited by a single dominant gene so that, after fusion, tolerant clones should be expected. Indeed, after fusion, all hybrids where one fusion parent was tolerant were tolerant with the same level of tolerance as the tolerant parent (Table 8.2). Similar results were obtained by the same authors when fusing dihaploids containing the *G. rostochiensis* resistance gene for *Ro 1* with susceptible clones. In the somatic hybrids of such combinations, all hybrids were scored as *Ro 1*-resistant (Table 8.2). For tuber shape, dominance of round tubers was observed; for flesh colour there was an intermediate inheritance (Möllers et al., 1991). At present, there are data on only a few plants and a few genotypes and with only limited repetitions under field conditions; but they demonstrate that for monogenic traits the expected gene expression can be observed.

There are, however, also some exceptions. As Table 8.2 demonstrates, after fusion of dihaploids with the gene *Rx* and other clones carrying the gene *Ry*, in most cases analysed so far, the expected combination of the resistances was observed. But in two hybrids either the PVY or both resistances were missing. Experiments are in progress to identify reasons for this behaviour (Thach et al., 1993).

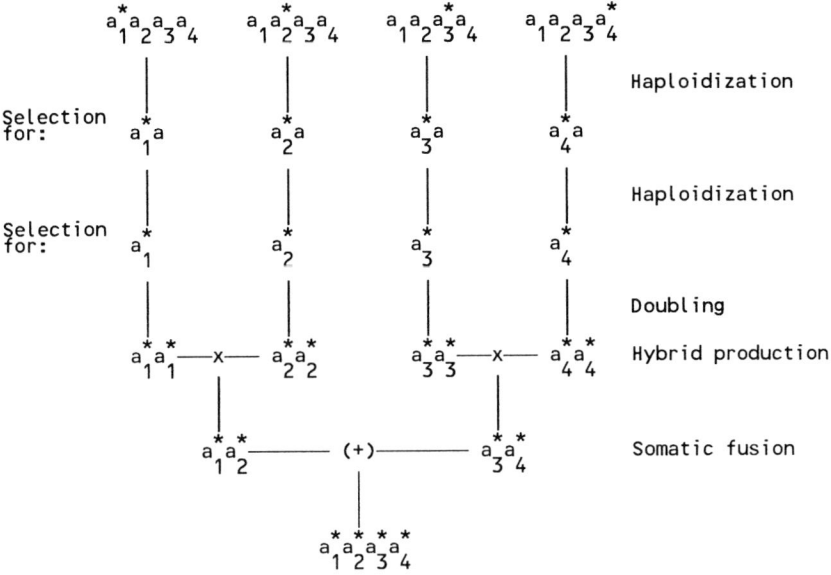

Fig. 8.3. Scheme for an analytical synthetic breeding procedure combining tissue culture techniques and conventional breeding steps (Wenzel et al., 1979).

Combined use of haploids and of somatic fusion

Wenzel *et al.* (1979) proposed the combination of haploidy, protoplast fusion and classical breeding steps for combining several traits of potato in an analytical synthetic manner (Figure 8.3). The monohaploid ploidy level is reached via two successive haploidization steps, and from these monohaploids homozygous dihaploids can be produced by spontaneous or induced doubling. In these clones, additive traits are also homozygous and will be inherited during a sexual combination to produce heterozygous interdihaploids. These interdihaploids can then be subjected to intensive sexual recombination and selection to give superior heterozygous interdihaploids expressing good agronomic traits. Finally, the somatic fusion of these hybrids combines several – at least four – characters and achieves, at the same time, the tetraploid level expressing maximum heterozygosity. It has taken more than 10 years from the publication of this scheme until the first analytical synthetic clones could be tested in the field. But it has now been shown that this scheme not only works for qualitative characters but also for quantitative ones. For recessive characters which are expressed in interdihaploids only when they are homozygous, there exists the possibility of expressing recessive alleles by maintaining homozygosity at the tetraploid level by fusing interdihaploids that both express the same recessive trait.

Many breeders possess interdihaploid populations, characterized for agronomically important traits, but many of which do not flower or are at least male-sterile. Now that fusion is becoming reproducible, all these clones can be transferred to the tetraploid level and their use in somatic hybrid breeding programmes estimated.

Conclusion

In discussing the use of tissue culture in potato genetics, it may be concluded that there is no doubt that the *in vitro* procedures of rapid propagation, coupled with meristem culture, offer advantages over the normal reproductive system and will aid the breeding of potatoes. Even without the means to entirely prevent genetic changes in these tissue cultures, it has become more reliable and cheaper to produce healthy breeding stocks. No other tissue culture techniques have yet made the jump from the experimental stage to economic success. Protoplast fusion is probably the next procedure which will pay, since it offers the opportunity of combining polygenic traits via asexual means. The fusion of primitive cultivars or wild species with *S. tuberosum* is probably much further away from application but will ultimately provide a means of integrating characters not available in a heritable form in the *S. tuberosum* gene pool. For this goal, asymmetrical fusion might have a better chance than the symmetrical fusion discussed here. There are already ideas on how to detect with repetitive DNA probes the desired chromosomes or chromosome fragments (Schweizer *et al.*, 1990). For the production of dihaploids, parthenogenetic techniques are available which avoid *in vitro* culture, but the production of monohaploids to achieve transient homozygosity appears to a practical breeder somewhat esoteric, as long as sufficient heterosis can be produced by crossing tetraploid heterozygotes. Breeders not yet willing to invest money in androgenetic procedures may change their minds when androgenesis can be reproducibly coupled to *in vitro* selection steps at the microspore level, as documented in barley (Ye *et al.*, 1987). Particularly in circumstances where agriculturally important characters are under polygenic control, molecular strategies for selection will be developed. This demands a better understanding of biochemical pathways, which is a prerequisite for finding out whether *in vitro* selection on more complex structures such as calli can be applied, and is also a prerequisite for the production of a broad spectrum of cheap and useful transgenic plants.

Finally, it should be said that it is necessary to grow all *in vitro*-produced plants in the field. Very often, a powerful *in vitro* trait does not show up at the whole-plant level under field conditions, even when calli subsequently induced for a second time from later generations express the character again, as shown for 2,4-D-resistant tobacco lines in the early work of Ono (1979). Miller *et al.* (1991) point to the important linkage of *in vitro* approaches and conventional breeding programmes, but it is the field test which will finally clarify whether tissue culture provides an opportunity for future potato genetics.

References

Austin, S., Baer, M., Ehlenfeldt, M., Kazmierczak, P.J. and Helgeson, J.P. (1985) Intra-specific fusions in *Solanum tuberosum*. *Theoretical and Applied Genetics* 71, 172–175.

Austin, S., Lojkowska, E., Ehlenfeldt, M.K., Kelman, A. and Helgeson J.P. (1988) Fertile interspecific somatic hybrids of *Solanum*: a novel source of resistance to *Erwinia* soft rot. *Phytopathology* 78, 1216–1220.

Bajaj, Y.P.S. (ed.) (1991) *High-Tech and Micropropagation*. Biotechnology in Agriculture and Forestry 17, Springer, Berlin.

Bajaj, Y.P.S. and Sopory, S.K. (1986) Biotechnology of potato improvement. In: Bajaj, Y.P.S. (ed.) *Biotechnology in Agriculture and Forestry*. Springer Verlag, Berlin, Vol. 2, pp. 429–454.

Berggren, B. and Sjödin, C. (1987) Sensitivity of potato protoplasts to *Phytophthora infestans* In: Foldö, N.E et al. (eds) *Proceedings of the 10th Triennial Conference of the European Association of Potato Research*. EAPR, Aalborg, p. 220.

Binding, H., Nehls, R., Schieder, O., Sopory, S.K. and Wenzel, G. (1978) Regeneration of mesophyll protoplasts isolated from dihaploid clones of *Solanum tuberosum*. *Physiologia Plantarum* 43, 52–54.

Chaput, M.-H., Sihachakr, D., Ducreux, G., Marie, D. and Barghi, N. (1990) Somatic hybrid plants produced by electrofusion between dihaploid potatoes: BF 15 (H1), Amica (H6) and Cardinal (H3). *Plant Cell Reports* 9, 411–414.

Clulow, S.A., Wilkinson, M.J., Waugh, R., Baird, E., De Maine, M.J. and Powell, W. (1991) Cytological and molecular observations on *Solanum phureja*-induced dihaploid potatoes. *Theoretical and Applied Genetics* 82, 545–551.

De Block, M. (1988) Genotype independent leaf disk transformation of potato (*Solanum tuberosum* L.) using *Agrobacterium tumefaciens*. *Theoretical and Applied Genetics* 76, 767–774.

Debnath, S.C. and Wenzel, G. (1987) Selection of somatic fusion products in potato by hybrid vigour. *Potato Research* 30, 371–380.

Dehmer, K.J., Graner, A. and Wenzel G. (1991) Screening for defined DNA sequences in minimal amounts of barley tissue by PCR. *Plant Breeding* 107, 70–72.

Deimling, S., Zitzlsperger, J. and Wenzel, G. (1988) Somatic fusion for breeding of tetraploid potatoes. *Plant Breeding* 101, 181–189.

De Jong, H. (1983) Inheritance of sensitivity to the herbicide metribuzin in cultivated diploid potatoes. *Euphytica* 32, 41–48.

de Vincente, M.C. and Tanksley, S.D. (1991) Genome-wide reduction in recombination of backcross progeny derived from male versus female gametes in an interspecific cross of tomato. *Theoretical and Applied Genetics* 83, 173–178.

Foroughi-Wehr, B. and Stolle, K. (1985) Resistenzselektion in vitro am Beispiel des Systems Kartoffel/*Phytophthora infestans* (Mont.) de Bary. *Nachrichtenblatt des Deutschen Pflanzenschutzdienstes* 37, 170–173.

Foroughi-Wehr, B. and Wenzel, G. (1990) Recurrent selection alternating with haploid steps – a rapid breeding procedure for combining agronomic traits in inbreeders. *Theoretical and Applied Genetics* 80, 564–568.

Foroughi-Wehr, B. and Wenzel, G. (1993) Andro- and parthenogenesis. In: Hayward, M.D., Bosemark, N.O. and Romagosa, I. (eds) *Plant Breeding – Principles and Prospects*. Chapman and Hall, London, pp. 261–277.

Foroughi-Wehr, B., Wilson, H.M., Mix, G. and Gaul, H. (1977) Monohaploid plants from anthers of a dihaploid genotype of *Solanum tuberosum* L. *Euphytica* 26, 361-367.

Garner, A. and Blake, J. (1989) The induction and development of potato microtubers *in vitro* on media free of growth regulating substances. *Annals of Botany* 63, 663-674.

Gibson, R.W., Jones, M.G.K. and Fish, N. (1988) Resistance to potato leaf roll virus and potato virus Y in somatic hybrids between *Solanum tuberosum* and *S. brevidens*. *Theoretical and Applied Genetics* 76, 113-117.

Glimelius, K., Fahlesson, J., Landgren, M., Sjödin, C. and Sundberg, E. (1991) Gene transfer via somatic hybridization in plants. *Trends in Biotechnology* 9, 24-30.

Haberlach, G.T., Cohen, B.A., Reichert, N.A., Baer, M.A., Towill, L.E. and Helgeson, J.P. (1985) Isolation, culture and regeneration of protoplasts from potato and several related *Solanum* species. *Plant Science* 39, 67-74.

Hein, T. and Schieder, O. (1986) An improved culture method of mechanically isolated heterokaryons of potato. *Plant Breeding* 97, 255-260.

Heinstein, P. and Emery, A. (1988) Processes with plant cell cultures. In: Rehm, H.J. (ed.) *Biotechnology*. VCH, Weinheim, Vol. 6b, pp. 213-248.

Hougas, R.W. and Peloquin, S.J. (1957) A haploid plant of the potato variety Katahdin. *Nature* 180, 1202-1210.

Irikura, Y. (1975) Induction of haploid plants by anther culture in tuber-bearing species and interspecific hybrids of *Solanum*. *Potato Research* 18, 133-140.

Jacobsen, E. and Sopory, S.K. (1978) The influence and possible recombination of genotypes on production of microspore embryoids in anther culture of *Solanum tuberosum* and dihaploid hybrids. *Theoretical and Applied Genetics* 52, 119-123.

Johansson, L. (1986) Improved method for induction of embryogenesis in anther cultures of *Solanum tuberosum*. *Potato Research* 29, 179-190.

Jongedijk, E., Ramanna, M.S., Sawor, Z. and Hermsen, J.G.Th. (1991) Formation of first division restitution (FDR) $2n$-megaspores through pseudohomotypic division in *ds-1* (desynapsis) mutants of diploid potato: routine production of tetraploid progeny from $2x$ FDR $\times$ $2x$ FDR crosses. *Theoretical and Applied Genetics* 82, 645-656.

Landsmann, J. and Uhrig, H. (1985) Somaclonal variation in *Solanum tuberosum* detected at the molecular level. *Theoretical and Applied Genetics* 71, 500-505.

Leifert, C. and Waites, W.M. (1990) Contaminations of plant tissue cultures. *IAPTC Newsletter* 60, 2-13.

Linsmaier, E.M. and Skoog, F. (1965) Organic growth requirements of tobacco tissue culture. *Physiologia Plantarum* 18, 100-127.

Marinus, J. (1990) *In vitro* tuberization of four potato cultivars under various day lengths and light intensities. In: Mackerron, D.K.L. *et al*. (eds) *Abstracts Eleventh Triennial Conference of the European Association for Potato Research*. EAPR, Edinburgh, pp. 107-108.

Masson, J., Lancelin, D., Bellini, C., Lecerf, M., Guerche, P. and Pelletier, G. (1989) Introduction of selectable markers by direct gene transfer and its application to somatic hybridization in potato (*Solanum tuberosum* L.). *Theoretical and Applied Genetics* 87, 153-159.

Matern, U., Strobel, G. and Shepard, J. (1978) Reaction to phytotoxins in a potato population derived from mesophyll protoplasts. *Proceedings of the National Academy of Sciences* 75, 4935-4939.

Meulemanns, M. and Fouarge, G. (1986) Regeneration of potato somaclones and *in vitro* selection for resistance to *Phytophthora infestans* (Mont.) de Bary. *Medelingen van de Faculteit Landbouwwetenschappen Rijksuniversiteit Gent* 51, 533-545.

Meyer, K. and Foroughi-Wehr, B. (1987) Possible ways to simplifying potato-meristem culture. *Plant Research and Development* 26, 12–18.

Meyer, R.C. (1991) Die genetische und zytologische Analyse der Kartoffel (*Solanum tuberosum* L.) nach Antherenkultur unter besonderer Berücksichtigung der Entwicklung reiner diploider Linien. Dissertation, University of Cologne.

Miller, D.R., Waskom, R.M., Brick, M.A. and Chapman, P.L. (1991) Transferring *in vitro* technology to the field. *Bio/technology* 9, 143–146.

Mix, G. (1983) Production of dihaploid plantlets from anthers of autotetraploid genotypes of *Solanum tuberosum*. *Potato Research* 26, 63–67.

Möllers, C. and Wenzel, G. (1992) Somatic hybridization of dihaploid potato protoplasts as a tool for potato breeding. *Acta Botanica* 105, 133–139.

Möllers, C., Hofferbert, R., Schriener, A. and Wenzel, G. (1991) Ergebnisse eines Feldanbaus tetraploider somatischer Kartoffelhybrieden. *Arbeitstagung der Saatzuchtleiter, Gumpenstein* 42, 129–135.

Möllers, C., Zhang, S. and Wenzel, G. (1992) The influence of silver thiosulfate on potato protoplast cultures. *Plant Breeding* 108, 12–18.

Möllers, C., Zitzlsperger, J. and Wenzel, G. (1992) The effect of toxin from *Phytophthora infestans* (Mont.) de Bary on potato protoplasts and microspores. *Physiological and Molecular Plant Pathology* 41, 427–435.

Murashige, T. and Skoog, F. (1962) A revised medium for rapid growth and bioassay with tobacco tissue culture. *Physiologia Plantarum* 15, 473–497.

Ono, H. (1979) Genetical and physiological investigations of a 2, 4-D resistant cell line isolated from tissue culture in tobacco. *Science Report Faculty of Agriculture Kobe University* 13, 273–277.

Perl, A., Aviv, D. and Galun, E. (1988) Ethylene and *in vitro* culture of potato: suppression of ethylene generation vastly improves protoplast yield, plating efficiency and transient expression of an alien gene. *Plant Cell Reports* 7, 403–406.

Saiki, R.K., Scharf, S., Faloona, F., Mullis, K.B., Horn, G.T., Erlich, H.A. and Arnheim, N. (1985) Enzymatic amplification of β-globin genome sequences and restriction site analysis for diagnosis of sickle-cell anemia. *Science* 230, 1350–1354.

Schieder, O. and Vasil, I.K. (1980) Protoplast fusion and somatic hybridization. *International Review of Cytololgy Supplement* 11B, 21–46.

Schilde-Rentschler, L., Boos, C. and Ninnemann, H. (1988) Somatic hybridization of diploid potato lines – a tool in potato breeding. In: Puite, K.J. Dons, J.J.M., Huizing, H.J., Kool, A.J., Koorneef, M. and Krens, F.A. (eds) *Progress in Plant Protoplast Research*. Kluwer, Dordrecht, pp. 195–196.

Schweizer, G., Stelzer, T. and Hemleben, V. (1990) RFLP-Analyse mit spezifischen Genomkomponenten der Kartoffel zur Identifizierung von symmetrischen und asymmetrischen Hybriden. *Vorträge für Pflanzenzüchtung* 18, 178–192.

Shahin, E.A. (1984) Isolation, culture and regeneration of potato leaf protoplasts from plants preconditioned *in vitro*. In: Vasil, I.K. (ed.) *Cell Culture and Somatic Cell Genetics of Plants*. Academic Press, Orlando, pp. 381–390.

Shepard, J.F. (1981) Protoplasts as sources of disease resistance in plants. *Annual Review of Phytopathology* 19, 145–166.

Singsit, C. and Veilleux, R.E. (1989) Intra-and interspecific transmission of androgenetic competence in diploid potato species. *Euphytica* 43, 105–112.

Sonnino, A., Tanaka, S., Iwanaga, M. and Schilde-Rentschler, L. (1989) Genetic control of embryo formation in anther culture of diploid potatoes. *Plant Cell Reports* 8, 105–107.

Sopory, S.K., Jacobsen, E. and Wenzel, G. (1978) Production of monohaploid embryoids and plantlets in cultured anthers of *Solanum tuberosum*. *Plant Science Letters* 12, 47–54.

Sree Ramulu, K. and Dijkhius, P. (1986) Flow cytometric analysis of polysomaty and *in vitro* genetic instability in potato. *Plant Cell Reports* 3, 234–237.

Taylor, R.J., Ruby, C.L. and Secor, G.A. (1988) Assessment of field performance and soft rot resistance in a population of protoplast-derived potato clones. *Phytopathology* 78, 1595–1603.

Thach, N.Q., Frei, U. and Wenzel, G. (1993) Somatic fusion for combining virus resistances in *Solanum tuberosum* L. *Theoretical and Applied Genetics* 85, 863–867.

Uhrig, H. (1985) Genetic selection and liquid medium conditions improve yield of androgenetic plants from diploid potatoes. *Theoretical and Applied Genetics* 71, 455–460.

Uhrig, H. and Salamini, F. (1987) Dihaploid plant production from $4x$ genotypes of potato by the use of efficient anther plant producing tetraploid strains ($4x$ EAPP-clones) – proposal of a breeding methodology. *Zeitschrift für Pflanzenzüchtung* 98, 228–235.

Uijtewaal, B.A., Suurs, L.C.M. and Jacobsen, E. (1987) Protoplast fusion of monohaploid ($2n = x = 12$) potato clones: identification of somatic hybrids using malate dehydrogenase as a biochemical marker. *Plant Science* 51, 277–284.

van Breukelen, E.W., Ramanna, M.S. and Hermsen, J.G.T. (1975) Monohaploids ($n = x = 12$) from autotetraploids *Solanum tuberosum* ($2x = 4x = 48$) through two successive cycles of female parthenogenesis. *Euphytica* 24, 567–574.

Vasil, I.K. and Vasil, V. (1980) Clonal propagation. *International Review of Cytology Supplement* 11A, 145–173.

Waara, S., Tegelström, H., Wallin, A. and Eriksson, T. (1989) Somatic hybridization between anther-derived dihaploid clones of potato (*Solanum tuberosum* L.) and identification of hybrid plants by isozyme analysis. *Theoretical and Applied Genetics* 77, 49–56.

Wenzel, G. (1980a) Recent progress in microspore culture of crop plants. In: Davies, D.R. and Hopwood, D.A. (eds) *The Plant Genome*. John Innes Charity, Norwich, pp. 185–196.

Wenzel, G. (1980b) Protoplast techniques incorporated into applied breeding programs. In: Ferenczy, L. and Farkas, L. (eds) *Advances in Protoplast Research*. Pergamon Press, London, pp. 327–340.

Wenzel, G. and Foroughi-Wehr, B. (1984) Anther culture of *Solanum tuberosum*. In: Vasil, I.K. (ed.) *Cell Culture and Somatic Cell Genetics of Plants*. Academic Press, Orlando, pp. 293–301.

Wenzel, G. and Foroughi-Wehr, B. (1990) Progeny tests of barley, wheat, and potato regenerated from cell cultures after *in vitro* selection for disease resistance. *Theoretical and Applied Genetics* 80, 359–365.

Wenzel, G. and Uhrig, H. (1981) Breeding for nematode and virus resistance in potato via anther culture. *Theoretical and Applied Genetics* 59, 333–340.

Wenzel, G., Schieder, O., Przewozny, T., Sopory, S.K. and Melchers, G. (1979) Comparison of single cell culture derived *Solanum tuberosum* L. plants and a model for their application in breeding programs. *Theoretical and Applied Genetics* 55, 49–55.

Wenzel, G., Bapat, V.A. and Uhrig, H. (1982) New strategy to tackle breeding problems of potato. In: Sen, S.K. and Giles, K.L. (eds) *Plant Cell Culture in Crop Improvement*. Plenum, New York, pp. 337–349.

Wenzel, G., Foroughi-Wehr, B., Frei, U., Graner, A., Kuhlman, U., Lind, V., Möllers, C., Schriener, A., Siedler, H. and Walther, H. (1991) Biotechnologische

Werge zur Resistenz bei Nutzpflanzen. *Vorträge für Pflanzenzüchtung* 19, 20-30.

Ye, J.M., Kao, K.N., Harvey, B.L. and Rossnagel, B.G. (1987) Screening salt tolerant barley genotypes via F_1 anther culture in salt stress media. *Theoretical and Applied Genetics* 74, 426-429.

Zitzlsperger, J. and Wenzel, G. (1990) Efficient production of androgenetic dihaploid potato plants from single anthers and isolated microspores via embryogenesis. In: *Abstracts VIIth International Congress on Plant Tissue and Cell Culture.* IAPTC, Amsterdam, p. 273.

9 Somaclonal Variation

A. KUMAR

Cell and Molecular Genetics Department, Scottish Crop Research Institute, Invergowrie, Dundee DD2 5DA, UK.

Introduction

Genetic variability is an essential component of any breeding programme designed to improve the characteristics of crop plants. Therefore, finding novel methods for producing genetic variability in plants is of immense importance, particularly if genetic variation is lacking. In recent years great interest has been generated by somaclonal variation, which provides an unexpectedly rich source of genetic variation among regenerated plants from explant cells, tissues and organ culture under *in vitro* conditions (Larkin and Scowcroft, 1981; Karp, 1990). In general, the *in vitro* plant regeneration process involves the establishment of dedifferentiated cells from tissue or organ culture under defined culture conditions, proliferation for a number of cell generations and subsequent regeneration of plants. In other words, a period of dedifferentiated cell proliferation is imposed between an explant and the next plant generation. It is now agreed that most of the somaclonal variation observed amongst the regenerated plants is generated during the *in vitro* tissue culture stages and in particular during the dedifferentiated culture stage. This can be seen by the changes in the frequency of chromosomal abnormalities with time in culture (Sree Ramulu, 1986). However, some of the variation arising in regenerated plants may reflect pre-existing heterogeneity in the explant. It is known that genetic changes, such as gene mutations, endoreduplication, aneusomaty and chimeras, also occur in plant tissues or cells *in vivo* (D'Amato, 1985). When such plant cells or tissues are cultured *in vitro*, genetic changes present in the explants can be separated and propagated in culture. Thus, genetic variation incorporated into regenerated plants is the sum of *in vivo* variation and variation that originates *in vitro*. The relative contribution of each of these may be different in different cases, depending upon factors such as genotype, type of culture, culture medium, culture age, etc.

The cultivated potato (*Solanum tuberosum* subsp. *tuberosum*) is a tetraploid in which heterozygosity seems essential in maximizing plant vigour. Due to occasional difficulties in performing sexual crosses (sterility and no flowering) and because of the tetrasomic nature of gene inheritance, improvement through conventional hybridization for disease resistance or for some other specific characters, without the loss of important agronomic properties, is often complex and difficult. The potato, however, is an ideal crop plant for improvement via somaclonal variation because it is highly amenable to various tissue culture techniques (see chapter 8 by Wenzel) and is normally propagated and established vegetatively from tubers. In recent years, a combination of cellular and molecular approaches, including exploitation of somaclonal variation, anther culture, somatic hybridization and genetic transformation, have been employed to improve existing potato cultivars (Kumar et al., 1992). However, the successes of these cellular and molecular techniques depend upon the availability of efficient, reliable and stable plant regeneration systems from explants. Somaclonal variation generated during the plant regeneration process could be a blessing or a curse depending upon the type of genetic manipulation method being used to improve the crop. For instance, production of a useful mutant line through somaclonal variation in a potato cultivar which confers resistance to a disease could be highly desirable (see chapter 16 by Umaerus and Umaerus and also chapter 19 by Elphinstone). On the other hand, it would be undesirable to generate somaclonal variation in a genetic manipulation programme based on the introduction of one or a few known isolated genes into a plant by a genetic transformation method. Similarly, it would also be undesirable to introduce somaclonal variation into micropropagation and cryopreservation programmes, since these techniques are essentially designed for true clonal propagation and conservation of an élite plant genotype. It is essential, therefore, that attempts should be made to understand more fully the mechanisms involved in the origin and genetic stability of somaclonal variation.

This chapter provides a brief but critical account of the origin and nature of somaclonal variation in potato and the factors influencing such variation, together with the role of somaclonal variation in potato breeding. More information on somaclonal variation can be found in the following articles: Bayliss (1980), Orton (1984), Karp and Bright (1985) and Lee and Phillips (1988).

Origin and Nature of Somaclonal Variation

Analyses of regenerated plants at the morphological, cytological, biochemical, molecular and genetic levels have revealed the extent of somaclonal variation and have also provided insight into the origin of somaclonal variation in plants (Sree Ramulu, 1986; Karp, 1990).

Several genetical changes have been reported to be responsible for the appearance of different types of somaclonal variation in plants (Peschke and

Phillips, 1992). These include karyotypic changes, gene mutations in the nuclear and cytoplasmic genomes, translocations, deletions, inversions, cryptic chromosome changes, gene rearrangements and non-conventional mutations involving gene amplifications and transposable elements. Most of the chromosomal abnormalities, at both nuclear and cytoplasmic levels, have been found in regenerated potato plants (Tables 9.1 and 9.2). Difficulties in performing sexual crosses (limited flowering and sterility) and the heterozygous nature of the tetraploid potato genome have hindered detailed genetic analyses of the somaclonal variant plants. However, in several other crop plants, such as tomato and maize (where genetic analyses are easier), Mendelian inheritance has been demonstrated for somaclonal variants and changes in characters controlled by multigene families and quantitative traits have also been shown. Additionally, single gene changes of dominant, semidominant and recessive natures have been detected among somaclonal variants of several crop plants (Evans and Sharp, 1986).

The start of chromosome instability has been well characterized in potato by several research groups (Sree Ramulu, 1986; Pijnacker and Sree Ramulu, 1990). It has been suggested that instability in the regulation of processes of maintenance and replication of DNA and of the mitotic process may lead to endoreduplication, differential DNA amplification, endomitosis and nuclear fragmentation, and result in one or more types of chromosomal aberration, such as polyploidy, aneuploidy, chromosome structural changes and gene mutation (Figure 9.1).

Sree Ramulu *et al.* (1985) and Pijnacker *et al.* (1986a, b) performed a detailed comparative study to follow the temporal sequence of the nuclear events occurring in cultures of protoplasts or shoot culture segments in tetraploid, dihaploid and monoploid potato lines, and were able to show that, within 7-14 days after culture, a considerable proportion of cells showed higher DNA contents, polyploidy and aneuploidy. Furthermore, the frequency of polyploidy and aneuploidy increased with the age of culture. In cultures of leaf segments of mono- and dihaploid genotypes, the results also revealed that interphase cells with higher DNA values, and mitotic metaphases with polyploid chromosome numbers and diplochromosomes occurred within 3-14 days after culture. With further growth of callus (72 days), the frequency and extent of polyploid and aneuploid cells increased. Interestingly, the explant material used was from juvenile shoot cultures consisting predominantly of young leaves, stems and shoot meristems and had normal DNA levels. However, a high degree of DNA and chromosome variation occurred during the initial stages of callus induction, suggesting that the source of genetic instability was inherent in the conditions of *in vitro* growth.

Thus, it is evident from various data in potato that polyploidy, aneuploidy and chromosome structural changes often occur in protoplast, cell or tissue culture. The data also suggest that endoreduplication is an important mechanism of polyploidization in potato. Similarly, chromosome lagging, bridges, dicentric and polycentric chromosomes contribute to the production of aneuploid cells.

In addition to the most commonly observed chromosomal abnormalities,

Table 9.1 Genomic changes observed in regenerated potato plants.

Genotype or cultivar	Explant source	Type of genomic variation	References
Protoplasts			
cv. Russet Burbank	Leaf protoplasts (*in vivo* plants)	Chromosomal structure Mitochondrial DNA	Gill et al., 1986 Kemble and Shepard, 1984
cv. Fortyfold	Leaf protoplasts (*in vivo* plants)	Chromosome number	Karp et al., 1982
Monohaploid PD 13-3-2	Leaf protoplasts (*in vitro* plants)	Chromosome number Chromosome structure	Sree Ramulu et al., 1986
Dihaploid H^2 140, 258, SH74-97-1168, SVP3:SH78-78-901, SVP4:PH75-1116-1544, SVP5:PH77-1445-2242, SVP7:PH77-1426-1872,	Leaf protoplast	Chromosome number Chromosome structure	Binding et al., 1978 Sree Ramulu, 1986 Sree Ramulu et al., 1989
PDH 40, 51	Leaf protoplast	Chromosome number	Wenzel et al., 1979 S. Cooper-Bland (unpublished)
Dihaploid $H^2$358, $H^2$260	Cell suspension	Chromosome number	Wenzel et al., 1979 Sree Ramulu et al., 1986
cv. Maris Bard	Leaf protoplast	Chromosome number	Karp et al., 1982 Thomas et al., 1982
cv. Bintje	Leaf protoplast	Chromosome number	Sree Ramulu et al., 1984a,b, 1986
cv. Majestic	Leaf protoplast	Chromosome number and chromosome structure (translocation and deletion)	Creissen and Karp, 1985
cv. Désiree	Leaf protoplast	Chromosome number	A. Kumar (unpublished)
Tetraploid line HH258 cv. Loreh	Leaf protoplast	Reduction in ribosomal RNA genes and/or changes in structure of ribosomal RNA gene Chromosome number	Landsmann and Uhrig, 1985 Potter and Jones, 1991
Solanum brevidens	Leaf protoplast	Chromosome number	Nelson et al., 1986
Tissue culture			
cv. Golden Wonder	Shoot meristem	Methylation of ribosomal RNA gene	Harding, 1991
cv. Désiree, Champion, Myatts Ashleaf, Bintje, Pentland Squire	Leaf, rachis, stem and tuber tissues	Chromosome number	Wheeler et al., 1985 Pijnacker and Ferwarda, 1990
Dihaploid lines PDH 40, 155, SVP11, PDH 40, 505, UP 88-147, UP 88-760, UP88-68	Leaf and stem tissues	Chromosome number	A. Kumar (unpublished) Jacobsen, 1981 Karp et al., 1984 Pijnacker et al., 1986a,b A. Kumar (unpublished)
Monoploid lines 7322,	Leaf tissues	Chromosome number	Tempelaar et al., 1985

Genotype or cultivar	Explant source	Type of genomic variation	References
PD 13-3-2			Sree Ramulu, 1986 Pijnacker and Sree Ramulu, 1990
Dihaploid clones	Anther culture	Chromosome number	Sopory and Tan, 1979 Wenzel and Uhrig, 1981
Hybrid plant of S. tuberosum × S. chacoense	Anther culture	Chromosome number	Cappadocia et al., 1984

Table 9.2. Genomic changes observed in regenerated potato somatic hybrid and transgenic plants.

	Explant source	Type of genomic variation	References
Somatic hybrids and cybrids			
Potato cv. Priekul (+) S. chacoense	Leaf protoplasts	Chromosome number	Butenko and Kuchko, 1980
Potato cv. Russet Burbank (+) S. chacoense	Leaf and cell suspension protoplasts	Chromosome number Mitochondrial recombination	Barsby et al., 1984 Fish and Karp, 1986
Diploid S. tuberosum (+) S. brevidens	Leaf protoplasts	Chromosome number	Austin et al., 1985 Masson et al., 1989
Potato cv. Atzimba (+) S. tuberosum	Leaf protoplasts	Chromosome number Mitochondrial recombination	Perl et al., 1990
Dihaploid (198.2) (+) Dihaploid (67.9) of S. tuberosum	Leaf protoplasts		Waara et al., 1992
Transgenic plants			
cv. Désiree	Tuber, leaf and stem tissues	Chromosome number	Stiekema et al., 1988 Barker et al., 1992 A. Kumar (unpublished)
cv. Bintje	Tuber and leaf tissues		
cv. Russet Burbank	Leaf and stem tissues		
cv. Berolina	Leaf tissues	Chromosome number	De Block, 1988
Homozygous diploid lines Mn 79.7322, amf-1 mutant 86.040	Leaf and stem tissues	Chromosome number	Visser et al., 1989
	Stem tissues	Chromosome number	Ooms et al., 1987

other non-conventional mutations, such as gene amplifications and diminution and activation of transposable elements, have also been observed in regenerated plants (Karp, 1990; Peschke and Phillips, 1992). For example, reduction in ribosomal RNA genes (rDNA) has been found in potato plants regenerated from protoplasts (Landsmann and Uhrig, 1985). Additionally, both structural rearrangement within rDNA and methylation of nucleotide

ANEUPLOIDY:
Changes in chromosome number
 – Monosomy
 – Trisomy
 – Tetrasomy
 – Nullisomy

POLYPLOIDY:
Changes in genome size
 – Monoploidy
 – Dihaploidy
 – Hexaploidy
 – Polyploidy

GENETIC VARIABILITY IN PLANTS PRODUCED BY SOMACLONAL VARIATION

KARYOTYPE:
Changes in chromosome structure
 – Translocation
 – Duplication
 – Deletion
 – Inversion
 – Dicentric chromosome
 – Iso- or heterodicentric
 – Telocentric chromosome
 – Isochromosomes

Changes in DNA structure
 – Differential DNA replication (amplification or diminution)
 – Gene mutations
 – Mitotic crossing over
 – Gene inactivation by methylation
 – Insertional mutations by transposable elements

Fig. 9.1. Various types of genomic changes which can occur during *in vitro* growth and differentiation of cells.

sequences of rDNA have been observed (Brown, 1989; Harding, 1991). Recently, genomic changes in a maize line and mutations in a tobacco line due to activation of transposable elements under *in vitro* tissue culture conditions have been demonstrated (Peschke and Phillips, 1992). However, such phenomena have not yet been observed in potato. It has also been shown that several types and copies of non-active transposable elements, including retro-transposable elements, are present in the genome of potato (Flavell *et al.*, 1992). Thus, it is likely that a few active transposable elements may also be present in potato and could be responsible for generating some degree of somaclonal variation in regenerated potato plants.

In addition to nuclear genomic changes, cytoplasmic organellar genomic (chloroplast and mitochondria) changes have also been detected in regenerated plants (Hanson, 1984). Molecular analyses based on restriction endonuclease digest patterns of mitochondrial and chloroplast genomes of regenerated potato plants derived from protoplasts have revealed a significant change in mitochondrial genomes but not of chloroplast genomes (Kemble and Shepard, 1984). It was suggested that the variation results from substantial DNA sequence rearrangements (e.g. deletion, addition and intramolecular recombinations) and cannot be explained by simple point mutations. However, point mutations can also occur in the genomes of organelles of regenerated plants. In contrast to mitochondrial genomes, chloroplast genomes appear to be more stable during *in vitro* culture. However, a striking exception has been reported in anther culture of cereals where, in wheat and barley, analysis of the chloroplast genome revealed that massive deletions of the chloroplast DNA had occurred (Day and Ellis, 1985).

Fig. 9.2. Plant regenerated from leaf tissues of potato cultivar Désiree showing albino phenotype due to mutation in the gene(s) involved with chlorophyll biosynthesis.

Morphological changes observed in regenerated plants vary from gross abnormalities to very little change in morphology. The most commonly observed morphological changes are dwarfness, male sterility, chlorophyll deficiency (Figure 9.2), changes in leaf, stem, flower and tuber sizes, shapes and colours, and variegation. It has not been possible to correlate morphological changes with chromosomal abnormalities observed in somaclonal variants. However, there is an indication that a high degree of polyploidy or aneuploidy in the regenerated plants correlates with gross morphological changes such as dwarfness and male sterility, etc. On the other hand, morphological changes have frequently been observed in regenerated plants which do contain normal numbers of chromosomes (Sree Ramulu, 1986). However, it is possible that these plants may contain gene mutations or slight structural rearrangements which it would not be possible to detect by cytological techniques. Such minor genomic changes could be detected by modern molecular techniques such as restriction fragment length polymorphism

(RFLP; Paterson et al., 1991) and randomly amplified polymorphic DNA (RAPD; Waugh and Powell, 1992). RAPD techniques have recently been used to assess the genomic variation at the molecular level of plants regenerated from potato protoplasts and fusion products (Baird et al., 1992). Lassner and Orton (1983) have demonstrated that isozymes are potentially powerful tools in studying and monitoring genetic variation in regenerated plants of in vitro tissue culture. Analysis of the total proteins from leaves of regenerated potato (cv. Désiree) plants using an isoelectric-focusing gel shows differences in band patterns between the control and some of the culture plants (Kumar and March, unpublished; Figure 9.3). In addition, advances in chromosome banding techniques (Pijnacker and Ferwarda, 1984; Waara et al., 1992; Figure 9.4) capable of detecting deletions, translocations and other structural alterations as well as molecular methods should help to elucidate the mechanism of somaclonal variation and hence help to show correlation between genomic changes and morphological changes.

Factors Influencing Somaclonal Variation

The important factors influencing genetic stability during plant regeneration under in vitro conditions are believed to be the source of explants, their genetic make-up and the composition of the culture medium used. Plant regeneration from various types of explant sources, such as leaf, stem, tuber, rachis and petiole tissues, and protoplasts isolated from leaves of in vitro shoots or glasshouse-grown plants or cell suspensions show different degrees of phenotypic and genotypic variations (Sree Ramulu, 1986). For instance, it has been demonstrated that protoplast-derived plants show a higher degree of chromosomal number variation than those derived from tissue or organ culture. It has been suggested that protoplasts are more prone to chromosome instability because they undergo a longer period of greater stress during the initial stages of cell division and dedifferentiation (Pijnacker and Sree Ramulu, 1990).

The ploidy and genetic make-up of the genotype used in plant regeneration have been shown to profoundly affect the type of variation observed in morphological characters and chromosome numbers among regenerated plants. For instance, plant regeneration from mono- and dihaploid potato genotypes often results in ploidy changes, but little aneuploidy or mixoploidy, whereas regenerated plants from tetraploid genotypes do display a wide range of aneuploidy and mixoploidy. It has been suggested that gross chromosomal and gene alterations can be tolerated to a greater extent because of the buffering capacity of the polyploid condition of the tetraploid cultivars (Sree Ramulu, 1986; Pijnacker and Sree Ramulu, 1990). Thus, phenotypic variation at the tetraploid level can be due to gross chromosomal and gene alterations which have a dominant phenotypic expression and will be otherwise deleterious in the great majority of cases. Furthermore, genetic make-up, especially heterozygosity of the tetraploid potato genome, might favour selective proliferation of chromosomally and genetically variant cells (Sree Ramulu, 1986).

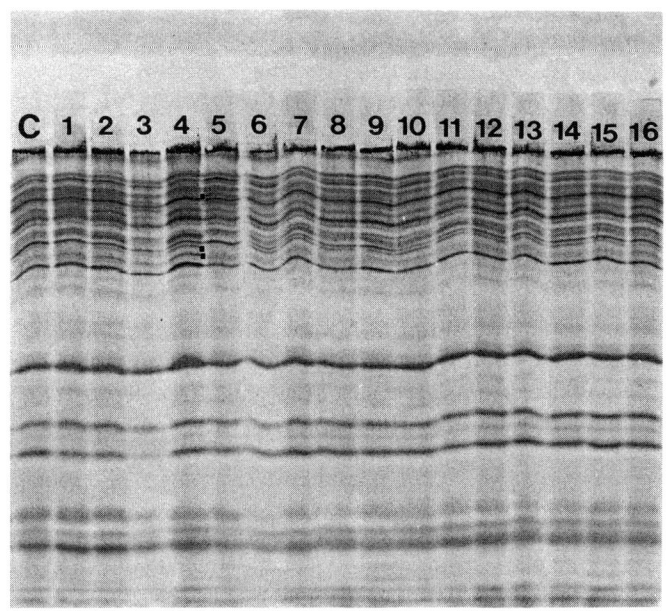

Fig. 9.3. Isoelectric-focusing profiles of total leaf proteins from a normal potato cultivar Désiree and from 16 regenerated plants from leaf tissues. Note that plant 4 shows a different protein profile from the control plant.

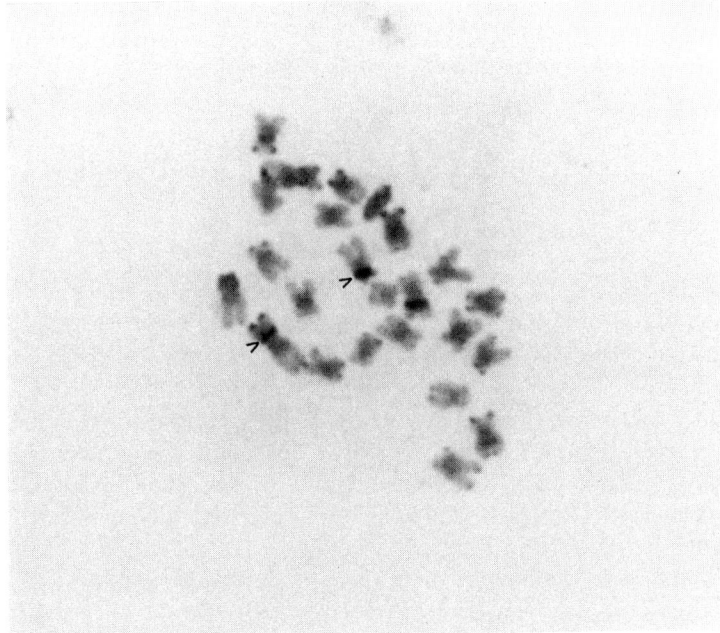

Fig. 9.4. C-banding pattern of the chromosomes from a normal leaf cell of dihaploid potato using a Giemsa-staining method. (Picture courtesy of Dr L.P. Pijnacker.)

Plant growth hormones, especially auxins and cytokinins, are essential for the induction of cell divisions in cultured cells and tissues, and their proper combinations in the culture media are required to induce regeneration of plants from explant cells and tissue. There is an indication that the concentration of auxins and cytokinins can influence changes in ploidy and phenotype of potato plants regenerated from callus or protoplast cultures (Jacobsen, 1981; Fish and Karp, 1986). For instance, in potato protoplast cultures, prolonged exogenous auxin concentration has been implicated in the occurrence of genotypic variations (Shepard, 1981). Hence, a decrease in the exogenous auxin concentration by dilution of the culture medium at 5 days after culture of protoplasts produced no significant changes in chromosome number variation (Nelson et al., 1986). In the case of callus cultures of dihaploid potato genotypes, there is an indication that 2,4-D increases the proportion of polyploid cells (Jacobsen, 1981). In fact, 2,4-D has been used intentionally to double the chromosome numbers of dihaploid lines (Jacobsen, 1978).

High concentrations of nutrient salts such as calcium ($CaCl_2$) and EDTA in the culture medium seem to increase the frequency of chromosomal abnormalities in cultured cells (Sree Ramulu, 1986). Recently, it has also been shown that high concentrations of sucrose (10 or 20 to $30\,g\,l^{-1}$) can induce polyploidization in callus cells derived from leaf segments of monohaploid lines but not in the callus cells of dihaploid and tetraploid lines (Pijnacker and Ferwarda, 1990). Thus, it can be concluded that somaclonal variations observed both at the phenotypic and genotypic levels could be the result of the types of culture media used. It is possible that some hormones, at certain concentrations or in certain combinations with other hormones, or in particular combinations with nutrients, may influence genetic stability in cultured cells by influencing their growth and metabolism.

Implications of Somaclonal Variation in Potato Breeding and Biotechnology

Any implications of somaclonal variation in potato breeding and biotechnology can only be realistically assessed by improving our understanding of the characteristics of somaclonal variation. In the past, these were not clear and therefore the potential usefulness of somaclonal variation in plant breeding was overestimated. We know now that somaclonal variation at every genetic level, from mutation of a single DNA base pair to the heritable alteration of traits encoded by multiple loci, can exist. The variation also encompasses heritable changes, non-heritable, epigenetic changes and changes that are both heritable and unstable. The use of somaclonal variation for crop improvement is dependent on a clear understanding of its advantages and disadvantages, as discussed by Karp (1990). Some of the advantages of somaclonal variation are that novel variants can arise, the frequencies of changes can be high, some useful changes in agronomic traits can arise, changes can occur in homozygous form and eventually new varieties may

Table 9.3 Agronomically useful somaclonal variation in potato.

Genotype or cultivar	Mode of propagation	Useful variants/ mutations	References
cv. Russet Burbank	Vegetative (tuber)	Yield, quality, uniformity and disease resistance (*Alternaria solani*)	Matern *et al.*, 1978 Secor and Shepard, 1981
cv. Fetwell, Foxton, Maris Piper	Vegetative (tuber)	Disease resistance (common scab *Streptomyces scabies*, potato virus Y and potato leafroll virus)	Thomson *et al.*, 1986
cv. Maris Piper	Vegetative (tuber)	Tuber shape, yield, maturity date, stem, leaf and flower morphology, early and late blight resistance	Thomas *et al.*, 1982 Bright *et al.*, 1983
Dihaploid lines	Vegetative (tuber)	Disease resistance (*Phytophthora infestans*, *Fusarium oxysporum*)	Behnke, 1979, 1980 Shepard, 1981
		Nematode and virus resistance, blackleg (*Erwinia carotovora*)	Wenzel and Uhrig, 1981 Wenzel, 1985

result. However, there are several disadvantages with somaclonal variation, including its uncontrollable and unpredictable nature and that most of the variation is of no apparent use, the variation is cultivar-dependent, not all changes are stable and heritable, changes occur at variable frequencies and most of the variation is not novel. Therefore, it is essential to devise methods which efficiently identify rare and useful somaclonal variants among a large population of useless variants. As pointed out earlier, the potato is especially suitable for investigating somaclonal variation due to its ease of regeneration, its asexual propagation and the difficulty of improving the potato by conventional methods. Several useful somaclonal variants have been identified, especially for resistance to diseases (Daub, 1986; Table 9.3). Some of these somaclonal variants have now been assessed under field conditions. Cassells *et al.* (1991) have reported a well-planned long-term experiment to evaluate the potential of somaclonal variation to correct character defects of potato cultivars, with particular reference to late blight of potato. They decided to screen adventitious regenerants for normal phenotype and vigour and then the latter were passed through one cycle of vegetative replication to break down chimeras, and following this were subjected to many years of field assessment for resistance to late blight. The selected lines were also screened for retention of the economically important traits of the parent cultivar. Their long-term field study has indicated that stability of somaclonal variants may not be a serious problem. Recently, Rietveld *et al.* (1991) have performed a detailed statistical study to assess quantitative, phenotypic somaclonal variability among adventitiously regenerated potato lines from tuber disc explants. This population was studied across five asexual generations, and the phenotypes were evaluated in three diverse locations. They observed that many somaclonal variants exhibited phenotypic stability and maintained the agronomically desirable characteristics inherent to the parent.

There has been no report of the production of a potato cultivar by somaclonal variation to date. However, in other crop plants, such as tomato, celery and carrot, somaclonal variation has resulted in cultivars (Evans and Sharp, 1986; Springen, 1987; Marcotrigiano and Jagannathan, 1988). It is also important to point out that somaclonal variation has resulted in several interesting biochemical mutants, which are being used successfully in plant metabolic pathways studies, i.e. amino acids and secondary metabolic pathways (Widholm, 1987; Weller *et al.*, 1987). Moreover, somaclonal variation may find its most useful application in the field of both basic and applied research in concert with selection for interesting mutations at the cellular level.

Somaclonal variation provides an excellent opportunity for studying mechanisms of genetic changes in plants. We have progressed well in our understanding of its causes and origins, but the question of whether the variation can be controlled in a meaningful way remains a challenge. With the increasing use of the biotechnological techniques in the horticulture and forest industries, including micropropagation, cryopreservation, anther culture, somatic hybridization and genetic transformation, the need to recognize and meet this challenge has never been more demanding. For example, it is essential that we explore all possible means to reduce or eliminate somaclonal variation in plants so that targeted genetic manipulation can effectively be used in producing useful transgenic plants. It remains to be seen whether the phenomenon of somaclonal variation in plant tissue culture will be viewed as a blessing or curse in the future.

Acknowledgements

I thank Professor A. Cassells, Dr A. Karp and Dr L.P. Pijnacker for providing their papers on somaclonal variation; Dr S. Cooper-Bland for her comments on this article; and Mrs E. Stewart for typing the manuscript.

References

Austin, S., Baer, M.A., Ehlenfeldt, M.K., Kazmiercgaak, P.J. and Helgesen, J.P. (1985) Intraspecific fusions in *Solanum tuberosum*. *Theoretical and Applied Genetics* 71, 172–175.

Baird, E., Cooper-Bland, S., Waugh, R., De Maine, M. and Powell, W. (1992) Molecular characterisation of inter- and intra-specific somatic hybrids of potato using randomly amplified polymorphic DNA (RAPD) markers. *Molecular and General Genetics* 233, 469–475.

Barker, H., Reavy, B., Kumar, A., Webster, K.D. and Mayo, M.A. (1992) Restricted virus multiplication in potato transformed with the coat protein gene of potato leafroll luteovirus: similarities with a type of host gene-mediated resistance. *Annals of Applied Biology* 120, 55–64.

Barsby, T.L., Shepard, J.F., Kemble, R.J. and Wong, R. (1984) Somatic hybridization in the genus *Solanum*: *S. tuberosum* and *S. brevidens*. *Plant Cell Reporter* 3, 165–167.

Bayliss, M.W. (1980) Chromosomal variation in plant tissues in culture. *International Review of Cytology Supplement* 11A, 113–144.

Behnke, M. (1979) Selection of potato callus for resistance to culture filtrates of *Phytophthora infestans* and regeneration of resistant plants. *Theoretical and Applied Genetics* 55, 69–71.

Behnke, M. (1980) General resistance to late blight of *Solanum tuberosum* plants regenerated from callus resistant to culture filtrates of *Phytophthora infestans*. *Theoretical and Applied Genetics* 56, 151–152.

Binding, H., Nehls, R., Schieder, O., Sopory, S.K. and Wenzel, G. (1978) Regeneration of mesophyll protoplasts isolated from dihaploid clones of *Solanum tuberosum*. *Physiologia Plantarum* 43, 52–54.

Bright, S.W.J., Jarrett, V., Nelson, R., Creissen, G., Karp, A., Franklin, J., Norburg, P., Kueh, J., Rognes, S. and Miflin, B. (1983) Modification of agronomic traits using *in vitro* technology. In: Mantell, S.H. and Smith, H. (eds) *Plant Biotechnology*. Cambridge University Press, London and New York, pp. 251–265.

Brown, P.T.H. (1989) DNA methylation in plants and its role in tissue culture. *Genome* 31, 717–729.

Butenko, R.G. and Kuchko, A.A. (1980) Somatic hybridization of *Solanum tuberosum* L. and *Solanum chacoense* Bitt. by protoplast fusion. In: Ferenczy, L., Farkas, G.L. and Lázár, G. (eds) *Advances in Protoplast Research*. Akadémiai Kiado, Budapest, pp. 293–300.

Cappadocia, M., Cheng, D.S.K. and Ludlum-Simonette, R. (1984) Plant regeneration from *in vitro* culture of anthers of *Solanum chacoense* Bitt. and interspecific diploid hybrids *S. tuberosum* L. × *S. chacoense* Bitt. *Theoretical and Applied Genetics* 69, 139–144.

Cassells, A.C., Deadman, M.L., Brown, C.A. and Griffin, E. (1991) Field resistance to late blight (*Phytophthora infestans* (Mont.) De Bary) in potato (*Solanum tuberosum* L.) somaclones associated with instability and pleiotropic effects. *Euphytica* 56, 75–80.

Creissen, G.P. and Karp, A. (1985) Karyotypic changes in potato plants regenerated from protoplasts. *Plant Cell, Tissue and Organ Culture* 4, 171–182.

D'Amato, F. (1985) Cytogenetics of plant cell and tissue cultures and regenerates. *CRC Critical Review of Plant Sciences* 3, 73–112.

Daub, M.E. (1986) Tissue culture and the selection of resistance to pathogens. *Annual Review of Phytopathology* 24, 159–186.

Day, A. and Ellis, T.H.N. (1985) Deleted forms of plastid DNA in albino plants from cereal anther culture. *Current Genetics* 9, 671–679.

De Block, M. (1988) Genotype-independent leaf disc transformation of potato (*Solanum tuberosum*) using *Agrobacterium tumefaciens*. *Theoretical and Applied Genetics* 76, 767–774.

Evans, D.A. and Sharp, W.R. (1986) Applications of somaclonal variation. *Bio/Technology* 4, 528–532.

Fish, N. and Karp, A. (1986) Improvements in regeneration from protoplasts of potato and studies on chromosome stability. 1. The effect of initial culture medium. *Theoretical and Applied Genetics* 72, 405–412.

Flavell, A.J., Smith, D.B. and Kumar, A. (1992) Extreme heterogeneity of Ty1-Copia group transposons in plants. *Molecular and General Genetics* 231, 233–242.

Gill, B.S., Kam, L.N.W. and Shepard, J.F. (1986) Origin of chromosomal and cytogenic variation in potato protoclones. *Journal of Heredity* 77, 13–16.

Hanson, M.R. (1984) Stability, variation, and recombination in plant mitochondrial genomes via tissue culture and somatic hybridisation. *Oxford Surveys of Plant Molecular and Cell Biology* 1, 33–52.

Harding, K. (1991) Molecular stability of the ribosomal RNA genes in *Solanum tuberosum* plants recovered from slow growth and cryopreservation. *Euphytica* 55, 141–146.

Jacobsen, E. (1978) Doubling dihaploid potato clones via leaf tissue culture. *Zeitschrift für Pflanzenzüchtung* 80, 80–82.

Jacobsen, E. (1981) Polyploidisation in leaf callus tissue and in regenerated plants of dihaploid potato. *Plant Cell, Tissue and Organ Culture* 1, 77–84.

Karp, A. (1990) On the current understanding of somaclonal variation. *Oxford Surveys of Plant Molecular and Cell Biology* 7, 1–58.

Karp, A. and Bright, S.W.J. (1985) On the causes and origins of somaclonal variation. *Oxford Surveys of Plant Molecular and Cell Biology* 2, 199–234.

Karp, A., Nelson, R.S., Thomas, E. and Bright, S.W.J. (1982) Chromosome variation in protoplast-derived potato plants. *Theoretical and Applied Genetics* 63, 265–272.

Karp, A., Risiott, R., Jones, M.G.K. and Bright, S.W.J. (1984) Chromosome doubling in monohaploid and dihaploid potatoes by regeneration from cultured leaf explants. *Plant Cell, Tissue and Organ Culture*, 3, 363–373.

Kemble, R.J. and Shepard, J.F. (1984) Cytoplasmic DNA variation in a protoclonal population. *Theoretical and Applied Genetics* 69, 211–216.

Kumar, A., Cooper-Bland, S. and Powell, W. (1992) Transfer of disease resistance genes in crop plants: the role of biotechnology. In: Wolf, J.N. (ed.) *Proceedings of Plant–Host Interactions, Senegal*. International Foundation of Sciences, Sweden, pp. 475–490.

Landsmann, J. and Uhrig, H. (1985) Somaclonal variation in *Solanum tuberosum* detected at the molecular level. *Theoretical and Applied Genetics* 71, 500–505.

Larkin, P.J. and Scowcroft, W.R. (1981) Somaclonal variation – a novel source of variability from cell cultures for plant improvement. *Theoretical and Applied Genetics* 60, 197–214.

Lassner, M.W. and Orton, T.J. (1983) Detection of somatic variation. In: Tanksley, S.D. and Orton, T.J. (eds) *Isozymes in Plant Genetics and Breeding, Part A*. Elsevier Science Publishers, Amsterdam, pp. 207–217.

Lee, M. and Phillips, R.L. (1988) The chromosomal basis of somaclonal variation. *Annual Review of Plant Physiology and Plant Molecular Biology* 39, 413–437.

Marcotrigiano, M. and Jagannathan, L. (1988) *Paulownia-tomentosa* cultivar somaclonal snowstorm. *Hortscience* 23, 226–227.

Masson, J., Lancelin, C., Bellini, C., Lecerf, M., Guerche, P. and Pelletier, G. (1989) Selection of somatic hybrids between diploid clones of potato (*Solanum tuberosum* L.) transformed by direct gene transfer. *Theoretical and Applied Genetics* 78, 153–159.

Matern, U., Strobel, G. and Shephard, J.F. (1978) Reaction to phytotoxins in a potato population derived from mesophyll protoplasts. *Proceedings of the National Academy of Sciences USA* 75, 4935-4939.

Nelson, R.S., Karp, A. and Bright, S.W.J. (1986) Ploidy variation in *Solanum brevidens* plants regenerated from protoplasts using an improved culture system. *Journal of Experimental Botany* 37, 253-261.

Ooms, G., Burrell, M.M., Karp, A., Bevan, M.W. and Hille, J. (1987) Genetic transformation in two potato cultivars with T-DNA from disarmed *Agrobacterium*. *Theoretical and Applied Genetics* 73, 744-750.

Orton, T.J. (1984) Genetic variation in somatic tissues: method or madness? *Advances in Plant Pathology* 2, 153-189.

Paterson, A.H., Tanksley, S.D. and Sorrells, M.E. (1991) DNA markers in plant improvement. *Advances in Agronomy* 46, 39-89.

Perl, A., Aviv, D. and Galun, E. (1990) Protoplast-fusion derived *Solanum* cybrids: application limitations; phylogenetic limitations. *Theoretical and Applied Genetics* 79, 632-640.

Peschke, V.M. and Phillips, R.L. (1992) Genetic variations of somaclonal variation in plants. *Advances in Genetics* 30, 41-75.

Pijnacker, L.P. and Ferwarda, M.A. (1984) Giemsa C-banding of potato chromosomes. *Canadian Journal of Genetics and Cytology* 26, 415-419.

Pijnacker, L.P. and Ferwarda, M.A. (1990) Effect of sucrose on polyploidization in early callus cultures of *Solanum tuberosum*. *Plant Cell, Tissue and Organ Culture* 21, 153-157.

Pijnacker, L.P. and Sree Ramulu, K. (1990) Somaclonal variation in potato: a karyotypic evaluation. *Acta Botanica Neerlandica* 39(2), 163-169.

Pijnacker, L.P., Hermelink, J.H.M. and Ferwarda, M.A. (1986a) Variability of DNA content and karyotype in cell cultures of an interdihaploid *Solanum tuberosum*. *Plant Cell Reports* 5, 43-46.

Pijnacker, L.P., Walch, K. and Ferwerda, M.A. (1986b) Behaviour of chromosomes in potato leaf tissue cultured *in vitro* as studied by Brd C-Giemsa labelling. *Theoretical and Applied Genetics* 72, 833-839.

Potter, R. and Jones, M.G.K. (1991) An assessment of genetic stability of potato *in vitro* by molecular and phenotypic analysis. *Plant Science* 76, 239-248.

Rietveld, R.C., Hasegawa, P.M. and Bressan, R.A. (1991) Somaclonal variation in tuber disc-derived populations of potato. *Theoretical and Applied Genetics* 82, 430-440.

Secor, G.A. and Shepard, J.F. (1981) Variability of protoplast-derived potato clones. *Crop Science* 21, 102-105.

Shepard, J.F. (1981) Protoplasts as sources of disease resistance in plants. *Annual Review of Phytopathology* 19, 145-166.

Sopory, S.K. and Tan, B.H. (1979) Regeneration and cytological studies of anther and pollen calli of dihaploid *Solanum tuberosum*. *Zeitschrift für Pflanzenzüchtung* 82, 31-35.

Springen, K. (1987) Improving on mother nature. *Newsweek* 109, 3.

Sree Ramulu, K. (1986) Case histories of genetic variability *in vitro*: potato. *Cell Culture and Somatic Cell Genetics of Plants* 3, 449-473.

Sree Ramulu, K., Dijkhuis, P. and Roest, S. (1984a) Genetic instability in protoclones of potato (*Solanum tuberosum* L. cv. 'Bintje'): new types of variation after vegetative propagation. *Theoretical and Applied Genetics* 68, 515-519.

Sree Ramulu, K., Dijkhuis, P., Roest, S., Bokelmann, G.S. and de Groot, B. (1984b) Early occurrence of genetic instability in protoplast cultures of potato. *Plant Science Letters* 36, 79-86.

Sree Ramulu, K., Dijkhuis, P., Hänisch ten Cate, Ch.H. and de Groot, B. (1985) Patterns of DNA and chromosome variation during *in vitro* growth in various genotypes of potato. *Plant Sciences* 41, 69-78.

Sree Ramulu, K., Dijkhuis, P., Roest, S., Bokelmann, G.S. and de Groot, B. (1986) Variation in phenotype and chromosome number of plants regenerated from protoplasts of dihaploid and tetraploid potato. *Plant Breeding* 97, 119-128.

Sree Ramulu, K., Dijkhuis, P. and Roest, S. (1989) Patterns of phenotypic and chromosome variation in plants derived from protoplast cultures of monohaploid, dihaploid and diploid genotypes and in somatic hybrids of potato. *Plant Sciences* 60, 101-110.

Stiekema, W.J., Heidekamp, F., Louwerse, J.D., Verhoeven, H.A. and Dijkhuis, P. (1988) Introduction of foreign genes into potato cultivars Bintje and Désiree using an *Agrobacterium tumefaciens* binary vector. *Plant Cell Reports* 7, 47-50.

Tempelaar, M.J., Jacobsen, E., Ferwarda, M.A. and Hartogh, M. (1985) Changes of ploidy levels by *in vitro* culture of monohaploid and polyploid clones of potato. *Zeitschrift für Pflanzenzüchtung* 95, 193-200.

Thomas, E., Bright, S.W.J., Franklin, J., Lancaster, V.A. and Miflin, B. (1982) Variation amongst protoplast-derived potato plants (*Solanum tuberosum* cv. 'Maris Bard'). *Theoretical and Applied Genetics* 62, 65-68.

Thomson, A.J., Gunn, R.E., Jellis, G.J. and Lacey, C.N.D. (1986) The evaluation of potato somaclones. In: *Symposium on Somaclonal Variation and Crop Improvement*. Gembloux, Belgium, pp. 233-240.

Visser, R.G.F., Jacobsen, E., Hesseling-Meinders, A., Schans, M.J., Witholt, B. and Feenstra, W.J. (1989) Transformation of homozygous diploid potato with an *Agrobacterium tumefaciens* binary vector system by adventitious shoot regeneration on leaf and stem segments. *Plant Molecular Biology* 12, 329-337.

Waara, S., Pijnacker, L., Ferwerda, M.A., Wallin, A. and Erikson, T. (1992) A cytogenetic and phenotypic characterization of somatic hybrid plants obtained after fusion of two different dihaploid clones of potato (*Solanum tuberosum* L.). *Theoretical and Applied Genetics* 85, 470-479.

Waugh, R. and Powell, W. (1992) Using RAPD markers for crop improvement. *Trends in Biotechnology* 10, 186-191.

Weller, S.C., Masiunas, J.B. and Gressel, J. (1987) Biotechnologies of obtaining herbicide tolerance in potato. In: Bajaj, Y.P.S. (ed.) *Biotechnology in Agriculture and Forestry 3*. Springer-Verlag, Berlin, pp. 281-297.

Wenzel, G. (1985) Strategies in unconventional breeding for disease resistance. *Annual Review of Plant Phytopathology* 23, 149-172.

Wenzel, G. and Uhrig, H. (1981) Breeding for nematode and virus resistance in potato via anther culture. *Theoretical and Applied Genetics* 59, 333-340.

Wenzel, G., Schieder, O., Przewozny, T., Sopory, S.K. and Melchers, G. (1979) Comparison of single cell culture derived *Solanum tuberosum* L. plants and a model for their application in breeding programs. *Theoretical and Applied Genetics* 55, 49-55.

Wheeler, V.A., Evans, N.E., Foulger, D., Webb, K.J., Karp, A., Franklin, J. and Bright, S.W.J. (1985) Shoot formation from explant cultures of fourteen potato cultivars and studies of the cytology and morphology of regenerated plants. *Annals of Botany* 55, 309-320.

Widholm, J.M. (1987) Potato improvement through *in vitro* selection for increased levels of free amino acids. In: Bajaj, Y.P.S. (ed.) *Biotechnology in Agriculture and Forestry 3*. Springer-Verlag, Berlin, pp. 268-280.

10 Molecular Genetics

K.N. WATANABE

Department of Plant Breeding and Biometry, 252 Emerson Hall, Cornell University, Ithaca, New York 14853-1902, USA.

Molecular Markers: Restriction Fragment Length Polymorphism (RFLP) and Randomly Amplified Polymorphic DNA (RAPD)

Introduction

Recent rapid developments in the area of molecular biology have led to new techniques for the revival and refreshment of classical plant science. Restriction fragment length polymorphisms (RFLP) and randomly amplified polymorphic DNA (RAPD) analyses, for example, are throwing light on the black box of plant genetics and plant breeding (Beckmann and Soller, 1986; Helentjaris and Burr, 1989; Tanksley *et al.*, 1989; Williams, J.G.K. *et al.*, 1990; Giovannoni *et al.*, 1991; Martin *et al.*, 1991; Michelmore *et al.*, 1991).

DNA markers generated from RFLP and RAPD can be used to assess and characterize genetic variation among plant genotypes of interest (Tanksley *et al.*, 1989; Dodds and Watanabe, 1990), and they have been extensively used to construct linkage maps in higher plants (Helentjaris and Burr, 1989; Tanksley *et al.*, 1989). Once plant chromosomes are well characterized by DNA markers, they can be used in breeding to monitor introgression and select for closely linked target traits. Monitoring RFLP or RAPD markers enables detection of variation at the DNA level, which, as DNA is the principal constituent of genes, allows detection of the plant genotypes of interest rather than monitoring plant phenotypes. The use of RFLP techniques in breeding offers the following advantages (Helentjaris and Burr, 1989; Tanksley *et al.*, 1989):

1. Identification of introgressed foreign genes is accurate and, following selection–recombination cycles, is very efficient.
2. It facilitates screening for target characters, even quantitatively inherited characters.

3. Early generation selection can be highly effective since it will allow direct selection of DNA types rather than indirect phenotypic selection.
4. Progeny testing is unnecessary to determine genotypes.
5. Recessive genes can be detected easily.

Both RFLP and RAPD techniques can provide an enormous number of DNA markers for breeding and genetics, whilst biochemical and morphological genetic markers are very scarce in potato. RFLP and RAPD markers also provide a highly accurate measure of genetic variation in contrast to isozyme or protein polymorphism since DNA is the physical component of alleles, and the allelic variation of these DNA markers is much greater than that of biochemical and morphological markers. RFLP or RAPD analyses should lead to the identification of new alleles or gene combinations that may be utilized effectively in potato breeding and genetics research.

Details of techniques and applications of RFLP are described elsewhere (Helentjaris and Burr, 1989; Tanksley *et al.*, 1989). RFLP markers are generated by the variable recognition sites of various restriction enzymes in the DNAs among plant genotypes. The procedure involves a series of rather tedious components (Figure 10.1): isolation of DNA from plant tissues, digestion of plant DNA by restriction enzymes, agarose gel electrophoresis of the digested DNAs, Southern blotting, hybridization with DNA probes, which are labelled either by radioactive isotopes (Feinberg and Vogelstein, 1983) or by a non-radioactive method (Kreike *et al.*, 1990), and radioautography in the case of radioactively labelled probes.

Randomly amplified polymorphic DNAs (RAPDs) generated *in vitro* by the polymerase chain reaction (PCR) also provide enormous potential as molecular markers in plant science (Welsh and McClelland, 1990; Williams, J.G.K. *et al.*, 1990; Giovannoni *et al.*, 1991; Klein-Lankhorst *et al.*, 1991; Martin *et al.*, 1991; Michelmore, *et al.*, 1991). Procedures for PCR are much simpler and more rapid than those of RFLP, and details can be obtained from Innis *et al.* (1990) (Figure 10.2). Segments of DNA are amplified by a thermostable enzymatic reaction called PCR, using the DNA from the target plant genotype as template and using random primers of common sizes ranging from 10 to 20 nucleotide sequences. The amplified DNAs are fractionated by agarose gel electrophoresis for analysis. The amplified products of PCR can be obtained in a few hours and the amount of DNA required for RAPD analysis is about one one-thousandth of that essential for RFLP analysis.

Genetic mapping and physical mapping

RFLP and RAPD markers segregate in a manner that is: (i) Mendelian and codominant; (ii) phenotype-neutral; (iii) free of epistatic interactions. Thus, these markers can be used to detect genetic linkage with major genes of interest (Figures 10.3 and 10.4).

The requirements for developing a DNA marker-based linkage map of an organism are: (i) sexual reproduction; (ii) a source of single-copy DNA clones

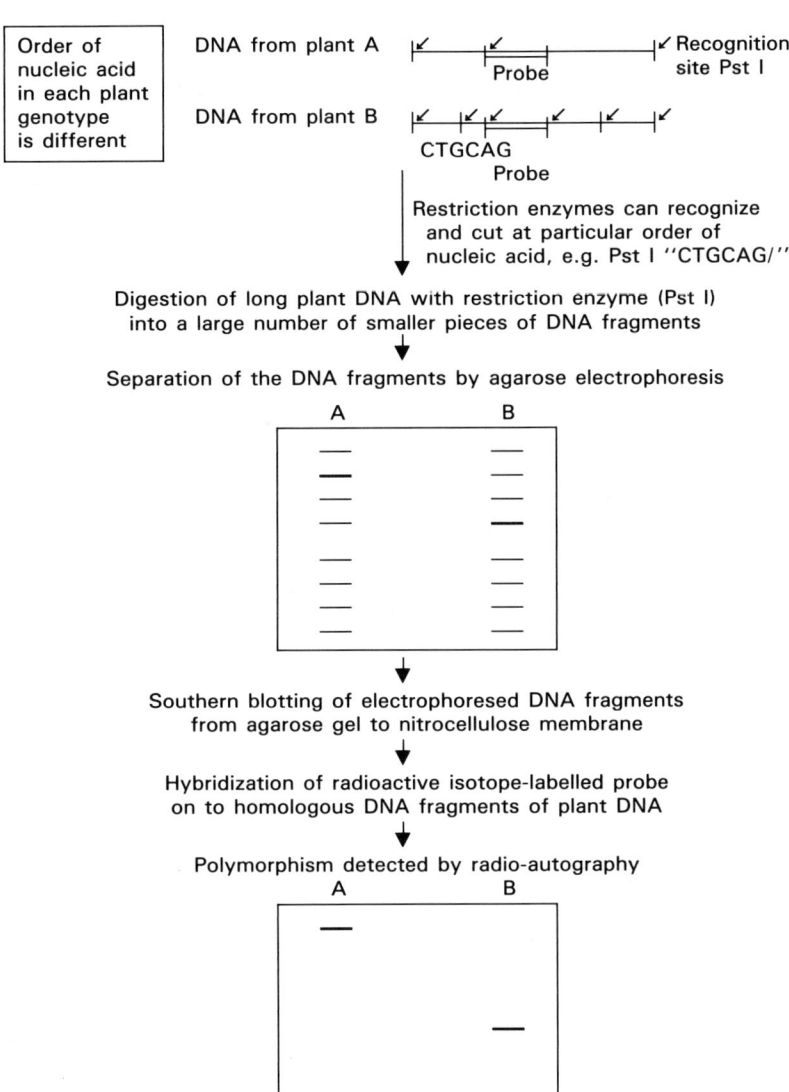

Fig. 10.1. A schematic diagram on generating restriction fragment length polymorphisms (RFLP) (Watanabe and Dodds, 1991).

(Tanksley *et al.*, 1989). Diploid potato species and breeding lines can easily be sexually hybridized to obtain a large number of progeny, and can be used to conduct molecular marker-based linkage mapping instead of tetraploid genotypes, with which the genetic analyses are far more complicated. *Solanum* section Petota, to which cultivated potatoes belong, exhibits a fair amount of RFLP polymorphisms (Bonierbale *et al.*, 1990; Debener *et al.*, 1990) and single-copy DNA clones can be easily identified (Bonierbale *et al.*, 1988; Gebhardt *et al.*, 1989a, b).

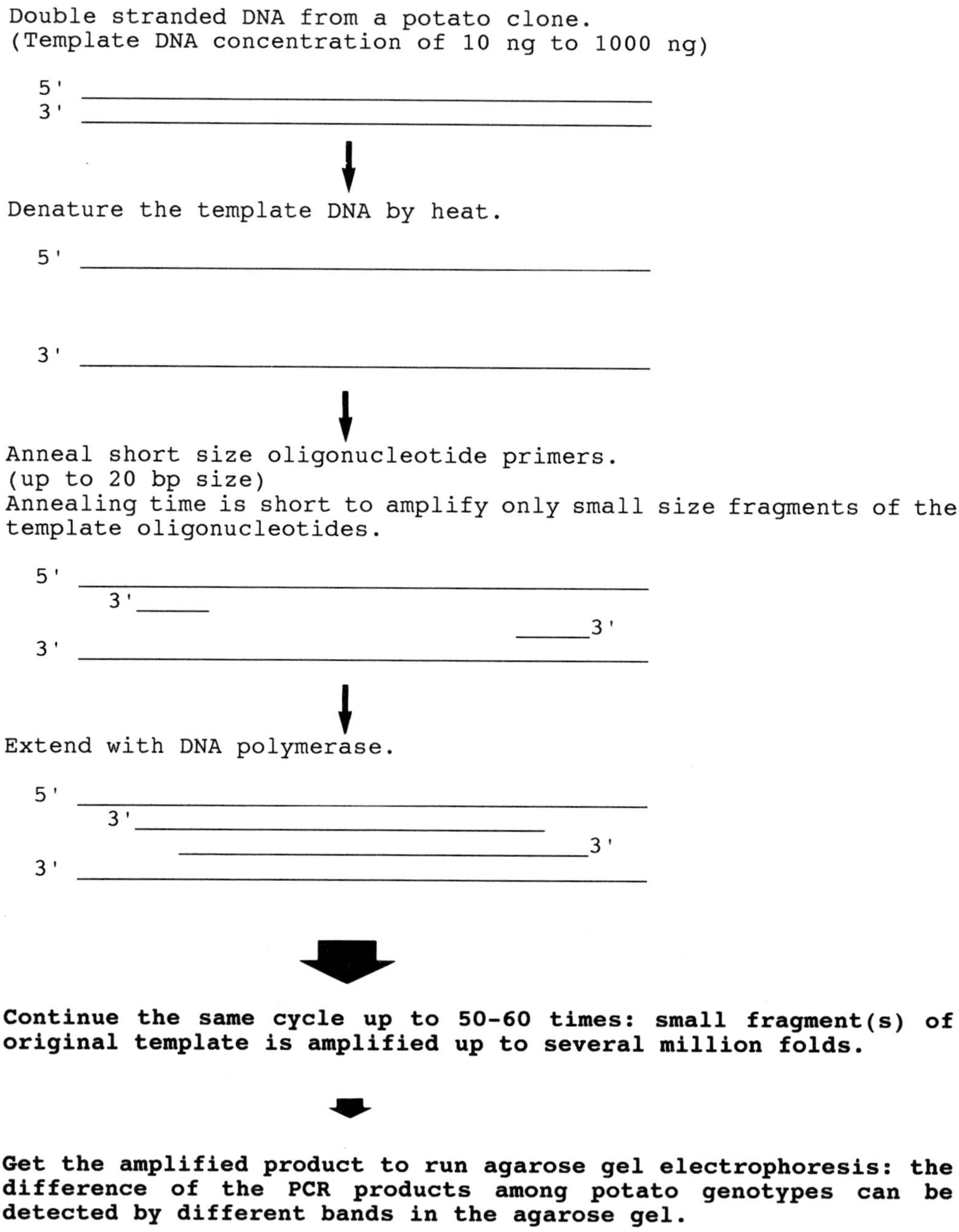

Double stranded DNA from a potato clone.
(Template DNA concentration of 10 ng to 1000 ng)

5' _____
3' _____

↓

Denature the template DNA by heat.

5' _____

3' _____

↓

Anneal short size oligonucleotide primers.
(up to 20 bp size)
Annealing time is short to amplify only small size fragments of the template oligonucleotides.

5' _____
   3'_____
                                    _____3'
3' _____

↓

Extend with DNA polymerase.

5' _____
   3'_____
         _____3'
3' _____

⬇

Continue the same cycle up to 50-60 times: small fragment(s) of original template is amplified up to several million folds.

⬇

Get the amplified product to run agarose gel electrophoresis: the difference of the PCR products among potato genotypes can be detected by different bands in the agarose gel.

Fig. 10.2. A schematic diagram on randomly amplified polymorphic DNAs (RAPD).

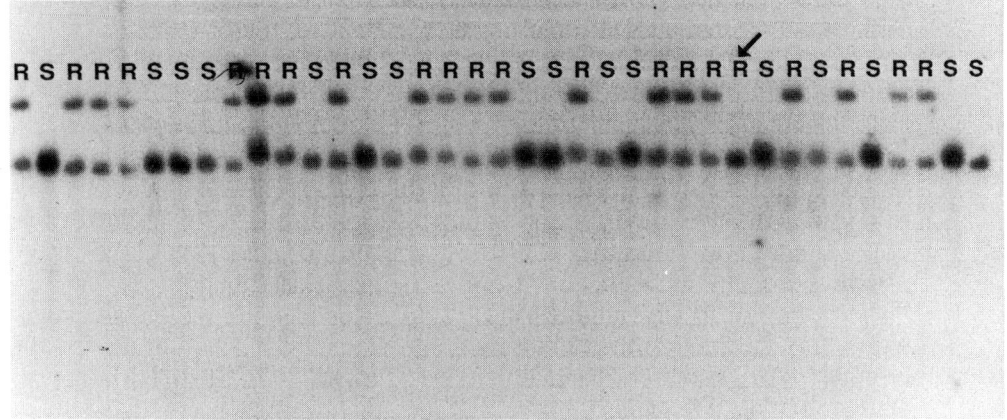

Fig. 10.3. RFLD marker CD 78 on chromosome V cosegragates with *H1* gene from CPC 1673, *S. tuberosum* subsp. *andigena*. R: resistant to *Globodera rostochiensis*, pathotype Ro 1; S: susceptible to Ro 1. (Courtesy of Pineda *et al.*, 1992.)

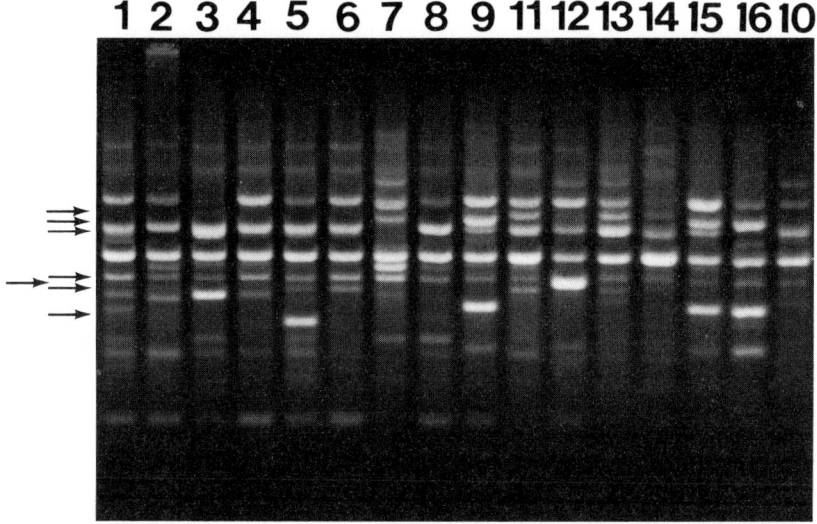

Fig. 10.4. DNA variation in potato cultivated and wild species generated by RAPD. Numbers on the top indicate different species. Arrows indicate mappable/scorable markers under different template concentrations. (Courtesy of Dr K. Hosaka, Kobe University, Japan, unpublished data.)

Genetic maps of molecular markers can be generated by using single-copy random genomic DNA clones or cDNAs (Tanksley *et al.*, 1989; Giovannoni *et al.*, 1991). Since probes from either animal or plant organisms can be used for molecular mapping (Tanksley *et al.*, 1989), enormous numbers of random genomic DNA clones and cDNAs are available to provide a vast amount of information on linkage between traits of interest and DNA markers in particular chromosome segments. DNA clones from either tomato or potato genomic and cDNA libraries have been used to construct genetic linkage maps in diploid potatoes (Bonierbale et al., 1988; Gebhardt et al., 1989b, 1991).

The RFLP data can be analysed with segregation of phenotypes of interest by using a computer program to generate a linkage map (Lander et al., 1987; Lander and Botstein, 1989). Most RFLP maps in plants are generated using F_2 or backcross generations derived from homozygous inbred lines. Potatoes, however, are highly heterozygous, are generally self-incompatible at the diploid level and suffer inbreeding depression, which may cause distortion in segregations. Generation of specific genetic stocks, such as inbred lines and near-isogenic lines (Young et al., 1988), for such molecular mapping is therefore highly unlikely in the immediate future. An alternative approach, using segregating F_1 populations from highly heterozygous parents, has been made to utilize existing genetic stocks efficiently (Bonierbale et al., 1988). The estimation of recombination frequencies in such F_1 populations was explained in detail by Ritter et al. (1990). Furthermore, bulked samples of plant DNA from individuals of a known phenotype can be used, instead of generating special genetic stocks, for genetic and cost efficiency (Giovannoni et al., 1991; Michelmore et al., 1991).

The first comprehensive RFLP linkage maps were generated by using two different types of segregating populations (Bonierbale et al., 1988; Gebhardt et al., 1989b). Bonierbale et al. (1988) used an interspecific cross from S. phureja × (S. tuberosum subsp. tuberosum × S. chacoense), which was mapped with tomato random genomic and cDNA clones, and found:

1. The alignment of tomato markers (135 loci) on the potato genome is principally the same as on tomato. Thus the nucleotide sequences are well conserved between these two genera and the karyotypes of these two related genera are very similar at the diploid level.
2. Five major paracentric-like inversions constitute the major difference between them.
3. The potato cross showed a reduced recombination frequency, resulting in a total map length of c. 700 cM (compared with 1600 cM in tomato interspecific crosses between L. esculentum × L. pennellii), while the genome size, estimated from nuclear DNA contents of diploid potato (1.9 pg) by flow cytometry, was close to that of tomato (2.0 pg) (Arumuganathan and Earle, 1991).

Gebhardt et al. (1989b) used intraspecific crosses among diploid potato breeding lines, mainly from S. tuberosum cultivars which had a broad genetic background tracing to several wild relatives. They used a backcross population for linkage mapping, which yielded a total map length of 690 cM, with 141 loci detected with potato random genomic clones and cDNAs.

The RFLP maps of Bonierbale et al. (1988) and Gebhardt et al. (1989b) were aligned using both common tomato and potato markers (Figure 10.5) (Gebhardt et al., 1991; Tanksley et al., 1992). Although the potato RFLP maps derived from different genetic backgrounds showed conservation of nucleotide sequences, as revealed by common tomato and potato markers, differences in chromosome and total map length were recognized. The total

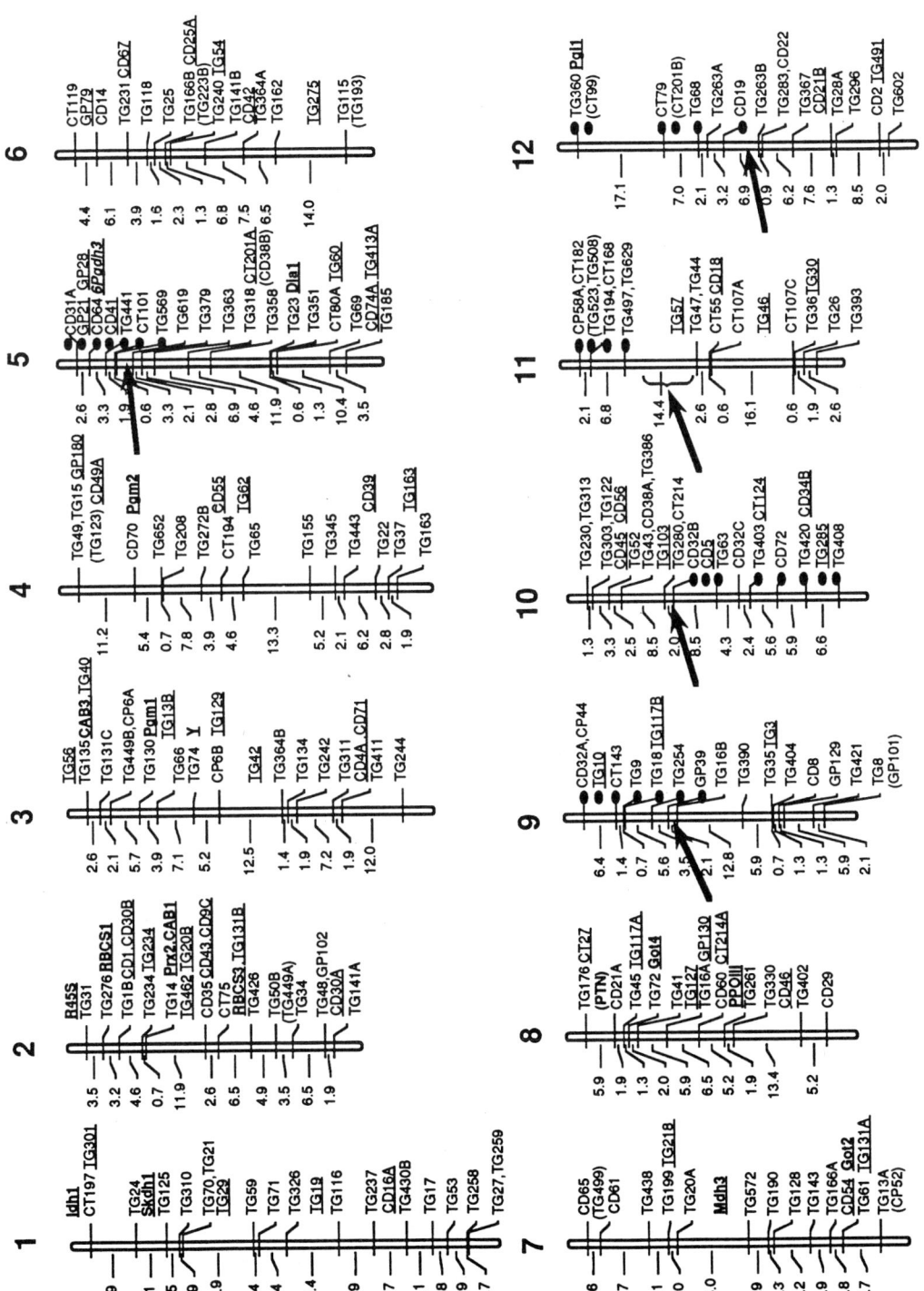

Fig. 10.5. A potato RFLP map based on a BC1 population from (2x *S. tuberosum* × *S. berthaultii*) × backcross to both parents. Arrows indicate paracentric inversion sites corresponding to tomato chromosomes. (Courtesy of Tanksley *et al.*, 1992.)

map length with the backcross population was 1034 cM, with 304 loci identified by 230 DNA markers and one morphological marker, whilst the map length in the interspecific cross was 670 cM, with 135 markers (Bonierbale et al., 1988). Therefore, a significant reduction in the map length was observed in the backcross which involved an interspecific cross-derived parent, compared with the intraspecific backcross (Gebhardt et al., 1991). It appeared that the reduction in map length was not random over the whole genome; thus it was attributed to the interspecific hybridity rather than population size, sampling error or large standard errors of marker positions due to distorted segregations (Gebhardt et al., 1991; Ritter et al., 1991). Sex-specific differences in recombination over the genome could also be an alternative reason for the reduction in the total map length in a population of interspecific origin (de Vicente and Tanksley, 1991). Gebhardt et al. (1991) took the view of Rick (1969) that recombination in an interspecific cross may not be impaired, but recombinant gametes or their zygotes may be preferentially eliminated. The reduction in recombination which was estimated by RFLP markers in populations of interspecific origin would thus be comparable with classical cytology in chromosome pairing and chiasma frequency (Singh et al., 1989; Matsubayashi, 1991).

The finding by Bonierbale et al. (1988) on the alignments of nucleotide sequences of the potato and tomato genomes confirmed that they are very conserved, as determined by reciprocal hybridizations of DNAs: tomato markers hybridized to the potato genome, and potato markers hybridized to the tomato genome (Gebhardt et al., 1991). The conservation of chromosome content allowed the designation of chromosome identification for the potato maps to be made based on tomato chromosome identification (Rick, 1974; Bonierbale et al., 1988; Tanksley and Mutschler, 1990; Gebhardt et al., 1991).

While tomato and potato differ by five major paracentric-like inversions (Bonierbale et al., 1992; Tanksley et al., 1992), the homosequential alignments of DNA markers between potato and tomato were recognized (Bonierbale et al., 1988; Gebhardt et al., 1991) and similar percentages of (G + C) contents in potato (37.4%) and tomato (41.4%) were observed (Messeguer et al., 1991); thus these inversions could be major causes in differentiation between tomato and potato (Figure 10.5). It appears that only small structural differences could be detected between potato and tomato: a small number of duplicated loci were observed with three tomato markers on the potato genome and six potato markers on the tomato genome; three markers did not locate at the corresponding homoeologous positions (Gebhardt et al., 1991). Cytosine methylation in tomato is, however, heritable and polymorphic, and it was pointed out that such a heritable polymorphism would have a potential role in the evolution of gene regulation (Messeguer et al., 1991). Although the chromosome inversions may have a major role in species differentiation between tomato and potato, gene regulation by altering the level and target sites of cytosine methylation may also have a role in the evolution of these two genera. However, this hypothesis must be confirmed in potato species by identifying the level, target sites and heritability of the cytosine methylation.

Distorted segregations were described in the current RFLP linkage analyses (Bonierbale *et al.*, 1988; Gebhardt *et al.*, 1989b, 1991; Ganal *et al.*, 1991; Tanksley *et al.*, 1992). While these may be attributed to the interspecific origin of parental mapping lines, another significant distortion of recombination was caused by the function of the self-incompatibility locus (Gebhardt *et al.*, 1991).

Gene–centromere distance is important to enable recombination for selection in potato breeding, as the position of a locus with respect to the centromere is the most critical factor on crossing over (Mendiburu and Peloquin, 1979; Peloquin *et al.*, 1989). When a target gene is proximal to the centromere the expected recombination frequency is low, whereas when it is distal to the centromere a higher probability of recombination between the centromere and the locus can be expected (Mendiburu and Peloquin, 1979). The centromere can be mapped on the genetic linkage map by using specific potato genotypes which confer $2n$ gametes (Mendiburu and Peloquin, 1979; Douches and Quiros, 1987; Rivard *et al.*, 1989; van Eck *et al.*, 1991; see chapter 5 by Tai). Furthermore, the distance between the target gene and centromere would provide an estimation of transmission rate of the gene to the progeny in germplasm enhancement, using $4x \times 2x$ crosses (Mendiburu and Peloquin, 1979). Isozyme loci have been employed to estimate map distances with corresponding centromeres (Douches and Quiros, 1987). Some knowledge of the position of centromeres in the DNA marker-based linkage map is now available (Tanksley *et al.*, 1992), but more effort to determine the chromosomal positions of centromeres must be encouraged for further characterization of the potato genome.

The DNA marker-based linkage maps may be used for cloning genes of interest. A strategy for the potato was postulated by Bonierbale *et al.* (1990), and this consists of four major steps. The first step is saturation of the potato genome by DNA markers, with a maximum gap size between markers of approximately 1.0 cM. The second step is the development of a physical map of the target chromosomal segment. Pulsed-field gel electrophoresis (PFGE) allows the separation of large DNA fragments of up to several megabase pairs (Mbp) (Ganal *et al.*, 1989). The patatin gene family was genetically and physically mapped by this approach (Bonierbale *et al.*, 1988; Ganal *et al.*, 1991): the chromosomal region containing the patatin gene family was approximately 1.4 Mbp long and carried a family of 10–15 patatin genes. Based on the physical map of the target region, in the third step the entire region of interest is cloned by chromosome walking for further characterization, with a system such as that based on the yeast artificial chromosome, which allows the cloning and amplification of DNA molecules of up to several hundred kilobases (Burke *et al.*, 1987). Finally, the cloned target DNA, which should contain the target gene, must be tested by a system such as *Agrobacterium*-mediated transformation to verify gene expression in the host plant.

Application of molecular markers for potato genetics and breeding

Genetic resources management and biosystematics

A serious bottleneck in genetic resources work on potatoes has been the lack of genetic markers. Isozyme and protein diversity, observed via starch or acrylamide gel electrophoresis, can provide some genetic markers. Douches and Quiros (1988) have established some allozyme loci in potato species, using the starch gel system, and this provides an opportunity to estimate genetic variation in germplasm collections. Cultivated potatoes can be clearly classified with respect to their parentage by the isozyme loci (Douches and Ludlam, 1991; Douches et al., 1991). Some additional information on genotypes at isozyme loci can also be obtained which may provide an estimate of homozygosity of chromosomal segments containing the isozyme loci (Douches and Quiros, 1987, 1988). Thus, isozyme loci can be used for fingerprinting potato genotypes to some extent.

Chloroplast DNA is much more conserved than the nuclear genome and can be used to estimate cytoplasmic differentiation, which provides information on the phylogenetic relationships among a wide range of taxa (Hosaka et al., 1984). Hosaka et al. (1984) generated a comprehensive phylogenetic tree of tuber-bearing *Solanum* species which has a close resemblance to the one from classical taxonomy. Buckner and Hyde (1985) observed chloroplast DNA variation between several genotypes of *S. tuberosum* subspp. *tuberosum* and *andigena*. Hosaka and Hanneman (1988) also made an extensive survey of the chloroplast DNA types in subsp. *andigena* collected throughout Latin America, where the cultivated potato originated. They classified these cultivated potatoes into five chloroplast DNA types, which is an unusually large amount of variation for a species. They also found that seven North American cultivars and one European cultivar have the same chloroplast DNA type as Chilean subsp. *tuberosum*. Kawagoe and Kikuta (1991) investigated further by sequencing chloroplast DNAs. They found a deletion of 241 bp which was not flanked by direct repeats in subsp. *tuberosum* compared with subsp. *andigena*. They also compared the sequences of two genes, rubisco large subunit (*rbcL*) and ATPase beta subunit (*atpβ*) among potato, tomato, tobacco and spinach: these two genes are highly conserved in their promoter regions but diverged to a high degree in the 5′ untranslated regions among these four species. The sequence data suggest that the rate of nucleotide substitution in potato chloroplast DNA is much slower than that in tomato (Kawagoe and Kikuta, 1991).

Nuclear genome-related DNA markers such as RFLPs can provide more refined information on genetic variation than allozyme markers and cytoplasmic DNA markers. RFLP and RAPD markers can be used for: (i) identifying cultivars by fingerprinting (Gebhardt et al., 1989a); (ii) establishment of representative 'core' collections in germplasm banks; and (iii) monitoring change of allelic frequencies in rejuvenations of true botanical seeds in germplasm collections. Hosaka and Spooner (1992) surveyed *S. acaule* germplasm extensively by nuclear RFLP analysis and clustered it into subspecies and several groups of particular accessions by principal component analysis. This may help to organize large germplasm collections into more manageable sizes at gene banks and achieve more efficient management. This approach needs to be examined further to increase cost efficiency and to decide whether isozyme loci or RFLP markers are used. RAPD markers would provide further ease in the generation of representative collections. Attempts have already been made with this type of marker to characterize dihaploids of tetraploid cultivars (Baird *et al.*, 1991) and several cultivated species and their closely related relatives (Cisneros and Quiros, 1991).

Potential in germplasm utilization and breeding

Although conventional potato breeding has demonstrably resulted in improving potato crops, there are several key bottlenecks (Watanabe, 1991).

1. The most important cultivated potato is a tetraploid with tetrasomic inheritance; thus segregation is more complicated than with disomic inheritance, and selection even for simply inherited characters is sometimes laborious and less cost-effective compared with that in diploids.
2. Major breeding targets tend to be quantitatively inherited, for example, resistances to bacterial wilt caused by *Pseudomonas solanacearum*, potato leafroll virus, potato tuber moth and root-knot nematodes, as well as tuber quality traits and yield components.
3. Genotype-environment interactions complicate the expression of the characters.
4. Identification of individual chromosomes by conventional cytogenetic methods is very difficult, and it is very hard to monitor introgression in hybridizations from wild species, which are valuable sources of resistances to major diseases and pests.

A major problem is the difficulty in monitoring introgression of target genes and elimination of undesirable characters, especially from wild species. Furthermore, confirmation of transmission of resistance(s) from progenitors by progeny testing is time-consuming but essential, since several genetic factors, such as cytoplasmic effects and female-male interactions, influence the expression of resistance in progenies. However, RFLP and RAPD markers are actual DNA; thus use of these markers could alleviate these bottlenecks greatly.

The cost of employing DNA markers, however, is an impediment to their use. Compared with the inherent difficulty in screening and selection for many target characters in potato breeding, the cost of DNA markers may nevertheless prove to be reasonably competitive. For example, present phenotypic screening for bacterial wilt resistance requires preparation of much plant material either by generation of seedlings or cuttings, and an incubation period of at least 2 weeks (Watanabe *et al.*, 1992). It also requires repeated testing and progeny testing to identify target genotypes. Thus, these procedures require enormous labour, materials and time. On the other hand, linked DNA markers would require a maximum of several days to identify the genotypes of interest, once small amounts of plant tissue were available (Giovannoni et al., 1991; Martin et al., 1991; Michelmore et al., 1991). Cost reductions could be made by simplifying several procedures in the utilization of DNA markers. Sample preparation is one step which can be simplified (Tai and Tanksley, 1990). Further simplification is being developed in certain cases by direct squash of plant tissue for assays (Pehu et al., 1990; Langridge et al., 1991). The employment of PCR-based RAPD markers could also facilitate the screening and selection procedures and increase efficiency in labour, cost and time (Martin et al., 1991).

Monitoring hybridity and introgression

Species-specific probes have been employed to select cell fusion products and subsequent backcross lines (Schwiezer et al., 1988; Pehu et al., 1990; Williams, C.E. et al., 1990). These probes can be used to confirm sexual and somatic hybridizations among distantly related species, where hybrid or heterokaryon identification is difficult due to lack of alternative appropriate markers. Also, introgression of chromosome segments from *S. brevidens* was monitored by using RFLP markers in backcross progeny from cell fusion products (Williams, C.E. et al., 1990).

Other series of such species-specific probes can also be obtained, since tuber-bearing *Solanum* spp. have extensive RFLP variations (Bonierbale et al., 1990; Debener et al., 1990). An example of species-specific DNAs is the small satellite repetitive sequences of 183 bp which are tandemly organized in *S. acaule* (Schwiezer et al., 1988). *S. tuberosum*- and *S. brevidens*-specific DNAs are also repetitive, and one *S. brevidens*-specific sequence is a satellite sequence (Visser et al., 1988; Pehu et al., 1990). These repetitive sequences can be used to characterize in detail telomeric and centromeric regions of the chromosomes (Pehu et al., 1990). These species-specific probes could be employed for the early phase of germplasm enhancement with wild species to eliminate genotypes with unnecessary chromosome segments, as well as to select for genotypes with target chromosome segments of wild species origin. Thus, the linkage drag (Zeven et al., 1983) in introgression would be minimized.

Utilization of DNA markers linked with economically important loci in screening and selection

Localization of economically important genes has already been demonstrated with some simply inherited loci which confer disease and pest resistances. For example, the $R1$ locus, which controls a race-specific resistance to late blight on chromosome 5, was mapped by using two F_1 populations (Leonards-Schippers et al., 1991).

Two loci which confer high levels of resistance to potato virus X were mapped to two independent chromosomal positions (Ritter et al., 1991). It appears that this finding corresponds to the classical genetic analyses by Cockerham (1970). These two loci originate from *S. tuberosum* subsp.

andigena and *S. acaule*, and were mapped by different approaches. The *Rx1* locus from subsp. *andigena* was mapped to the distal end of chromosome 12 by using random mapping of probes, revealing polymorphic loci spread through the genome. The *Rx2* locus from *S. acaule* was mapped to chromosome 5 by using target-mapping of polymorphic restriction fragments, one of which was linked with the locus.

A simply inherited locus *Gro1* for resistance to cyst nematode (*Globodera rostochiensis*) pathotype Ro1 was mapped to chromosome 7 with a close marker locus *TG20* (Barone et al., 1990). Jacobs et al. (1991) found that the resistant locus from *S. vernei* does not have a linkage relationship with the locus *TG20*, which was mapped by Barone et al. (1990). Furthermore, the simply inherited *H1* gene from CPC 1673, *S. tuberosum* subsp. *andigena*, which shows resistance to the same Ro1, was located on chromosome 5 (Pineda et al., 1992; Figure 10.3). Thus the loci which confer resistance to pathotype Ro1 of *G. rostochiensis* occupy different chromosomal positions among different sources of the resistance from several wild species.

Once DNA markers linked with target traits are available, screening for the trait should be quickly but accurately achieved, compared with conventional screening procedures. Since availability of pathogens/pests of interest is not essential for resistance screening by molecular markers, testing for resistances to quarantined pathogens and pests can be facilitated. Selection of a genotype with respect to the entire genome and particular chromosomal segments of interest can be conducted based on a linkage map by a graphic computer program (Young and Tanksley, 1989).

QUANTITATIVELY INHERITED TRAITS

Attempts to identify molecular markers linked to genes affecting quantitatively inherited traits which are major constraints on potato breeding are under way (M.W. Bonierbale, pers. comm.; D.S. Douches, pers. comm.). It appears that some isozyme loci have significant associations with loci controlling tuber specific gravity and dormancy in diploid interspecific populations (Freyre and Douches, 1991). However, the complicated nature of the observed genotype–environment interactions must be elucidated in detail before such isozyme loci would be very useful for screening for such tuber characters. On the other hand, DNA markers could help to resolve the complexity of the control of quantitative traits by identifying the genes involved and their interactions with environments. Thus, the genetics of quantitative traits might be elucidated clearly and the DNA markers would then be applied to screening without the complications of environmental and other external factors.

Some achievements have already been demonstrated in mapping quantitative trait loci (QTL) in several species (see Helentjaris and Burr, 1989; Paterson et al., 1990). QTL mapping can be achieved by using 'selected overlapping recombinant chromosomes' (Lander and Botstein, 1989; Paterson et al., 1990). Several disease and pest resistances which are quantitatively inherited are targeted for mapping by DNA markers, such as glandular trichomes, which confer multiple insect resistances in foliage (Bonierbale et al., 1992), and multiple resistances to cyst nematodes (Jacobs et al., 1991).

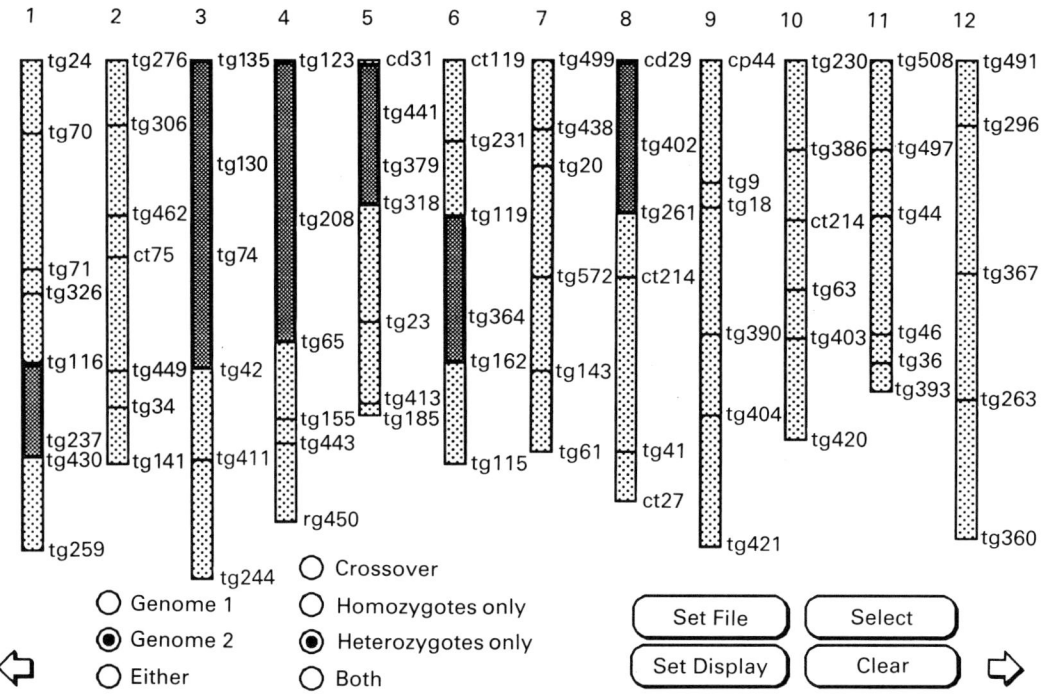

Fig. 10.6. A chromosomal idiotype on quantitative traits which confer insect resistances, which is based on RFLP map information from a BC1 generation to cultivated potato. Dark marked chromosomal regions are from wild species donor of the resistance, and dotted regions are from cultivated potato (M.W. Bonierbale *et al.*, unpublished data).

An attempt has been made to select potato clones for quantitative traits with a chromosomal idiotype which was generated based on a RFLP map (Figure 10.6) (M.W. Bonierbale *et al.*, unpublished data).

Gene Expression: Transposons and Self-incompatibility

The potato has been used as a model plant species in plant cell and molecular biology due to its unique reproductive characters and ease of handling *in vitro*. Gene organization and expression have been extensively studied in several stress-induced genes, such as wound-inducible genes, especially in promoter regions (see references in Stanford *et al.*, 1989; Davis *et al.*, 1990; Keil *et al.*, 1990; Crosby and Vayda, 1991; Ricky and Belknap, 1991). Gene characterization and tissue- and developmental-specific expressions have been intensively investigated in the patatin gene family (see further references in Blundy *et al.*, 1991; Ganal *et al.*, 1991; Liu *et al.*, 1991). Some aspects of carbohydrate metabolism have been elucidated and the genes encoding corresponding enzymes cloned and characterized (see Müller-Röber *et al.*, 1990; Koßmann *et al.*, 1991). As much recent literature covers these subjects, this section deals only with transposons and the molecular characterization of the self-incompatibility gene as unique topics in potato molecular biology.

Transposons

While molecular marker-based linkage and physical maps can be used to clone genes of interest, transposable elements are alternative tools for isolating target genes. The strategy is called transposon tagging (Schmidt *et al.*, 1987; Wienand and Saedler, 1988). The technique of transposon tagging has an advantage when the target gene does not produce abundant transcripts or products or when the gene product has not been identified due to its regulatory nature (Schmidt *et al.*, 1987). The target gene of interest is marked by the integration of a transposon causing interruption of expression in the target gene. The mutated gene can then be isolated by DNA hybridization, using the transposon as the probe. Several maize genes were isolated by transposon targeting (see Wienand and Saedler, 1988).

Maize transposable elements can be transferred to and integrated into potato. The *Ac* element was found to transpose by introducing it into a diploid potato genotype via tDNA of *Agrobacterium tumefaciens* (Knapp *et al.*, 1988), and the *En-1/Spm* element revealed excision–integration events in a dihaploid from a 4*x* potato cultivar (Frey *et al.*, 1989). *Ac/Ds* elements from maize are being employed in diploid potatoes to isolate genes of economic importance (Blok *et al.*, 1991; Pereira *et al.*, 1991).

Transposons occur extensively in the plant kingdom (Nevers *et al.*, 1985), and transposon-like elements occur in potatoes (Camirand *et al.*, 1990; Köster-Töpfer *et al.*, 1990; Kumar *et al.*, 1991). Köster-Töpfer *et al.* (1990) found an insertion in front of the putative transcription start site in an inactive patatin gene belonging to the class II subfamily. The insertion contained a structure resembling a transposable element: an 11 bp inverted repeat at the termini and an 8 bp duplication flanking the insertion site. The element contains a terminal, nearly perfect, inverted repeat of 11 bp, which has a significant homology to the *Ac/Ds* family of maize, the *Tam3* element of snapdragon and the *Tpc1* element of parsley. It is likely that this inactivation of the patatin gene is due to a transposable element.

Camirand *et al.* (1990) compared differences between the potato starch phosphorylase gene and the human glycogen phosphorylase gene. They found differences in the positions of 14 introns and also found that the fifth intron of potato starch phosphorylase contained a *copia*-like transposable element inserted in the opposite orientation to that of the phosphorylase gene. They designated the element *Tst1*. It contains long terminal repeats of 285 bp at the 5′ end and of 283 bp at the 3′ end, with an internal domain of 4492 bp, which is characterized by four open reading frames that encode protein domains for a reverse transcriptase, an integrase, an RNA-binding site and a protease. It appears that *copia–Ty* family retrotransposons commonly occur in solanaceous species, and that *copia-Ty* family elements occur repeatedly in potatoes but with highly heterogeneous sequences (Kumar et al., 1991).

Reproductive biology: self-incompatibility

Gametophytic self-incompatibility exists in tuber-bearing *Solanum*, and is a common feature of solanaceous species, where it is controlled by a single S locus with many alleles (Hermsen, 1978; Clarke et al., 1989; Rivard et al., 1989; Xu et al., 1990). Self-incompatibility is one of the unique features of reproductive biology in potato species with respect to species differentiation, evolution and potential uses in breeding and seed production. The pollen–pistil interactions and further pollination events are key issues in utilizing valuable wild germplasm, and understanding self-incompatibility at the molecular level would be a breakthrough for comprehensive potato reproductive biology.

The self-incompatibility locus of *S. chacoense* was characterized by pistil-specific proteins (Xu et al., 1990). Homology of amino acid sequence between the S_2 and S_3 proteins was 41.5%, which indicates a high level of sequence variability between alleles. On the other hand, some specific amino acid residues were very conserved between the two alleles, and furthermore common residues were recognized with those of S proteins from *Nicotiana alata*, S_2, S_3 and S_6 alleles (Clarke et al., 1989).

Kaufmann et al. (1991) confirmed allelic diversity and gene structure of the S locus in *S. tuberosum*. They isolated and characterized genomic and cDNA clones encoding self-incompatible pistil-specific proteins of S_1, S_{r1} and S_2. The S_{r1} allele is derived from cultivar Roxy, of unknown S phenotype, and is highly homologous to the standard S_1 allele. A high degree of difference in sequence homology (66%) was seen between the S_1 and S_2 alleles. The S_1 and S_{r1} alleles were distinct in phenotype, although protein sequence homology between S_1 and S_{r1} was 95%, which would indicate that small sequence variability is enough for altering the function/phenotype of S alleles. It appears that the gene coding for S_1- and S_2-specific proteins has a simple structure: two exons are separated by a small intron of 113 (S_1) and 117 (S_2) bp.

Judging from the above reports, it seems that the gene coding for S-specific proteins is structurally organized into conserved and hypervariable regions, similar to the S genes coding *Nicotiana alata* S proteins (Clarke et al., 1989; Kaufmann et al., 1991). Thompson et al. (1991) proposed a structure for the S locus which consists of two tightly linked genes: a style-specific gene encoding the S-linked glycoprotein (SLG) and a pollen-specific gene, but the idea must be proved by isolating and characterizing a pollen-specific protein which is as yet undetermined.

Gebhardt et al. (1991) mapped the self-incompatibility locus by using a cDNA clone of S_{r1}, which was characterized by Kaufmann et al. (1991). The locus was closely linked with loci *CP100* and *Gp128(a)* on chromosome I. These two flanking RFLP markers are 1.2 cM apart. Furthermore, the chromosomal position of the S locus is very similar to that of tomato, judging from a marker locus *TG24* which is commonly mapped on potato and tomato (Tanksley and Loaiza-Figueroa, 1985; Bonierbale et al., 1988; Gebhardt et al., 1991; Tanksley et al., 1992).

A self-compatible (SC) mutation was recognized in dihaploids ($2x$) of *S. tuberosum* (Hermsen, 1978; Thompson et al., 1991). Hermsen (1978) proposed that this SC mutation of the S_1 allele could be attributed to a translocation of the S_1 allele to a new chromosomal location. This hypothesis is supported by the molecular characterization of the S locus by Thompson et al. (1991). The latter, however, proposed two possibilities for the origin of the SC mutation, either a translocation of the S allele or an S allele inhibitor mutation at a distinct locus. It should be noted that one of the pollen-expressed unilateral incompatibility loci derived from wild tomato species was mapped very close to the S locus of tomato (Chetelat and DeVarna, 1991), while the map position of the S locus in the potato corresponds well with the position of the tomato S locus (Gebhardt et al., 1991). This would suggest another possibility of S locus mutation in potato and would provide an alternative hypothesis for species differentiation by incompatibility–compatibility mutation.

Conclusion

Potato molecular genetics is providing basic knowledge to help interpret classical potato genetics. Although the application of the knowledge and techniques is not yet established in potato breeding, further efforts and knowledge should facilitate potato breeding, especially by the application of DNA markers. Although the potato is highly heterozygous, and as a rule the self-incompatibility system precludes the generation of pure breeding lines by selfing in diploid species, it is ideal experimental plant material for the study of basic molecular genetics and biology in such areas as gene cloning, gene expression, transposons, carbohydrate metabolism and plant–microbe interactions.

Acknowledgements

The author would like to thank Drs M.W. Bonierbale and S.D. Tanksley, Department of Plant Breeding and Biometry, Cornell University, New York, USA, for their productive suggestions and reviews of this manuscript. This work was supported by the International Potato Center.

References

Arumuganathan, K. and Earle, E.D. (1991) Nuclear DNA content of some important plant species. *Plant Molecular Biology Reporter* 9(3), 208–218.
Baird, E., Cooper-Bland, S., Waugh, R. and Powell, W. (1991) The development and use of randomly amplified polymorphic DNA (RAPD) markers in potato. *Second*

International Potato Molecular Biology Symposium, St Andrews, Scotland, 11–15 August 1991. Abstract.

Barone A., Ritter, E., Schachtschabel, U., Debener, T., Salamini, F. and Gebhardt, C. (1990) Localization by restriction fragment length polymorphism in potato of a major gene conferring resistance to the potato cyst nematode *Globodera rostochiensis*. *Molecular and General Genetics* 224, 177–182.

Beckmann, J.S. and Soller, M. (1986) Restriction fragment length polymorphisms and genetic improvement of agriculture species. *Euphytica* 35, 111–124.

Blok, K.M., van Vark, A., Faber, N.M. and Hille, J. (1991) Towards isolation of disease-resistance genes from potato using transposon-tagging. *Second International Potato Molecular Biology Symposium*, St Andrews, Scotland, 11–15 August 1991. Abstract.

Blundy, K.S., Blundy, M.A.C., Carter, D., Wilson, F., Park, W.D. and Burrell, M.M. (1991) The expression of class I patatin gene fusions in transgenic potato varies with both gene and cultivar. *Plant Molecular Biology* 16, 153–160.

Bonierbale, M.W., Plaisted, R.L. and Tanksley, S.D. (1988) RFLP maps based on a common set of clones reveal modes of chromosomal evolution in potato and tomato. *Genetics* 120, 1095–1103.

Bonierbale, M.W., Ganal, M.W. and Tanksley, S.D. (1990) Application of restriction fragment length polymorphisms and genetic mapping in potato breeding and molecular genetics. In: Vayda, M.E. and Park, W.D. (eds) *The Molecular and Cellular Biology of the Potato*. CAB International, Wallingford, UK, pp. 13–24.

Bonierbale, M.W., Plaisted, R.L. and Tanksley, S.D. (1992) Application of genome mapping in potato. *Joint EAPR-Breeding EUCARPIA-Potato Sections Conference*, INRA, Landerneau, France, 12–17 January 1992. Abstract.

Buckner, B. and Hyde, B.B. (1985) Chloroplast DNA variation between the common cultivated potato (*Solanum tuberosum* subsp. *tuberosum*) and several South American relatives. *Theoretical and Applied Genetics* 71, 527–531.

Burke, D., Carle, G. and Olson, G. (1987). Cloning of large segments of exogenous DNA into yeast by means of artificial chromosome vectors. *Science* 236, 806–812.

Camirand, A., St-Pierre, B., Marineau, C. and Brisson, N. (1990) Occurrence of a copia-like transposable element in one of the introns of potato starch phosphorylase gene. *Molecular and General Genetics* 224, 33–39.

Chetelat, R.T. and DeVarna, J.W. (1991) Expression of unilateral incompatibility in pollen of *Lycopersicon pennellii* is determined by major loci on chromosomes 1, 6, 10. *Theoretical and Applied Genetics* 82, 704–712.

Cisneros, P. and Quiros, C.F. (1991) Evolutionary studies of potatoes using RAPD markers. *Second International Potato Molecular Biology Symposium*, St Andrews, Scotland, 11–15 August 1991. Abstract.

Clarke, A.E., Anderson, M.A., Atkinson, A., Bacic, A., Ebert, P.R., Jahnen, W., Lush, W.M., Mau, S.L. and Woodward, J.R. (1989) Recent developments in the molecular genetics and biology of self-incompatibility. *Plant Molecular Biology* 13, 267–271.

Cockerham, G. (1970) Genetical studies on resistance to potato viruses X and Y. *Heredity* 25, 309–348.

Crosby, J.S. and Vayda, M.E. (1991) Stress-induced translational control in potato tubers may be mediated by polysome-associated proteins. *Plant Cell* 3, 1013–1023.

Davis, M.C., Butler, W. and Vayda, M.E. (1990) Molecular responses to environmental stresses and their relationship to soft rot. In: Vayda, M.E. and Park, W. (eds) *Molecular and Cellular Biology of the Potato*. CAB International, Wallingford, UK, pp. 71–87.

Debener, T., Salamini, F. and Gebhardt, C. (1990) Phylogeny of wild and cultivated *Solanum* species based on nuclear restriction fragment length polymorphism (RFLP). *Theoretical and Applied Genetics* 79, 360-368.

de Vicente, M.C. and Tanksley, S.D. (1991) Genome-wide reduction in recombination of backcross progeny derived from male versus female gametes in an interspecific cross of tomato. *Theoretical and Applied Genetics* 83, 173-178.

Dodds, J.H. and Watanabe, K. (1990) Plant genetics resources management and biotechnology. *Diversity* 6(3-4), 26-28.

Douches, D.S. and Ludlam, K. (1991) Electrophoretic characterization of North American potato cultivars. *American Potato Journal* 68, 767-780.

Douches, D.S. and Quiros, C.F. (1987) Use of $2x$-$4x$ crosses to determine gene-centromere map distances of isozyme loci in *Solanum* species. *Genome* 29, 519-527.

Douches, D.S. and Quiros, C.F. (1988) Additional loci in tuber-bearing Solanums: inheritance and linkage relationships. *Journal of Heredity* 79, 377-384.

Douches, D.S., Ludlam, K. and Freyre, R. (1991) Isozyme and plastid DNA assessment of pedigrees of nineteenth century potato cultivars. *Theoretical and Applied Genetics* 82, 195-200.

Feinberg, A.P. and Vogelstein, B. (1983) A technique for radiolabelling DNA restriction fragments to a high specific activity. *Annals of Biochemistry* 132, 6-13.

Frey, M., Tavantzis, S.M. and Saedler, H. (1989) The maize *Em/Spm* element transposes in potato. *Molecular and General Genetics* 217, 172-177.

Freyre, R. and Douches, D.S. (1991) Use of molecular markers for the study and analyses of quantitative tuber traits in diploid potatoes. *Second International Potato Molecular Biology Symposium*, St Andrews, Scotland, 11-15 August 1991. Abstract.

Ganal, M.W., Young, N.D. and Tanksley, S.D. (1989) Pulse field gel electrophoresis and physical mapping of large DNA fragments in the *Tm-2a* region of chromosome 9 in tomato. *Molecular and General Genetics* 215, 395-400.

Ganal, M.W., Bonierbale, M.W., Roeder M.S., Park, W.D. and Tanksley, S.D. (1991) Genetic and physical mapping of the patatin genes in potato and tomato. *Molecular and General Genetics* 225, 501-509.

Gebhardt, C., Blomendahl, C., Schachtschnabel, U., Debener, T., Salamini, F. and Ritter, E. (1989a) Identification of $2n$ breeding lines and $4n$ varieties of potato (*Solanum tuberosum* subsp. *tuberosum*) with RFLP fingerprints. *Theoretical and Applied Genetics* 78, 16-22.

Gebhardt, C., Ritter, E., Debener, T., Schachtschnabel, U., Walkemeier, B., Uhrig, H. and Salamini, F. (1989b) RFLP analysis and linkage mapping in *Solanum tuberosum*. *Theoretical and Applied Genetics* 78, 65-75.

Gebhardt, C., Ritter, E., Barone, A., Debener, T., Walkemeier, B., Schachtschnabel, U., Kaufmann, H., Thompson, R.D., Bonierbale, M.W., Ganal, M.W., Tanksley, S.D. and Salamini, F. (1991) RFLP maps of potato and their alignment with the homoeologous tomato genome. *Theoretical and Applied Genetics* 83, 49-57.

Giovannoni, J.J., Wing, R.A., Ganal, M.W. and Tanksley, S.D. (1991) Isolation of markers from specific chromosomal intervals using pools from existing mapping populations. *Nucleic Acid Research* 19(23), 6553-6558.

Helentjaris, T. and Burr, B. (1989) *Development and Application of Molecular Markers to Problems in Plant Genetics*. Cold Spring Harbor Laboratory, New York, USA.

Hermsen, J.G.Th. (1978) Genetics of self-compatibility in dihaploids of *Solanum tuberosum* L. 2. Detection and identification of all possible incompatibility and compatibility genotypes in six F1's from interdihaploid crosses. *Euphytica* 27, 1-11.

Hosaka, K. and Hanneman, R.E., Jr (1988). The origin of the cultivated tetraploid potato based on chloroplast DNA. *Theoretical and Applied Genetics* 76, 172-176.

Hosaka, K. and Spooner, D.M. (1992) RFLP analysis of the wild potato species, *Solanum acaule* Bitter (*Solanum* sect. *Petota*). *Theoretical and Applied Genetics* 84, 851-858.

Hosaka, K., Ogihara, Y., Matsubayashi, M. and Tsunewaki, K. (1984) Phylogenetic relationship between the tuberous *Solanum* species as revealed by restriction endonuclease analysis of chloroplast DNA. *Japanese Journal of Genetics* 59, 349-369.

Innis, M.A., Gelfand, D.H., Sninsky, J.J. and White, T.J. (1990) *PCR Protocols - a Guide to Methods and Applications.* Academic Press, San Diego, California, USA.

Jacobs, J.M.E., Kreike, N.C.M., Arens, P.F.P. and Stiekema, W.J. (1991) RFLP mapping of the potato cyst nematode resistance Ro1 from *S. vernei*. *Second International Potato Molecular Biology Symposim*, St Andrews, Scotland, 11-15 August 1991. Abstract.

Kaufmann, H., Salamini, F. and Thompson, R.D. (1991) Sequence variability and gene structure at the self-incompatibility locus of *Solanum tuberosum*. *Molecular and General Genetics* 226, 457-466.

Kawagoe, Y. and Kikuta, Y. (1991) Chloroplast DNA evolution in potato (*Solanum tuberosum*). *Theoretical and Applied Genetics* 81, 13-20.

Keil, M., Sánchez-Serrano, J., Schell, J. and Willmitzer, L. (1990) Localization of elements important for the wound-inducible expression of a chimeric potato proteinase inhibitor II-CAT gene in transgenic tobacco plants. *Plant Cell* 2, 61-70.

Klein-Lankhorst, R.M., Vermunt, A., Weide, R., Liharska, T. and Zabel, P. (1991) Isolation of molecular markers for tomato (*L. esculentum*) using random amplified polymorphic DNA (RAPD). *Theoretical and Applied Genetics* 83, 108-114.

Knapp, S., Coupland, G., Uhrig, H., Starlinger, P. and Salamini, F. (1988) Transposition of the maize transposable element *Ac* in *Solanum tuberosum*. *Molecular and General Genetics* 213, 285-290.

Koßmann, J., Visser, R., Müller-Röber, B., Willmitzer, L. and Sonnewald, U. (1991) Cloning and expression analysis of a potato cDNA that encodes branching enzyme: evidence for co-expression of starch biosynthetic genes. *Molecular and General Genetics* 230, 39-44.

Köster-Töpfer, M., Frommer, W.B., Rocha-Sosa, M. and Willmitzer, L. (1990) Presence of a transposon-like element in the promoter region of an inactive patatin gene in *Solanum tuberosum* L. *Plant Molecular Biology* 14, 239-247.

Kreike, C.M., de Koning, J.R.A. and Krens, F.A. (1990) Non-radioactive detection of single-copy DNA-DNA hybrids. *Plant Molecular Biology Reporter* 8(3), 172-179.

Kumar, A., Smith, D. and Flavell, A. (1991) Extreme heterogeneity of *Copia-Ty* family retrotransposons in potato. *Second International Potato Molecular Biology Symposium*, St Andrews, Scotland, 11-15 August 1991. Abstract.

Lander, E.S. and Botstein, D. (1989) Mapping Mendelian factors underlying quantitative traits using RFLP linkage maps. *Genetics* 121, 185-191.

Lander, E.S., Green, P., Abrahamson, J., Barlow, A., Daly, M.J., Lincoln, S.E. and Newburg, L. (1987) MAPMAKER: an interactive computer package for constructing primary genetic linkage maps of experimental and natural populations. *Genomics* 1, 174-181.

Langridge, U., Schwall, M. and Langridge, P. (1991) Squashes of plant tissues as substrate for PCR. *Nucleic Acid Research* 19(24), 6954.

Leonards-Schippers, C., Gieffers, W., Ritter, E., Salamini, F. and Gebhardt, C. (1991) RFLP marker based localization of the gene R1 conferring vertical resistance to *Phytophthora infestans* in potato. *Second International Potato Molecular Biology Symposium*, St Andrews, Scotland, 11–15 August 1991. Abstract.

Liu, X.Y., Rocha-Sosa, M., Hummel, S., Willmitzer, L. and Frommer, W.B. (1991) A detailed study of the regulation and evolution of the two classes of patatin genes in *Solanum tuberosum* L. *Plant Molecular Biology* 17, 1139–1154.

Martin, G.B., Williams, J.G.K. and Tanksley, S.D. (1991) Rapid identification of markers linked to a *Pseudomonas* resistance gene in tomato by using random primers and near-isogenic lines. *Proceedings of the National Academy of Sciences USA* 88, 2336–2340.

Matsubayasbi, M. (1991) Phylogenetic relationships in the potato and its related species. In: Tsuchiya, T. and Gupta, P.K. (eds) *Chromosome Engineering in Plants: Genetics, Breeding, Evolution. Part B.* Elsevier, Amsterdam, The Netherlands, pp. 93–118.

Mendiburu, A.O. and Peloquin, S.J. (1979) Gene-centromere mapping by $4x \times 2x$ matings in potatoes. *Theoretical and Applied Genetics* 54, 177–180.

Messeguer, R., Ganal, M.W., Steffens, J.C. and Tanksley, S.D. (1991) Characterization of the level, target sites and inheritance of cytosine methylation in tomato nuclear DNA. *Plant Molecular Biology* 16, 753–770.

Michelmore, R.W., Paran, I. and Kesseli, R.V. (1991) Identification of markers linked to disease-resistance genes by bulked segregant analysis: a rapid method to detect markers in specific genomic regions by using segregating populations. *Proceedings of the National Academy of Sciences USA* 88, 9828–9832.

Müller-Röber, B.T., Koßmann, J., Hannah, L.C., Willmitzer, L. and Sonnewald, U. (1990) One of two different ADP-glucose pyrophosphorylase genes from potato responds strongly to elevated levels of sucrose. *Molecular and General Genetics* 224, 136–146.

Nevers, P., Shepherd, N.S. and Saedler, H. (1985) Plant transposable elements. *Advances in Botanical Research* 12, 102–203.

Paterson, A.H., DeVerna, J.W., Lanini, B. and Tanksley, S.D. (1990) Fine mapping of quantitative trait loci using selected overlapping recombinant chromosomes, in an interspecific cross of tomato. *Genetics* 124, 735–742.

Pehu, E., Thomas, M., Poutala, T., Karp, A. and Jones, M.G.K. (1990) Species specific sequences in the genus *Solanum*: identification, characterization, and application to study somatic hybrids of *S. brevidens* and *S. tuberosum*. *Theoretical and Applied Genetics* 80, 693–698.

Peloquin, S.J., Yerk, G.L., Werner, J.E. and Darmo, E. (1989) Potato breeding with haploids and 2n gametes. *Genome* 31, 1000–1004.

Pereira, A., van Agtmaal, S., te Lintel-Hekkert, W., Aarts, M., Stiekema, W., Laschuit, J. and Jacobsen, E. (1991) Transposon targeting in potato. *Second International Potato Molecular Biology Symposium*, St Andrews, Scotland, 11–15 August 1991. Abstract.

Pineda, O., Bonierbale, M.W., Plaisted, R.L., Brodie, B.B. and Tanksley, S.D. (1992) Identification of RFLP markers linked to the *H1* gene conferring resistance to the potato cyst nematode (*Globodera rostochiensis*). *Genome* 36, 152–156.

Rick, C.M. (1969) Controlled introgression of chromosomes of *Solanum pennellii* into *Lycopersicon esculentum*: segregation and recombination. *Genetics* 62, 753–768.

Rick, C.M. (1974) The tomato. In: King, R.C. (ed.) *Handbook of Genetics*. Plenum Press, New York, Vol. 2, pp. 247-280.

Ricky, T.M. and Belknap, W.R. (1991) Comparison of the expression of several stress-responsive genes in potato tubers. *Plant Molecular Biology* 16, 1009-1018.

Ritter, E., Gebhardt, C. and Salamini, F. (1990) Estimation of recombination frequencies and construction of RFLP linkage maps in plants from crosses between heterozygous parents. *Genetics* 125, 645-654.

Ritter, E., Debener, T., Barone, A., Salamini, F. and Gebhardt, C. (1991) RFLP mapping on potato chromosomes of two genes controlling extreme resistance to potato virus X (PVX). *Molecular and General Genetics* 227, 81-85.

Rivard, S.R., Cappadocia, M., Vincent, G., Brisson, N. and Landry, B.S. (1989) Restriction fragment length polymorphic (RFLP) analyses of plants produced by *in vitro* anther culture. *Theoretical and Applied Genetics* 78, 49-56.

Schmidt, R.J., Burr, F.A. and Burr, B. (1987) Transposon tagging and molecular analysis of the maize regulatory locus *opaque-2*. *Science* 238, 960-963.

Schwiezer, G., Ganal, M., Ninnemann, H. and Hemleben, V. (1988) Species-specific DNA sequences for identification of somatic hybrids between *Lycopersicon esculentum* and *Solanum acaule*. *Theoretical and Applied Genetics* 75, 679-684.

Singh, A.K., Salamini, F. and Uhrig, H. (1989) Chromosome pairing in 14 F_1 hybrids among 11 diploid potato species. *Journal of Genetics and Breeding* 43, 1-5.

Stanford, A., Bevan, M. and Northcote, D. (1989) Differential expression within a family of novel wound-induced genes in potato. *Molecular and General Genetics* 215, 200-208.

Tai, T.H. and Tanksley, S.T. (1990) A rapid and inexpensive method for isolation of total DNA from dehydrated plant tissue. *Plant Molecular Biology Report* 8, 297-303.

Tanksley, S.D. and Loaiza-Figueroa, F. (1985) Gametophytic self-incompatibility is controlled by a single locus on chromosome 1 in *Lycopersicon peruvianum*. *Proceedings of National Academy of Sciences USA* 82, 5093-5096.

Tanksley, S.D. and Mutschler, M.A. (1990) Linkage map of the tomato (*Lycopersicon esculentum*) ($2n = 24$). In: O'Brien, S.J. (ed.) *Genetic Maps*, 5th edn. Cold Spring Harbor Laboratory Press, Cold Spring Harbor, NY, USA, pp. 6.3-6.15.

Tanksley, S.D., Young, N.D., Paterson, A.H. and Bonierbale, M.W. (1989) RFLP mapping in plant breeding: new tools for an old science. *Bio/Technology* 7, 257-264.

Tanksley, S.D., Ganal, M.W., Prince, J.P., de Vicente, M.C., Bonierbale, M.W., Broun, P., Fulton, T.M., Giovanonni, J.J., Grandillo, S., Martin, G.B., Messeguer, R., Miller, J.C., Miller, L., Paterson, A.H., Pineda, O., Roder, M.S., Wing, R.A., Wu, W. and Young, N.D. (1992) High density molecular linkage maps of the tomato and potato genomes. *Genetics* 132, 1141-1160.

Thompson, R.D., Uhrig, H., Hermsen, J.G.Th., Salamini, F. and Kaufmann, H. (1991) Investigation of a self-compatible mutation in *Solanum tuberosum* clones inhibiting S-allele activity in pollen differentially. *Molecular and General Genetics* 226, 283-288.

van Eck, H.J., Ramanna, M.S. and Jacobsen, E. (1991) Use of RFLPs for centromere mapping in diploid potato. *Second International Potato Molecular Biology Symposium*, St Andrews, Scotland, 11-15 August 1991. Abstract.

Visser, R.G.F., Hoekstra, R., Leji, F.R. van der, Pijnacker, L.P., Witholt, B. and Feenstra, W.J. (1988) *In situ* hybridization to somatic metaphase chromosomes of potato. *Theoretical and Applied Genetics* 76, 420-424.

Watanabe, K. (1991) Bottlenecks in germplasm enhancement and application of the biotechnology. In: Dodds, J.H. (ed.) *Proceedings of Planning Conference on the Application of Biotechnology to Germplasm Enhancement of Potatoes*. International Potato Center, Lima, Peru, pp. 135-140.

Watanabe, K. and Dodds, J.H. (1991) Gene mapping to improve potatoes: tools of molecular biology expand use of CIP genetic resources. *CIP Circular* 18(2), 1-5.

Watanabe, K., El-Nashaar, H. and Iwanaga, M. (1992) Transmission of bacterial wilt resistance by FDR $2n$ pollen via $4x \times 2x$ crosses in potatoes. *Euphytica* 60, 21-26.

Welsh, J. and McClelland, M. (1990) Fingerprinting genomes using PCR with arbitrary primers. *Nucleic Acid Research* 18(24), 7213-7218.

Wienand, U. and Saedler, H. (1988) Plant transposable elements: unique structures for gene tagging and gene cloning. In: Hohn, Th. and Schell, J. (eds) *Plant DNA Infectious Agents*. Springer, Vienna, New York, pp. 205-228.

Williams, C.E., Hunt, G.J. and Helgeson, J.P. (1990) Fertile somatic hybrids of *Solanum* species: RFLP analysis of a hybrid and its sexual progeny from crosses with potato. *Theoretical and Applied Genetics* 80, 545-551.

Williams, J.G.K., Kubelik, A.R., Livak, K.J., Rafalski, J.A. and Tingey, S. (1990) DNA polymorphisms amplified arbitrary primers are useful as genetic markers. *Nucleic Acid Research* 18(22), 6531-6535.

Xu, B., Mu, J., Nevins, D.L., Grun, P. and Kuo, T. (1990) Cloning and sequencing of cDNAs encoding two self-incompatibility associated proteins in *Solanum chacoense*. *Molecular and General Genetics* 224, 341-346.

Young, N.D. and Tanksley, S.D. (1989) Restriction fragment length polymorphism maps and the concept of graphical genotypes. *Theoretical and Applied Genetics* 77, 95-101.

Young, N.D., Zamir, D., Ganal, M.W. and Tanksley, S.D. (1988) Use of isogenic lines and simultaneous probing to identify DNA markers tightly linked to the Tm-$2a$ gene in tomato. *Genetics* 120, 579-585.

Zeven, A.C., Knott, D.R. and Johnson, R. (1983) Investigation of linkage drag in near-isogenic lines of wheat by testing for seedling reaction to races of stem rust, leaf rust and yellow rust. *Euphytica* 32, 319-327.

IV ENVIRONMENTAL STRESS, MORPHOLOGY AND QUALITY

11 Environmental Stress and Its Impact on Potato Yield

M.E. VAYDA

Department of Biochemistry, Microbiology and Molecular Biology, University of Maine, Orono, Maine 04469, USA.

Introduction

The harvest index of potato is not genetically limited, so that the realized yield is greatest under growing conditions with adequate light, water and cool temperatures. In experiments where there is no stress, the observed variation between potato cultivars in tuber dry-weight yield appears to be directly related to cumulative light interception (van der Zaag and Doornbos, 1987). However, in the field environmental stress conditions do vary dramatically and affect yield adversely. Potato yield is especially sensitive to heat and drought. Even a short period of acute stress can cause a substantial decrease in total and marketable yield. In addition, postharvest stresses to the tuber crop can further reduce marketable yield. Most notably, prolonged cold storage leads to the phenomenon of cold-induced sweetening, and low oxygen conditions in storage render tubers susceptible to rot by opportunistic bacterial pathogens. In general, these stress conditions impair the plant's ability to function at the molecular level, resulting in either poor photosynthetic efficiency, inadequate partitioning of photosynthate to the tuber sink, or, in postharvest stress, altered metabolic processes.

The molecular basis for these physiological responses is not clearly understood and little is known concerning the genetics of stress tolerance: genes for resistance to environmental stresses have not been identified. However, there are differences in the degree to which individual cultivars are affected by these stresses, and strategies for selection of tolerant cultivars in breeding programmes have been developed. The effort to identify genes or chromosomal segments associated with stress tolerance may be assisted by the new technique of restriction fragment length polymorphism (RFLP) mapping (see chapter 10 by Watanabe). Furthermore, gene products that accumulate in several plant species during temperature and water stress conditions have been isolated and cloned and their putative cellular functions have been identified. It is hoped that understanding the molecular responses to acute stress may lead to identification and manipulation of gene products that mediate stress tolerance. However, it has not yet been demonstrated that the expression and accumulation of these products confers stress tolerance at either the cell or the whole-plant level.

This chapter reviews the effect on yield and quality of four environmental stress conditions: heat, drought, cold and oxygen deficiency. It tries to explain the basis for these effects in physiological and molecular terms, and discusses the prospects for identification of genetic factors involved in stress tolerance.

Heat Stress and Thermotolerance

The commercial potato, *Solanum tuberosum* L. subsp. *tuberosum*, has been bred as a crop acclimatized to northern Europe and North America. As such, optimal growth and yield are obtained in temperate climates with cool temperatures and long days. Growth, tuber yield and quality are adversely affected by high temperatures and water deficit (Levy, 1986a). The optimum temperature for growth of most cultivars is 17–20°C, with every 5°C rise above 20°C effecting a 25% decrease in the photosynthetic rate (Burton, 1981; Demagante and van der Zaag, 1988). Prolonged high temperatures lead to an underdeveloped canopy which is less capable of supporting tuber growth (van der Zaag, 1984). Heat-stressed plants are less likely to tuberize (Ewing, 1981; Ben Khedher and Ewing, 1985), and they exhibit continued foliage production after the initiation of tuberization, diverting photosynthate from the tuber sink (van der Zaag, 1984). For these reasons, temperatures above 20°C severely depress both tuber initiation and bulking such that temperatures above 29°C effectively stop tuber production and reduce dry-matter accumulation (Marinus and Bodlaender, 1975; Ewing, 1981; Ben Khedher and Ewing, 1985). Even short periods of stress cause malformed and less marketable tubers that exhibit a knobby or bottleneck shape, secondary growth cracks, heat necrosis and premature sprouting (Iritani and Weller, 1973; Burton, 1981; Iritani, 1981; Levy, 1985).

Despite these problems, potato production and utilization are increasing faster than any other crop in the semitropical and tropical areas of the world. In hot climates such as found in Israel or the Philippines, where daily temperatures reach 40°C, expected yields may be only 23 tonnes per hectare as compared with optimal yields in temperate climates of up to 90 tonnes per hectare (Burton, 1981). To compensate for these conditions growers have adopted two strategies: intercropping and selection of heat-tolerant potato clones.

Intercropping with 'shade' crops such as maize, sugarcane and bean can cool the soil temperature 5–10°C (Levy *et al.*, 1986; Demagante and van der Zaag, 1988; Struik *et al.*, 1989b). Although air temperatures may remain high, reduction of the root and stolon temperatures is sufficient to promote stolon growth, a prerequisite for tuber formation (Struik *et al.*, 1989a, b, c).

Potato is a C3 plant and thus does not need high solar energy input for carbon assimilation. However, shading does have an adverse effect in that it delays the onset of the rapid tuber bulking period and slows the overall bulking rate because of continued haulm growth (Demagante and van der Zaag, 1988).

Breeding for heat tolerance has been difficult because heat stress affects at least three distinct physiological processes of the potato plant: photosynthetic efficiency and haulm growth, tuber initiation and photosynthate partitioning. High ambient temperatures cause a decrease in photosynthetic efficiency (Sipos and Prange, 1986; Prange *et al.*, 1990). This results in decreased photosynthate production and a reduced haulm. A well-developed and photosynthetically active canopy is necessary for tuber initiation and bulking (van der Zaag, 1984). High temperatures inhibit the accumulation of the 'tuberization stimulus' (Ewing, 1981), so that the period of foliage growth is prolonged and the timing of tuber initiation is delayed, causing a reduction in the total number of tubers produced. High temperatures are thought to cause the production of gibberellins, which promote shoot growth, stolon formation and stolon elongation but delay tuber formation (Midmore, 1984; Struik *et al.*, 1989b). High air temperatures also cause potato plants to partition to shoot synthesis photosynthate that would be utilized for tuber dry matter under optimal conditions (Struik *et al.*, 1989c; Prange *et al.*, 1990; Gawronska *et al.*, 1992). This complexity compels breeders to assess germplasm for two independent characteristics: foliage resistance to heat stress and the ability to tuberize under high-temperature conditions (Reynolds and Ewing, 1989; Levy *et al.*, 1991).

Reynolds and Ewing (1989) demonstrated that these two characteristics are not correlated. Numerous accessions of wild *Solanum* species exhibit foliage heat tolerance, including *S. berthaultii*, *S. chacoense*, *S. demissum*, *S. jamesii*, *S. kurtzianum*, *S. papita*, *S. spegazzinii*, *S. stoloniferum* and *S. sucrense*. The high rate of photosynthesis in these species correlates with their heat tolerance. However, only a subset of the heat-tolerant accessions are able to tuberize well at high temperature (Reynolds and Ewing, 1989). Similarly, the ability to initiate tuberization at elevated temperature differs among commercial potato cultivars (Ben Khedher and Ewing, 1985). Conversely, attempts to identify clones that tuberize well at elevated temperatures without first assessing shoot tolerance to heat stress have found little correlation with heat tolerance in the field (Harvey *et al.*, 1988; Nowak and Colborne, 1989).

The primary effect of acute heat stress to foliage appears to be an alteration of photosynthetic efficiency (van der Zaag and Doornbos, 1987; Bhagsari *et al.*, 1988; Demagante and van der Zaag, 1988). This reduction in photosynthetic efficiency appears to be due to an impaired function of photosystem II (PS II), as measured by chlorophyll fluorescence yield *in vivo* (Sipos and Prange, 1986; Prange *et al.*, 1990) or in isolated chloroplasts (Hetherington *et al.*, 1983). There is no apparent effect on the enzymes of the dark reactions, because the activity of these enzymes *in vitro* increases with increased temperature (Sharpe, 1983). Thus, the plant is unable to process the energy of intercepted radiation: energy absorbed by PS II is not transferred to PS I with subsequent reduction of CO_2, but rather is emitted as fluorescence.

It is not clear whether this dysfunction results from thermal denaturation of the proteins comprising the PS II complex or indirectly by changes in thylakoid membrane fluidity that lead to PS II destabilization. The ability of thylakoid membranes to resist heat damage varies among potato cultivars (Hetherington et al., 1983) and thermotolerance could involve the ability to adapt thylakoid membrane composition to resist heat stress, as appears to be the case in other plant species (Martineau et al., 1979; Blum and Ebercon, 1981; Nagarajan and Bansal, 1986). In contrast, thermal stability of the photosynthetic apparatus in several plant species is thought to be closely related to the appearance, accumulation and function of heat-shock proteins (HSPs) (Key et al., 1985; Schoffl et al., 1988; Nover et al., 1989; Ellis, 1990; Vierling, 1991).

The HSPs are expressed in virtually every organism subjected to acute heat stress, typically upon an elevation in temperature of 10°C above the usual growth temperature (Nover et al., 1989). In angiosperms, there are three classes of HSPs: the HSP70 class, exhibiting molecular weights of 70–80 kD, the HSP60 class, exhibiting molecular weights of 55–60 kD, and the low-molecular-weight (LMW) class of HSPs, exhibiting molecular weights of 15–22 kD. The steady-state levels of the mRNAs encoding these species accumulate within minutes of acute heat shock and are efficiently translated *in vivo*. The translation of other mRNA species is suspended until the heat stress is relieved. Thus, one effect of prolonged heat stress could be decreased synthesis and turnover of cellular proteins under high-temperature conditions.

The recently elucidated functions of the HSPs suggest that HSPs protect PS II or other macromolecular assemblies from irreversible denaturation at high temperatures (Schuster et al., 1988). HSP70 exhibits an ATPase activity in the presence of denatured protein (Schoffl et al., 1988; Nover et al., 1989; Ellis, 1990). The HSP70 proteins are thought to bind to denatured proteins and unfold them. Upon hydrolysis of ATP, HSP70 releases the bound protein and aids its folding into a native configuration. Thus, HSP70s have been termed 'molecular chaperones' or 'chaperonins' for their ability to catalytically aid in the refolding of heat-denatured proteins. There are two principal forms of HSP70: one cytoplasmic and the other associated with the endoplasmic reticulum (ER). Both forms have been demonstrated to be present in unstressed cells, and at least the ER-associated form has been shown to have an important function in non-stressed cells: this protein is necessary for the proper folding of secreted proteins (Mason et al., 1984; Wu et al., 1988; Pelham, 1989; Ellis and Hemmingsen, 1989). Several forms of HSP70 have been identified in potato that are specifically expressed in response to either heat shock, wounding or developmental cues (Belknap and Rickey, 1990; Rickey and Belknap, 1991; W. Belknap, pers. comm.). Thus, expression of the HSP proteins under heat-stress conditions appears to be an amplification of a basal cell function to cope with denatured protein complexes.

The HSP60 class appears to play a similar role in the assembly of organellar macromolecular complexes. The plant HSP60 proteins bear sequence similarity to the *gro*EL and *gro*ES proteins of *E. coli*. These bacterial proteins were first identified as gene products necessary for the assembly of several types of bacteriophage, and were subsequently also observed to accumulate to high levels in response to heat shock. The plant homologues are targeted to chloroplasts and mitochondria, and have been demonstrated to partake in assembly of ribulose-1,5-bisphosphate carboxylase subunits and other macromolecular assemblies (Lubben et al., 1989; Prasad and Hallberg, 1989; Roy, 1989). Accordingly, the HSP60 class of proteins is likely to be involved in the stabilization or renaturation of PS II and other thylakoid complexes during heat stress.

The LMW HSPs are a complex family of 12 to 27 abundant polypeptides (Nagao et al., 1985; Mansfield and Key, 1987; Helm et al., 1989; DeRocher et al., 1991). One class of LMW HSPs is localized in the chloroplast whereas the other two families are localized in the cytoplasm (Vierling et al., 1986, 1988, 1989; Chen et al., 1990). The LMW HSPs persist long after relief from heat stress, and for this reason are thought to be involved in thermoadaptation. However, neither the molecular nor the physiological function of these proteins is known.

The HSPs are thought to be involved in the phenomenon of acquired thermotolerance because they accumulate to levels proportional to the applied thermal stress, and exhibit exceptionally long half-lives of 50 days or more (Chen et al., 1990). The HSPs persist after periods of transient heat shock and, accordingly, have been detected in field-grown plants with midday temperatures ranging from 34 to 40°C (Burke et al., 1985; Kimpel and Key, 1985; Chen et al., 1990). Plants in which HSPs have accumulated, either by previous heat shock or by arsenite treatment, are more resistant to cellular damage upon a subsequent heat shock (Lin et al., 1984; Altschuler and Mascarenhas, 1985; Key et al., 1985; Schuster et al., 1988), presumably because the presence of the long-lived HSPs prevents or limits damage to cellular complexes. However, although HSPs are circumstantially implicated in thermotolerance, a protective role for the HSPs has not been demonstrated directly. For example, expression of HSPs in transgenic plants has not yielded increased thermotolerance. One study following the segregation of a specific LMW HSP in maize found no correlation between the presence of this gene and thermotolerance (Ottaviano et al., 1991). This is not entirely unexpected, however, because the LMW HSPs appear to be encoded by a multigene family with redundant functions (Nagao et al., 1985; DeRocher et al., 1991), such that the loss of one particular member is not significant. Indeed, RFLP and classical genetic analysis of maize and soyabean indicates that shoot thermotolerance is a complex character affected by at least six quantitative trait loci (Martineau et al., 1979; Ottaviano et al., 1991).

Thus, it is not yet clear if shoot thermotolerance can be obtained by optimization of the heat-stress response, despite considerable effort to understand the system at the molecular level. In contrast, the molecular and genetic factors underlying the initiation of tuberization and partitioning of photosynthate to the tuber sink are poorly understood. Therefore, at present, tolerance of these physiological processes to heat stress cannot be addressed in any way other than by empirical selection.

Breeding for potato clones that can initiate and bulk tubers well during periods of heat stress is complicated by the fact that breeders have selected clones for optimal yield as a first priority. As a result, some of the cultivars that produce the highest yields in hot climates do not exhibit true heat tolerance when it is defined as the ratio of yield obtained when grown under heat stress relative to yield obtained when grown under temperate conditions. For example, the yield of cv. Norchip grown under heat stress was 66% that of the yield obtained when grown under temperate conditions, whilst clones that produce well in Mediterranean climates, such as DTO-28, LT-1 and cv. Desiree, exhibited yields under heat-stress conditions that were less than 25% of the yield obtained when grown in a cool greenhouse (Ben Khedher and Ewing, 1985). DTO-28 and LT-1 have been demonstrated at high temperatures to partition ^{14}C photoassimilate more efficiently to tuber dry matter than Russet Burbank and Desiree (Gawronska et al., 1992; Malik et al., 1992). However, cultivars such as Desiree are so vigorous that their absolute yield under heat stress, even reduced by 75% or more, is greater than that of the truly heat-tolerant clones. Thus, clones selected for their performance in the tropics may not be truly heat-tolerant as much as exhibiting general good yielding ability and vigour. This is further evidence that breeders for heat tolerance must first identify heat-tolerant germplasm and then breed for increased yield.

Heat tolerance of *S. tuberosum* cultivars has long been associated with early maturity. It is likely that some of these early-maturing cultivars are not truly heat-tolerant, but rather, escape the stress conditions by accumulating yield quickly prior to onset of the hottest weather, and thus are exposed to fewer and shorter periods of heat stress (Levy, 1986a, b). For example, DTO-28 exhibits high tuber yield and relatively early tuber initiation and bulking (Malik et al., 1992). However, the presence of heat tolerance in medium–late-maturing cultivars, such as Diamant, indicates that true tolerance is separable from earliness (Levy et al., 1990). This was further demonstrated by crossing the late-maturing heat-sensitive cultivar Cara, with Blanka, an early-maturing heat-tolerant cultivar. The result was hybrid progeny of the medium–late class which exhibited heat tolerance (Levy et al., 1991). These efforts demonstrate that it appears possible to genetically separate the character of heat tolerance from the early-maturing phenotype. However, the number, identity and characteristics of genes involved in heat resistance are unknown.

Water Stress and Drought Resistance

Water stress is the most important physiological stress to potato production in most areas of the world. The potato is very sensitive to soil water conditions, much more sensitive than most other crop species (Begg and Turner, 1976). It exhibits morphological changes at -0.4 bar (Shalhevet et al., 1983; Shimshi and Susnoschi, 1985), that is, when soil moisture only drops to 70–85% of field capacity, depending on the relative humidity (Burton, 1981;

van Loon, 1981; Stark and Wright, 1985). Stress periods as short as 1 day can cause visible effects (Burton, 1981; MacKerron and Jefferies, 1988). Thus, water availability is not only an extreme problem in arid regions, but frequently is a problem in prime growing areas such as Maine, USA, where irrigation is not a practical option and fluctuating water availability can have severe effects on both total and marketable yield.

There is a linear relationship between the reduction in tuber yield and the amount of soil moisture when the water applied is less than that lost daily by evapotranspiration (Shimshi and Susnoschi, 1985). However, this apparently simple relationship disguises a complex set of responses. At all stages of growth, water stress reduces photosynthetic efficiency (Burton, 1981; van Loon, 1981), but drought during the period of tuber initiation and bulking has the most drastic effect on yield. Tuber initiation is blocked during the interval of water stress (MacKerron and Jefferies, 1988), as is the initiation of stolons (Haverkort et al., 1990). Thus, drought reduces the number of tuber initiation events in a manner proportional to the duration of the stress. Further, the bulking of tubers initiated prior to the onset of stress is dramatically decreased during drought periods, effecting loss in dry matter that is proportional to both the severity and duration of the stress (van Loon, 1981; MacKerron and Jefferies, 1988; MacKerron et al., 1988).

Even relief of drought stress can have adverse effects. When tuber growth is inhibited for periods of several days, the tubers become 'set', that is, the basal portion ceases to grow (Iritani, 1981). When adequate soil water is resumed, the apical end of the tuber resumes growth, yielding malformed pear-shaped, dumb-bell-shaped or knobby tubers, such as those shown in Figure 11.1, which reduce the marketable potential of the crop. Prolonged periods of water stress during early tuber development cause depletion of starch in the basal end, leading to translucent sugar or jelly ends, low in starch and high in reducing sugars, which cause browning during cooking (Iritani and Weller, 1973; Sowokinos et al., 1985). Furthermore, the rapid tuber growth that often occurs upon relief of water stress causes growth cracks and contributes to other maladies, such as hollow heart, shown in Figure 11.2 (Iritani, 1981; MacKerron and Jefferies, 1985). Thus, even short periods of water stress have dramatic effects on the marketable yield.

It is generally accepted that there are cultivar differences in susceptibility to water stress, although all potato cultivars are affected to significant degrees (Shalhevet et al., 1983; Wolfe et al., 1983; Susnoschi and Shimshi, 1985; van der Zaag and Doornbos, 1987). The North American standard, cv. Russet Burbank, is notoriously sensitive to water stress, as is the Dutch cultivar Veenster (Miller and Martin, 1987; Schapendonk et al., 1989), whereas cvs Nooksack, Lemhi, Spunta and Desiree are more tolerant. The basis for drought tolerance is not clear, because diffusive resistance, photosynthetic efficiency and partitioning of assimilates are all adversely affected by water deficit (Ewing, 1981; Shimshi et al., 1983; Wolfe et al., 1983; Shimshi and Susnoschi, 1985; Levy et al., 1988). For example, cv. Spunta, which exhibits an early and rapid bulking period, maintains a high relative-yield rating under drought conditions, although it is not the highest-yielding variety in irrigated

Fig. 11.1. Selected Russet Burbank tubers exhibiting extreme effects of secondary growth, resulting from a growing season that was cool but with two prolonged periods without rainfall.

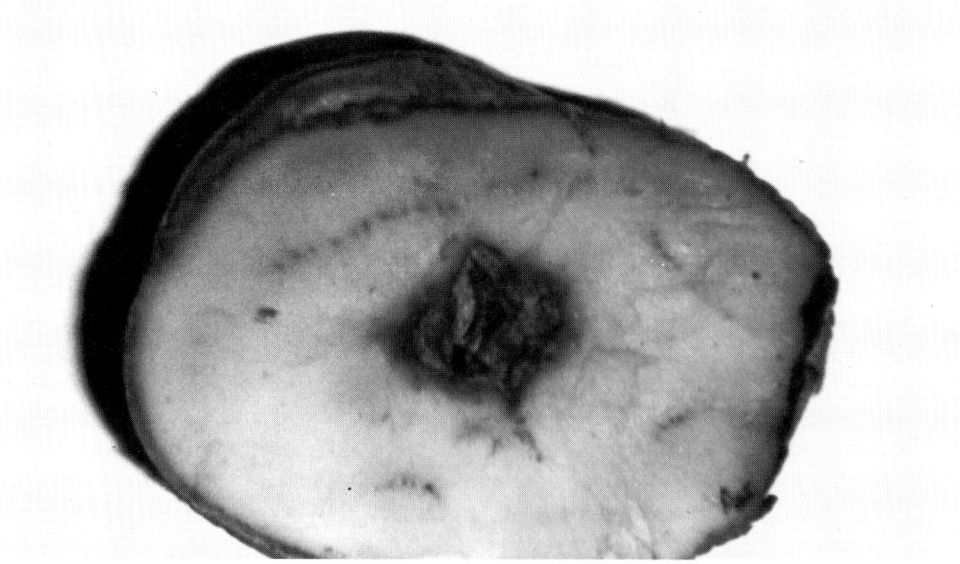

Fig. 11.2. An FL-1607 tuber exhibiting hollow heart, a cultivar-specific flaw that is exacerbated by periods of rapid growth following relief of drought stress.

rows. The persistence of foliage during water stress also does not correlate, because cv. Spunta exhibits early haulm senescence whereas cv. Alpha, which maintains its foliage during water stress, exhibits low relative yields. Similarly, high water potential does not correlate with high yield during drought stress: cv. Up-to-Date maintains a high water potential during stress but yields lower than cv. Spunta (Shimshi and Susnoschi, 1985). In agreement with this finding, antitranspirant treatment has no effect on tuber yield, although it does improve water-use efficiency (Stark and Dwelle, 1989). The cultivar Desiree yields well during water stress, apparently because it exhibits a high photosynthetic rate in both irrigated and water-stressed conditions, and, although decreased by stress, the amount of photosynthate accumulating in both the haulm and the tuber sink is substantial (Shimshi and Susnoschi, 1985; Susnoschi and Shimshi, 1985). Thus, several of the cultivars suitable for growth in the Mediterranean region may escape the most severe stress by early tuber initiation and bulking (Levy *et al.*, 1990). However, other cultivars, such as cv. Cara and cv. Up-to-Date, exhibit tolerance to water-stressed conditions.

The physiological basis for differences in cultivar tolerance is not known. One explanation of the acute sensitivity of potato to drought is its relatively shallow root system, and the inability of potato roots to penetrate the plough pan (DeRoo and Waggoner, 1961; Bishop and Grimes, 1978; Miller and Martin, 1987). The practice of subsoiling allows potato roots to penetrate an additional 0.5 metres into the soil, and affords up to 4 days of tolerance to insufficient water supply (Miller and Martin, 1987).

The acute molecular effect of water stress appears to be a decrease in photosynthetic efficiency. But, unlike heat stress, fluorescence yield analysis reveals no decrease in function of either PS II or PS I, but rather indicates an inhibition of the Calvin-cycle enzymes (Schapendonk *et al.*, 1989). Water stress causes immediate stomatal closure, which decreases water loss via evapotranspiration but also shuts out the CO_2 that is necessary for assimilation (Burton, 1981; van Loon, 1981; Bansal and Nagarajan, 1987). However, within 3 days, stomata open partially and internal CO_2 levels approximate prestress levels. Under these conditions, an apparent inhibition of the Calvin-cycle enzymes persists, and the degree of this inhibition is greatest in cultivars such as Veenster, which correlates with their sensitivity to drought (Schapendonk *et al.*, 1989). The inhibition of dark-cycle reactions may be due to photoinhibition or leaf damage, although the precise cause is not known.

The molecular basis for resistance to water stress is also not known, in either potato or other plant species. Drought could effect a loss of water from the cytoplasm, which may lead to an electrolyte imbalance, but the cytoplasm is certainly far from dried out. The phenomenon has been studied in several other plant species, notably soyabean. In this case, plants accumulate proline and betaine, which are thought to alter the osmotic potential of the cytoplasm (Hanson and Hitz, 1982). Growth inhibition in soyabeans is due in part to the rise in abscisic acid (ABA) and decrease in gibberellic acid (GA) that accompanies transfer to low water potential (Bensen *et al.*, 1988, 1990). Changes in the polysome-associated mRNA populations indicate that gene expression is modulated by water deficit, but only a subset of the mRNAs that accumulate have been identified to date (Mason et al., 1988; Creelman et al., 1990; Creelman and Mullet, 1991). Water-stressed soyabean plants accumulate vegetative storage proteins (Mason and Mullet, 1990). Proteins of related amino acid sequence, termed 'dehydrins', accumulate in monocots subjected to water stress (Close et al., 1989). Cotton, *Brassica* spp. and other plant species also accumulate similar proteins when water-stressed. The latter were first identified as polypeptides or mRNAs which accumulated during seed desiccation ('LEA', late embryogenesis abundant) or in response to ABA treatment ('Rab', responsive to ABA) (Dure et al., 1989). ABA levels rise during water stress and seed desiccation and may mediate expression or accumulation of these gene products (Galau et al., 1987; Bray, 1988; Creelman et al., 1990). The nucleic acids encoding these species have been cloned and sequenced, and found to specify very similar proteins that are extremely hydrophilic, resistant to denaturation and high in residues with hydroxyl groups, that is, serine and threonine (Dure et al., 1989). These proteins are postulated to bind and retain water in the cytoplasm despite the draw of transpiration during drying conditions. However, expression of these gene products has not been directly demonstrated to confer resistance to drought stress. One attempt to increase drought tolerance has yielded inconclusive results. The gene encoding a Rab protein from the desiccation-tolerant resurrection plant (*Ceratostigma plantagineum*) has been isolated, cloned and introduced into tobacco by *Agrobacterium*-mediated transformation. Although the transgenic tobacco plants expressed the gene product, increased drought tolerance was not observed (Iturriaga et al., 1991). The molecular responses of potato to water stress have not been investigated. Specifically, the level of expression of genes encoding dehydrin-like proteins has not been assessed in potato cultivars subjected to water stress. Studies to date have only assessed the physiological responses of potato at the whole-plant level, usually by field assay.

The resistance of potato to drought conditions has not been clearly identified at either the genetic or molecular level. This is partly because sensitivity operates on at least three levels: photosynthetic efficiency, initiation of stolon and tuber development, and carbon partitioning and growth deformities. Furthermore, water stress is often accompanied by heat stress, which complicates field studies and cultivar assessment. As a result, the molecular response of potato to drought and the genetics of resistance to water stress are poorly understood, except to say that it is a whole-plant phenomenon and probably a polygenic character.

Cold Stress

Cold stress has two major effects on the potato, frost-killing of the haulm and cold-induced sweetening in the tuber, although tubers can also freeze in a hard frost. These are two distinct phenomena; one affects the foliage and limits the growing season, while the other is a severe postharvest problem that affects the quality of potato tubers for processing.

Frost tolerance

Cultivated potato plants can survive temperatures down to −3°C. Below −4°C, ice crystals form in the extracellular spaces of plant tissues and cause cell death (Li et al., 1981). It is the formation of these ice crystals in the extracellular space and not the cold temperatures *per se* that cause damage, because supercooling of potato leaf tissues to −6.9°C without ice formation does not cause tissue death (Rajashekar et al., 1983). Ice crystals form from water that is slowly extruded from the cytoplasm during the process of freezing. Even if ice crystals do not form, the leaf tissue may exhibit irreversible damage that is the result of cold-induced dehydration (Sukumaran and Weiser, 1972).

All cultivars of *S. tuberosum* are frost-sensitive, as are the majority of tuber-bearing *Solanum* species. However, some *Solanum* species exhibit frost tolerance. For example, *S. acaule*, *S. commersonii*, *S. multidissectum*, *S. chomatophilum*, *S. boliviense*, *S. megistacrolobum* and *S. sanctae-rosae* can withstand temperatures of −4.0 to −6.0°C, with *S. acaule* the most resistant to frost (Chen and Li, 1980a). A number of *Solanum* species subjected to short days and cool temperatures exhibit cold acclimatization, which enables them to withstand temperatures 3 to 7°C below the usual killing temperature (Li and Fennell, 1985). After acclimatization to cold temperatures, *S. oplocense* and *S. polytrichon* can withstand temperatures of −8°C and −6°C, respectively, whereas both are otherwise frost-sensitive. The frost-tolerant species *S. acaule*, *S. multidissectum* and *S. chomatophilum* exhibit acclimatization that affords them resistance to −9°C, and cold-acclimatized *S. commersonii* can tolerate temperatures as low as −11.5°C. However, *S. tuberosum* is a frost-sensitive species that does not exhibit cold acclimatization.

The molecular basis of frost tolerance is not understood. In a variety of plant species, proline or monomeric sugars have been demonstrated to accumulate in the cytoplasm during the acclimatization process (Cox and Levitt, 1976; Levitt, 1980; Stout, 1980). Chen and Li (1980b) reported that, in *Solanum* species, both free sugars and starch increase during acclimatization to cold. The increase in free sugars upon cold acclimatization was greatest in *S. commersonii*. However, accumulated sugars alone cannot account for acclimatized tolerance, because species such as *S. tuberosum* also accumulated sugars when subjected to short days and cool temperatures but remained sensitive to temperatures below −3°C. Furthermore, the amount of sugars accumulated by species that can acclimatize does not correlate with the temperature those species can withstand, nor is the increase in sugars of *S. commersonii* proportional to the degree of frost hardiness it exhibits. Thus, some other factor induced by the cold-acclimatization process must mediate the induced tolerance.

A class of mRNAs is known to accumulate in several plant species during cold acclimatization (Lee et al., 1991), and *Solanum* species which can acclimatize to cold exhibit an increase in soluble proteins which is proportional to their degree of induced frost hardiness (Li and Fennell, 1985). Hormonal levels have been observed to fluctuate during the cold-acclimatization process: in species that can acclimatize to cold, ABA rises and GA declines. In those *Solanum* species that can acclimatize, the accumulation of soluble proteins can be elicited by exogenously applied ABA alone. Such ABA treatment induces cold tolerance in these *Solanum* species (Chen *et al.*, 1983). Cold acclimatization of several other plant species includes accumulation of the soluble dehydrin/LEA/Rab proteins (Haskell *et al.*, 1991; Quigley *et al.*, 1991; Thomashow *et al.*, 1991). Thus, it is interesting to speculate that homologues of these gene products may also mediate the induced frost tolerance of cold-acclimatizing *Solanum* species, although this has not been demonstrated. If this is so, it may be possible to introduce some degree of induced frost tolerance into *S. tuberosum* commercial cultivars by *Agrobacterium*-mediated transformation and expression of these gene products.

Cold-induced sweetening during storage

Of great importance to the potato processing industry is the phenomenon of cold-induced sweetening. This topic is dealt with at length in the chapter by Dale and Mackay on the inheritance of table and processing quality, and so only a summary of the biochemical basis of this phenomenon is included here. Briefly, the storage of harvested tubers at cool temperatures (3°C to 6°C) to maintain dormancy induces a physiological change that is detrimental to processing quality. Primarily, starch is degraded and monomeric reducing sugars accumulate. These sugars interact with amino groups during cooking, yielding an undesirable brown colour (the Maillard reaction) and imparting a sweet taste (Shallenberger *et al.*, 1959). Thus, cold stress in storage renders the crop less marketable to the processing industry. The degree to which stored tubers exhibit cold-induced sweetening is cultivar-specific, although all potato cultivars will exhibit sweetening if stored at cool temperatures for long periods.

The biochemical basis for cold-induced sweetening appears to be a flux of inorganic phosphate (Pi) from the vacuole to the cytosol (Shekar and Iritani, 1978; Sowokinos *et al.*, 1985). The accumulation of Pi allows phosphorolytic breakdown of starch to yield glucose-1-phosphate, which can be efficiently transported from the amyloplast to the cytoplasm via a hexose-phosphate/Pi transporter (Entwistle and ap Rees, 1988; Keeling *et al.*, 1988; Tyson and ap Rees, 1988). Glucose-1-phosphate is converted into sucrose-6-phosphate via the action of UDP-glucose pyrophosphorylase (UPPL), sucrose-6-phosphate synthase (SPS) and two hexose phosphate isomerases. Sucrose is generated by the action of sucrose-6-phosphatase, and the reducing sugars that accumulate are produced by the action of any of a number of invertases present in the cytoplasm (Sowokinos, 1990a, b).

The flux of Pi into the cytosol and amyloplast may result from damaged or impaired membranes: in the cold, membranes become more rigid, that is, less fluid and more leaky to ions (Shekar *et al.*, 1978; Workman *et al.*, 1979;

Walker et al., 1991). Indeed, the cultivars most sensitive to cold-induced sweetening exhibit high levels of electrolyte leakage at 3°C compared with resistant cultivars (Sowokinos, 1990a). Thus, genetic factors which affect membrane composition and the ability to maintain membrane integrity at low temperatures may dictate the sensitivity of specific cultivars to cold-sweetening.

The levels of UPPL and SPS directly correlate with the susceptibility of cultivars to cold-induced sweetening. SPS and UPPL activities increase in response to stresses that cause sweetening. Furthermore, cultivars with high sugar levels exhibit the largest cold-induced increases in these enzymatic activities, whereas the most cold-resistant potato clones exhibit only modest increases (Sowokinos, 1990a, b). Thus, the susceptibility of cultivars to cold-induced sweetening may reside in the induction of the two enzymes, SPS and UPPL. However, it is not clear whether the inducibility of the genes encoding these enzymes is due to allelic variance of genes encoding SPS and UPPL, or the presence of some effectors that regulate expression of these genes.

The genetic basis for resistance to cold-induced sweetening is unknown. The two resistant clones, ND 651-9 and ND 860-2, contain *S. phureja* germplasm, which itself exhibits cold resistance (Sowokinos, 1990a). However, the genetic characters associated with the resistance of *S. phureja* have not yet been identified. The construction of genetic maps of *S. tuberosum*, *S. phureja* and their close relatives using RFLPs is in progress and may help to localize these elusive genetic factors (see chapter by Watanabe on molecular genetics). Furthermore, the gene sequences encoding SPS, UPPL and invertases have been, or are in the process of being, cloned. In the near future, it will be possible to correlate susceptibility to cold-sweetening with specific alleles of SPS and UPPL, using molecular probes, and to test directly in transgenic potato clones whether the expression of modified SPS or UPPL sequences or antisense constructs has any effect on reducing cold-induced sweetening.

Hypoxic Stress and Postharvest Losses

Hypoxia is a postharvest stress that can cause severe reduction in marketable yield. Low-oxygen conditions favour tuber infection by opportunistic pathogens in the soil (*Erwinia* spp.), resulting in the disease known as soft rot (Perombelon and Kelman, 1980; Davis et al., 1990; Figure 11.3). Low-oxygen conditions can also prevail during storage, in transportation and upon planting of seed tubers when condensation forms a water film on the tuber surface, which drastically reduces gas diffusion (Perombelon and Kelman, 1980). The pathogenic *Erwinia* species secrete hydrolytic enzymes that degrade plant tissues, but these bacteria only establish an infection in tubers when plant cell functions are impaired. Such an impairment occurs when tubers are subjected to hypoxic conditions.

Potato tubers exhibit resistance to bacterial infection when maintained in adequately aerated conditions. Tubers of the cultivar BelRus, however, are

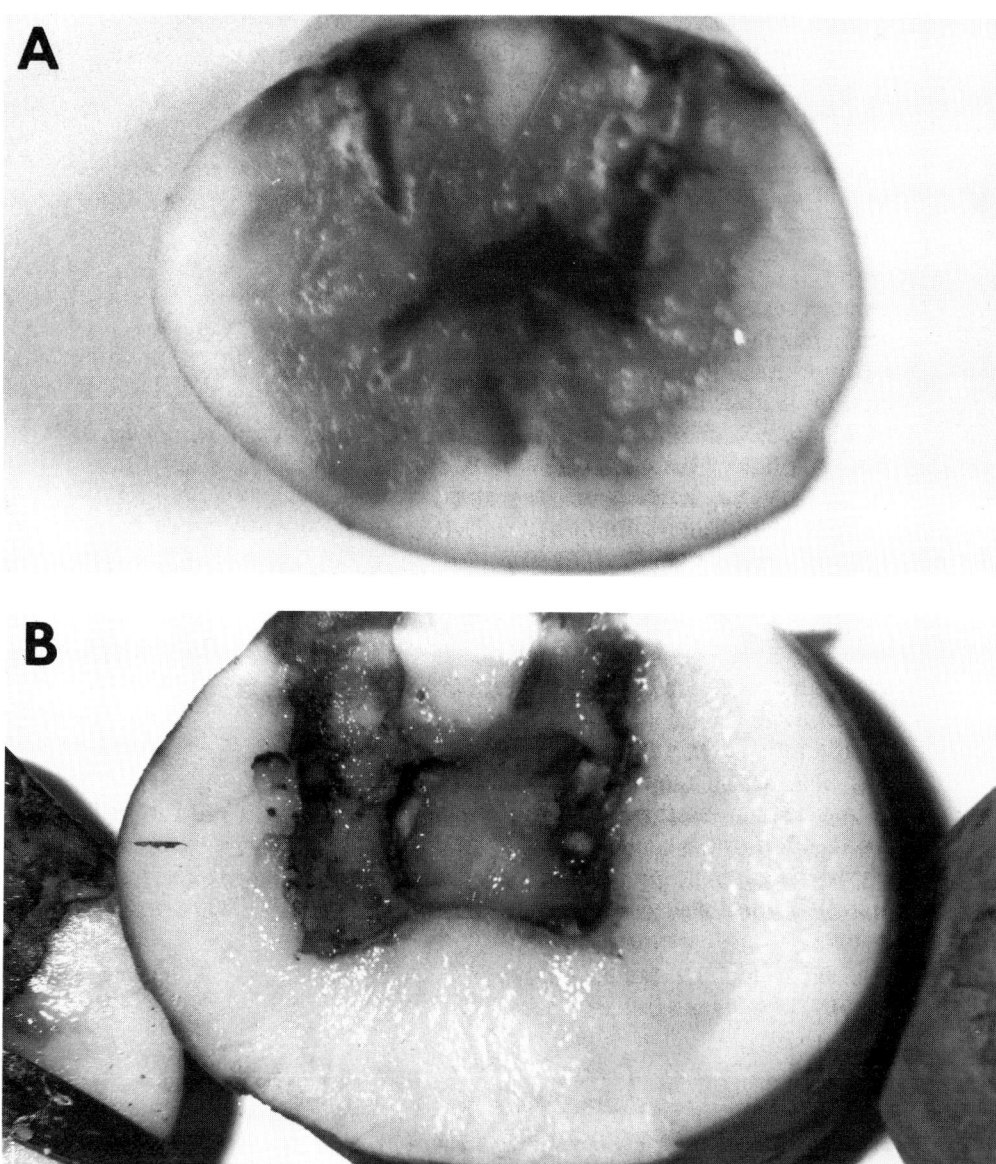

Fig. 11.3. Russet Burbank tubers decayed by bacterial soft rot. Tubers were inoculated with *Erwinia carotovora* subsp. *atroseptica* and incubated for 72 hours in a low-oxygen atmosphere (see Vayda *et al.*, 1992) A. Cross-section of tuber through inoculation sites showing extensive tissue maceration. B. Extent of rot is evident by cavity remaining after removal of the macerated tissue.

more resistant than those of cultivars Superior and Katahdin (Davis et al., 1990; Vayda et al., 1992). Under aerobic conditions, 10^6 to 10^7 bacteria per wound site are required to initiate maceration of tuber tissues. In contrast, all commercial cultivars exhibit increased susceptibility under hypoxic conditions. For example, only approximately 10^2 bacteria per wound site are required to initiate maceration of tubers maintained in hypoxic conditions (Perombelon and Lowe, 1975; DeBoer and Kelman, 1978; Maher and Kelman, 1983). Aerobic tubers wounded several hours prior to inoculation with *Erwinia caratovora* subsp. *caratovora* exhibit less rot than tubers inoculated at the time of wounding (Vayda et al., 1992), indicating that at least one component of aerobic resistance is inducible by wounding. This resistance is lost by treatments which inhibit tuber protein synthesis. For example, tubers incubated with cycloheximide prior to inoculation with pathogenic bacteria exhibit susceptibility to rot that is comparable to that of hypoxic tubers (Zucker and El Zayat, 1968; Zucker and Hanklin, 1970; Vayda et al., 1992). Thus, resistance to bacterial infection appears to be mediated by factors that are expressed in aerobic tubers, and rot occurs when the expression of these gene products is inhibited.

Hypoxic stress causes a global inhibition of tuber protein synthesis which may explain the loss of resistance to bacterial pathogens (Vayda and Schaeffer, 1988). A cessation of aerobic tuber gene expression is apparent within 30 minutes of the onset of hypoxia, which is mediated by inhibition of both transcriptional and translational processes (Butler et al., 1990). However, it is the inhibition of the translational machinery that appears to be the most serious consequence of hypoxic stress because aerobically expressed mRNAs persist for many hours after the onset of hypoxia (Crosby and Vayda, 1991). Prolonged hypoxic stress of 14 hours or more induces the accumulation of mRNAs encoding alcohol dehydrogenase and other genes whose products are involved in fermentative metabolism. These mRNAs become polyribosome-associated and are translated under hypoxic conditions, although the wound-response mRNAs present prior to the onset of hypoxia are not translated. Thus, the response of tubers to hypoxic stress appears to be an alteration in metabolism to allow survival under this adverse condition. However, this alteration precludes expression of putative factors that may confer resistance to bacterial pathogens.

The factors involved in aerobic resistance are not known. Only a small number of gene products induced upon wounding in potato tubers have been identified. These include proteins involved in phenylpropanoid metabolism, cell wall biogenesis and cell replication (Vayda and Schaeffer, 1988; Butler et al., 1990; Rumeau et al., 1990; Vayda et al., 1992). Accordingly, it is reasonable to believe that the processes of suberization and wound-healing are important for limiting or preventing bacterial proliferation. However, expression of these gene products was detected solely because molecular probes were available, and there is no direct evidence that expression of any of the genes identified to date confer resistance to bacterial pathogens. Numerous other unidentified polypeptides are synthesized by tubers in response to wounding (Vayda and Schaeffer, 1988), any of which might exhibit antibacterial activities. Thus, the nature of aerobic resistance to soft rot remains undefined. Efforts to identify such factors genetically are reviewed by Elphinstone in chapter 19 on the inheritance of resistance to bacterial diseases.

Several laboratories are in the process of attempting to introduce novel resistance into potato cultivars by genetic engineering. To date, transgenic potatoes expressing three different foreign genes have been obtained. Potatoes expressing silk moth cecropin and chicken lysozyme (Vayda and Belknap, 1992; W. Belknap and M.E. Vayda, unpublished results), as well as T4 lysozyme (During et al., 1991), exhibited enhanced resistance to bacterial infection in preliminary laboratory testing but this resistance is not apparent in subsequent field testing (W. Belknap, D. Corsini and M.E. Vayda, unpublished). These gene constructs may be more effective if they can be engineered to be expressed specifically under hypoxic conditions. To this end, Vayda and collaborators have been trying to isolate the promoter elements of genes expressed during hypoxia stress, and to understand the mechanism which allows translation of hypoxia-induced mRNAs while inhibiting translation of wound-response mRNAs.

Summary and Conclusions

Various environmental stresses reduce the absolute and marketable yield of potato plants. Stress during the growing season exerts three effects: reduced photosynthetic efficiency and function, inhibition of stolon and tuber initiation, and altered partitioning of photosynthate, which is usually limited under stress conditions. Furthermore, stresses, both during the growing season and after harvest, can affect the marketable quality of the tuber crop. This adverse effect is the result of the temporary suspension of basal tuber metabolic processes, causing either irregular starch deposition, accumulation of reducing sugars or loss of resistance factors. However, the molecular mechanisms and genetics of these physiological processes are not clearly understood.

The genetics and molecular basis of resistance to environmental stresses are ill-defined. Breeders have selected potato clones that are suited for growth under stress conditions, for example, potatoes that yield well in hot, arid growing areas, or that retain their marketable quality during cold storage. However, it is not clear what genetic factors are involved in such tolerance. The activity of specific enzymes has been correlated with stress tolerance or susceptibility in specific cases, for example, HSPs in stress tolerance, dehydrin-like LEA proteins in drought and frost tolerance, and SPS and UPPL in cold-sweetening. However, direct participation of these factors in stress tolerance remains to be demonstrated.

Nevertheless, efforts to understand the physiology, biochemistry and molecular genetics of stress responses have provided insight into these processes. Biochemical and molecular analyses have identified gene products that may mediate stress responses. The role of these gene products and their effectiveness in stress tolerance will be tested directly now that transgenic potato plants can be routinely produced. Furthermore, RFLP mapping of the potato may help identify genetic factors that comigrate with resistant phenotypes. Thus, there is reason to be optimistic that genes mediating tolerance to environmental stress conditions will be identified in the near future.

References

Altschuler, M. and Mascarenhas, J.P. (1985) Transcription and translation of heat shock and normal proteins in seeds of soybean exposed to a gradual temperature increase. *Plant Molecular Biology* 5, 291-297.

Bansal, K.C. and Nagarajan, S. (1987) Reduction of leaf growth by water stress and its recovery in relation to transpiration and stomatal conductance in some potato (*Solanum tuberosum* L.) genotypes. *Potato Research* 30, 497-506.

Begg, J.E. and Turner, N.C. (1976) Crop water deficits. *Advances in Agronomy* 28, 161-217.

Belknap, W. and Rickey, T. (1990) Physiological and stress-induced changes in potato tuber gene expression. In: Vayda, M.E. and Park, W.D. (eds) *The Molecular and Cellular Biology of the Potato*. CAB International, Wallingford, UK, pp. 89-95.

Ben Khedher, M. and Ewing, E.E. (1985) Growth and analysis of eleven potato cultivars grown in the greenhouse under long photoperiods with and without heat stress. *American Potato Journal* 62, 537-554.

Bensen, R.J., Boyer, J.S. and Mullet, J.E. (1988) Water-deficit-induced changes in abscisic acid, growth, polysomes, and translatable RNA in soybean hypocotyls. *Plant Physiology* 88, 289-294.

Bensen, R.J., Beall, F.D., Mullet, J.E. and Morgan, P.W. (1990) Identification of endogenous gibberellins and their relationship to hypocotyl elongation in soybean seedlings. *Plant Physiology* 94, 77-94.

Bhagsari, A.S., Webb, R.E., Phatak, S.C. and Jaworski, C.A. (1988) Canopy photosynthesis, stomatal conductance and yield of *Solanum tuberosum* grown in a warm climate. *American Potato Journal* 65, 393-406.

Bishop, J.C. and Grimes, D.W. (1978) Precision tillage effects on potato root and tuber production. *American Potato Journal* 55, 65-71.

Blum, A., and Ebercon, A. (1981) Cell membrane stability as a measure of drought and heat tolerance in wheat. *Crop Science* 21, 43-47.

Bray, E.A. (1988) Drought-and ABA-induced changes on polypeptide and mRNA accumulation in tomato leaves. *Plant Physiology* 88, 1210-1214.

Burke, J.J., Hatfield, J.L., Klein, R.R. and Mullet, J.E. (1985) Accumulation of heat shock proteins in field-grown cotton. *Plant Physiology* 78, 394-398.

Burton, W.G. (1981) Challenges for stress physiology in potato. *American Potato Journal* 58, 3-14.

Butler, W., Cook, L. and Vayda, M.E. (1990) Hypoxic stress inhibits multiple aspects of the potato tuber wound response. *Plant Physiology* 93, 265-270.

Chen, H.H. and Li, P.H. (1980a) Characteristics of cold acclimation and deacclimation in tuber-bearing *Solanum* species. *Plant Physiology* 65, 1146-1148.

Chen, H.H. and Li, P.H. (1980b) Biochemical changes in tuber-bearing *Solanum* species in relation to frost hardiness during cold acclimation. *Plant Physiology* 66, 414-421.

Chen, H.H., Li, P.H. and Brenner, M.L. (1983) Involvement of abscisic acid in potato cold acclimation. *Plant Physiology* 71, 362-365.

Chen, Q., Lauzon, L.M., DeRocher, A.E. and Vierling, E. (1990) Accumulation, stability and localization of a major chloroplast heat-shock protein. *Journal of Cell Biology* 110, 1873–1883.

Close, T.H., Kortt, A.A. and Chandler, P.M. (1989) A cDNA-based comparison of dehydration-induced proteins (dehydrins) in barley and corn. *Plant Molecular Biology* 13, 95–108.

Cox, W. and Levitt, J. (1976) Interrelations between environmental factors and freezing resistance of cabbage leaves. *Plant Physiology* 57, 553–555.

Creelman, R.A. and Mullet, J.E. (1991) Water deficit modulates gene expression in growing zones of soybean seedlings: analysis of differentially expressed cDNAs, a new β-tubulin gene, and expression of genes encoding cell wall proteins. *Plant Molecular Biology* 17, 591–608.

Creelman, R.A., Mason, H.S., Bensen, R.J., Boyer, J.S. and Mullet, J.E. (1990) Water deficit and abscisic acid cause differential inhibition of shoot versus root growth in soybean seedlings: analysis of growth, sugar accumulation and gene expression. *Plant Physiology* 92, 205–214.

Crosby, J.S. and Vayda, M.E. (1991) Stress-induced translational control in potato tubers may be mediated by polysome-associated proteins. *Plant Cell* 3, 1013–1023.

Davis, M.C., Butler, W. and Vayda, M.E. (1990) Molecular responses to environmental stresses and their relationship to soft rot. In: Vayda, M.E. and Park, W.D. (eds) *The Molecular and Cellular Biology of the Potato*. CAB International, Wallingford, UK, pp. 71–87.

DeBoer, S.H. and Kelman, A. (1978) Influence of oxygen concentration and storage factors on the susceptibility of potato tubers to soft rot. *Potato Research* 21, 65–80.

Demagante, A.L. and van der Zaag, P. (1988) The response of potato (*Solanum* spp.) to photoperiod and light intensity under high temperatures. *Potato Research* 31, 73–83.

DeRocher, A.E., Helm, K.W., Lauzon, L.M. and Vierling, E. (1991) Expression of a conserved family of cytoplasmic low molecular weight heat shock proteins during heat stress and recovery. *Plant Physiology* 96, 1038–1047.

DeRoo, H.C. and Waggoner, P.E. (1961) Root development of potatoes. *Agronomy Journal* 53, 15–17.

Dure, L., Crouch, M., Harada, J., Ho, T.-H., Mundy, J., Quatrano, R., Thomas, T. and Sung, Z.R. (1989) Common amino acid sequence domains among the LEA proteins of higher plants. *Plant Molecular Biology* 12, 475–486.

During, K., Fladung, M. and Lorz, H. (1991) Anti-bacterial resistance of transgenic potato plants producing T4 lysozyme. *Third International Congress of the International Society for Plant Molecular Biology*, Tucson, Arizona, p. 1270. Abstract.

Ellis, R.J. (1990) Molecular chaperones: the plant connection. *Science* 250, 954–959.

Ellis, R.J. and Hemmingsen, S.M. (1989) Molecular chaperones: proteins essential for the biogenesis of some macromolecular structures. *Trends in Biochemical Sciences* 14, 339–342.

Entwistle, G. and ap Rees, T. (1988) Enzymic capacities of amyloplasts from wheat (*Triticum aestivum*) endosperm. *Biochemical Journal* 255, 391–396.

Ewing, E.E. (1981) Heat stress and the tuberization stimulus. *American Potato Journal* 58, 31–49.

Galau, G.A., Bijaisoradat, N. and Hughes, D.W. (1987) Accumulation kinetics of cotton late embryogenesis-abundant mRNAs and storage protein mRNAs: coordinate regulation during embryogenesis. *Developmental Biology* 123, 198–212.

Gawronska, H., Thorton, M.K. and Dwelle, R.B. (1992) Influence of heat stress on dry matter production and photoassimilate partitioning by four potato clones. *American Potato Journal* 69, 653–664.

Hanson, A.D. and Hitz, W.D. (1982) Metabolic responses of mesophytes to plant water deficits. *Annual Reviews of Plant Physiology* 33, 163–203.

Harvey, B.M.R., Lee, H.C. and Susnoschi, M. (1988) Assessment of heat tolerance in potato (*Solanum tuberosum* ssp. *tuberosum*) under controlled environmental conditions: the 'cutting technique'. *Potato Research* 31, 659–666.

Haskell, D., Guy, C., Neven, L. and Smeiser, C. (1991) Spinach leaf proteins accumulating during cold stress also accumulate in response to small changes in water potential. *Third International Congress of the International Society for Plant Molecular Biology*, Tucson, Arizona, p. 1553. Abstract.

Haverkort, A.J., van de Waart, M. and Bodlaender, K.B.A. (1990) The effect of early drought stress on numbers of tubers and stolons of potato in controlled and field conditions. *Potato Research* 33, 89–96.

Helm, K.W., Petersen, N.S. and Abernethy, R.H. (1989) Heat shock response of germinating embryos of wheat. *Plant Physiology* 90, 598–605.

Hetherington, S.E., Smillie, R.M., Malagamba, P. and Huaman, Z. (1983) Heat tolerance and cold tolerance of cultivated potatoes measured by chlorophyll fluorescence method. *Planta* 159, 119–124.

Iritani, W.M. (1981) Growth and preharvest stress and processing quality of potatoes. *American Potato Journal* 58, 71–80.

Iritani, W.M. and Weller, L. (1973) The development of translucent end tubers. *American Potato Journal* 50, 223–233.

Iturriaga, G., Salamini, F. and Bartels, D. (1991) Expression of drought-induced genes in transgenic tobacco. *Third International Congress of the International Society for Plant Molecular Biology*, Tucson, Arizona, p. 1521. Abstract.

Keeling, P.L., Wood, J.R., Tyson, R.H. and Bridges, I.G. (1988) Starch biosynthesis in developing wheat grain: evidence against the direct involvement of triose phosphates in the metabolic pathway. *Plant Physiology* 87, 311–319.

Key, J.L., Kimpel, J.A., Vierling, E., Lin, C.Y., Nagao, R.T., Czarnecka, E. and Schoffl, F. (1985) Physiological and molecular analysis of the heat shock response in plants. In: Atkinson, B. and Walden, D. (eds) *Changes in Eukaryotic Gene Expression in Response to Environmental Stress*. Academic Press, New York, pp. 327–347.

Kimpel, J.A. and Key, J.L. (1985) Presence of heat shock mRNA in field grown soybeans. *Plant Physiology* 79, 672–678.

Lee, S.P., Chen, H.H. and Fuchigami, L.H. (1991) Changes in the translatable RNA population during abscisic acid induced freezing tolerance in bromegrass suspension culture. *Plant Cell Physiology* 32, 45–56.

Levitt, J. (1980) *Responses of Plants to Environmental Stresses*. Academic Press, New York.

Levy, D. (1985) The response of potatoes to a single transient heat or drought stress imposed at different stages of tuber growth. *Potato Research* 28, 415–424.

Levy, D. (1986a) Tuber yield and tuber quality of several potato cultivars as affected by seasonal high temperatures and by water deficit in a semi-arid environment. *Potato Research* 29, 95–107.

Levy, D. (1986b) Genotypic variation on the response of potatoes to high ambient temperatures and water deficit. *Field Crops Research* 15, 85–96.

Levy, D., Livesku, L. and van der Zaag, D.E. (1986) Double cropping of potatoes in a semi-arid environment: the association of ground cover with tuber yields. *Potato Research* 29, 437–449.

Levy, D., Pehu, E. and Veilleux, R.E. (1988) Variability of diffusive leaf resistance and transpiration rates of various potato genotypes and its possible contribution to water economy. *Potato Research* 32, 275-282.

Levy, D., Genizi, A. and Goldman, A. (1990) Compatibility of potatoes to contrasting seasonal conditions, to high temperatures and to water deficit: the association with time of maturation and yield potential. *Potato Research* 33, 325-334.

Levy, D., Kastenbaum, E. and Itzhak, Y. (1991) Evaluation of parents and selection for heat tolerance in the early generations of a potato (*Solanum tuberosum* L.) breeding program. *Theoretical and Applied Genetics* 82, 130-136.

Li, P.H. and Fennell, A. (1985) Potato frost hardiness. In: Li, P.H. (ed.) *Potato Physiology*. Academic Press, New York, pp. 457-479.

Li, P.H., Huner, N.P.A., Toivio-Kinnucan, M., Chen, H.H. and Palta, J.P. (1981) Potato freezing injury and survival, and their relationships to other stress. *American Potato Journal* 58, 15-29.

Lin, C.Y., Roberts, J.K. and Key, J.L. (1984) Acquisition of thermotolerance in soybean seedlings. *Plant Physiology* 74, 152-160.

Lubben, T.H., Donaldson, G.K., Viitanen, P.V. and Gatenby, A.A. (1989) Several proteins imported into chloroplasts form stable complexes with the *groEL*-related chloroplast molecular chaperone. *Plant Cell* 1, 1223-1230.

MacKerron, D.K.L. and Jefferies, R.A. (1985) Observations on the effects of relief of late water stress in potato. *Potato Research* 28, 349-359.

MacKerron, D.K.L. and Jefferies, R.A. (1988) The distributions of tuber sizes in droughted and irrigated crops of potato. I. Observations on the effect of water stress on graded yields from differing cultivars. *Potato Research* 31, 269-278.

MacKerron, D.K.L., Marshall, B. and Jefferies, R.A. (1988) The distributions of tuber sizes in droughted and irrigated crops of potato. II. Relation between size and weight of tubers and the variability of tuber-size distributions. *Potato Research* 31, 279-288.

Maher, E.A. and Kelman, A. (1983) Oxygen status of potato tuber tissue in relation to maceration by pectic enzymes of *Erwinia carotovora*. *Phytopathology* 73, 536-539.

Malik, N.J., Dwelle, R.B., Thorton, M.K. and Pavek, J.J. (1992) Dry matter accumulation in potato clones under seasonal high temperature conditions in Pakistan. *American Potato Journal* 69, 667-676.

Mansfield, M.A. and Key J.L. (1987) Synthesis of the low-molecular weight heat shock proteins in plants. *Plant Physiology* 84, 1007-1017.

Marinus, J. and Bodlaender, K.B.A. (1975) Response of some potato varieties to temperature. *Potato Research* 18, 189-204.

Martineau, J.R., Williams, J.H. and Specht, J.E. (1979) Temperature tolerance in soybeans. II. Evaluation of segregating population for membrane thermostability. *Crop Science* 19, 79-82.

Mason, H.S. and Mullet, J.E. (1990) Expression of two soybean vegetative storage protein genes during development and in response to water deficit, wounding and jasmonic acid. *Plant Cell* 2, 569-579.

Mason, H.S., Mullet, J.E. and Boyer, J.S. (1988) Polysomes, messenger RNA, and growth in soybean stems during development and water deficit. *Plant Physiology* 86, 725-733.

Mason, P.J., Hall, L.M.C. and Gausz, J. (1984) The expression of heat shock genes during normal development in *Drosophila melanogaster*. *Molecular and General Genetics* 194, 73-78.

Midmore, D.J. (1984) Soil temperature effects on emergence, plant development and yield. *Field Crops Research* 8, 255–271.

Miller, D.E. and Martin, M.W. (1987) The effect of irrigation regime and subsoiling on yield and quality of three potato cultivars. *American Potato Journal* 64, 17–25.

Nagao, R.T., Czarnecka, E., Gurley, W.B., Schoffl, F. and Key, J.L. (1985) Genes for low molecular weight heat shock proteins of soybeans: sequence analysis of a multigene family. *Molecular and Cellular Biology* 5, 3417–3428.

Nagarajan, S. and Bansal, K.C. (1986) Measurement of cellular membrane thermostability to evaluate foliage heat tolerance of potato. *Potato Research* 29, 163–167.

Nover, L., Neumann, D. and Scharf, K.D. (1989) *Heat Shock and Other Stress Response Systems of Plants*. Springer-Verlag, Berlin.

Nowak, J. and Colborne, D. (1989) *In vitro* tuberization and tuber proteins as indicators of heat stress tolerance in potato. *American Potato Journal* 66, 35–45.

Ottaviano, E., Sari Gorla, M., Pe, E. and Frova, C. (1991) Molecular markers (RFLPs and HSPs) for the genetic dissection of thermotolerance in maize. *Theoretical and Applied Genetics* 81, 713–719.

Pelham, H.R.B. (1989) Heat shock and the sorting of luminal ER proteins. *EMBO Journal* 8, 3171–3176.

Perombelon, M.C. and Kelman, A. (1980) Ecology of the soft rot Erwinias. *Annual Reviews of Phytopathology* 18, 361–387.

Perombelon, M.C. and Lowe, R. (1975) Studies on the initiation of bacterial soft rot in potato tubers. *Potato Research* 18, 64–82.

Prange, R.K., McRae, K.B., Midmore, D.J. and Deng, R. (1990) Reduction in potato growth at high temperature: role of photosynthesis and dark respiration. *American Potato Journal* 67, 357–369.

Prasad, T.K. and Hallberg, R.L. (1989) Identification and metabolic characteristics of the *Zea mays* mitochondrial homolog of the *Escherichia coli groEL* protein. *Plant Molecular Biology* 12, 609–618.

Quigley, A.S., Ougham, H.J., Eagles, C.F. and Schunmann, P.H.D. (1991) Homologies between cold hardening and other stresses in winter oats (*Avena sativa* L.). *Third International Congress of the International Society for Plant Molecular Biology*, Tucson, Arizona, p. 1567. Abstract.

Rajashekar, C., Li, P.H. and Carter, J.V. (1983) Frost injury and heterogeneous ice nucleation in leaves of tuber-bearing *Solanum* species. *Plant Physiology* 71, 749–755.

Reynolds, M.P. and Ewing, E.E. (1989) Heat tolerance in tuber bearing *Solanum* species: a protocol for screening. *American Potato Journal* 66, 63–74.

Rickey, T.M. and Belknap, W.R. (1991) Comparison of the expression of several stress-responsive genes in potato tubers. *Plant Molecular Biology* 16, 1009–1018.

Roy, H. (1989) Rubisco assembly: a model system for studying the mechanism of chaperonin action. *Plant Cell* 1, 1035–1042.

Rumeau, D., Maher, E.A., Kelman, A. and Showalter, A.M. (1990) Extensin and phenylalanine ammonia-lyase gene expression altered in potato tubers in response to wounding, hypoxia, and *Erwinia carotovora* infection. *Plant Physiology* 93, 1134–1139.

Schapendonk, A.H.C.M., Spitters, C.J.T. and Groot, P.J. (1989) Effects of water stress on photosynthesis and chlorophyll fluorescence of five potato cultivars. *Potato Research* 32, 17–32.

Schoffl, F., Baumann, G. and Raschke, E. (1988) The expression of heat shock genes: a model for the environmental stress response. In: Goldberg, B. and Verna, D.P.S.

(eds) *Temporal and Spatial Regulations of Plant Genes*. Springer-Verlag, Berlin, pp. 253-273.

Schuster, G., Even, D., Kloppstech, K. and Ohad, J. (1988) Evidence for protection by heat-shock proteins against photoinhibition during heat shock. *EMBO Journal* 7, 1-6.

Shalhevet, J., Shimshi, D. and Meir, T. (1983) Potato irrigation requirements in a hot climate using sprinkler and drip methods. *Agronomy Journal* 75, 13-16.

Shallenberger, R.S., Smith, O. and Treadway, R.H. (1959) Role of the sugars in the browning reaction in potato chips. *Journal of Agriculture and Food Chemistry* 7, 274-277.

Sharpe, P.J.H. (1983) Responses of photosynthesis and dark respiration to temperature. *Annals of Botany* 52, 325-343.

Shekar, V.C. and Iritani, W.M. (1978) Starch to sugar interconversion in *Solanum tuberosum* L.: influence of inorganic ions. *American Potato Journal* 55, 345-350.

Shekar, V.C., Iritani, W.M. and Magnuson, J.R. (1978) Starch-sugar interconversion in *Solanum tuberosum* L.: influence of membrane permeability and fluidity. *American Potato Journal* 55, 394-407.

Shimshi, D. and Susnoschi, M. (1985) Growth and yield studies of potato development in a semi-arid region. 3. Effect of water stress and amounts of nitrogen top dressing on physiological indices and on tuber yield and quality of several cultivars. *Potato Research* 28, 177-191.

Shimshi, D., Shalhevet, J. and Meir, T. (1983) Irrigation regime effects on some physiological responses of potato. *Agronomy Journal* 75, 262-267.

Sipos, J. and Prange, R.K. (1986) Response of ten potato cultivars to temperature as measured by chlorophyll fluorescence *in vivo*. *American Potato Journal* 63, 683-694.

Sowokinos, J. (1990a) Stress-induced alterations in carbohydrate metabolism. In: Vayda, M.E. and Park, W.D. (eds) *The Molecular and Cellular Biology of the Potato*. CAB International, Wallingford, UK, pp. 137-158.

Sowokinos, J. (1990b) Effect of stress and senescence on carbon partitioning in stored potatoes. *American Potato Journal* 67, 849-857.

Sowokinos, J.R., Lulai, E.C. and Knoper, J.A. (1985) Translucent tissue defects in *Solanum tuberosum* L. I. Alterations in amyloplast membrane integrity, enzyme activities, sugars and starch content. *Plant Physiology* 78, 489-493.

Stark, J.C. and Dwelle, R.B. (1989) Antitranspirant effects on yield, quality and water use efficiency of Russet Burbank potatoes. *American Potato Journal* 66, 563-574.

Stark, J.C. and Wright, J.L. (1985) Relationship between foliage temperature and water stress in potatoes. *American Potato Journal* 62, 57-68.

Stout, D.G. (1980) Alfalfa water status and cold hardiness as influenced by cold acclimation and water stress. *Plant, Cell and Environment* 3, 237-241.

Struik, P.C., Geertsema, J. and Custers, C.H.M.G. (1989a) Effects of shoot, root and stolon temperatures on the development of the potato (*Solanum tuberosum* L.) plant. I. Development of the haulm. *Potato Research* 32, 133-141.

Struik, P.C., Geertsema, J. and Custers, C.H.M.G. (1989b) Effects of shoot, root and stolon temperatures on the development of the potato (*Solanum tuberosum* L.) plant. II. Development of stolons. *Potato Research* 32, 142-149.

Struik, P.C., Geertsema, J. and Custers, C.H.M.G. (1989c) Effects of shoot, root and stolon temperatures on the development of the potato (*Solanum tuberosum* L.) plant. III. Development of tubers. *Potato Research* 32, 151-158.

Sukumaran, M.P. and Weiser, C.J. (1972) Freezing injury in potato leaves. *Plant Physiology* 50, 564-567.

Susnoschi, M. and Shimshi, D. (1985) Growth and yield studies of potato development in a semi-arid region. 2. Effect of water stress and amounts of nitrogen top dressing on growth of several cultivars. *Potato Research* 28, 161-176.

Thomashow, M., Gilmour, S., Hajela, R., Horvath, D., Lin, C. and Guo, W. (1991) Cold acclimation in *Arabidopsis thaliana*. *Third International Congress of the International Society for Plant Molecular Biology*, Tucson, Arizona, p. 149. Abstract.

Tyson, R.H. and ap Rees, T. (1988) Starch synthesis by isolated amyloplasts from wheat endosperm. *Planta* 175, 33-38.

van der Zaag, D.E. (1984) Reliability and significance of a simple method of estimating the potential yield of the potato crop. *Potato Research* 27, 51-53.

van der Zaag, D.E. and Doornbos, J.H. (1987) An attempt to explain difference in the yielding ability of potato cultivars based on differences in cumulative light interception, utilization efficiency of foliage and harvest index. *Potato Research* 30, 551-568.

van Loon, C.D. (1981) The effect of water stress on potato growth, development, and yield. *American Potato Journal* 58, 51-69.

Vayda, M.E. and Belknap, W.R. (1992) The emergence of transgenic potatoes as commercial products and tools for basic science. *Transgenic Research* 1, 149-163.

Vayda, M.E. and Schaeffer, H.J. (1988) Hypoxic stress inhibits the appearance of wound-response proteins in potato tubers. *Plant Physiology* 88, 805-809.

Vayda, M.E., Antonov, L.S., Yang, Z., Butler, W.O. and Lacy, G.H. (1992) Hypoxic stress inhibits aerobic wound-induced resistance and activates hypoxic resistance to bacterial soft rot. *American Potato Journal* 68, 239-253.

Vierling, E. (1991) The roles of heat shock proteins in plants. *Annual Reviews of Plant Physiology and Plant Molecular Biology* 42, 579-620.

Vierling, E., Mishkind, M.L., Schmidt, G.W. and Key, J.L. (1986) Specific heat shock proteins are transported into chloroplasts. *Proceedings of the National Academy of Sciences USA* 76, 361-365.

Vierling, E., Nagao, R.T., DeRocher, A.E. and Harris, L.M. (1988) A heat shock protein localized to chloroplasts is a member of a eukaryotic superfamily of heat shock proteins. *EMBO Journal* 7, 575-581.

Vierling, E., Harris, L.M. and Chen, Q. (1989) The major low-molecular-weight heat shock protein in chloroplasts shows antigenic conservation among diverse higher plant species. *Molecular and Cellular Biology* 9, 461-468.

Walker, M.A., McKersie, B.D. and Peter Pauls, K. (1991) Effects of chilling on the biochemical and functional properties of thylakoid membranes. *Plant Physiology* 97, 663-669.

Wolfe, D.W., Ferreres, E. and Voss, R.E. (1983) Growth and yield responses of two potato cultivars at various levels of applied water. *Irrigation Science* 3, 211-222.

Workman, M., Cameron, A. and Twomey J. (1979) Influence of chilling on potato tuber respiration, sugar, O-dihydroxyphenolic content and membrane permeability. *American Potato Journal* 56, 277-288.

Wu, C.H., Caspar, T., Browse, J., Lindquist, S. and Somerville, C. (1988) Characterization of an hsp70 cognate gene family in *Arabidopsis*. *Plant Physiology* 88, 731-740.

Zucker, M. and El Zayat, M.M. (1968) The effect of cycloheximide on the resistance of potato tuber discs to invasion by a fluorescent pseudomonad. *Phytopathology* 58, 339-344.

Zucker, M. and Hanklin, L. (1970) Physiological basis for cycloheximide-induced soft rot of potatoes by *Pseudomonas fluorescens*. *Annals of Botany* 34, 1047-1062.

12 Inheritance of Morphological and Tuber Characteristics

R. ORTIZ[1] and Z. HUAMAN[2]

[1]*Department of Genetics, University of Wisconsin-Madison, 1575 Linden Drive, Madison, Wisconsin 53706, USA.* [2]*Department of Genetic Resources, International Potato Center (CIP), Apartado Postal 5969, Lima, Peru.*

Despite the tremendous amount of phenotypic variability that exists in morphological and tuber characters of the tuber-bearing *Solanum* species, knowledge on the mode of inheritance of several of these traits is still scanty. The two main obstacles have been: (i) the tetrasomic inheritance of the cultivated potato (Cadman, 1942); and (ii) the strong inbreeding depression and gametophytic self-incompatibility at the diploid level, which makes the production of inbred lines difficult. Many morphological mutants useful in genetic studies result in weak and sterile plants (De Jong, 1991).

The inheritance of the most important morphological and tuber traits of the potato were first reviewed by Swaminathan and Howard (1953); later on Howard (1960, 1970) updated the information. Since then many papers related to these traits have been published. The general morphology of the potato plant is shown in Figure 12.1.

Anthocyanin Pigmentation in the Potato

Pink, red, blue and purple pigmentation in flowers, sprouts, stems and tubers are due to anthocyanins. A wider range of anthocyanins has been found in cultivated potatoes than in its wild relatives (Dodds and Long, 1955, 1956; Harborne, 1960; Dodds and Paxman, 1962), which have only an acylated derivative of petunidin. Similarly, malvidin has been found only in the cultivated tetraploids and not in diploid cultivated potatoes (Simmonds and Harborne, 1965).

The most recent review of the inheritance of anthocyanin pigmentation in potatoes was made by De Jong (1991). Howard (1970) considered that there were major differences in the anthocyanin pigmentation systems between Tuberosum and cultivated diploids. However, De Jong (1991) considers that

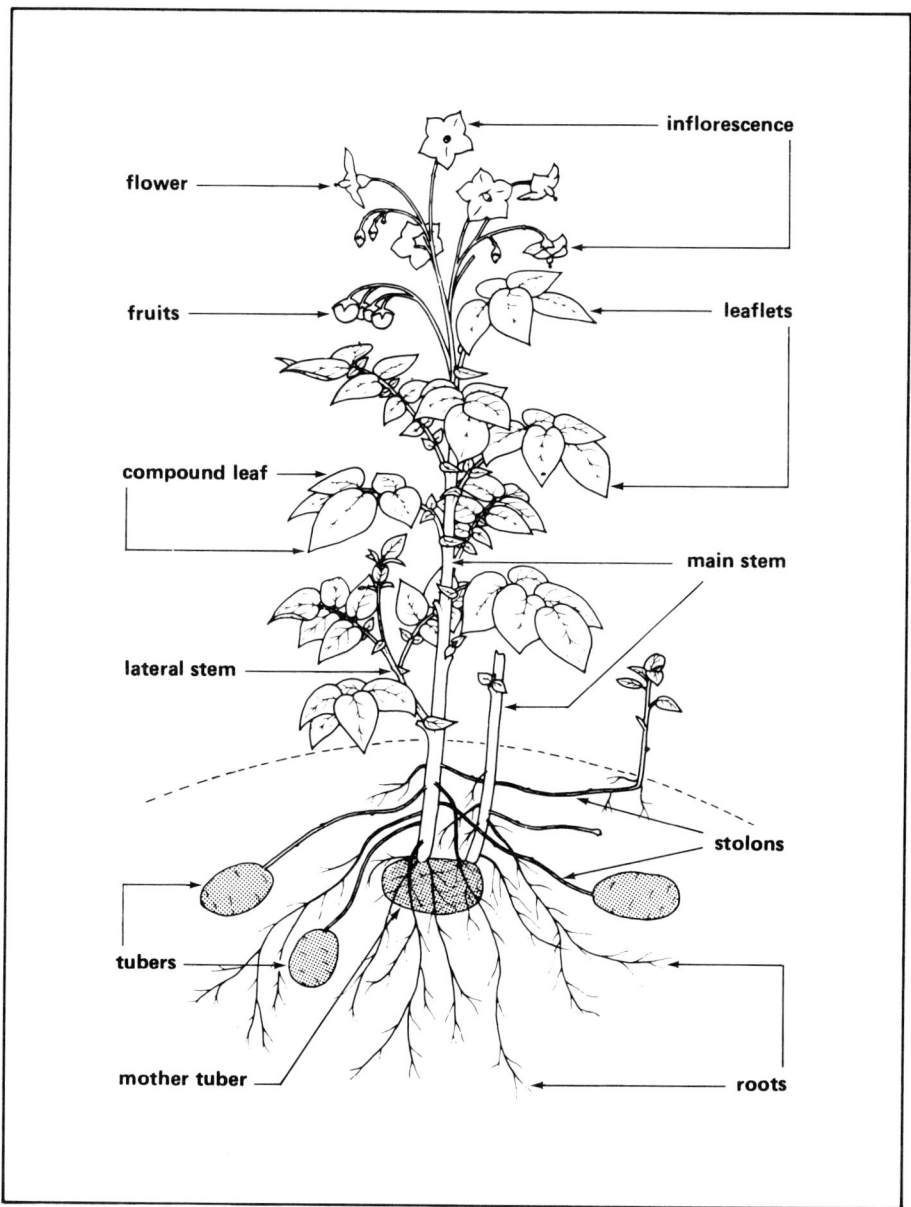

Fig. 12.1. Morphology of the potato plant.

these differences may be due more to the differences in alleles which have thus far been identified rather than differences in the presence or absence of different loci between Tuberosum and the cultivated diploids.

The locus D carries the basic gene for pigmentation throughout the plant. The P and R loci control the synthesis of anthocyanins and are responsible for production of purple (delphinidin and petunidin derivatives) and red pigments (cyanidin and pelargonidin derivatives), respectively. The locus Ac controls the acylation of anthocyanins.

Several loci also affect the distribution of anthocyanin in different plant parts (Dodds and Long, 1956; De Jong and Rowe, 1972; Kessel and Rowe, 1974b; Garg et al., 1981; De Jong, 1987). The locus B is a multi-allelic locus, with at least five alleles in a hierarchical system of dominance ($Bd > Bc > Bb > Ba > b$), controlling pigmentation at the base of organs that are homologous to leaves, e.g. floral abscission layer, tuber eyebrow, cotyledonary node of the embryo (seed spot, allele Bd) and nodal band at the base of the petiole. The accumulation of anthocyanins in the youngest leaves of the plants (pigmented whorl) is controlled by the Pw locus. The gene F acts as an intensifier of pigmentation in the flower and requires D and R (or P) for expression (De Jong, 1991). The pigmentation of the ovary wall is controlled by locus Ow. The I locus controls the distribution of the pigments on the tuber skin and the stem. The gene M restricts tuber periderm pigmentation to areas around the eyes (Kelly, 1924; Howard, 1962). Another locus, E, has also been postulated for the red pigmentation of the periderm of tubers, eyes, stems and flowers (Lunden, 1937). Howard (1966) found that loci E and M were closely linked, and that genes E and R, which together with D produced red anthocyanin pigments in the periderm and phelloderm of tubers, could be allelomorphic (Howard, 1964). The loci for pigmented whorl (Pw) and ovary wall (Ow) were found to be linked to the F-I-B linkage group (De Jong and Rowe, 1972; Kessel and Rowe, 1974b).

Plant Type (see Figure 12.2)

Salaman and Lesley (1920) found in $4x$ Tuberosum that the normal upright habit of the potato plant was dominant to prostrate. They observed upright to prostrate ratios of 63:1 in the segregating F_2 generations which they (probably incorrectly) interpreted as indicating disomic inheritance involving three loci. The procumbent type, an intermediate growth habit between upright and prostrate, was found to be recessive to upright but its relationship with prostrate was not clear.

Some mutations are known to produce some variations in plant type that differ from the normal type. Thus, the topiary mutant was identified in the diploid wild species *S. infundibuliforme* (den Nijs et al., 1980). It is a pleiotropic trait characterized by its compact plant type at the early seedling stage, which develops into a spherical shape; by tubers that are clustered around the stem; and by extremely knobby tubers with poor apical dominance during sprouting. The pleiotropic action of the mutant was found to be inherited as a recessive (tp) affecting cytokinin activity.

Dwarfism was found in Andigena and it is characterized by producing a compact, dark green, rosette plant. The dwarf phenotype is conferred by a nulliplex status at a single locus. Tuberosum cultivars such as Superior and

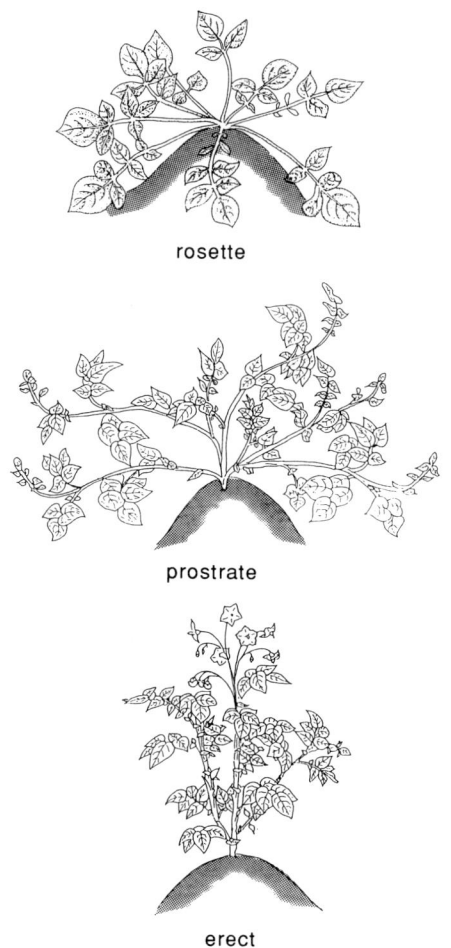

Fig. 12.2. Plant forms.

its mutant New Superior were found to be triplex at this locus (Bamberg and Hanneman, 1987).

Lunden (1937) found white (albinism) and light green (virescens) seedlings in a selfed progeny of an F_1 hybrid between the Tuberosum cultivars Parnassia and Louis Botha. Their segregation in the F_2 fit a 35:1 ratio, for both normal:white and normal:virescens, which was expected after selfing an individual duplex for both loci (*WWww VVvv*) with chromosome segregation. Estrada (1960) found albino seedlings in cultivated diploid and tetraploid potatoes, and in the disomic tetraploid *S. acaule*. Lam and Erickson (1971) indicated that the albino character in *S. chacoense* is controlled by a single recessive gene, *a*, with normal green (*A*−) being completely dominant over albino (*aa*). The monogenic recessive mutant virescens (*v*) was also reported in Tuberosum dihaploids (Hermsen *et al.*, 1978). The character is observed most clearly in the young seedling stages, which are characterized by their slower growth and lighter green colour than normal plants ($V-$) At later stages, these differences decrease but the virescent plant (vv) rarely becomes completely normal.

Plaisted and Peterson (1967) observed a mutant in *S. vernei* which, in its homozygous recessive condition (gg), produced plants with chlorotic foliage 6 weeks after planting. The *G* locus producing normal green plants was found to be linked to a factor controlling anthocyanin pigmentation.

The dark green mutant seedling is inherited as a monogenic recessive character, *ds* (Hermsen *et al.*, 1978). The mutant is characterized by both its dark green colour and its overall slow growth. The *dsds* genotype is not lethal but plants have variable viability.

Stem Characters

Pigmentation

Stem colour is due to the accumulation of anthocyanin pigments in the hypodermal layer, which traces back to the L2 histogenic layer. It is commonly observed in the leaf axis. Stem pigmentation is probably due to a single dominant gene (Kumikura, 1967).

Pubescence

The stems of the potato plant can be glabrous (without hairs) or pubescent. A clear expression of this trait is observed in young plants about 3 weeks after emergence. Matsubayashi (1979) studied 65 dihaploids of cv. Chippewa with stem pubescence ranging from dense to almost none (glabrous). The observed ratio for pubescent to glabrous stems was 48 to 17, which fits an 11 : 3 ratio. Based on these results, it therefore appears that a single gene controls this trait, with pubescent stem dominant to glabrous stem, and that the locus displays chromatid segregation. A duplex genotype was assigned to Chippewa.

Stem wing type

Choudhuri (1944) observed that all plants had a broad stem wing in the F_1 of a cross between *S. phureja* (rudimentary stem wing) and *S. simplicifolium* (broad stem wing). The observed phenotypes in the F_2, backcross to *S. phureja* and backcross to *S. simplicifolium*, fitted 3 : 1, 1 : 1 and 1 : 0 broad : rudimentary stem wing ratios, respectively. These results indicated that a single major gene, *W*, was controlling the inheritance of the trait. Furthermore, Taylor (1978) found that wing stem type was controlled by one gene with crenulate ($C-$) dominant to straight (cc).

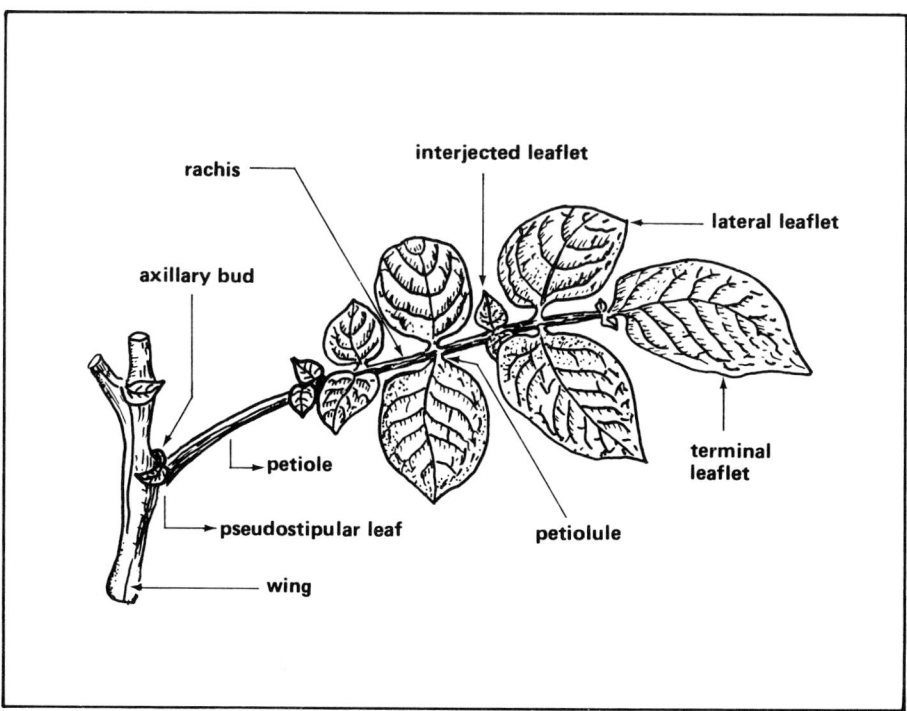

Fig. 12.3. Leaf of potato plant.

Leaf Characters (see Figure 12.3)

Dark heart of leaves (pigmented whorl)

Some potato plants produce a dark heart in the youngest leaves, which disappears as they mature, due to an accumulation of anthocyanins at different intensities. Koopmans (1951) studied this trait in F_2 progenies of *S. rybinii* × *S. chacoense* and *S. rybinii* × *S. commersonii* and postulated that this trait was controlled by a recessive gene. However, Kessel and Rowe (1974b) studied the same trait in progenies involving *S. phureja* ($\equiv$ *S. rybinii*), *S. stenotomum* and dihaploid Tuberosum and suggested that a single dominant gene, *Pw*, controlled pigmented whorl.

Underleaf pigmentation

The abaxial or underside of the leaflets of some genotypes is totally pigmented, while in others only small areas of the leaflet are pigmented. However, this characteristic is not found on all leaves of a single plant, probably due to incomplete penetrance. Kessel and Rowe (1974b) postulated that a single dominant gene, *Ul*, determined underleaf pigmentation.

Nodal band pigmentation

The pigmented ring that covers the base of the leaf petioles and the stem nodes of some potato genotypes is known as nodal band. It is one manifestation of the complementary action of the genes *B* and *P* (Dodds and Long, 1956).

Leaf shape

All cultivated potatoes produce compound leaves that comprise a terminal leaflet and several pairs of primary lateral leaflets. The degree of leaf dissection can be increased by the presence of secondary folioles, which are located between the lateral leaflets and sometimes even on their stalks or petiolules, which join the leaflet to the rachis. Leaves also differ in the number, size and breadth of leaflets and folioles, and in the length of petiolules of the lateral leaflets. Simple leaves are found only in some wild potato species.

Choudhuri (1943, 1944, 1948a, b) studied the inheritance of leaf shape in both cultivated and wild species. He found that the compound leaf of *S. phureja* was dominant to the simple leaf of *S. simplicifolium*. All F_1 plants had compound leaves and the segregation ratios of the F_2 and the backcrosses to *S. simplicifolium* and *S. phureja* fitted 3 : 1, 1 : 1 and 1 : 0 ratios, respectively. These results indicated a simple monofactorial genetic system, with *L* assigned to the allele for compound leaf (Choudhuri, 1944). Moreover, Emme (1937) indicated that the narrow leaflets and more divided leaf of *S. stenotomum* were dominant in the F_1 to the leaf types of both *S. phureja* and *S. goniocalyx*. Koopmans (1951) concluded that leaf shape was determined by multiple genetic factors. The aberrant types have been reported as being dominant over normal (Heiken and Ewertson, 1962; Howard, 1967b).

Terminal leaflet shape

In general, the terminal leaflet shape can be broad or narrow. The variation of this trait in a dihaploid population from cv. Chippewa was continuous but followed a bimodal pattern. The dihaploids were grouped as narrow (mean 49.4) and broad (mean 61.4), expressed in percentage of the width/length ratio of mature leaves (Matsubayashi, 1979). The segregation was 56 narrow : 9 broad which fits a 5 : 1 ratio, indicating one gene with chromosomal segregation controlling the phenotypic expression of the trait.

Primary lateral leaflets

Narrow leaflets have been reported to be recessive (*nlnl*) to broad in *S. phureja* (Dodds and Paxman, 1962). Furthermore, an 11 broad : 1 narrow segregation ratio in a cross between two Tuberosum tetraploid cultivars was observed (Kumikura, 1972). This is the expected ratio for a cross between a duplex × simplex parent with the locus close to the centromere (chromosomal segregation).

The number of lateral leaflets differs with the position of the leaves. However, there is little variation in the leaves at the base of the stem. A clear 13 : 15 segregation ratio of more than three pairs of leaflets to less than three pairs was observed in a cross between the tetraploid cultivars Spartan and Maris Piper (Kumikura, 1972). These results suggested that one dominant gene was responsible for less than three pairs of leaflets and that the locus was located far away from the centromere (chromatid segregation). Koopmans (1951), however, postulated a more complicated genetic system, because her results were explained only by assuming that multiple factors determined leaflet numbers.

Ivy leaf

Van Harten *et al.* (1973) produced the mutant ivy leaf through induced mutations in Tuberosum. They indicated that this characteristic leaf shape was inherited as dominant. Their results were confirmed by Pongsupasamit (1990), who indicated that this leaf shape was mainly determined by the constitution of the cells in the L2 histogenic layer because those of L1 and L3 have little effect.

Mutations affecting leaf characters

More than 26 other mutants have been reported that could be distinguished on a morphological (somatic) or physiological basis (Heiken, 1958; Simmonds, 1965). Simmonds (1965) indicated that the following mutants were inherited as recessives in *S. phureja*: curled leaf (*cl*), crumpled leaf (*cr*), droopy plants (*dr*), defective xylem (*dx*), hairy dwarf (*hd*) and plants with leaves resembling leafroll virus infection (*lr*). He also reported other recessive mutants, including plants in Andigena with leaflets which were pink underneath and had yellow margins (*py*), short plants in *S. stenotomum* (*s*) and veinal chlorosis in *S. goniocalyx* (*rc*).

Simmonds (1965) reported a mutant leaf margin characterized by small roundish leaflets with yellow or reddish margins. The mutant plants may tuberize but they do not produce flowers (Hermsen, 1978). It is inherited as a simple recessive, *ym* (Dodds and Paxman, 1962).

Flower Characters

Number of inflorescences and flowers per inflorescence

Pineda-Colorado (1990) applied generation mean analysis to study the inheritance of both total number of inflorescences and number of flowers per inflorescence. He obtained five populations derived from intermating five inbred lines of *S. chacoense*. Both parents and the F_1, F_2 and backcrosses to each parent were included in the analysis of each population. He indicated that his results did not fit an additive–dominance model and that genotype–environment interaction was significant in the expression of both traits. These results could be an indication that different major genes were turned on under different environmental conditions. However, other genes with minor additive effects and epistasis cannot be discarded for the genetic control of these traits.

Inflorescence type

Different types of inflorescence can be observed in the potato, mainly due to the number of flower clusters. Taylor (1978) found a simple genetic system with one locus having chromatid segregation, where the type 1 allele with two flower clusters was dominant to the type 2 allele with three flower clusters.

Calyx shape

Hawkes (1956) presented preliminary results indicating that the regular form of calyx with short sepals of *S. sparsipilum* was dominant to that found in the cultivated *S. stenotomum*, which is characterized by its irregular form with relatively long sepals. The gene symbols *Ca* (regular form of calyx) and *ca* (irregular) were suggested by Howard (1973) for the dominant and recessive alleles, respectively. An irregular form of calyx never segregates out in tetraploid progenies from tetraploid cultivars, which have dihaploid populations with both irregular (*caca*) and regular (*Ca*−) calyxes (Howard, 1973). However, observed ratios reported by Taylor (1978) did not fit any expected ratio in selfed tetraploid progeny, parthenogenic tetraploids derived from second-division restitution 2*n* eggs, and dihaploids from the same tetraploid parent. Howard (1973) also suggested a close linkage between the locus controlling calyx shape and the S-incompatibility locus. However, Woodcock and Howard (1975) found no evidence to support this hypothesis and suggested that more than one gene may be involved in controlling the different states of calyx irregularity.

Calyx tube length

Taylor (1978) reported one locus with chromatid segregation controlling calyx tube length, where short is dominant to long. This study was made using a ratio between the length from the base of the calyx tube to the sepal acumen tip and the calyx tube length.

Corolla pigmentation

Potato flower colour is a character with a great deal of variation, and Black (1933) postulated that it is controlled by several genetic factors. Hermsen (1978) found that two complementary dominant genes could explain his flower colour results when he selfed two self-compatible Tuberosum dihaploids from cultivars Black 4495 and Ginecke. Furthermore, Taylor (1978) concluded that corolla pigmentation is under the control of a two-gene system, with both loci having chromatid segregation. Studies with chimeras demonstrated that corolla pigmentation was due to pigments in tissues which trace back to the L1 histogenic layer at the stem growing point (Howard, 1970).

Corolla shape

Corolla shape varies from stellate, through semistellate, pentagonal and rotate, to very rotate. These shapes can also be differentiated by using a corolla index, such as the one used by Matsubayashi (1979), where I is the ratio between the length from the centre of the flower to the intersection of the petals (B) and the length to the petal acumen tip (A).

This trait has been studied extensively in crosses between wild and cultivate diploid species (Emme, 1938; Propach, 1940; Koopmans, 1951). However, there are different genetic explanations for its inheritance. In the F_1 from *S. phureja* (stellate) × *S. jamesii* (rotate), the corolla shape was intermediate (Emme, 1938). However, Propach (1940) observed a 3 rotate : 1 stellate ratio in an F_2 from a cross between *S. verrucosum* and *S. chacoense*. The results indicated that rotate was due to a dominant gene, *Rot*, and that the genotype for stellate corolla should be *rot/rot*. In addition, Matsubayashi (1979) reported that the tetraploid cultivar Chippewa, with subrotate corolla shape ($I = 0.70$), showed a large variation in its derived dihaploids: 16 were stellate ($I = 0.52$), 28 were pentagonal ($I = 0.62$) and 12 were subrotate ($I = 0.73$). The observed segregation in this dihaploid population fits a 3 : 8 : 3 ratio and could indicate that one locus, with alleles having incomplete dominance and chromatid segregation, controls the trait.

Stigma type

The main types of stigma in the potato are the notched stigma, with snicked and large-sized lobes, and the entire stigma, with smooth-faced and small-sized lobes. This character was studied in a Chippewa dihaploid population (Matsubayashi, 1979), which had 47 individuals with notched and 9 with entire stigma. The results fitted a 5 : 1 ratio, which indicates that notched is dominant to entire stigma, that the locus displays chromosomal segregation, and that Chippewa had a duplex genotype for the notched stigma locus.

Another type of stigma, the split stigma, has been frequently found in aneuploid plants from interspecific $3x \times 2x$ crosses, but was not related to the aneuploidy of these plants. Kessel and Rowe (1974a) provided evidence, using sib mating and backcross generations, that one or two independent loci in the homozygous recessive state were responsible for the expression of this type of stigma.

Ovary wall pigmentation

This phenotype was first reported by Pushkarnath (1960), who did not indicate the inheritance of the trait, which was later on found to be inherited by a single dominant gene, *Ow* (De Jong and Rowe, 1972).

Flower abnormalities

Deformed flowers, characterized as having short anthers, absence of anthers or lack of dehiscent pores in the anthers, were observed in F_2 generations of *S. chacoense* by Koopmans (1951, 1952). Koopmans (1951) found that her results fitted a 3 : 1 or 13 : 1 ratio of normal to abnormal flowers in the F_2 of a cross between *S. phureja* and *S. chacoense*, and a 5 : 3 ratio in the F_2 derived from a cross between *S. phureja* and *S. commersonii*. She also found cytoplasmic effects. The interaction between a recessive gene, *df*, and a sensitive cytoplasm has been reported to result in short anthers (Grun, 1970; Lee and Rowe, 1976), and an interaction between a dominant gene, *In*, and a sensitive cytoplasm to result in indehiscence of anthers (Grun, 1970).

The anther–style fusion abnormality was found to be controlled by a dominant gene *AF* in tetraploid Tuberosum (Grun, 1970).

Seed embryo spot

Embryo spot is observed as a deep purple or reddish pigmentation at the base of the embryo's cotyledon on both sides of the flat seeds (Hermsen and Verdenius, 1973). Dominant genes with pleiotropic effects on plant organs

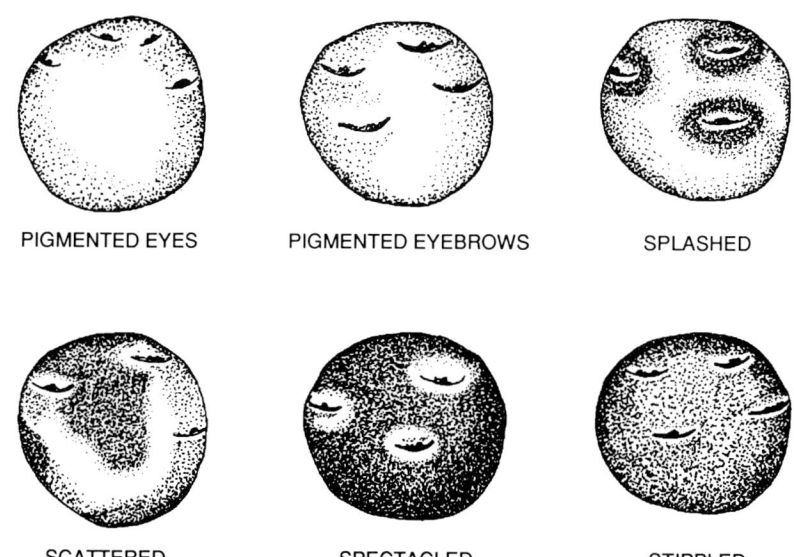

Fig. 12.4. Distribution of tuber skin colour.

which are homologous to cotyledons (leaves and leaflets, scale leaves of stolons, tuber eyebrows, floral abscission layers) determine the expression of the character (Dodds and Long, 1955, 1956). The action of the genes for anthocyanin formation, *P* (purple) and *R* (red), in combination with the allele *Bd* cause the seed embryo spot. The desired genotype for embryo spot is *BdBdPP--*, because *P* is epistatic to *R*; i.e. the genotype *BdBdppRR* has an uncoloured hypocotyl at the young seedling stage.

Tuber Characteristics

Tuber periderm pigmentation

Burton (1966) indicated that the presence of anthocyanin dissolved in the cell sap of the periderm (or peripheral cortical cells) is the basis of skin pigmentation in potatoes, with brown tuber skin dominant to colourless (Kumikura, 1972; Verma, 1972). A one-gene system with chromatid segregation was postulated to explain the inheritance of the trait (Taylor, 1978).

Splashed, hidden and spectacled tubers (Figure 12.4)

Splashed tubers have pigmentation confined to areas around the eyes and are due to the action of the gene *M*, which restricts tuber periderm pigmentation to areas around the eyes (Howard, 1967a).

Hidden pigmented tubers are characterized by uniform pigmentation except around the eyes, where deeper pigmentation is observed. Spotted pink tubers are produced by three combinations of genes, *ED*, *REMD* and *EEMD* (Howard, 1964, 1967a). The gene *P* acts epistatically to these genotypes, converting them into hidden spotted purple tubers. Carson and Howard (1945) and Howard (1964) indicated that not all the *ED* genotypes are hidden spotted. *REMD* genotypes have splashed pink periderm and full pink phelloderm. Furthermore, the locus *R* (red pigmentation of tuber phelloderm) has been reported to be possibly linked to locus *F* (Lunden, 1960).

Spectacled tubers have areas of periderm and phelloderm around the eyes without pigment, but the remaining area of the tuber is intensively pigmented. Different genetic systems between diploid and tetraploid cultivated species have been reported for this trait. The interaction of *Iisp* with either *P* or *R*, or both, produces spectacled tubers in diploid cultivated potatoes (Dodds and Paxman, 1962; Simmonds, 1969). However, it is more complicated at the tetraploid level, e.g. individuals with hidden spotted or spectacled tubers are either *REMD* or *EEMD* (Asseyeva, 1931; Howard, 1964, 1967a). The locus *M* has the alleles responsible for the pigment restriction and is linked to the *R* locus in Tuberosum. However, the allele *isp*, responsible for the restriction of pigmentation at the diploid level and located in the *I* locus, is not linked to *R*.

Flesh colour

Tuber flesh colour is generally white or yellow. Dihaploid × species hybrids between these flesh colours were studied by Smiley (1963). He found that yellow-fleshed F_1 individuals segregated 3 yellow : 1 white upon selfing and that the F_1s with white flesh bred true. Based on these results, yellow flesh was postulated to be due to a monogenic allele which was dominant to that for white flesh (Fruwirth, 1912; Simmonds, 1964). Moreover, the intermediate intensity of different yellow colours has been reported to be due to minor modifier genes controlling the degree of yellowing (Black, 1930; Schick, 1956; Engel, 1957; Howard, 1978).

Some potato cultivars also produce tubers with anthocyanin pigmentation in the flesh. Thus, the purple flesh of Congo black was found by Krantz (1922, 1926) to be dominant over white flesh and due to a single gene. The *Pf* (pigmented flesh) locus has been proposed to explain tuber flesh pigmentation in cultivated diploids (De Jong, 1987) and tetraploids (Krantz, 1922). The gene *I*, which controls the distribution of pigment in the tuber skin, is necessary for the expression of the *Pf* allele and is closely linked to the *Pf* locus. Furthermore, a two-gene system, with both loci having chromatid segregation, has been postulated for the genetic control of the trait in Andigena (Taylor, 1978).

Eye colour

The pigmentation of the eye periderm area was found by Masson (1985) to be under the genetic control of two complementary loci. The locus D controlling eye colour, was considered to be distal to the centromere. The locus M, proximal to the centromere, restricts the pigmentation to the eye area. It was reported to be tightly linked to the E locus (Howard, 1966) and was treated as the EM locus for mapping purposes.

Sprout colour

Sprout colour results from an accumulation of anthocyanins in the hypodermal layer of the skin, which traces back to the L2 histogenic layer at the growing point. Four dominant genes (D, R, E and P) acting independently lead to pigmented sprouts (Matsubayashi, 1979).

Russet tuber skin

The inheritance of russet skin in diploid potatoes was studied by Pavek and Corsini (1981). They indicated that three independent complementary genes controlled russeting in russet × non-russet crosses and from intermating russet parents. The extreme russeting appeared to be due to homozygosity at one or more of the three loci.

Tuber shape (Figures 12.5 and 12.6)

Although tuber shape is mainly a varietal characteristic, it can be influenced by environmental conditions and cultural practices. This complicates the evaluation of this trait.

The shape of the potato tuber is easy to score by determining the length/breadth (I) ratio (Winiger and Ludwig, 1974). Salaman (1910, 1911, 1912, 1926) indicated that tuber shape depended essentially on the presence or absence of a single gene for tuber length. Tuber shape ranged from compressed (longitudinal axis shorter than the diameter axis) to very long (tubers are often curved or hooked). The most common types were round ($I < 1.4$), oval ($1.5 < I < 1.9$) and long ($I > 2.0$).

Contradictory reports about the inheritance of the trait have been published. Some authors (Salaman, 1926; Rudorf and Baerecke, 1974) considered long tubers as being dominant to round tubers, but others (Schick, 1956; De Jong and Rowe, 1972) suggested that round shape was dominant to either oval or long oval. However, in a recent review, De Jong and Burns (1993) indicate that, although there has been disagreement over which tuber

Inheritance of Morphological and Tuber Characteristics

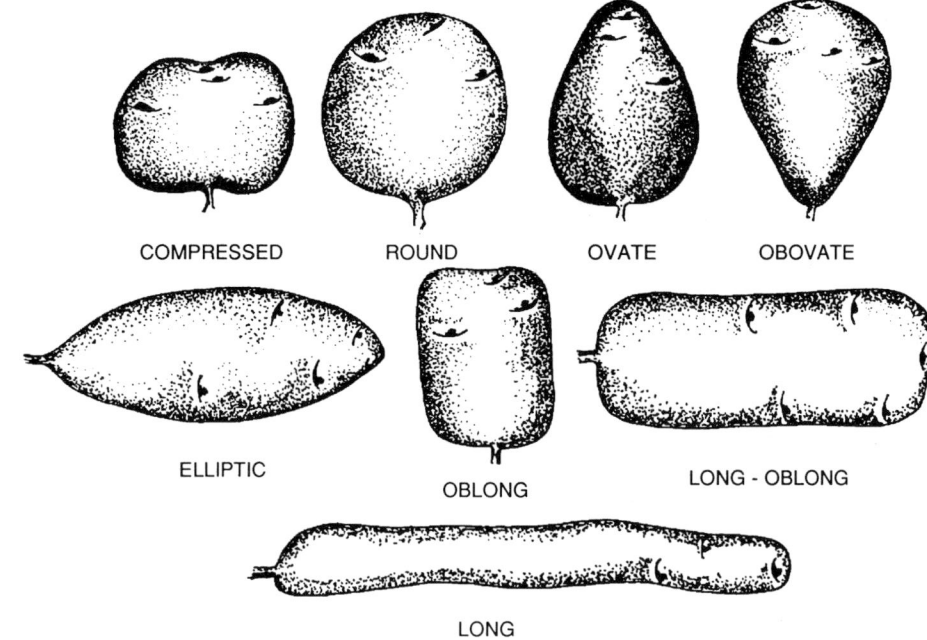

Fig. 12.5. Tuber shapes.

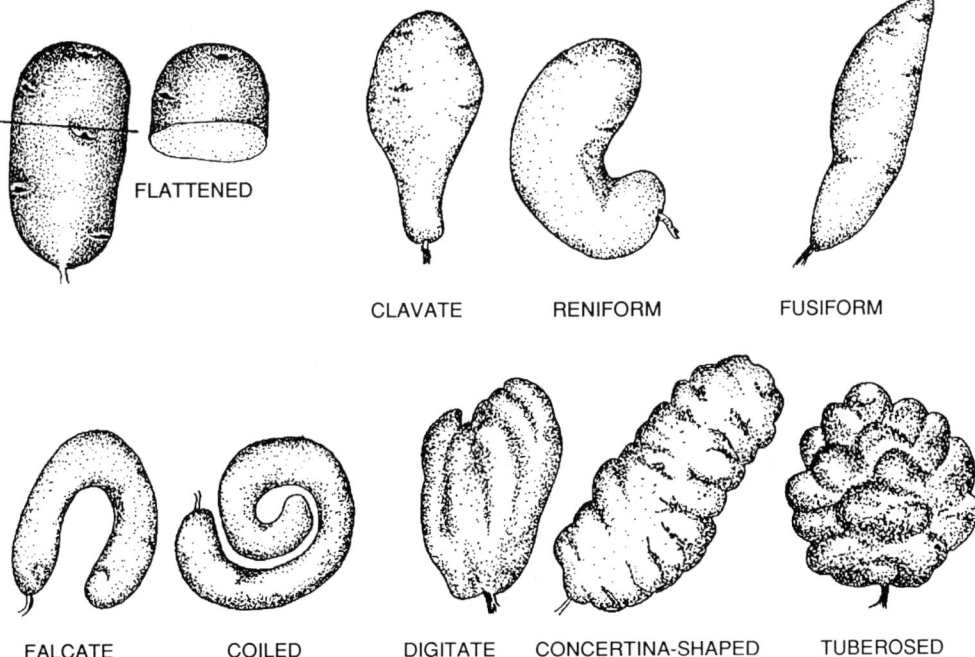

Fig. 12.6. Unusual tuber shapes.

shape is dominant, most authors agree that this trait is relatively simply inherited.

Matsubayashi (1979) classified the tuber shapes of a dihaploid population derived from the cultivar Chippewa as long-ovate, oval and round, using a transverse/long axis ratio. The cultivar Chippewa had tubers ranging from oval to round and the observed segregation in the dihaploid population was 10 long-ovate, 42 oval and 13 round, which fits a 1:4:1 ratio and could indicate that one locus, with alleles showing partial dominance and chromosomal segregation, was responsible for the genetic control of the trait.

In diploid cultivated potatoes, Emme (1937) found evidence that tuber shape is controlled by one major gene and that round is dominant over long. Later on, De Jong and Rowe (1972) observed that long shapes bred true upon selfing, but segregation occurred in selfed progeny of round tubers, which indicated that the long shape was recessive. They also found that this major gene for tuber shape was associated with the *B–I–F* linkage group. Masson (1985) proposed the symbol *Ro* for this gene. More recently, De Jong and Burns (1993) studied a population derived from hybrids between *S. phureja* or *S. stenotomun* adapted to long days and dihaploid Tuberosum. Their results indicated that, in addition to the *Ro* gene, other genes, some with modifying effects, were also operating. They also confirmed that *Ro* is linked with *I*. However, Masson (1985) placed the *Ro* (tuber shape) locus with the linked loci controlling tuber eye pigmentation.

Okwuagwu (1981) analysed a large population of a round × oval cross between *S. phureja* – Tuberosum dihaploid hybrids. She classified the resulting tubers of the progeny as round (more than 50%), oval, long, pointed, compressed and flat. Besides confirming that round is dominant over long, she also postulated that flat and compressed tuber shapes are recessive in their expression. However, the results obtained by De Jong and Burns (1993) did not support this hypothesis.

Some studies using tetraploids that were reported in 1930 by J. Bartosch (in De Jong and Burns, 1993) postulated that tuber shape was controlled by three or four genes. However, B. Maris in 1962 (cited in De Jong and Burns, 1993) considered that this trait was governed by one major gene with a cumulative effect and an unknown number of minor genes. Furthermore, Maris also postulated that the major gene controlling eye depth was linked to the gene controlling tuber shape. Taylor (1978) also analysed a population derived from an Andigena parent and concluded that round is dominant to oval and possibly under the control of one gene.

Information is scarce on the unusual tuber shapes found in Andean potato cultivars and described by Huaman et al. (1977). All of these 'odd' tuber shapes are genetically inherited since they are observed in sexually derived progenies. Huber (1930) studied progenies obtained by selfing cultivars with 'sausage' (concertina) and 'pinecone' (tuberosed) tuber shapes. He developed a hypothesis of three genes, at least two of which are dominant for round. Plaisted and Peterson (1972) found that duplicate recessive epistasis produced the 'knobby tubers' derived from Katahdin and that this pattern of inheritance was unaffected by Potato Spindle Tuber Viroid (*PSTV*) in the parents and offspring. Some of the tuber shapes produced in these progenies resembled some tuberosed tuber shapes that are common in Peruvian cultivars.

Depth of eyes

East (1910) and Rudorf and Baerecke (1974) reported shallow eyes as being dominant to deep eyes. Black (1930) also suggested that deep eyes should be recessive and controlled by more than one factor. Huber (1930) also postulated that at least two complementary genes were necessary for shallow eyes. However, other authors found deep eyes dominant to shallow (Heribert-Nilsson, 1913; Howard, 1978). Salaman (1910, 1911, 1912) observed that varieties with deeply incised eyes bred true for this character but plants with both shallow and deep eyes were found in their hybrid progeny, which indicates complete dominance for shallow eyes (Black, 1933). Ross (1986) mentioned that a parent with deep eyes always shifts the progeny towards deeper eyes.

Stolon length

Matsubayashi (1979) studied this character by classifying the progenies into long stolons (more than 15 cm) or short ones (less than 15 cm). He observed that 46 Chippewa dihaploids were of short type and the remaining 19 were of long type. This result fits an 11 short : 3 long segregation ratio and indicates that Chippewa could be duplex for the locus controlling short stolons (dominant to long). The locus is very distal to the centromere and has chromatid segregation.

References

Asseyeva, T. (1931) Bud mutations in the potato. *Trudy po Prikladnoi Botanike, Genetike i Selektsii* 27, 135-217. (In Russian with long summary in English.)

Bamberg, J.B. and Hanneman, R.E., Jr (1987) Characterization of a gibberellic acid related dwarfing locus in *Solanum tuberosum. American Potato Journal* 64, 431.

Black, W. (1930) Notes on the progenies of various potato hybrids. *Journal of Genetics* 22, 27-43.

Black, W. (1933) Studies of the inheritance of tuber colour in potatoes. *Journal of Genetics* 27, 319-339.

Burton, W.G. (1966) *The Potato: a Survey of Its History and of the Factors Influencing Its Yield, Nutritive Value, Quality and Storage*. European Association of Potato Research, Longman, Wageningen.

Cadman, C.H. (1942) Autotetraploid inheritance in the potato: some new evidence. *Journal of Genetics* 44, 33-52.

Carson, G.P. and Howard, H.W. (1945) Notes on the inheritance of the King Edward type of colour in potatoes. *Journal of Genetics* 46, 358-360.

Choudhuri, H.C. (1943) Cytological studies in the genus *Solanum*. 1. Wild and native cultivated diploid potatoes. *Transactions of the Royal Society of Edinburgh* 61, 113–135.

Choudhuri, H.C. (1944) Cytological studies in the genus *Solanum*. 2. Wild and native cultivated diploid potatoes. *Transactions of the Royal Society of Edinburgh* 61, 199–219.

Choudhuri, H.C. (1948a) Genetic studies in wild and cultivated potatoes. *Proceedings of the Indian Science Congress* 3(10), 226.

Choudhuri, H.C. (1948b) III. Genetic studies in wild and cultivated potatoes. *Bulletin of the Botanical Society of Bengal* 2, 186–209.

De Jong, H. (1987) Inheritance of pigmented tuber flesh in cultivated diploid potatoes. *American Potato Journal* 64, 337–343.

De Jong, H. (1991) Inheritance of anthocyanin pigmentation in the cultivated potato: a critical review. *American Potato Journal* 68, 585–593.

De Jong, H. and Burns, V.J. (1993) Inheritance of tuber shape in cultivated diploid potatoes. *American Potato Journal* 70, 267–283.

De Jong, H. and Rowe, P.R. (1972) Genetic markers in inbred clones of cultivated diploid potatoes. *Potato Research* 15, 200–208.

den Nijs, T.P.M., Leue, E.F. and Peloquin, S.J. (1980) Topiary, a mutant character in *Solanum infundibuliforme*. *Journal of Heredity* 71, 57–60.

Dodds, K.S. and Long, D.H. (1955) The inheritance of color in diploid potatoes. I. Types of anthocyanidins and their genetic loci. *Journal of Genetics* 53, 136–149.

Dodds, K.S. and Long, D.H. (1956) The inheritance of color in diploid potatoes. II. A three factor linkage group. *Journal of Genetics* 54, 27–41.

Dodds,, K.S. and Paxman, G.J. (1962) The genetic system of cultivated potato. *Evolution* 16, 154–167.

East, E.M. (1910) Inheritance in potatoes. *American Naturalist* 44, 424–430.

Emme, H. (1937) Genetik der kartofell. III. *Solanum rybinii* Juz. et Buk., *S. stenotomum* Juz. et Buk. and *S. goniocalyx* – inte Bedeuntung fur die Zuchtung and phylogenotiche Beziehungen. *Biologicheskii Zhurnal* 6, 787–796.

Emme, H. (1938) Studies on interspecific hybridization of tuber bearing potatoes section Tuberarium Bitt. Genus *Solanum* L. *Biologicheskii Zhurnal* 7, 1093–1104.

Engel, K.H. (1957) Grunlegende Fragen zu einem schema fur arbeiten mit inzuchten bei kartoffeln. *Zuchter* 27, 98–124.

Estrada, N. (1960) Herencia del albinismo en especies de papas diploides y tetraploides. *Agricultura Tropical* 16, 348–352.

Fruwirth, C. (1912) Zur Zuchtung der Kartoffel. *Deutsche Landwirtschaftliche Presse* 39, 551–552, 565–567.

Garg, K.C., Tiwari, S.P. and Sharma, K.P. (1981) Inheritance of leaf pigmentation in dihaploid-Phureja hybrids of potatoes. *Journal of Indian Potato Association* 8, 31–34.

Grun, P. (1970) Cytoplasmic sterilities that separate the cultivated potato from its putative ancestors. *Evolution* 24, 750–758.

Harborne, J.B. (1960) Plant polyphenols. I. Anthocyanin production in cultivated potato. *Biochemical Journal* 74, 262–269.

Hawkes, J.G. (1956) Taxonomic studies on the tuber-bearing Solanums. I. *Solanum tubersum* and the tetraploid species complex. *Proceedings of the Linnean Society, London* 166, 97–144.

Heiken, A. (1958) Aberrant types in the potato. *Acta Agriculturae Scandinavica* 8, 319–358.

Heiken, A. and Ewertson, G. (1962) The chimerical structure of a somatic *Solanum* mutant revealed by ionizing irradiation. *Genetica* 33, 88-94.

Heribert-Nilsson, N. (1913) Potatisforadling och Potatisbedoming. *Weibulls Aarsbok* 8, 4-31.

Hermsen, J.G.Th. (1978) Genetics of self-incompatibility in di-haploids of *Solanum tuberosum* L. 4. Linkage between an S-bearing translocation and a locus for virescens. *Euphytica* 27, 381-384.

Hermsen, J.G.Th. and Verdenius, J. (1973) Selection from *Solanum tuberosum* group Phureja of genotypes combining high-frequency haploid induction with homozygosity for embryo-spot. *Euphytica* 22, 244-259.

Hermsen, J.G.Th., Taylor, L.M., van Breukelen, E.W.M. and Lipski, A. (1978) Inheritance of genetic markers from two potato dihaploids and their respective parent cultivar. *Euphytica* 27, 681-688.

Howard, H.W. (1960) Potato cytology and genetics, 1952-1959. *Bibliographia Genetica* 19, 87-216.

Howard, H.W. (1962) Experiments with potatoes on the effect of the pigment-restricting gene *M*. *Heredity* 17, 145-156.

Howard, H.W. (1964) Further effects on the pigment-restricting gene, *M*, in potatoes: hidden spotted and spectacled. *Heredity* 19, 349-356.

Howard, H.W. (1966) Recombination values for genes *E* and *M* in potatoes. *Heredity* 21, 313-315.

Howard, H.W. (1967a) Differentiation in potatoes: hidden-spotted and spectacled. *Heredity* 22, 57-64.

Howard, H.W. (1967b) Further experiments on the use of X-rays and other methods in investigating potato chimeras. *Radiation Botany* 7, 389-399.

Howard, H.W. (1970) *Genetics of Potato*. Springer Verlag, New York.

Howard, H.W. (1973) Calyx form in dihaploids in relation to the origin of *Solanum tuberosum*. *Potato Research* 16, 43-46.

Howard, H.W. (1978) The production of new varieties. In: Harris, P.M. (ed.), *The Potato Crop*. Chapman and Hall, London, pp. 607-646.

Huaman, Z., Williams, J.T., Salhuana, W. and Vicent, L. (1977) *Descriptors for the Cultivated Potato and for the Maintenance and Distribution of Germplasm Collections*. International Board for Plant Genetic Resources, Rome, Italy.

Huber, J.A. (1930) Genetische Versuche mit Salatkartoffeln. *Pflanzenbau, Pflanzenschutz, Pflanzenzucht* 15, 75-85.

Kelly, J.P. (1924) Seed progeny of a potato with faintly colored tubers. *Journal of Genetics* 14, 197-199.

Kessel, R. and Rowe, P.R. (1974a) The inheritance of the split stigma character in diploid *Solanum* species. *Potato Research* 17, 227-233.

Kessel, R. and Rowe, P.R. (1974b) Inheritance of two qualitative traits and a proposed map for their linkage group in diploid potatoes. *Potato Research* 17, 283-295.

Koopmans, A. (1951) Cytogenetics studies of *Solanum tuberosum* L. and some of its relatives. *Genetica* 25, 193-337.

Koopmans, A. (1952) Changes in sex in the flowers of the hybrid *Solanum rybinii* × *S. chacoense*. *Genetica* 26, 359-380.

Krantz, F.A. (1922) The application of genetic principles to potato breeding. *Proceedings American Society for Horticultural Science* 19, 124-129.

Krantz, F.A. (1926) Genetic studies in potato. II. The inheritance of red cortical colour in tubers. *Proceedings Potato Association of America* 13, 52-55.

Kukimura, H. (1967) Intervarietal differences of radiosensitivity of potato plants. *Annual Report Institute of Radiation Breeding* (Japanese).

Kukimura, H. (1972) Effects of gamma-rays on segregation ratios in potato families. *Potato Research* 15, 106–116.

Lam, S.L. and Erickson, H.T. (1971) Location of a mutant gene causing albinism in a diploid potato. *Journal of Heredity* 62, 207–208.

Lee, H.K. and Rowe, P.R. (1976) Genetic segregation of the deformed flower in trisomics of *Solanum chacoense*. *Euphytica* 25, 313–320.

Lunden, A.P. (1937) Arvelighetsundersokelser i potet (Inheritance studies in the potato). *Meldinger fra Norges Landbrukshøishkole* 17, 1–156.

Lunden, A.P. (1960) Some more evidence of autotetraploid inheritance in the potato (*Solanum tuberosum*). *Euphytica* 9, 225–234.

Masson, M.F. (1985) Mapping, combining abilities, heritabilities and heterosis with $4x \times 2x$ crosses in potato. Unpublished PhD thesis, University of Wisconsin-Madison.

Matsubayashi, M. (1979) Genetic variation in dihaploid potato clones, with special reference to phenotypic segregations in some characters. *Science Report of Agriculture Kobe University* 13, 185–192.

Okwuagwu, C.O. (1981) Phenotypic evaluation and cytological analysis of 24-chromosome hybrids for analytical breeding in potatoes. Unpublished PhD thesis, University of Wisconsin-Madison.

Pavek, J.J. and Corsini, D.L. (1981) Inheritance of russet skin in diploid potatoes. *American Potato Journal* 58, 515–516.

Pineda-Colorado, R. (1990) Quantitative and genetic analysis in *Solanum chacoense* Bitt. using inbred lines. Unpublished PhD thesis, University of Wisconsin-Madison.

Plaisted, R.L. and Peterson, L.C. (1967) Linkage of genes affecting tuber color and chlorotic foliage. *American Potato Journal* 44, 411–414.

Plaisted, R.L. and Peterson, L.C. (1972) Inheritance of the Katahdin 'Knobby Tuber'. *American Potato Journal* 49, 285–290.

Pongsupasamit, S. (1990) Comparative study of mutation methods in potatoes, *Solanum tuberosum* L. Unpublished PhD thesis, School of Agriculture and Forestry, University of Melbourne, Australia.

Propach, H. (1940) Cytogenestiche untersuchungen in der gattung *Solanum* sect. Tuberarium. V. Diploide artbastarde. *Zeitschrift für Induktiun Abstammungs und Vererbungslehre* 78, 115–128.

Pushkarnath (1960) Potato flower biology series. I. Flower and flowering. *Indian Potato Journal* 2, 12–29.

Ross, H. (1986) *Potato Breeding – Problems and Perspectives*. Advances in Plant Breeding 13, Verlag Paul Parey, Berlin and Hamburg.

Rudorf, W. and Baerecke, M.L. (1974) Methoden der qualitatsveurteilungen bei kartoffeln fur den menschlichen konsum. *Potato Research* 17, 434–465.

Salaman, R.N. (1910) The inheritance of colour and other characters in the potato. *Journal of Genetics* 1, 7–46.

Salaman, R.N. (1911) Studies in potato breeding. *Proceedings of the IVth International Conference on Genetics*, Paris 1911, pp. 397–405.

Salaman, R.N. (1912) Hereditary changes in the potato. *Journal of the Royal Horticultural Society* 38, 34–39.

Salaman, R.N. (1926) *Potato Varieties*. Cambridge University Press, Cambridge.

Salaman, R.N. and Lesley, J.W. (1920) Genetic studies in potatoes for the inheritance of abnormal haulm type. *Journal of Genetics* 10, 21–37.

Schick, R. (1956) Methoden und problems der kartoffelzuchtung. *Sitzungsberichte der Deutschen Akademie der Landwirtschaftswissenschaften, Berlin* 5(29), 1–40.

Simmonds, N.W. (1964) The genetics of seed tuber dormancy in the cultivated potatoes. *Heredity* 19, 489-504.

Simmonds, N.W. (1965) Mutant expression in diploid potatoes. *Heredity* 20, 65-72.

Simmonds, N.W. (1969) Genetics of spectacle in diploid potatoes. *Heredity* 24, 487-490.

Simmonds, N.W. and Harborne, J.B. (1965) Control of malvidin synthesis in the cultivated potatoes. *Heredity* 20, 315-318.

Smiley, J.H. (1963) Behaviour of self-fertility and its utilization in *Solanum* diploid species – haploid *Solanum tuberosum* hybrids. Unpublished PhD thesis, University of Wisconsin-Madison.

Swaminathan, M.S. and Howard, H.W. (1953) The cytology and genetics of the potato (*Solanum tuberosum* L.) and related species. *Bibliographia Genetica* 16, 1-192.

Taylor, L.M. (1978) Variation patterns of parthenogenetic plants derived from 'unreduced' embryo-sacs of *Solanum tuberosum* spp. *andigena* (Juz. et Buk.) Hawkes. *Theoretical and Applied Genetics* 52, 241-249.

Van Harten, A.M., Bouter, H. and Schut, B. (1973) Ivy leaf of potato (*Solanum tuberosum*), a radiation-induced dominant mutation for leaf shape. *Radiation Botany* 13, 287-292.

Verma, S.M. (1972) Note on the genetics of skin color of tuber in the first inbred population of 'Plulwa' potato. *Indian Journal of Agricultural Science* 42, 114.

Winiger, F.A. and Ludwig, J.W. (1974) Methoden der qualitätsbeurteilung bei kartoffeln für den menschlinchen konsum. *Potato Research* 17, 434-465.

Woodcock, K.M. and Howard, H.W. (1975) Calyx types in *Solanum tuberosum* dihaploids, *S. stenotomum*, *S. sparsipilum* and their hybrids. *Potato Research* 18, 460-465.

13 Inheritance of Table and Processing Quality

M.F.B. Dale and G.R. Mackay

Crop Genetics Department, Scottish Crop Research Institute, Invergowrie, Dundee DD2 5DA, UK.

Introduction

Quality is perhaps one of the most important characteristics of such an important human food source as the potato, and yet it is probably also the most poorly defined and genetically least researched. The potato is the fourth most important human food crop in the world after wheat, maize and rice and produces more value per unit area than each of these three cereal crops (Anon., 1985; Horton, 1988). It is perhaps a reflection of the generally high nutritional value of all potato varieties that so little research has been directed towards differences between cultivars, i.e. between genotypes. Many research publications on aspects of nutritional quality also tend to ignore potential genetic variation and, although discussing a feature or features of the quality of 'potatoes', often refer to data gathered from a single cultivar (= genotype). In this chapter, therefore, the authors have assumed evidence for genetically based variation, or for heritable variation, when reviewing research on more than one genotype, which indicates differences between these genotypes.

In contrast to the major cereal crops, the vast bulk of the potato crop is presented to the consumer in its raw state. Consequently characters such as tuber appearance, size, shape, colour and skin finish, etc., which may influence consumer choice, e.g. a preference for yellow flesh versus white, may be deemed 'quality' characters. However, the true quality of a potato is reflected in its culinary or nutritional value after cooking or processing. Indeed, in most developed and developing countries, an increasing proportion of the crop is now processed prior to consumption. In a sophisticated market such as that of the UK, Potato Marketing Board statistics now show that nearly 30% of the crop is processed (Anon., 1992a), while in America it is estimated that more than 50% of the crop is processed. In this case certain very specific quality traits are essential to the manufacture of (primarily) fried products

Table 13.1 Quality characteristics of importance to processor and consumer and relative importance to each: H = high, M = medium, L = low priority.

Character	Processing	Table use
Tuber defects	H	H
Damage - external	M	H
Damage - internal	H	M
Glycoalkaloids	H	H
Greening	H	H
Nutritional value	H	H
After-cooking blackening	M	H
Texture	M	M
Enzymic browning	H	M
Sugar content	H	L
Dry matter	H	L
Flavour	M	H

such as crisps (= chips) and chips (= french fries or *pommes frites*). For these reasons this chapter will concentrate on those quality traits of relevance to usage, i.e. consumption and processing, rather than tuber difference, which is dealt with in Chapter 12. Readers are referred to Horton and Anderson (1992) for fuller detail on world potato production figures and patterns in consumption.

Initially, the genetics of those traits which generally affect quality in its widest sense are addressed, particularly those defects in the tuber which detract from the quality of the raw product. Aspects of quality which are common for both home consumption (table use) and manufacture (processing) are then examined, and then traits which are more important specifically to either one or the other. Table 13.1 lists the quality traits to be reviewed and their importance to the user. First we consider tuber defects which are of interest to all users, since each of these detract from the saleable or usable yield for whatever use to which the crop is to be put. An observation common to practically all characters affecting quality, noted in early reviews more than 80 years ago, is that, whereas there are clear varietal differences indicating genetic variation, environmental conditions can give rise to differences within varieties as great as those between them (East, 1908), perhaps itself a reason why the genetics is so little understood.

Tuber Defects

Growth cracking

Tuber growth cracking is a serious defect, sometimes leading to substantial marketable yield losses. The causes of growth cracking are poorly understood, being attributed in part to fluctuating water stress (Iritani, 1981; van Loon,

1981) or to changes in tuber growth rates (Gray and Hughes, 1978). However, Jefferies and MacKerron (1987) reported that conditions which produce very rapid tuber growth, which may or may not accompany the relief of water stress, can produce large, often deep, cleavage cracks, as observed in some varieties such as cv. Guardian, but relief of such water stress can also result in shallower, more frequent, 'jagged tearing' cracks in other cultivars, such as cv. Record. The two quite distinct types of growth cracking are not mutually exclusive. At present, breeders know little of the factors influencing growth cracking. From the reports on the influence of variation in tuber growth patterns and the different responses to water stress, there is undoubtedly a significant environmental component. From observations on material in the early stages of a breeding programme, there is also a heritable component which does not result in clearly delineated groups, suggesting that a number of minor genes may be implicated. Until breeders have access to reliable testing procedures for this, character selection will have to rely on observations in various environments over a number of years, as is presently practised.

Hollow heart

Hollow heart in the potato tuber is a physiological defect resulting in an internal cavity of varying dimensions. It may be preceded by the appearance of brown centre, a necrosis of the pith cells. Generally, hollow heart is found more frequently in larger tubers, though this is not always the case (Nelson and Thoreson, 1986; Rex and Mazza, 1989). Development of hollow heart is often associated with periods of rapid tuber growth, which may have been preceded by a period of moisture or nutritional stress. Further details on the physiological and nutritional factors which influence the degree of hollow heart are given elsewhere (Rex and Mazza, 1989; Storey and Davies, 1992).

The expression of hollow heart is affected by various environmental and genetic factors. Cultivars demonstrate variation for the degree of observed hollow heart and their susceptibility varies between environments.

Veilleux and Lauer (1981) in studies of *S. tuberosum* ($4x$) × *S. phureja* Juz et Buk ($2x$) and *S. tuberosum* × *S. tuberosum* crosses found that 10% of all genotypes expressed severe hollow heart symptoms but that 46% of these were derived from one parent, 'Norgold Russet', which was in the parentage of only 16% of the seedlings. The cultivar 'Shoshomi' and genotype 148-17 were also responsible for transmitting a high degree of susceptibility to hollow heart to their offspring when used as pollen parents. The authors found high GCA estimates for hollow heart and less, but still significant, SCA effects, though only in the $4x$–$4x$ cross. Jansky and Thompson (1990) studied hollow heart in segregating tetraploid potato populations in a breeding programme. They found positive correlations between hollow heart and both mean yield and tuber size and, as might be expected, a negative correlation between hollow heart and tuber number. The authors found the highest levels of hollow heart in families with susceptible parents, although susceptible genotypes were observed in the progeny of resistant parents.

Given the nature of this defect, the environmental component and the lack of a rapid and reliable test, it is advisable to avoid the use of susceptible parents. Breeders can routinely assess the larger tubers of genotypes throughout their programmes over a number of environments and years, selecting against those exhibiting symptoms.

Brown centre

Brown centre, as mentioned previously, is often, though not always, a precursor to hollow heart and is sometimes referred to as incipient hollow heart. Brown centre is characterized by browning due to cell death in the pith area of the tuber. Cool temperatures during tuber initiation are reported to induce brown centre (Hiller *et al.*, 1979). Tuber growth rate is an important factor affecting symptoms, which can be found in both large and small tubers. Little is known of the causes of the disorder and, as such, control is difficult. As with hollow heart, breeders need to assess material over a number of years and locations, selecting against genotypes whenever they express symptoms.

Internal rust spot

Internal rust spot (IRS) is a physiological disorder of potatoes characterized by rust-coloured spots or blotches distributed irregularly throughout the tubers. Its development is reported to be induced by a number of environmental factors, including restricted water supply and high temperatures during the growing season and by uneven growth rates (O'Brien and Rich, 1976; Hooker, 1981). Its development is associated with a low concentration of calcium in the affected tubers (Collier *et al.*, 1980). Several environmental factors can increase the incidence of IRS (Davies and Ross, 1986), including soil moisture and temperature fluctuations (i.e. factors which lead to variable tuber growth rates) and soil type, with sandy soils tending to exacerbate symptoms, although these should not be confused with tobacco rattle virus symptoms. Planting date, with later planting reducing IRS symptoms, and nutrient treatment, with high N causing excessive foliage growth and increased IRS symptoms, represent two environmental factors operating in opposite directions. Davies and Ross (1986) suggest that such factors can adversely affect the calcium content of the tubers, leading to increased IRS symptoms.

There is variation for the character between cultivars (Collier *et al.*, 1980; Davies and Ross, 1986), and a wider range of variation is frequently observed in breeding material at earlier stages of a programme. Without access to a reliable test that can assess large numbers of genotypes, breeders must rely, as with hollow heart and growth cracking, on assessing material over a number of sites and seasons by cutting a number of large tubers of individual genotypes from each plot and discarding those with symptoms.

Damage Resistance

Mechanical damage and bruising are major causes of yield loss in potatoes. In a survey of 665 farms in the UK, 23% of potatoes had sustained visible flesh damage and a further 13% had internal damage (Anon., 1974). Smittle *et al.* (1974) in the USA estimated that 26% of farmers' potential gross income was lost due to damage. A more recent UK survey (Balls *et al.*, 1982) found 9% of tubers severely damaged and 26% bruised.

Damage to tubers is generally classified into two groups, external and internal. External damage ranges from slight scuffing to splits, cuts or cracks, while internal damage includes shattering and blackspot with localized discoloration due to melanin formation. Tuber condition at the time of impact affects the degree of damage susceptibility; internal blackspot damage susceptibility is lower in turgid tubers and increases with dehydration, while external shatter-type damage decreases with tuber dehydration (Kunkel and Gardner, 1965; Smittle *et al.*, 1974; Hiller *et al.*, 1985). A close relationship between blackspot susceptibility and specific gravity has been observed (Kunkel and Gardner, 1965; Smittle *et al.*, 1974). Storey and Davies (1992) suggest that specific gravity is not important in determining blackspot susceptibility between cultivars, but may be an important factor within a cultivar, with increasing specific gravity generally resulting in increased bruising. Marked increases in both types of damage are observed when tubers are harvested or handled at low temperatures, although response varies between genotypes (Smittle *et al.*, 1974; Gray and Hughes, 1978; McRae, 1979). Larger tubers tend to be damaged to a greater extent than smaller tubers (Carruthers, 1982). Both Munzert (1987) and Cole (1980) describe a positive correlation between late maturity, higher starch content, tuber size and increased damage susceptibility. McRae (1986) and McRae *et al.* (1976, 1986) have done much to assess causes of damage due to mechanized harvesting and handling and to reduce subsequent losses. Further details of how various factors affect the degree of damage in a crop can be found elsewhere (Storey and Davies, 1992).

Field evaluation of material is difficult due to variation between locations, e.g. soil types, temperature, harvest techniques and seasons. To facilitate testing and breeding of new cultivars with resistance to damage, it is important to have improved and reliable methods of assessment. Different tests have therefore been designed, ranging from shaker tests which simulate riddling conditions to various types of drop tests, in which tubers are impacted under a range of conditions, often involving using a bolt or pendulum type of weight. The latter system, developed at the Food Research Institute, Norwich, UK, inflicts damage over a range of forces and measures the potato's dynamic properties, such as tuber firmness or turgidity and the energy absorbed during impact. The various tests utilized are assessed and reviewed by Umaerus (1978), Skrobacki *et al.* (1989) and Storey and Davies (1992).

Due to the complex nature of damage resistance, there is limited information on the inheritance of resistance to damage. Hunnius and Munzert (1976) found variation due to heritable factors for damage resistance between 52 and 56% and to environmental factors between 20 and 22%. Based on studies of 2x-4x crosses, De Maine (1988) identified significant variation between dihaploid genotypes derived from two tetraploids for tuber deformation resistance. The results indicated that dihaploids with good levels of damage resistance could be used to produce tetraploids with high levels of resistance. De Maine et al. (1992), in a detailed study involving progeny testing of material derived from eight cultivars, intercrossed in a half-diallel design, including selfs, found that general combining ability accounted for all of the progeny variation and that there was no significant specific combining ability. These results indicate that there is a significant heritable component and that the phenotype of the parents gives a good indication of their breeding value with respect to this character. It is evident that a number of genetic factors are involved in damage resistance; these are probably inherited in a polygenic manner. Breeders have reviewed strategies (Cole, 1980; Munzert, 1987; Scholtz, 1987) and generally proceed by selecting against susceptible genotypes over a number of years trialling.

Glycoalkaloids

Potato tubers contain small quantities of naturally occurring chemicals called steroidal glycoalkaloids, a class of potentially toxic compounds which are found throughout the family Solanaceae. Approximately 95% of the total glycoalkaloids present in potatoes are accounted for by α-solanine and α-chaconine, both of which are structurally similar, being differently glycosylated forms of the aglycone, solanidine (Maga, 1980). While glycoalkaloids are found throughout the potato plant, the highest levels are reported in those parts with high metabolic rates, such as flowers, unripe berries, young leaves and sprouts (Van Gelder, 1990). Tubers generally have a much lower glycoalkaloid content, although their distribution is not uniform, with higher concentrations found in the periderm and cortex, decreasing markedly towards the pith. However, in tubers with high contents, the glycoalkaloids diffuse towards the centre (Van Gelder, 1990). Cultivars vary with regard to their inherent glycoalkaloid content; at low levels it is suggested that they may enhance potato flavour, but at higher concentrations (above $15 \text{ mg } 100 \text{ g}^{-1}$ fresh weight) they impart bitterness (Maga, 1980). Levels above $20 \text{ mg } 100 \text{ g}^{-1}$ fresh weight are considered unsuitable for human consumption (Jadhav and Salunkhe, 1975), resulting in various symptoms typically associated with food poisoning (for a fuller description, see Parnell et al., 1984; Van Gelder, 1990).

It is generally thought that certain glycoalkaloids may confer a degree of protection to the plant against various pathogens. Glycoalkaloids have been implicated in resistance to early blight, *Alternaria solani* (Sinden et al., 1973), to the Colorado beetle (Shreiber, 1957; Sturckow and Lou, 1961) and to the spores of *Fusarium coerulum* (McKee, 1959). However, Grassert and Lellbach (1987) found no relationship between resistance to *Globodera rostochiensis* or *G. pallida* and glycoalkaloid content, and Deahl et al. (1973) found no correlation between blight resistance and glycoalkaloid content.

A number of factors can influence the level of glycoalkaloids in potato tubers, including cultivar, climate, storage environment, maturity, damage, temperature and exposure to light (Maga, 1980; Jadhav et al., 1981). High glycoalkaloid levels occur following damage and severe bruising (Fitzpatrick et al., 1978; Olsson, 1986). Generally, high correlations are observed between the initial glycoalkaloid contents and the increases following damage. Significantly increased glycoalkaloid levels are also observed following exposure to light, where the duration, intensity and wavelength are all important (Gull and Isenberg, 1960; Liljcmark and Widoff, 1960). However, the question of whether increases in glycoalkaloid synthesis always parallels light-induced greening and chlorophyll synthesis remains unresolved, and, indeed, this may itself be genotype-dependent (De Maine et al., 1988; Kozukue and Mizuno, 1990; Dale et al., 1992). Ramaswamy et al. (1976) and Nair et al. (1981) demonstrated an association between chlorophyll synthesis and the synthesis of glycoalkaloids in that formate, glycine and pyruvate were intermediates to the mevalonic acid pathway, these being common precursors to chlorophyll and glycoalkaloids.

It is desirable that potato breeders select genotypes with low glycoalkaloid contents. The glycoalkaloid contents of potato cultivars vary quite widely. In a large survey of 58 German varieties (Lepper, 1949) from six environments, the glycoalkaloid contents ranged from 2 to 22 mg $100\,g^{-1}$ fresh weight. A survey of glycoalkaloid content of 32 American cultivars ranged from 2 to 13 mg $100\,g^{-1}$ fresh weight (Wolf and Duggar, 1946). A study of 13 UK cultivars (Bintcliffe et al., 1982), grown over three sites and two years, indicated that the glycoalkaloid levels ranged from 3.6 to 14.2 mg $100\,g^{-1}$ fresh weight and were significantly affected by cultivar, site and season and that there were significant cultivar–site and cultivar–season interactions. Although these interaction effects were significant, the ranking order of glycoalkaloid content of genotypes remained comparatively constant.

In a study of 27 genotypes grown over four seasons at two locations, a significant correlation ($r = 0.86$) was found between the genotype averages of the glycoalkaloid contents and the coefficients of variation calculated for each genotype (Van Gelder and Dellaert, 1988). Sinden and Webb (1972) found that cultivars with high mean glycoalkaloid contents were more likely to produce excessive glycoalkaloid content compared with other cultivars with low mean glycoalkaloid contents when subjected to environmental stress or damage. A well-documented example is that of an American cultivar, Lenape, which had a high average TGA content of 29 mg $100\,g^{-1}$ over 39 locations in the US (Sinden and Webb, 1972) and which rose to as high as 65 mg $100\,g^{-1}$ when grown in stressed conditions (Sinden and Webb, 1974). Lenape was removed from commerce in 1970 due to its high glycoalkaloid levels. It is believed that the high glycoalkaloid levels were due to *Solanum chacoense* in its immediate ancestry (Zitnak and Johnston, 1970).

Beyond such surveys examining the range of levels and distribution of glycoalkaloids within cultivars in various countries, there has been limited effort directed at investigating the inheritance of this character. Early work by Ross (1966) suggested that suppression of glycoalkaloid synthesis is inherited in a dominant manner, with the F_1 hybrid between a wild species and *S. tuberosum* generally exhibiting lower solanine content. He further stated that most cultivars and breeding hybrids appear to have two or more dominant alleles for low solanine synthesis. However, other workers, including Georgieva and Ronkov (1954) and Schwarze (1962), have found that the F_1 hybrids between wild species and *S. tuberosum* result in genotypes that are generally high in glycoalkaloid, while Prokoshev *et al.* (1952) reported that interspecific hybrids using various wild species contained glycoalkaloid levels intermediate between the two parents.

A more detailed study by Sandford and Sinden (1972) found significant differences in glycoalkaloid content between parents and between the progeny means, based on a 2-year study of ten tetraploid *S. tuberosum* crosses. The glycoalkaloid contents of the parents ranged from 3.6 to 36 mg $100\,g^{-1}$ fresh weight. The observed variation within the progenies was generally continuous and the authors concluded that glycoalkaloid content was inherited in a polygenic manner, with heritability ranging from 86% to 89% in the broad sense and from 66% to 84% in the narrow sense. Sinden *et al.* (1984) presented a detailed review of research on the inheritance of glycoalkaloid synthesis, not only in *S. tuberosum* subsp. *tuberosum* but also in a large number of wild species. The authors reported studies indicating inheritance in a quantitative manner and also cases of major genes. McCollum and Sinden (1979) reported simple segregation ratios of 3:1 in an F_2 and 1:1 in a backcross generation for the presence : absence of solanine, chaconine and commersonine in a study of glycoalkaloid inheritance in *S. chacoense*.

Given the potentially toxic nature of the various glycoalkaloids, it is therefore important that initial content of parental material be assessed within a breeding programme, particularly when utilizing wild species as a source of a desired character. An increasing body of data on different glycoalkaloids contained in various wild species is becoming available and breeders should therefore assess their options prior to initiating crossing/backcrossing programmes (Van Gelder *et al.*, 1988).

Greening

Potato tubers exposed to periods of light before, during or after harvest develop a green pigmentation, initially at the surface and subsequently throughout the whole tuber. This greening condition is caused by the formation of chlorophylls a and b, which are harmless and tasteless to humans. Greened potatoes are often associated with increased levels of glycoalkaloids, which are poisonous. The relationship between the synthesis of toxic glycoalkaloids and of chlorophyll was discussed in the previous section on glycoalkaloids. The extent to which tubers are exposed to different quantities and qualities of light, including daylight and sunlight during the growing season and at harvest and ultraviolet, fluorescent or incandescent light during subsequent storage and in retail outlets, will vary with a number of environmental and marketing factors. However, light intensity as low as 53.8 lux can produce greening, which increases with increasing light intensity (Sharma and Salunkhe, 1989). Extensive greening can dramatically reduce the marketability of affected tubers, resulting in substantial losses when tubers are prepared for sale. With reported yield losses of up to 10% due to greening in some cultivars, e.g. cv. King Edward (Brown and Riley, 1976), it is evident that there is scope for considerable savings, in terms of both yield and economics. In the field preharvest, the major causes of greening are insufficient soil cover either at planting or from subsequent heavy rainfall. The greening problem is, however, often exacerbated after harvest when tubers are stored and displayed in retail outlets in bright conditions. Liljemark and Widoff (1960) found when studying the effects of light that daylight-type fluorescent tubes were most effective at inducing chlorophyll synthesis. Temperature was also found to influence the rate of greening under light conditions (Larsen, 1949), with little response at 5°C but extensive greening at 2°C. Possible solutions to this problem include improved agronomic practices to protect the tubers from light in field conditions, better control of storage and retail conditions and breeding for resistance to greening.

There are a number of reports on the differences between cultivars with respect to rates of greening. Brown and Riley (1976) found consistent significant differences between those cultivars studied for their rates of greening. Reeves (1988) examined the greening reaction of 144 potato cultivars and found that depth of greening was less in the russeted cultivars and that the red-skinned cultivars generally showed less greening. The study demonstrated differences between the cultivars for greening and depth of greening and suggested that these may be independent of each other.

There are, however, few detailed studies of the inheritance of greening factors. Akeley *et al.* (1962) examined several cultivars and their progeny when exposed to different light regimes. On examining five tuber progenies of four diverse parents, significant differences were evident. The family segregations and comparisons of the means for the greening classes indicated that a number of factors were involved and that dominance was incomplete. It was evident that selection could be effective in identifying genotypes with reduced greening on exposure to light. Parfitt and Peloquin (1981) examined diploid material and found no evidence of any qualitative inheritance for tuber greening, with indications that at least several genes were involved, with most of the variation being additive. The authors surprisingly found no dominance or maternal effects, given that one might expect greening to be chloroplast-dependent and so might expect indications of maternal inheritance (Von Wettstein, 1967). Their studies supported the conclusions of Akeley *et al.* (1962) that tuber greening is inherited in a quantitative manner. The broad-sense heritability identified for greening was sufficiently large to permit effective selection within potato breeding programmes.

Nutritional Value

The potato is a proved source of good-quality protein and a rich source of energy, with a favourable ratio of protein calories to total calories. It is also an important source of vitamins, particularly ascorbic acid (vitamin C), and minerals, such as calcium, potassium and phosphorus. Its value within human diets is often underestimated, if not ignored.

Proteins

On a dry-weight basis, potatoes can have a protein content (total N $\times$ 6.25) in the order of 10%, which is comparable to that of wheat and higher than most rice or maize. Potatoes can produce more protein per unit area than any other crop except soyabeans (Smith, 1984). Potato protein notably contains a higher proportion of the essential amino acid lysine than most cereals and, as a consequence, it is often used to fortify various cereal products, e.g. rice or pasta. However, it contains a lower concentration of the sulphur-containing amino acids (methionine, cysteine) than cereals.

The total tuber nitrogen content consists of soluble, coagulable protein, insoluble protein and soluble non-protein N. The latter is composed of free amino acids, amides (asparagine and glutamine), small amounts of nitrate N, basic nitrogen compounds, nucleic acids and alkaloids. Fuller accounts of the nutritional value of potato proteins are given elsewhere (Markakis, 1975; Horton and Sawyer, 1985; Woolfe, 1987).

There are two principal reasons for seeking to breed potatoes with higher protein contents. While many of the developed countries have ample alternative sources of protein, in many of the developing or Third World countries there is a general lack of protein and in some of these the potato is an important constituent of the diet and consequently breeding for increased protein levels is an important objective.

The second reason is the use of the potato as an industrial starch source, as in The Netherlands, where about 50% of potato production is for the starch industry. During processing, the industry recovers protein as a valuable by-product for use as feed supplements. An increase in the content of coagulable protein would be of value (Miedema *et al.*, 1976).

In order to breed higher protein contents, it is necessary to have available genetic variation, a suitable screening technique which can assess reasonably large numbers of genotypes, preferably some indication of the inheritance of the important genetic factors and also how these relate to other important characteristics.

Varietal differences in the protein content of potatoes have been reported by a number of workers. (Note: all figures are given on a dry-weight basis.) Schwimmer and Burr (1967) found potato protein contents ranging from 3.5% to a remarkably high 23%. Kaldy and Markakis (1972) reported on six cultivars with protein levels ranging from 8.1% to 12.3%. Fitzpatrick *et al.*

(1969) found protein levels in 83 unselected seedlings ranging from 8.75% to 17.75%. When examining 34 cultivars, Miedema et al. (1976) identified a range of protein contents from 4.8% to 10.1%.

To assess protein contents, rapid methods which can screen moderate numbers of genotypes for coagulable protein levels have been developed (Miedema et al., 1976). Other methods have also been described (Stegemann et al., 1973; Hoff, 1975; Snyder and Desborough, 1978).

Various factors influence protein levels, principally genotype, year, location and soil type effects (Desborough and Weiser, 1974; Hunnius et al., 1976; Hoff et al., 1978). As has been reported many times with potatoes and other crops, a positive correlation exists between the degree of nitrogen fertilizer applied and the total nitrogen content, though with diminishing returns as nitrogen levels increase (see Augustin, 1975). Hoff et al. (1971) applied nitrogen at rates up to 336 lb. acre^{-1} (378 kg ha^{-1}) and found that the total amino acid pool almost doubled, indicating that the excess nitrogen was not incorporated into protein. Miedema et al. (1976) found that the coagulable protein content in fresh material correlated significantly with dry matter ($r = 0.756$), with total yield ($r = 0.615$) and with earliness ($r = -0.361$), but not significantly with dry-matter yield ($r = 0.309$). From their analyses the authors concluded that a high coagulable protein content can be combined with early maturity and high dry-matter yield and that coagulable protein levels in the dry matter appeared to be the most suitable selection criteria. Hunnius et al. (1976) found a strong relationship between starch and protein contents.

There have been few attempts to elucidate the inheritance of protein levels. From the continuous range of variation observed in the numerous cultivars assessed for protein content and from detailed studies by Fitzpatrick et al. (1969) and Veilleux et al. (1981), it is apparent that levels are influenced by many genes. Veilleux et al. (1981) identified parents with good general combining ability for tuber protein content. The authors found that S. phureja contained individuals with notably high protein levels but that there was wide variation between plants of the same genotype and that the high protein levels were poorly transmitted to the 4x-2x hybrids. However, the transmission of higher protein contents to the 4x S. tuberosum-4x S. tuberosum progeny was more effective, with no indication of heterosis, with mean protein contents of the hybrids (5.3%) falling short of the midparental value (5.7%).

The most prominent tuber protein is patatin, a glycoprotein with a molecular weight of 40,000, accounting for up to 40% of the total soluble proteins (Park, 1983). Rosahl et al. (1986) described the isolation and physical characterization of the patatin-encoding cDNA, as well as of one genomic clone. Jervis et al. (1986) and De Maine and Jervis (1989) described variation in the tuber proteins, including patatin, of dihaploids derived from the cultivar Pentland Crown. The authors suggested that, through the use of dihaploids and subsequent somatic chromosome doubling, it could be possible to fix such variation in heterozygous tetraploid potatoes. However, subsequent work by Clulow et al. (1992) has reported that S. phureja genes can be incorporated into the dihaploid genome following hybridization and are expressed. It is therefore possible that the previously reported variation in the patatin protein within *S. tuberosum* dihaploids may in fact have been introduced from *S. phureja*, and this requires further investigation.

Vitamins

Potatoes are a substantial source of vitamins, including ascorbic acid (vitamin C), thiamine, niacin and pyridoxine and its derivatives (vitamin B_6 group), folic acid, pantothenic acid (vitamin B_5) and riboflavin. Woolfe (1987) and Storey and Davies (1992) reviewed the various vitamin contents, nutritional details and factors influencing contents in detail. Vitamin C is the main vitamin in potatoes, with a range of 15 to 25 mg $100\,g^{-1}$ fresh weight, and Augustin (1975) demonstrated a large range of variation between potato varieties from 84 to 145 mg $100\,g^{-1}$ dry weight basis, while also noting environmental influences, including site and storage conditions.

There has been no investigative work to assess the inheritance of factors influencing the content of any of the vitamins, although there are undoubtedly varietal differences which are heritable. Sinden et al. (1978) found significant differences between 98 *S. tuberosum* genotypes for ascorbic acid content, even after 6 months storage. The authors also reported contents in 10 different *Solanum* species, ranging from 5.3 mg $100\,g^{-1}$ fresh weight in *S. bulbocastanum* to 25.4 mg $100\,g^{-1}$ in *S. stoloniferum*. The authors found that losses of ascorbic acid over a 3-month storage period were highly correlated ($r = 0.80$, $P = 0.01$) with the initial content, with genotypes having high initial contents losing a greater percentage during storage than those with lower contents. Based on a sample of ten biparental families, the authors estimated narrow-sense heritability at 0.45, suggesting that it is possible to increase ascorbic acid content of potatoes by breeding and subsequent selection. There are no apparent major genetic factors and it is probable that inheritance is polygenic, with final levels also subject to environmental factors.

After-Cooking Blackening

The development of after-cooking blackening (ACB) in susceptible potato cultivars is due to the formation of a colourless iron–chlorogenic acid complex which, on exposure to air, oxidizes to the bluish-grey compound ferridichlorogenic acid (Hughes and Swain, 1962a,b). This defect is most noticeable in susceptible cultivars following boiling or steaming, though it may also occur following baking, frying or dehydration. Its occurrence is worldwide and, while it has no known detrimental effect on taste or nutritive value, the quality of affected tubers is reduced due to darkened and unsightly flesh colour after cooking. Considerable effort has been directed to investigate the causes of, and factors affecting, after-cooking blackening. The formation of ferridichlorogenic acid is affected by citric acid levels, with high levels reducing the extent of ACB by acting as an alternative substrate for the iron present in the tuber to form a colourless complex, the citric acid sequestering the iron more efficiently than the chlorogenic acid (Mulder, 1949). Malic acid and phosphate, while present in lesser quantities, can also act in a similar manner. The important factors affecting the degree of ACB in the tuber are the quantities of chlorogenic acid and citric acid. Hughes and Swain (1962a) demonstrated that the distribution of ACB within the tuber was dependent on the ratio of both acids, the level of citric acid being lower at the stem end, where most ACB is observed (Shekhar and Iritani, 1979). The ratio of chlorogenic acid to citric acid was also demonstrated as being the principal factor accounting for the differences in ACB within individual tubers and between cultivars at different sites and in different years (Hughes and Evans, 1967).

Chlorogenic acid levels vary from cultivar to cultivar, while citric acid levels are more influenced by environmental factors. Environmental factors affecting ACB can be summarized as follows. Potatoes grown in soils high in organic matter with low potassium (K) tend to blacken more, are lower in citric acid and sometimes higher in chlorogenic acid (Hughes and Evans, 1967). In some soils iron (Fe) uptake may be important, being affected by soil pH or high calcium levels. Potatoes grown in cooler, wetter conditions often exhibit more blackening due to having higher chlorogenic acid levels. More comprehensive details of factors affecting ACB are given by Storey and Davies (1992). However, considering the importance of ACB, very little research has been directed to examining the inheritance of factors which influence varietal responses. Dalianis *et al.* (1966) demonstrated that there are genetic differences responsible for the development of ACB in different progenies, observing wide variation amongst ten parental lines used in a diallel cross. Their evidence indicated considerable variation attributable to general combining ability, that the correlation between phenotypic performances and the resultant progeny justified selection of parents for their freedom from ACB and that selection for non-blackening could be effective early in a breeding programme.

In a detailed study of a number of important traits within a half-diallel set of crosses using six parents, Killick (1977) identified significant SCA effects. This is at variance with those results reported by Dalianis *et al.* (1966), although different parental material was used. Killick (1977) also reported that ACB did not appear to be associated with any of the other characters examined in the study, including specific gravity, maturity, texture and yield. Pika *et al.* (1984) top-crossed three commercial cultivars and two parental genotypes on to two test clones and assessed the progeny in their second and third year in the field for ACB. They discerned significant differences between the parental material and attributed most of the differences to additive gene effects, with heritability in the broad sense ranging between 0.60 and 0.68 and in the narrow sense from 0.33 to 0.63. Caligari (1992), discussing unpublished work, mentioned the successful adoption of cross-prediction techniques for some quality characters, notably ACB, crisp colour and sloughing, with reasonable agreement in the rankings of six progenies for observed and predicted degree of ACB and crisp colour.

It is evident that ACB is a complex character with a heritable component which is influenced by various environmental factors. However, little is known of the detailed inheritance of this important quality character, although what evidence there is would indicate that it is inherited in a polygenic manner. With selection of parents expressing little, if any, ACB and the use of effective progeny testing and subsequent clonal selection early in a breeding programme, it is possible to ensure the quality of selected material.

Texture

There is a consensus amongst potato researchers that the texture of cooked tubers is extremely important in terms of the organoleptic properties of the consumed product. Despite the importance of this trait and observed varietal differences, very little is known about its heritability or genetics. This is probably because in most breeding programmes texture tends to be regarded as a descriptive trait rather than a selection criterion. Varieties tend to be classified by degrees, from extremely waxy to extremely floury in texture, and, whilst there may be strong local preferences for either side of the median, there is insufficient preference for either 'waxy' or 'floury' potatoes to exert selection in favour of the one type versus the other.

One texture trait, disintegration on boiling or sloughing, is undesirable in a variety for table use but is often a feature of high dry-matter clones suitable for processing, although the association between sloughing and dry matter is not absolute and other factors may be involved (Linehan and Hughes, 1969). Nevertheless, environmental factors which enhance dry-matter content may lead to sloughing problems in a variety which is quite acceptable for boiling purposes (Holden, 1980). Texture is clearly a complex trait determined by many biochemical and structural characteristics of the tuber, and has been deemed to be of such import that European breeders have collaborated to agree a common classification system (Lugt, 1960). Whilst influenced by environment, genotype is believed to play the more significant role (for review, see Storey and Davies, 1992). The association between texture and dry matter is a strong one, as reported independently by several authorities (Faulks and Griffiths, 1983; Sato et al., 1991), but factors such as cell adhesion and starch structure have also been implicated (Barrios et al., 1963; Linehan and Hughes, 1969; Warren and Woodman, 1974; Shomer and Levy, 1988).

Whilst most of this research stresses or highlights the varietal (i.e. genotypic) component of variation for texture, there have been few attempts to partition the total variation into its genetic and environmental components, although Killick (1977) was able to deduce that specific combining ability (SCA) was important. The conclusion, therefore, is that texture is a complex trait, determined by various factors, of which the most important is dry-matter content. The classification of varieties (genotypes) into limited categories is somewhat arbitrary, and texture more probably exhibits continuous variation from extreme waxiness to extreme flouriness. It is probably under complex polygenic control, which only further research can elucidate. It is possible that more basic research into starch structure and composition such as amylose : amylopectin ratios, may provide a clearer insight into the genetics of texture, but cell wall structure and adhesion are also likely to be important, possibly independently inherited, factors, which will complicate genetic analyses of texture *per se*.

Enzymic Browning

Discoloration of tuber flesh is an important problem to growers and processors and can involve increased costs through losses, further labour required for sorting and preventive measures during processing. Enzymic browning is found in peeled or cut potatoes or in tubers suffering from external or internal bruising caused by mechanical damage. The causes of enzymic browning are well documented (Muneta, 1977; Walker, 1977; Rhodes and Wooltorton, 1978), being the oxidation of tyrosine and other orthodihydric phenols by polyphenoloxidase (PPO). Tyrosine oxidation initiates the subsequent formation of the dark or black melanin pigment (Joslyn and Ponting, 1951; Storey and Davies, 1992).

To control enzymic browning, processors utilize various chemical means; these have been reviewed by Smith (1977). Chemical control can be effected by inhibiting PPO activity by altering the pH (Alma and Francis, 1961), adding chelating agents, adding reducing agents, or, most commonly and most effectively, using bisulphites or sulph-hydryl compounds (Lerner, 1953). However, there is increasing concern about additives to food and their continued use may be subject to question.

Various environmental factors influence the observed degree of enzymic browning, although there is limited information on which factors are important. Mapson *et al.* (1963) identified a good relationship in some cases between rainfall and the amount of tyrosine in the tubers. Umaerus and Olsson (1974) found that the tyrosine content in the tuber had a high predictive value with regard to the degree of enzymic discoloration and that the content of chlorogenic acid influenced the speed of the reaction. The authors suggested that clones with high tyrosine contents could therefore be discarded from a breeding programme, although some of the low-tyrosine clones still gave a fairly high discoloration.

Variation in degree of observed enzymic browning is exhibited between varieties (Anon., 1992b) and appears to be continuous in nature, suggesting that there are no clear major genetic factors operating. Gubb *et al.* (1989) assessed 50 genotypes representing accessions of *Solanum hjertingii* and five commercial varieties, and found that the wild species possesses a non-browning character and that reduced PPO activity limited browning. Other species in the series Longipedicellata were shown by Woodwards and Jackson (1985) to possess varying degrees of reduced enzymic browning, with *S. polytrichon* and *S. papita* showing the least browning. These authors suggested, from limited data, that non-browning appeared to be a dominant character governed by a small number of genetic factors.

Breeding work to specifically reduce the degree of enzymic browning in tetraploid cultivars has a relatively low priority and will depend on how important the objective is to the market in future in comparison with other characters. At present there appears to be adequate genetic variation within *Solanum tuberosum* without resort to introgression from wild species, which may prove difficult and possibly ineffective in an *S. tuberosum* genetic background.

Sugar Content

Whilst starch is the major carbohydrate source in potato tubers, all contain small but varying amounts of sugars, namely sucrose, glucose and fructose. In terms of quality for processing, these sugar levels are critical, particularly the hexose sugars, glucose and fructose. The vast bulk of processed potatoes are manufactured into fried products, namely chips (*pomme frites* or french fries) and crisps (chips), and a proportion of products reconstituted from dehydrated potato are also fried. For an extensive review, the reader is referred to Talburt and Smith (1987). The principal reason for the critical role of hexose reducing sugars in potato quality is the fact that frying at high temperature results in a typical Maillard reaction between these sugars and the α-amino acid groups of nitrogenous compounds also present in the tuber (Schallenberg *et al.*, 1959). This results in a dark-coloured, bitter-tasting product. A reducing sugar content of 0.1% fresh weight is ideal for the manufacture of crisps and higher than 0.33% is unacceptable (for review, see Storey and Davies, 1992).

Sugar content of tubers varies considerably from season to season, from site to site and between varieties. Cunningham and Stevenson (1963) carried out an extensive study of the inheritance of chip (= crisp) colour on 236 genotypes (varieties and seedlings) over 3 years and concluded that heritability for chip colour rating after cold storage (3°C) and reconditioning was high and that progeny means were fairly reliable for predicting parental breeding values. The low-temperature storage regime employed in this study reflects the very important influence of storage temperature on sugar content. Many of the problems associated with high reducing sugar levels in potato tubers can be avoided or ameliorated by storage at high (8–10°C) temperature, as is current practice by most major processors. However, storage at such temperatures, particularly for prolonged periods, is costly and requires the use of sprout-suppressant chemicals. Storage at lower temperatures (4°C) results in cold-induced sweetening, a stress-induced factor (see chapter 11 by Vayda), which is probably the most important problem faced by potato processors. Cold-induced sweetening is not peculiar to the potato, but because of its extreme economic importance it has been the subject of much study (for reviews, see ap Rees *et al.*, 1988; Sowokinos, 1990). The biochemistry of cold-induced sweetening in potato tubers is complex and the regulation of starch mobilization is not yet fully understood. There are, for example, contradictory reports regarding the presence or absence of amylolitic enzymes (α- and β-amylose) in cold-stored tubers, although it is now generally accepted that they are present. However, despite differences in amylolitic activity between tubers under different storage regimes and between clones exhibiting different sugar stabilities during low-temperature storage, the precise role of these enzymes in genetic differences in cold-induced sweetening is unclear (Cochrane et al., 1991). Phosphorolytic enzymes, in particular phosphofructokinase (PFK), which is known to be cold-labile, are believed to play a major role in cold-induced sweetening. Differences between forms of PFK isozymes extracted from the cultivar Record (normal type) and cultivar Brodick (sugar-stable at low temperature) would appear to provide a genetic basis for this (ap Rees et al., 1988). As sucrose is a precursor of glucose and fructose, the regulatory mechanisms of sucrose synthesis may also be involved. However, differences in rate of hexose sugar accumulation were not reflected in genotypic differences in total invertase activity (Richardson et al., 1990).

Until recently, whilst it was recognized that some varieties (= genotypes) were to a degree more sugar-stable at low temperature than others and conventional studies have provided evidence for the heritability of fry colour (O'Keefe, 1974; Lynch et al., 1992), there was little evidence for genetic variation for 'resistance' to cold-induced sweetening in the cultivated potato, *Solanum tuberosum*. No varieties were capable of being stored at low temperature (3–4°C) for prolonged periods without accumulating excessive levels of hexose reducing sugars. Consequently, a great deal of effort has been directed towards the molecular approach, which might enable sugar levels to be manipulated in existing varieties (Davies and Viola, 1992), or towards incorporating low-temperature sugar stability into *S. tuberosum* from related wild species or primitive forms where it has been reported (Lauer and Shaw, 1970; Accatino et al., 1973; Colon et al., 1989). Recent studies at the SCRI have shown that variation for low-temperature sugar stability does exist between tetraploid *S. tuberosum* clones (Brown et al., 1990; Mackay et al., 1990). Some of these clones were capable of storage for more than 5 months at 4°C without accumulating excessive levels of hexose sugars and were capable of producing an acceptable pale-coloured fry product direct from storage at 4°C without reconditioning. Two of these clones have now been admitted to the UK National List (Anon., 1990, 1991) and are undergoing commercial evaluation as named varieties in the UK and elsewhere.

Crosses have been made between some of these sugar-stable clones and normal varieties. Subsequent studies on the tubers of progenies from these crosses, stored at 10°C and 4°C, have illustrated that the trait is heritable but complex (Mackay, 1992). The glucose content of tubers seems to be the most important factor in determining fry colour, but the latter can be more accurately predicted when fructose content is taken into account also, in accordance with previous published work (Brown et al., 1990). Sucrose levels do not correlate well with either of the hexose sugar levels or with fry colours (Figure 13.1). In so far as fry colour is itself measured only on a visual scale of 1–9 (dark to pale) in this type of work (Mackay et al., 1990), one must interpret

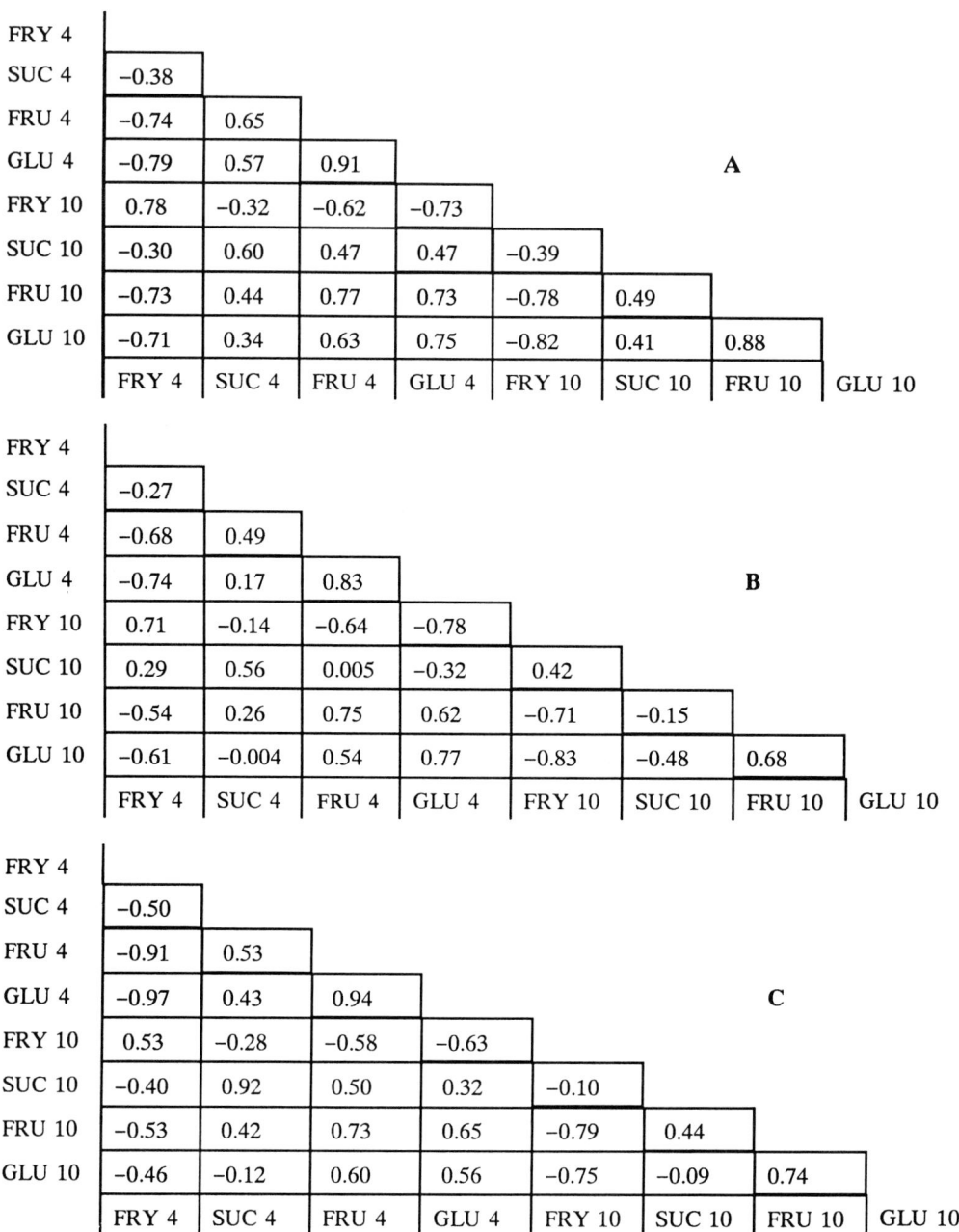

Fig. 13.1. Correlation matrices between fry colours after storage at 4°C (FRY 4) and 10°C (FRY 10) and sugar contents of the same tuber samples. SUC = sucrose; FRU = fructose; GLU = glucose; 4 = 4°C storage and 10 = 10°C storage. A) 22 clones and cultivars (see Mackay et al., 1990; Brown et al., 1990). B) Between means of 30 progenies from a half-diallel (see Mackay, 1992). C) Between parents of B.

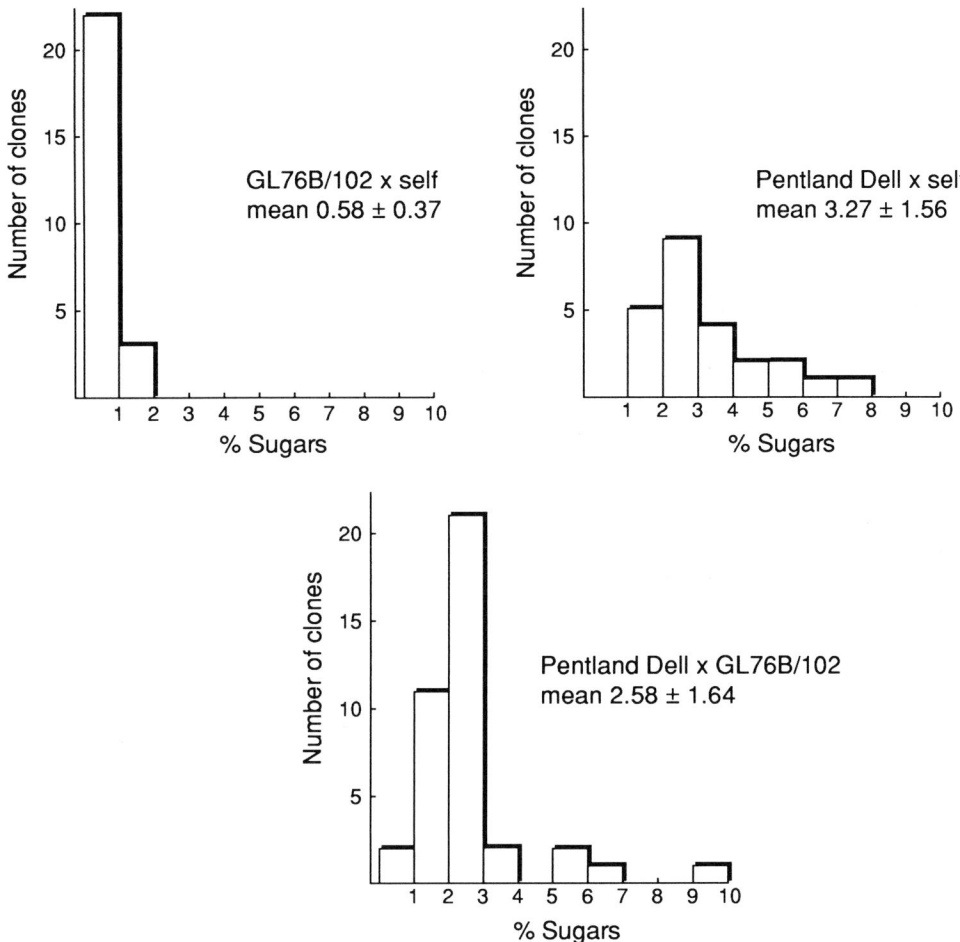

Fig. 13.2. Segregation of progenies by selfing and crossing of a low-temperature sugar-stable clone (GL76B/102) and a normal variety (Pentland Dell) for total hexose (glucose and fructose) sugars, in tubers stored at 4°C for 3 months. Sugars expressed as percentage dry matter from a three-tuber sample of each clone.

these relationships with caution, but the fact that the glucose and fructose contents do not entirely explain the variation in fry colour would support the argument that other factors such as amino acid balance and composition may be important (Storey and Davies, 1992). Statistical analyses of an incomplete half-diallel set of crosses between four phenotypically sugar-stable clones and two normal varieties have shown that glucose levels after storage at 10°C (for 3 months) have a significant GCA component but insignificant SCA, whilst glucose levels from 4°C have significant SCA (G.R. Mackay, unpublished).

Examination of the segregation pattern from progenies produced by selfing and by pair-crossing a sugar-stable clone with a normal variety illustrated the complexity of inheritance (Figure 13.2). In this instance the fact that the phenotypically sugar-stable clone bred relatively true on selfing whilst the normal cultivar segregated would suggest that genetic factors influencing low-temperature sugar stability tend to be recessive to the normal 'wild type'. In considering this hypothesis, reference to the breeding history of this particular clone may be relevant. GL76B/102 was identified in a population of Neotuberosum derived by cyclical recurrent selection from Andigena (see Chapter 21). It is now believed that in the course of development of this Neotuberosum population, in which true seed was produced by natural open pollination of selected clones, quite a high degree of self pollination (inbreeding) occurred. GL76B/102 was itself obtained from a seed of a berry obtained by open pollination. The fact that a small proportion of the selfed progeny of the non-sugar-stable variety Pentland Dell were more sugar-stable than their parent lends credence to the view that inbreeding is exposing recessive factors involved in low-temperature stability (Figure 13.2). The tendency for most potato breeders to treat *S. tuberosum* as an obligate outbreeder, to avoid inbreeding and to maximize heterozygosity, in the belief that this enhances yield potential, may be one reason for the paucity of variation for this trait in most selected populations of *S. tuberosum*. Further research is needed before hard conclusions can be drawn but the indications are that there may be benefits in a degree of mild inbreeding in the breeders' pursuit of low-temperature sugar-stable varieties.

Dry Matter

Starch is the major component of the dry-matter (DM) content of potato tubers, accounting for approximately 70% of the total solids. It is usually determined by estimating the specific gravity of a tuber sample, either directly, by use of a potato hydrometer, or indirectly, by the difference between the weight of a sample in air and the weight of the same sample in water (underwater weight). The relationships between these various characteristics were reviewed by Simmonds (1977). The DM content of potato tubers has been known to vary between varieties for some considerable time and the fact that a proportion of this variation is heritable is also recognized (Akeley and Stevenson, 1943). However, there is considerable variation within varieties and dry-matter content is subject to considerable influence of the environment, both the growing crop and the stored tubers, (for reviews, see Smith, 1967; Storey and Davies, 1992). Early attempts to examine the inheritance of DM content concluded that the varieties used as parents were heterozygous with respect to the factors controlling DM content (tuber density) and that the trait was polygenically controlled; but there was some suggestion that high DM content (tuber density) seemed dominant to low (Akeley and Stevenson, 1944). Phenotypic recurrent selection, based on single plants ('hills'), elicited a small but consistent response, confirming a heritable component to variation for DM content (Plaisted and Peterson, 1963). Singh (1969) obtained heritability estimates from 5 to 60% in progenies from five pair crosses. He concluded that, whilst genetic gain may be achieved by selection amongst crosses for high specific gravity, the trait was also considerably affected by environment. He was unable to confirm the observations of Akeley and Stevenson (1944), for in his experiments a cross between a low-specific-gravity parent and a medium-specific-gravity parent produced the highest specific gravity progeny. Moreover, there appeared to be no relationship between parental phenotype and offspring, but he concurred with Akeley and Stevenson (1943) that there was genetic variation with a large environmental effect. Singh's (1969) study confirmed the conclusion of Cunningham and Stevenson (1963) that specific gravity was a quantitatively inherited trait with a low heritability, and that recessive genes might also be involved, as reported by Borger et al. (1954).

Further studies on the genetic control of DM content have confirmed the complexity of the expression of this important trait over sites and seasons (Killick and Simmonds, 1974; MacArthur and Killick, 1976). Analysis of data from an intervarietal diallel found very little evidence for any significant genetic variation for DM content (specific gravity) (Killick, 1977). This may have been in part due to the rather narrow genetic base of the six varieties used as parents in this study, but as Killick (1977) concluded, significant replicate (block)–progeny interactions were indicative of a substantial environmental effect on expression of DM content, which would require extensive studies over a wide range of environments to resolve. This conclusion is supported by studies in Poland, which have also shown that the repeatability of starch yield versus tuber yield over years was very low; on average only 9% of the variation for starch yield in the second year could be accounted for by that in the first year (Swiezynski et al., 1977), although this contrasts with Kaminski (1977), who obtained high estimates of heritability for DM and starch content. This may be partly explained by the fact that Kaminski (1977) was able to clonally replicate his experiment and that his DM data were collected from relatively large samples (five-plant plots) so that intraclonal variation was minimized. Estimates of narrow-sense heritability for specific gravity in diploid potatoes have ranged from 27 to 74% (Ruttencutter et al., 1979). However, this material was drawn from a broader genetic base than the parents used by Killick (1977), and there was also evidence of substantial genotype–environment interactions, which may have affected estimates of heritability obtained by realized selection response and parent–offspring regression, due to experimental design.

The DM content of potato tubers is an extremely important character, which has been associated with many other quality traits, such as texture, suitability for processing and susceptibility to mechanical damage or bruising (for review, see Storey and Davies, 1992), although in the latter instance Killick and MacArthur (1980) were unable to demonstrate any significant correlation between specific gravity and bruise volume.

It is therefore somewhat surprising that so little is known of the genetic architecture of DM content. However, it clearly has a heritable component and will respond to selection, but the large environmental effects are such that much effort would be required to achieve significantly higher levels than are currently available in known varieties that have a higher than average dry matter. Too high a DM content also detrimentally affects certain culinary properties, such as disintegration on boiling (sloughing), so that selection for higher DM content *per se* is seldom a breeding object; perhaps it will only become so if the potato were to become more important as a source of starch solely for industrial use rather than consumption or processing as food. In this respect, further research into specific characteristics of potato starch which render it more suitable for certain industrial processes than (say) cereal starches might be more immediately useful than raising the content *per se*. That such variation can be found may be well illustrated by the recent identification of an amylose-free mutant (chapter 7 by Jacobsen and Ramanna).

Flavour

The flavour of potatoes is perhaps one of the more important characteristics for the consumer, and yet comparatively little is known about it. Important factors influencing overall flavour include aroma, taste and texture. The latter character, which affects the 'mouth-feel', was dealt with separately and is associated with dry-matter content. The aroma of boiled potatoes is due largely to a mixture of volatile compounds with relatively low boiling-points (Self *et al.*, 1963). The majority of the volatile compounds can be produced *in vitro* by boiling solutions of amino acids with sugars and pectin. Potatoes of different varieties or of different age, if they contain different levels of the precursors, may be expected to have different aromas (Casey *et al.*, 1963). Nursten and Sheen (1974) identified 35 components in the essence of cooked potatoes. Fuller details of the many components implicated in the flavour and aroma of cooked potatoes are described elsewhere (Nursten and Sheen, 1974; Smith, 1977).

Variation for flavour exists between varieties currently available, although there is little published research. McBean and Coote (1974) assessed 262 varietal samples of 117 different varieties over a 13-year period in Australia, finding significant differences in flavour. The authors found colour, texture and flavour to be correlated (Kendall's concordance coefficient $W = 0.71$, $P < 0.001$), suggesting that the determination of texture could be of use when assessing quality early in a breeding programme. Neenan *et al.* (1967) found that differences in flavour due to soil type were small in relation to differences between varieties, and that the flavour characteristic of varieties remained relatively stable across years and sites.

Testing for flavour in a breeding programme is normally restricted to limited numbers being subjected to organoleptic tests. Unless more reliable testing procedures can be developed, such as the identification of the principal chemical components using an easy and rapid test, there would appear to be limited opportunity for significant advances. A further problem confronting the breeder is that different consumers in different regions or countries often prefer different products with different flavours, and, as a result, the probability of identifying one flavour which is universally preferred appears remote and is not necessarily desirable.

References

Accatino, P.I., Peloquin, S.J. and Cipar, M.S. (1973) Inheritance of potato chip color at the diploid and tetraploid levels of ploidy. *American Potato Journal* 50, 335-336 (Abstract).

Akeley, R.V. and Stevenson, F.J. (1943) Yield, specific gravity and starch content of tubers in a potato breeding program. *American Potato Journal* 20, 203-217.

Akeley, R.V. and Stevenson, F.J. (1944) The inheritance of dry-matter content in potatoes. *American Potato Journal* 21, 83-89.

Akeley, R.V., Houghland, G.V.C. and Schark, A.E. (1962) Genetic differences in potato tuber greening. *American Potato Journal* 39, 409-417.

Alma, B.L. and Francis, F.J. (1961) Effect of dipping solution on the quality of prepeeled potatoes. *American Potato Journal* 38, 121-130.

Anon. (1974) *Report on a National Damage Survey, 1973*. Potato Marketing Board, Oxford.

Anon. (1985) *Food and Agricultural Organization Yearbook for 1985* Series 39, FAO, Rome, p. 331.

Anon. (1990) *Scottish Crop Research Institute Annual Report 1990*. Dundee, p. 23.

Anon. (1991) *Scottish Crop Research Institute Annual Report 1991*. Dundee, p. 25.

Anon. (1992a) *Potato Statistics in Great Britain 1987-1991*. Potato Marketing Board, Oxford, p. 28.

Anon. (1992b) *Potato Variety Handbook*. NIAB, Cambridge.

ap Rees, T., Burrell, M.M., Entwistle, T.G., Hammond, J.B.W., Kirk, D. and Kruger, N.J. (1988) Effects of low temperature on the respiratory metabolism of carbohydrates by plants. In: Long, S.P. and Woodward, F.I. (eds) *Plants and Temperature*. For the Society of Experimental Biology by the Company of Biologists, Cambridge, pp. 377-393.

Augustin, J. (1975) Variations in the nutritional composition of fresh potatoes. *Journal of Food Science* 40, 1259-1299.

Balls, R.C., Gunn, J.S. and Stirling, A.J. (1982) *The National Potato Damage Awareness Campaign*. Potato Marketing Board and Agricultural Development and Advisory Service, Oxford, 32 pp.

Barrios, E.P., Newsom, D.W. and Miller, J.C. (1963) Some factors influencing the culinary quality of Irish potatoes. II. Physical characters. *American Potato Journal* 40, 200-208.

Bintcliffe, E.J.B., Clydesdale, A. and Draper, S.R. (1982) Effects of genotype, site and season on the glycoalkaloid content of potato tubers. *Journal of the National Institute of Agricultural Botany* 16, 86-91.

Borger, H. Von, Kohler, D. and Sengbusch, R.V. (1954) Untersuchungen uber die Zuchtung von Kartoffeln mit hohem Starkeertrag. *Zuchter* 24, 273-276.

Brown, E. and Riley, W. (1976) Greening of potato tubers: varietal response to controlled exposure to light. *Journal of the National Institute of Agricultural Botany* 14, 70-76.

Brown, J., Mackay, G.R., Bain, H., Griffith, D.W. and Allison, M.J. (1990) The processing potential of tubers of the cultivated potato, *Solanum tuberosum* L., after storage at low temperatures. 2. Sugar concentration. *Potato Research* 33, 219-227.

Caligari, P.D.S. (1992) Breeding new varieties. In: Harris, P.M. (ed.) *The Potato Crop*, 2nd edn. Chapman and Hall, London, pp. 334-372.

Carruthers, J. (1982) A comparison of impact levels from two shapes of potato harvester web rods. Department Note SIN/329, Scottish Institute of Agricultural Engineering, Penicuik, Edinburgh (unpublished).

Casey, J.C., Self, R. and Swain, T. (1963) Origin of methanol and dimethyl sulphide from cooked potatoes. *Nature* 200, 885.

Clulow, S.A., Wilkinson, M.J. and De Maine, M.J. (1992) Dihaploid formation - a new hypothesis. In: Rouselle-Bourgeois, F. and Rouselle, P. (eds) *Proceedings of the Joint Conference of the EAPR Breeding and Varietal Section and the EUCARPIA Potato Section*, Landerneau, France, pp. 165-169.

Cochrane, M.P., Duffus, C.M., Allison, M.J. and Mackay, G.R. (1991) Amylolytic activity in stored potato tubers. 2. The effect of low temperature storage on the activities of α- and β-amylase and α-glycosidase in potato tubers. *Potato Research* 34, 333-341.

Cole, C.S. (1980) Breeders' problems. *Annals of Applied Biology* 96, 354-357.

Collier, G.F., Wurr, D.C.E. and Huntington, V.C. (1980) The susceptibility of potato varieties to internal rust spot. *Journal of Agricultural Science* 94, 407-410.

Colon, L.T., Sijpkes, L. and Hartmans, K.J. (1989) The cold stability of *Solanum goniocalyx* and *S. phureja* can be transferred to adapted diploid and tetraploid *S. tuberosum* germplasm. In: Louwes, K.M., Toussaint, H.A.J.M. and Dellaert, L.M.W. (eds) *Parental Line Breeding and Selection in Potato Breeding*. PUDOC, Wageningen, The Netherlands, pp. 76-80.

Cunningham, C.E. and Stevenson, F.J. (1963) Inheritance of factors affecting potato chip colour and their associations with specific gravity. *American Potato Journal* 40, 253-265.

Dale, M.F.B., Griffiths, D.W. and Bain, H. (1992) Glycoalkaloids in potatoes-shedding light on an important problem. *Aspects of Applied Biology* 33, 221-227.

Dalianis, C.D., Plaisted, R.L. and Peterson, L.C. (1966) Selection for freedom from after cooking darkening in a potato breeding program. *American Potato Journal* 43, 207-215.

Davies, H.V. and Ross, H.A. (1986) The development of internal rust spot in potato tuber. *Aspects of Applied Biology* 13, 433-435.

Davies, H.V. and Viola, R. (1992) Regulation of hexose accumulation in potato tubers: possibilities for molecular manipulation. *Post Harvest News and Information* 3(5), 97-100.

Deahl, K.L., Young, R.J. and Sinden, S.L. (1973) A study of the relationship of late blight resistance to glycoalkaloid content in fifteen potato clones. *American Potato Journal* 50, 248-253.

De Maine, M.J. (1988) Testing the tuber deformation resistance of tetraploid and dihaploid potatoes. *Journal of Agricultural Science* 110, 445-449.

De Maine, M.J. and Jervis, L. (1989) The use of dihaploids in increasing the homozygosity of tetraploid potatoes. *Euphytica* 44, 37-42.

De Maine, M.J., Bain, H and Joyce, J.A.L. (1988) Changes in the total tuber glycoalkaloid content of potato cultivars on exposure to light. *Journal of Agricultural Science* 111, 57-58.

De Maine, M.J., Bradshaw, J.E. and Caligari, P.D.S. (1992) Inheritance of the external mechanical damage resistance of potato cultivars. *Annals of Applied Biology* 121, 379-384.

Desborough, S. and Weiser, C.J. (1974) Improving potato protein. I. Evaluation of selection techniques. *American Potato Journal* 51, 185-196.

East, E.M. (1908) A study of the factors influencing the improvement of the potato. *Bulletin Illinois Agricultural Experiment Station* 127, 375-456.

Faulks, R.M. and Griffiths, N.M. (1983) Influence of variety, site and storage on physical, sensory and compositional aspects of mashed potato. *Journal of the Science of Food and Agriculture* 34, 979–986.

Fitzpatrick, T.J., Akeley, R.V., White, J.W., Jr, and Porter, W.L. (1969) Protein, nonprotein and total nitrogen in seedlings of potato. *American Potato Journal* 46, 237–284.

Fitzpatrick, T.J., McDermott, J.A. and Osman, S.F. (1978) Evaluation of injured commercial potato samples for total glycoalkaloid contents. *Journal of Food Science* 43, 1617–1618.

Georgieva, R. and Ronkov, B. (1954) Investigating the inheritance of the solanine type of glycoalkaloids in some interspecific hybrids of the potato. *Izvestiya na Institute po Rastenievudstvo (Sofiya)* 12, 225–240.

Grassert, V. and Lellbach, H. (1987) Investigations of glycoalkaloid content of potato hybrids resistant to potato cyst eelworms, *G. rostochiensis* and *G. pallida*. *Biochemie und Physiologie der Pflanzen* 182, 473–479.

Gray, D. and Hughes, J.C. (1978) Tuber quality. In: Harris, P.M. (ed.) *The Potato Crop: the Scientific Basis for Improvement.* Chapman and Hall, London, pp. 504–544.

Gubb, I.R., Hughes, J.C., Jackson, M.T. and Callow, J.A. (1989) The lack of enzymic browning in the wild potato species *Solanum hjertingii* Hawkes compared with commercial *Solanum tuberosum* varieties. *Annals of Applied Biology* 114, 579–586.

Gull, D.D. and Isenberg, F.M. (1960) Chlorophyll and solanine content and distribution in four varieties of potato tubers. *Proceedings of the American Society for Horticultural Science* 75, 545–556.

Hiller, L.K., Koller, D.C. and Van Denburgh, R. (1979) Brown centre of potatoes - what have we learned? *Proceedings of the 18th Annual Washington State Potato Conference*, USA, pp. 21–27.

Hiller, L.K., Koller, D.C. and Thornton, R.E. (1985) Physiological disorders of potato tubers. In: Li, P.H. (ed.) *Potato Physiology.* Academic Press, London, pp. 389–455.

Hoff, J.E. (1975) A simple method for the approximate determination of soluble protein in potato tubers. *Potato Research* 18, 428–432.

Hoff, J.E., Jones, C.M., Wilcox, G.E. and Castro, M.D. (1971) The effect of nitrogen fertilization on the composition of the free amino acid pool of potato tubers. *American Potato Journal* 48, 390–394.

Hoff, J.E., Lam, S.L. and Erickson, H.T. (1978) Breeding for high protein and dry matter in the potato at Purdue University. *Research Bulletin - Purdue University Agricultural Experiment Station*, No: 953, 23 pp.

Holden, J.H.W. (1980) Problems in breeding for tuber quality. *Proceedings of Joint Meeting of EAPR/EUCARPIA*, Edinburgh, June 1980 (Abstract).

Hooker, W.J. (ed.) (1981) *Compendium of Potato Diseases.* St Paul, Minnesota, USA.

Horton, D.E. (1988) Potatoes: truly a world crop. *SPAN* 30(3), 116–118.

Horton, D.E. and Anderson, J.L. (1992) Potato production in the context of the world and farm economy. In: Harris, P.M. (ed.) *The Potato Crop*, 2nd edn. Chapman and Hall, London, pp. 794–815.

Horton, D.E. and Sawyer, R.L. (1985) The potato as a world food crop, with special reference to developing areas. In: Li, P.H. (ed.) *Potato Physiology.* Academic Press, London, pp. 1–34.

Hughes, J.C. and Evans, J.L. (1967) Studies on after-cooking blackening in potatoes. IV Field experiments. *European Potato Journal* 10, 16–36.

Hughes, J.C. and Swain, T. (1962a) After-cooking blackening in potatoes. II.

Core experiments. *Journal of the Science of Food and Agriculture* 13, 229–236.

Hughes J.C. and Swain, T. (1962b) After-cooking blackening in potatoes. III. Examination of the interaction of factors by *in vitro* experiments. *Journal of the Science of Food and Agriculture* 13, 358–363.

Hunnius, W. and Munzert, M. (1976) Der Einfluss der Jahres Und Erntewitterung auf die Knollenbeschadigung von Kartoffelsorten. *Zeitschrift fur Acker und Pflanzenbau* 142, 237–247.

Hunnius, W., Fritz, A. and Munzert, M. (1976) On the effect of year and course of weather on protein content of potatoes. *Landwirtschaftliche Forschung* 29, 141–148.

Iritani, W.M. (1981) Growth and preharvest stress and processing quality of potatoes. *American Potato Journal* 58, 71–80.

Jadhav, S.J. and Salunkhe, D.K. (1975) Formation and control of chlorophyll and glycoalkaloids in tubers of *Solanum tuberosum* L. and evaluation of glycoalkaloid toxicity. *Advances in Food Research* 21, 307–354.

Jadhav, S.J., Sharma, R.P. and Salunkhe, D.K. (1981) Naturally occurring toxic alkaloids in foods. *CRC Critical Reviews in Toxicology* 9, 21–104.

Jansky, S.H. and Thompson, D.M. (1990) Expression of hollow heart in segregating tetraploid potato families. *American Potato Journal* 67, 695–703.

Jefferies, R.A. and MacKerron, D.K.L. (1987) Observations on the incidence of tuber growth cracking in relation to weather patterns. *Potato Research* 30, 613–623.

Jervis, L., Shepherd, A.L. and De Maine, M.J. (1986) Variations in soluble tuber proteins in tetraploid, dihaploid and doubled dihaploid potato (*S. tuberosum*). *Biochemical Society Transactions* 14, 1986–1987.

Joslyn, M.A. and Ponting, J.D. (1951) Enzyme-catalysed oxidative browning of fruit products. *Advances in Food Research* 3, 1–44.

Kaldy, M.S. and Markakis, P. (1972). Amino acid composition of selected potato varieties. *Journal of Food Science* 37, 375–377.

Kaminski, R. (1977) Variability and heritability of morphological and physiological characters of potato. *Genetica Polonica* 18, 115–123.

Killick, R.J. (1977) Genetic analysis of several traits in potatoes by means of a diallel cross. *Annals of Applied Biology* 86, 279–289.

Killick, R.J. and MacArthur, A.W. (1980) The relationship between bruising and specific gravity in some potato varieties. *Potato Research* 23, 457–461.

Killick, R.J. and Simmonds, N.W. (1974) Specific gravity of potato tubers as a character showing small genotype–environment interactions. *Heredity* 32, 109–112.

Kozukue, N. and Mizuno, S. (1990) Effects of light exposure and storage temperature on greening and glycoalkaloid content in potato tubers. *Journal of the Japanese Society for Horticultural Science* 59, 673–677.

Kunkel, R. and Gardner, W.H. (1965) Potato tuber hydration and its effect on blackspot of Russet Burbank in the Columbia Basin of Washington. *American Potato Journal* 42, 109–124.

Larsen, E.C. (1949) Investigations on cause and prevention of greening in potato tubers. *Idaho Agricultural Experiment Station Research Bulletin* 16, 32 pp.

Lauer, F. and Shaw, R. (1970) A possible genetic source for chipping potatoes from 40°F storage. *American Potato Journal* 47, 275–278.

Lepper, W. (1949) Solaningehalt von 58 Kartoffelsorten. *Zeitschrift fur Lebensmitteluntersuchung und Forschung* 89, 264–273.

Lerner, A.B. (1953) Metabolism of phenylalanine and tyrosine. *Advances in Enzymology* 14, 73–128.

Liljemark, A. and Widoff, E. (1960) Greening and solanine development of white potato in fluorescent light. *American Potato Journal* 37, 379–389.

Linehan, D.J. and Hughes, J.C. (1969) Texture of cooked potato. *Journal of the Science of Food and Agriculture* 20, 110–112.

Lugt, C. (1960) 1 Results of the assessment of the cooking quality of internationally exchanged potato samples. *Proceedings of 1st EAPR Triennial Conference*, Braunschweig, The Netherlands, pp. 321–323.

Lynch, D.R., Tai, G.C.C. and Coffin, R.H. (1992) Genetic components of potato chip quality evaluated in three environments and under various storage regimes. *Canadian Journal of Plant Science* 72, 535–543.

MacArthur, A.W. and Killick, R.J. (1976) Environmental and genetic variation in some economically important traits in potatoes. *Journal of Agricultural Science* 8, 39–43.

McBean, D.McG. and Coote, G.G. (1974) Quality tests of potato varieties in Australia. *American Potato Journal* 51, 165–169.

McCollum, G.D. and Sinden, S.L. (1979) Inheritance study of tuber glycoalkaloids in a wild potato, *S. chacoense* Bitter. *American Potato Journal* 56, 95–113.

Mackay, G.R. (1992) Selecting for low temperature sugar-stability in potatoes. Rouselle-Bourgeois, F. and Rouselle, P. (eds) *Proceedings of the Joint Conference of the EAPR Breeding and Varietal Assessment Section and the Eucarpia Potato Section*, Landerneau, France, pp. 143–148.

Mackay, G.R., Brown, J. and Torrance, C.J.W. (1990) The processing potential of tubers of the cultivated potato, *Solanum tuberosum* L., after storage at low temperature. 1. Fry colour. *Potato Research* 33, 211–218.

McKee, R.K. (1959) Factors affecting the toxicity of solanine and related compounds to *Fusarium coeruleum*. *Journal of General Microbiology* 20, 686–698.

McRae, D.C. (1979) Potatoes mechanical handling damage project 55. *Central Council for Agricultural Engineering Research* 22, 229–245.

McRae, D.C. (1986) Mechanical damage to potatoes, causes, measurement and remedies. *Aspects of Applied Biology* 13, 383–392.

McRae, D.C., Carruthers, J. and Porteous, R.L. (1976) The effect of drop height on damage sustained by some main crop varieties. Department Note SIN/202, Scottish Institute of Agricultural Engineering, Penicuik, Edinburgh.

McRae, D.C., Hutchison, P.S. and Carruthers, J. (1986) Sieving control and horizontal agitation of potato harvester chains. *Transactions – American Society of Agricultural Engineers* 29, 366–369.

Maga, J.A. (1980) Potato glycoalkaloids. *Critical Reviews in Food Science and Nutrition* 12, 371–405.

Mapson, L.W., Swain, T. and Tomalin, A.W. (1963) Influence of variety, cultural conditions and temperature of storage on enzymic browning of potato tubers. *Journal of the Science of Food and Agriculture* 14, 673–684.

Markakis, P. (1975) The nutritive quality of potato protein. In: Friedman, M. (ed.) *Protein Nutritional Quality of Foods and Feeds, Part 2. Quality Factors – Plant Breeding, Composition, Processing and Anti-nutrients*. Marcel Dekker, New York, pp. 471–488.

Miedema, P., van Gelder, W.M.J. and Post, J. (1976) Coagulable protein in potato: screening method and prospects for breeding. *Euphytica* 25, 663–670.

Mulder, E.G. (1949) Mineral nutrition in relation to the biochemistry and physiology of potatoes. *Plant and Soil* 2, 59–121.

Muneta, P. (1977) Enzymic blackening in potatoes: influence of pH on dopachrome oxidation. *American Potato Journal* 54, 387–393.

Munzert, M. (1987) Potato breeding strategy in the Federal Republic of Germany. In: Jellis, G.J. and Richardson, D.E. (eds) *The Production of New Potato Varieties*. Cambridge University Press, Cambridge, pp. 38-44.

Nair, P.M., Behere, A.G. and Ramaswamy, N.K. (1981) Glycoalkaloids of *S. tuberosum* Linn. *Journal of Scientific and Industrial Research* 40, 529-535.

Neenan, M., Mulqueen, J. and Franklin, A.A. (1967) Influence of soil type on certain quality characteristics of potatoes. *European Potato Journal* 10, 167-179.

Nelson, D.C. and Thoreson, M.C. (1986) Relationships between tuber size and time of harvest to hollow heart initiation in dryland Norgold Russet Potatoes. *American Potato Journal* 63, 155-161.

Nursten, H.E. and Sheen, M.R. (1974) Volatile flavour components of cooked potato. *Journal of the Science of Food and Agriculture* 25, 643-663.

O'Brien, M.J. and Rich, A.E. (1976) *Potato Diseases*. Agricultural Handbook, US Department of Agriculture, Washington, D.C.

O'Keefe, R.B. (1974) Heritability and repeatability for quality factors in potatoes. *American Potato Journal* 51, 295.

Olsson, K. (1986) The influence of genotype on the effects of impact damage on the accumulation of glycoalkaloids in potato tubers. *Potato Research* 29, 1-12.

Parfitt, D.E. and Peloquin, S.J. (1981) The genetic basis for tuber greening in 24 chromosome potatoes. *American Potato Journal* 58, 299-304.

Park, W.D. (1983) Tuber proteins of potato - a new and surprising molecular system. *Plant Molecular Biology Reporter* 1, 61-66.

Parnell, A., Bhur, V.S. and Bintcliffe, E.J.B. (1984) The glycoalkaloid content of potato varieties. *Journal of the National Institute of Agricultural Botany* 16, 535-541.

Pika, N.A., Tarasenko, V.A and Mitsko, V.N. (1984) Combining ability of potato varieties and hybrids for tuber flesh blackening after cooking. *Selektsiia-i-Semenorodstro, USSR* 7, 16-17.

Plaisted, R.C. and Peterson, L.C. (1963) Two cycles of phenotypic recurrent selection for high specific gravity. *American Potato Journal* 4, 396-402.

Prokoshev, S.M., Petrochenko, E.I. and Baranova, V.Z. (1952) Inheritance of glycoalkaloids by interspecific hybrids of the potato (in Russian). *Doklady Academii Nauk, USSR* 83, 457-460.

Ramaswamy, N.K., Behere, A.G. and Nair, P.M. (1976) A novel pathway for the synthesis of solanidine in the isolated chloroplasts from greening potatoes. *European Journal of Biochemistry* 67, 275-282.

Reeves, A.F. (1988) Varietal differences in potato tuber greening. *American Potato Journal* 65, 651-658.

Rex, B.L. and Mazza, G. (1989) Cause, control and detection of hollow heart in potatoes: A review. *American Potato Journal* 66, 165-183.

Rhodes, J.M.C. and Wooltorton, L.S.C. (1978) The biosynthesis of phenolic compounds in wounded plant storage tissues. In: Kahl, G. (ed.) *Biochemisty of Wounded Plant Tissues*. Walter de Gruyer, Berlin and New York, pp. 243-286.

Richardson, D.L., Davies, H.V., Ross, H.A. and Mackay, G.R. (1990) Invertase activity and its relation to hexose accumulation in potato tubers. *Journal of Experimental Botany* 41 (222), 95-99.

Rosahl, S., Schmidt, R., Schell, J. and Willmitzer, L. (1986) Isolation and characterisation of a gene from *S. tuberosum* encoding patatin, the major storage protein of potato tubers. *Molecular and General Genetics* 203, 214-220.

Ross, H. (1966) The use of wild *Solanum* species in German potato breeding of the past and today. *American Potato Journal* 43, 63-80.

Ruttencutter, G., Haynes, F.L. and Mull, J.H. (1979) Estimation of narrow-sense heritability for specific gravity in diploid potatoes (*Solanum tuberosum* ssp. *phureja* and *stenotomum*). *American Potato Journal* 56, 447–453.

Sandford, L.L. and Sinden, S.L. (1972) Inheritance of potato glycoalkaloids. *American Potato Journal* 49, 209–217.

Sato, H., Takano, K., Mitsuura, N., Tanimura, W. and Kamoi, I. (1991) Texture profiles of potatoes with different specific gravities. *Journal of the Japanese Society of Food* 38(12), 1134–1136.

Schallenberg, R.S., Smith, O. and Treadaway, R.H. (1959) Role of sugars in the browning reaction in potato chips. *Journal of Agricultural and Food Chemistry* 7, 274–277.

Scholtz, M. (1987) Potato breeding strategy in the German Democratic Republic. In: Jellis, G.J and Richardson, D.E. (eds) *The Production of New Potato Varieties*. Cambridge University Press, Cambridge, pp. 32–37.

Schwarze, P. (1962) Methods for identification and determination of solanine in potato breeding material. *Zuchter* 32, 155–160.

Schwimmer, S. and Burr, H.K. (1967) Structure and chemical composition of the potato tubers. In: Talbert, W.F. and Smith, O. (eds) *Potato Processing*, 2nd edn. Avi Publishing Co., Westport, Connecticut, pp. 12–43.

Self, R., Rolley, H.L.J. and Joyce, A.E. (1963) Some volatile components from cooked potatoes. *Journal of the Science of Food and Agriculture* 14, 8–14.

Sharma, R.P. and Salunkhe, D.K. (1989) *Solanum* Glycoalkaloids. In: Cheeke, P.R. (ed.) *Toxicants of Plant Origin*. CRC Press, Boca Raton, USA, pp. 179–236.

Shekhar, V.C. and Iritani, W.M. (1979) Changes in malic and citric acid contents during growth and storage of *Solanum tuberosum* L. *American Potato Journal* 56, 87–94.

Shomer, I. and Levy, D. (1988) Cell wall mediated bulkiness as related to the texture of potato (*Solanum tuberosum* L.) tuber tissue. *Potato Research* 31, 321–334.

Shreiber, K. (1957) Naturally occurring plant resistance factors against the Colorado potato beetle and their possible mode of action. *Zuchter* 27, 289–299.

Simmonds, N.W. (1977) Relations between specific gravity, dry matter content and starch content of potatoes. *Potato Research* 20, 137–140.

Sinden, S.L. and Webb, R.E. (1972) Effect of variety and location on the glycoalkaloid content of potatoes. *American Potato Journal* 49, 334–338.

Sinden, S.L. and Webb, R.E. (1974) Effect of environment on glycoalkaloid content of six potato varieties at 39 locations. *US Department of Agricultural Technology Bulletin* 1472.

Sinden, S.L., Goth, R.W. and O'Brien, M.J. (1973) Effect of potato glycoalkaloids on the growth of *A. solani* and their possible role as resistance factors in potatoes. *Phytopathology* 63, 303–307.

Sinden, S.L., Webb, R.E. and Sandford, L.L. (1978) Genetic potential for increasing ascorbic acid content in potatoes. *American Potato Journal* 55, 394–395.

Sinden, S.L., Sandford, L.L. and Webb, R.E. (1984) Genetic and environmental control of potato glycoalkaloids. *American Potato Journal* 61, 141–156.

Singh, K.B. (1969) Inheritance of specific gravity in potato tuber. *Indian Journal of Genetics and Plant Breeding* 29, 433–437.

Skrobacki, A., Halderson, J.L., Pavek, J.J. and Corsini, D.L. (1989) Determining potato tuber resistance to impact damage. *American Potato Journal* 66, 401–415.

Smith, M.A. (ed.) (1984) *Encyclopedia Americana*. Grolier Inc., Danbury, Connecticut, Vol. 22, p. 464.

Smith, O. (1967) Factors affecting specific gravity or dry matter of potatoes. In:

Talburt, W.F. and Smith, O. (eds) *Potato Processing*. Avi Publishing Co., Westport, Connecticut, pp. 69-88.

Smith, O. (1977) Flavour and odour of potatoes. In: Smith, O. (ed.) *Potatoes: Production, Storing and Processing*, 2nd edn. Avi Publishing Co., Westport, Connecticut, pp. 635-637.

Smittle, D.A., Thornton, C.L., Peterson, C.L. and Dean, B.B. (1974) Harvesting potatoes with minimum damage. *American Potato Journal* 51, 152-164.

Snyder, J.C. and Desborough, S.L. (1978) Protein, starch and non-protein nitrogen accumulation in high protein hybrids and low protein cultivars. *American Potato Journal* 55, 453-465.

Sowokinos, J. (1990) Stress induced alterations in carbohydrate metabolism. In: Vayda, V.E. and Park, W.D. (eds) *The Molecular and Cellular Biology of the Potato*. CAB International, Wallingford, pp. 137-157.

Stegemann, H., Francksen, H. and Macko, V. (1973) Potato proteins: genetic and physiological changes, evaluated by one- and two-dimensional PAA-gel techniques. *Zeitschrift fur Naturfurschung C* 28, 722-732.

Storey, R.M.J. and Davies, H.V. (1992) Tuber quality. In: Harris, P.M. (ed.) *The Potato Crop*, 2nd edn. Chapman and Hall, London, pp. 507-569.

Sturckow, B. and Lou, I. (1961) The effects of some *Solanum* alkaloid glycosides on the potato beetle. *Entemologia Experimenta et Applicata* 4, 133-142.

Swiezynski, K.M., Kocyk, B., Kuzminska, E. and Wojcik, R. (1977) Evaluation of tuber yield, starch content in the tuber and some other characters in potato breeding. *Genetica Polonica* 18, 1-13.

Talburt, W.F. and Smith, O. (1987) *Potato Processing*, 4th edn. Van Nostrand Reinhold, New York.

Umaerus, M. (1978) Report of survey of methods for screening susceptibility to mechanical tuber damage. *7th Triennial Conference of the European Association for Potato Research, Warsaw*, Poland, pp. 117-118.

Umaerus, M. and Olsson, K. (1974) Varietal differences in tyrosine and chlorogenic acid in relation to enzymic discoloration of potato tubers. *Potato Research* 17, 157-158.

Van Gelder, W.M.J. (1990) Chemistry, toxicology and occurrence of steroidal glycoalkaloids: potential contaminants of the potato. In: Rizk, A.F. (ed.) *Poisonous Plant Contamination of Edible Plants*. CRC Press, Boca Raton, USA, pp. 117-156.

Van Gelder, W.M.J. and Dellaert, L.M.W. (1988) Alkaloids in potatoes. *Prophyta* 42(9), 236-238.

Van Gelder, W.M.J., Vinke, J.H. and Scheffer, J.J.C. (1988) Steroidal glycoalkaloids in tubers and leaves of *Solanum* species used in potato breeding. *Euphytica* 39, 141-145.

Van Loon, C.D. (1981) The effect of water stress on potato growth, development and yield. *American Potato Journal* 58, 51-69.

Veilleux, R.E. and Lauer, F.I. (1981) Breeding behaviour of yield components and hollow heart in tetraploid-diploid vs. conventionally derived potato hybrids. *Euphytica* 30, 547-561.

Veilleux, R.E., Lauer, R.E. and Desborough, S.L. (1981) Breeding behaviour for tuber protein in *S. tuberosum* and Tuberosum-Phureja hybrids. *Euphytica* 30, 563-577.

Von Wettstein, D. (1967) Chloroplast structure and genetics. In: San Petro, A., Greer, F.A. and Army, T.J. (eds) *Harvesting the Sun*. Academic Press, New York, pp. 153-190.

Walker, J.R.L. (1977) Enzymic browning in foods, its chemistry and control. *Food Technology in New Zealand* 12, 19-25.

Warren, S. and Woodman, J.S. (1974) The texture of cooked potatoes: a review. *Journal of the Science of Food and Agriculture* 25, 129-138.

Wolf, M.J. and Duggar, B.M. (1946) Estimation and physiological role of solanine in the potato. *Journal of Agricultural Research* 73, 1-32.

Woodwards, L. and Jackson, M.T. (1985) The lack of enzymic browning in wild potato species, Series Longipedicellata, and their crossability with *S. tuberosum*. *Zeitschrift fur Planzenzuchtung* 94, 278-287.

Woolfe, J.A. (1987) *The Potato in the Human Diet*. Cambridge University Press, Cambridge.

Zitnak, A. and Johnston, G.R. (1970) Glycoalkaloid content of B5141-6 potatoes. *American Potato Journal* 47, 256-260.

V Inheritance of Resistance to Pests and Diseases

14 Inheritance of Resistance to Nematodes

M.S. PHILLIPS

Zoology Department, Scottish Crop Research Institute, Invergowrie, Dundee DD2 5DA, UK.

Introduction

Nematodes comprise a large group of worm-like animals, many of which are parasitic on plants. Plant-parasitic nematodes are usually small (0.2–10 mm in length) and inhabit the soil, from where they most frequently attack roots, tubers and occasionally aerial parts of plants. Species parasitic on plants are characterized by the possession of a mouth stylet or spear, which is used to penetrate plant cells and extract cell contents. The stylet is hollow in all parasitic species, except the Trichodoridae. Some species feed directly on the cytoplasm, whilst others modify plant cells to increase the supply of nutrients. The latter nematodes become sedentary and lose their vermiform shape; their reproductive rate is usually high and they include the most damaging species (*Globodera*, *Meloidogyne* and *Nacobus*). The harm caused to plants is usually direct but may be exacerbated by interactions with other pathogenic bacteria or fungi. Some species act as virus vectors.

The plant-parasitic nematodes that attack potatoes have been identified by Jensen *et al.* (1979), who lists 67 species associated with potatoes, though many of them are of little or no importance regarding crop production. Brodie (1984) and Evans and Trudgill (1992) identify the cyst nematodes, *Globodera rostochiensis* and *G. pallida*, as the most damaging species on a worldwide scale. Other significant species with an extensive distribution are the root-knot nematodes, *Meloidogyne* spp., the stubby root nematodes, *Trichodorus* and *Paratrichodorus* spp., the root lesion, *Pratylenchus* spp., and the potato rot nematodes, *Ditylenchus* spp. Other species that are of local importance are the false root-knot nematode, *Nacobuss aberrans*, and the sting nematode, *Belonolaimus longicaudatus*.

This relative importance of each nematode group is reflected in the numbers of papers published on these species between 1986 and 1991 (Table 14.1).

Table 14.1 The number of references to nematodes and potatoes together with inheritance studies 1986-1991 CAB Abstracts.

Species	Name	Number of references	
		Nematodes and potatoes	Inheritance and nematodes and potatoes
Globodera spp.	Cyst nematodes	379	42
Meloidogyne spp.	Root-knot nematodes	93	9
Ditylenchus spp.	Tuber-rot nematodes	41	0
Pratylenchus spp.	Root-lesion nematodes	71	0
Trichodorus spp. and Paratrichodorus spp.	Stubby root nematodes	24	0
Nacobbus aberrans	False root-knot nematodes	4	0
Belonolaimus	Sting nematode	0	0

However, since reported studies on the inheritance of resistance are restricted to *Globodera* spp. and *Meloidogyne* spp., further discussion in this chapter is confined to them. For the purposes of the discussion, the term resistant genotype describes a plant that restricts or prevents nematode multiplication and does not describe the expression of the disease. The terms tolerance and intolerant relate to the ability of a host genotype to withstand or recover from the damaging effects of a nematode attack and are independent of resistance (Trudgill, 1991).

Potato Cyst Nematodes

There are two species of potato cyst nematode (PCN), *Globodera rostochiensis* and *G. pallida*, which originate from the Andean region of South America (Evans and Trudgill, 1992), where they have coevolved with their preferred hosts (Stone, 1985). Both species spread to Europe, which in turn probably acted as a secondary centre of distribution to many other countries. PCN was first recorded in Germany in 1881 but was thought to be a strain of the sugarbeet cyst nematode, *Heterodera schachtii*. Wollenweber (1923) described these nematodes as *H. rostochiensis* and Stone (1973) described a second species, *H. pallida*. Later, the subgenus, *Globodera*, into which the round cyst nematodes were placed (Behrens, 1975), was elevated to generic rank, and thus potato cyst nematodes have assumed their current names. Both species are diploid outbreeders and thus show variation both within and between populations.

Pathotype scheme

Both species of PCN can be divided into pathogenic variants, designated pathotypes. With studies of variation in PCN populations and the discovery of different sources of resistance in different countries, the designation of pathotypes became confused. Kort *et al.* (1977) brought together European interests and published an 'international' pathotype scheme that sought to produce a unified approach. At the same time a new system for classifying 'races' or pathotypes was produced from South America (Canto-Saenz and de Scurrah, 1977). The two systems are very similar and are shown in Table 14.2.

There are limitations to the Kort *et al.* (1977) scheme in that the differentials used to identify pathotypes are of different kinds. To differentiate between some of the *G. rostochiensis* pathotypes, a differential clone with the major *H1* resistance gene identifies, qualitatively and genotypically, PCN populations that are avirulent to this gene, such as pathotypes Ro1 and Ro4. Likewise, within *G. pallida*, a clone with the *H2* resistance gene identifies the Pa1 populations, which are avirulent on genotypes with this gene. To distinguish between the other pathotypes, differential clones bred from polygenic sources of resistance are used and identification is based on whether or not

Table 14.2 Potato cyst nematode pathotype schemes proposed by Kort *et al.* (1977) (A) and Canto-Saenz and de Scurrah (1977) (B), where + signifies a virulent pathotype.

Differential clone	A	Ro1	Ro2	Ro3	Ro4	Ro5	Pa1	Pa2	Pa3			
	B	R1A	R2A	R3A	R1B		P1A	P4A	P5A	P1B	P3A	P2A
S. tuberosum subsp. *tuberosum*		±	+	+	+	+	+	+	+	+	+	+
S. tuberosum subsp. *andigena* CPC 1673		–	+	+	–	+	+	+	+	+	–	+
S. kurtzianum 60.21.19		–	–	–	+	+	+	–	+	+	–	–
S. vernei GLKS 58.1642/4		–	–	–	–	+	+	+	+	+	+	–
S. vernei 62.33.3		–	–	–	–	+	–	–	+	+	–	–
S. vernei 65.346/19		–	–	–	–	–	+	+	+	+	–	–
S multidissectum		+	+	+	+	+	–	+	+	+	–	–

a population is able to multiply on these clones. Phillips *et al.* (1979a) and Phillips (1984) have shown that multiplication rates of PCN on clones with partial or quantitative resistance are affected by environmental factors. This being the case, the use of absolute multiplication rates to determine pathotypes becomes arbitrary and inappropriate. Trudgill (1985) argues that pathotypes such as Ro2, Ro3, Ro5, Pa2 and Pa3 are not pure pathotypes but are heterogeneous for virulence genes and it has been proposed that these should be called virulence groups (Anon., 1985) and that Ro1 and Ro4 be amalgamated into one pathotype.

Sources of resistance

Ellenby (1948, 1952, 1954) screened the Commonwealth Potato Collection (CPC) with *H. rostochiensis* (now *G. rostochiensis* Ro1) and identified resistance in the wild, diploid species *Solanum vernei* (with accession numbers CPC 105, 2413 and 2414), in a single triploid clone (CPC 1647), described as having affinities to *S. tuberosum* subsp. *andigena*, and in five *S. tuberosum* subsp. *andigena* accessions (CPC 1595, 1673, 1685, 1690 and 1692).

Mai and Peterson (1952), Goffart and Ross (1954) and Rothacker (1958) also reported on resistance to PCN from *S. vernei*. In 1960 Huijsman reported two resistance genes from *S. kurtzianum* and Dunnett (1960) one gene from *S. multidissectum*. Ross (1962) examined resistance from *S. spegazzinii* and identified two genes conferring resistance to *G. rostochiensis*. De Scurrah *et al.* (1973b) also reported on resistance from this source as well as from *S. santa-rosae*. Howard *et al.* (1970) and Fuller and Howard (1974), working with *S. tuberosum* subsp. *andigena* CPC 2802, identified a second gene (*H3*), from this source, conferring resistance to *G. pallida*. Uhrig and Wenzel (1981) also identified *S. gourlayi* as a source of resistance to this species.

In addition, several other workers have screened potato collections to identify other sources of resistance to both *G. rostochiensis* and *G. pallida*: Rothacker *et al.* (1966), Ross and Huijsman (1969), Deshmukh and Weischer (1970), Van Soest *et al.* (1983), Stelter (1987), Chavez *et al.* (1988a, b), Dellaert *et al.* (1988) and Turner (1989). Table 14.3 indicates the most commonly used sources of resistance in breeding programmes.

Inheritance of major gene resistance to G. rostochiensis

Gene *H1*

The first report of the mode of inheritance of a PCN resistance gene was made by Toxopeus and Huijsman (1953), who studied the resistance to what is now recognized as *G. rostochiensis* Ro1 from *S. tuberosum* subsp. *andigena* CPC 1673 and CPC 1685. They concluded that the gene, which they termed *H*, was a single dominant gene inherited tetrasomically. These results were confirmed and clarified by Huijsman (1955). This gene, which was later designated *H1*

Table 14.3 The main sources of resistance to potato cyst nematodes.

Source	Genes	Author
S. tuberosum subsp. andigena CPC 1673	H1	Ellenby (1952); Toxopeus and Huijsman (1953); Dunnett (1963)
S. tuberosum subsp. andigena CPC 2802	Polygenic	Dale and Phillips (1982)
S. multidissectum	H2	Dunnett (1963)
S. kurtzianum	A (K1), B	Huijsman (1960); Kort et al. (1977)
S. spegazzinii	Fa, Fb	Ross (1962)
S. vernei	Polygenic	Ellenby (1948); Goffart and Ross (1954); Rothacker (1958); Dale and Phillips (1982)

on the discovery of other resistance genes, confers resistance to *G. rostochiensis* Ro1 and Ro4 (Kort *et al.* 1977). Cole and Howard (1957) later described an identical gene from CPC 1670.

Studies on the inheritance of virulence to the *H1* gene in the nematode have shown that there is a gene-for-gene relationship, the virulence gene being recessive. This was originally suggested by Parrott (1981), as a result of findings from experiments conducted on the basis of mass matings of virulent and avirulent nematode populations. Janssen *et al.* (1991) have confirmed this, using inbred lines of nematodes and controlled single matings.

Recently Gebhardt *et al.* (1993) have reported identifying RFLP markers closely linked to the gene and have mapped the *H1* gene to a distal position on chromosome 5. A diploid population, derived from a cross between a dihaploid line Amaryl H5 carrying *H1* and a susceptible $2n$ *S. phureja*, was used for the mapping. Of the hybrid lines, 44 were resistant and 47 susceptible. Linkage was detected with six markers on chromosome 5, with one marker, CP113, found to be linked with 0% recombination. Fifty-three tetraploid potato cultivars, 20 of which have the *H1* gene, were screened for the presence of the RFLP allele, but it was not detected. Pineda *et al.* (1993) have also identified seven RFLP markers from chromosome 5 linked to the *H1* gene and have mapped the gene to a similar position (see Chapter 10, Figure 10.3). They too used a dihaploid population, but initially used DNA bulked from either resistant or susceptible lines. Rather than test a range of cultivars to examine linkage in tetraploids, Pineda *et al.* examined a progeny from a tetraploid cross and found that four of the markers cosegragated with the *H1* gene. The mapping of this gene is the first step in developing a molecular probe for it, which could provide a rapid screen, and which could also lead to isolation of the gene and consequently an understanding of the biochemical basis of this resistance mechanism.

Gene *K1*

Huijsman (1960) reported on studies of resistance from the diploid species *S. kurtzianum*, which he had intercrossed with cultivated diploid *S. tuberosum* genotypes. From these progenies, he selected resistant seedlings and backcrossed them to susceptible *S. tuberosum* to provide material for inheritance studies. With the test population of nematodes used (*G. rostochiensis* Ro1), he found, on examination of the root ball, that the progeny were either very resistant, with less than five cysts per plant, or were susceptible, with many cysts. He concluded that the observed segregations could be explained by two dominant genes.

A clone (KTT 60.21.19) with one resistance gene (*K1*) was derived from this material and incorporated into the pathotype schemes mentioned above. This clone is used to differentiate Ro1 and Ro2 (avirulent) from Ro3, Ro4 and Ro5 (virulent). Examination of the data from Kort *et al.* (1977) indicates that this clone confers some level of resistance to pathotypes (Ro3–Ro5), though not to any of their *G. pallida* pathotypes. The former observation, however, can be explained in terms of Ro3–Ro5 populations having differing and higher proportions of the virulence gene(s) necessary to overcome the resistance.

Genes *Fa* and *Fb*

Ross (1962) was the first to report resistance to *G. rostochiensis* in *S. spegazzinii* and found that it was resistant to several populations of PCN, some of which were *G. rostochiensis* and others *G. pallida*. Examining F_2 to F_4 families, he postulated that there were two equally effective, independent and dominant genes, which he called *Fa* and *Fb*. He considered that *Fa* was identical with the *H1* gene, as it gave resistance to the same populations, and that the *Fb* gene gave resistance to all the other populations.

Further testing by Ross and Huijsman (1969) confirmed that resistance derived from *S. spegazzinii* was effective against both species. De Scurrah *et al.* (1973a), testing *S. spegazzinii* with three South American populations of *G. pallida*, found resistance but reported no inheritance studies. Momeni *et al.* (1969) studied *S. spegazzinii* × *S. tuberosum* crosses, and tested them with *G. rostochiensis* Ro1. Segregation ratios in the F_2 suggested that the resistance found was the result of a single dominant gene.

Barone *et al.* (1990) used RFLPs to map resistance genes in a progeny derived from *S. spegazzinii*. They did so for a gene that confers resistance to *G. rostochiensis* Ro1 and Ro5. The segregation of resistant and susceptible plants in the progeny they obtained was again compatible with there being one dominant resistance gene. They located the gene to the potato chromosome 7, which is homoeologous with the tomato chromosome 7 (Gebhardt *et al.*, 1991). Based on the resistance expressed by this material in relation to the *G. rostochiensis* pathotypes, they consider that it is most likely to be the *Fb* gene as described by Ross (1962).

Inheritance of major gene resistance to **G. pallida**

Gene *H2*

Following the discovery of populations that were virulent to the *H1* gene, Dunnett (1957) tested the wild diploid *S. multidissectum* for resistance and found a resistance gene effective against some of these virulent populations. This gene was later designated *H2* (Dunnett, 1963). Dunnett (1960) examined an F_1 (*S. tuberosum* × *S. multidissectum*) and two backcross generations and used populations of PCN which are now known to have been *G. pallida*, one of which was pathotype Pa1. He concluded that the wild species possessed a major dominant resistance gene to the Pa1 population, but he could not be sure whether the original breeding line contained one or two independently acting genes. He also concluded that there were supplementary and probably polygenic resistance genes present as well as the major gene(s), because some breeding lines, which lacked a major gene, did give a moderate level of resistance to one of his test populations.

As with virulence to the *H1* gene in *G. rostochiensis*, Parrott (1981) also suggested that there was a complementary recessive virulence gene in *G. pallida* to the *H2* gene.

Gene *H3*

Howard et al. (1970) attempted to identify sources of resistance from within cultivated Andigena, as this was likely to lead to more rapid progress in incorporating resistance into new cultivars than would be achieved by using resistance genes from wild diploid species. The first screening of the CPC had been done only with *G. rostochiensis* Ro1 and they saw no reason why Andigena should only possess resistance to that pathotype. They identified five lines for study from CPC 2775 selfed, 2802 selfed, 2805 selfed, 2774 × 2775 and 2793 × 2775. These lines were tested with Pa1 and Pa2/3 populations. One family from the 2774 × 2775 cross segregated 15 resistant to 24 susceptible when tested with *G. pallida* Pa2/3. Eleven of the resistant lines were then tested with *G. rostochiensis* Ro1 and segregated six resistant and five susceptible. The authors suggested, from a small sample, that two genes were probably involved. One was identical with or similar to the *H1* gene; the second (which gave resistance to *G. pallida*) they designated *H3*. Another family, derived from CPC 2802, segregated eight resistant and four susceptible when tested with the Pa2/3 population. Testing with Ro1 showed there was no resistance and it was concluded that this accession had only the *H3* gene.

Fuller and Howard (1974) reported on further studies where lines with this resistance were crossed with parents having the *H1* gene alone or in combination with the *H2* gene. The data they presented indicated that many of the progeny (c. 30%) were very resistant (< 4% susceptible) but that c. 50% were more than 20% susceptible, with no clear segregation between 'resistant' and susceptible. The authors suggested that this was due to variations in penetrance of the gene, which is not always expressed as completely as the *H1* gene.

Franco and Evans (1978) compared the performance of 44 South American and nine European populations of PCN, on clones possessing the *H1*, *H2* or *H3* genes. They found that the *H3* gene was effective against the European *G. pallida* populations but that its resistance to many South American *G. pallida* populations was only partial. They argued that it was unlikely that the populations they used were mixed, but rather that the designated *H3* gene may involve several genes.

The experience of using this resistance in breeding programmes in the UK also suggested that the inheritance of this resistance was not simple and led Dale and Phillips (1982) to compare the distribution of resistance in progenies derived from *S. vernei* and from *S. tuberosum* subsp. *andigena* CPC 2802 and CPC 1675 (H1). Their results showed that the distributions from ex *S. vernei* and CPC 2802 were continuous and different from the discrete partitioning shown by the progenies derived from parents with the *H1* gene. They concluded that the resistance from CPC 2802 is complex and probably polygenic.

Phillips and Dale (1982) conducted progeny tests on crosses between susceptible parents and those derived from CPC 2802 (in a North Carolina 2-mating scheme) and intercrosses between the latter (in a half-diallel). The results from the susceptible × resistant experiment indicated that GCA effects accounted for all the observed genetic variation. Partitioning of these effects showed that both sets of parents contributed, but the resistant one more so. In the half-diallel (resistant × resistant), whilst GCA effects were observed, there were also significant and equally important SCA effects, giving further evidence of the complexity of this source of resistance and its inheritance.

Inheritance of quantitative resistance

Solanum vernei

An important source of resistance to both species of PCN comes from the wild diploid species *S. vernei*. This resistance was first reported by Ellenby (1948) and subsequently by Mai and Peterson (1952). Goffart and Ross (1954) screened 21 wild species of *Solanum*, including *S. vernei*, and produced selfs and F_1 generations. They assessed resistance to *G. rostochiensis* in the *S. vernei* progenies and observed a wide range of responses, from very resistant to susceptible. They considered that there was some dominance but that the resistance was inherited polygenically.

Rothacker (1958) studied intercrosses between *S. vernei* and other *Solanum* species and likewise concluded that the resistance was polygenic.

Plaisted et al. (1962) conducted a series of experiments on the resistance to *G. rostochiensis* Ro1 in diploid plants derived from intercrosses between different *S. vernei* lines to produce a genetic model of resistance. The intercrosses were made between 'susceptible' and resistant lines, although intercrosses between the former gave progenies with a range of resistances. The progeny from the susceptible/resistant crosses also gave a range of resistances in the progeny, though skewed towards resistance. The authors decided to delineate susceptible plants from resistant ones on the basis of the number of cysts observed on the root ball of each plant (susceptible > 20 cysts) and then to classify the resistant plants into four groups within the range 0 to 20 cysts. On this basis, they suggested an explanation of the inheritance by proposing two loci, *B* and *C*, where *B- C-* were most resistant, *B- cc* and *bb C-* were less resistant and *bb cc* were susceptible. This explanation did not adequately explain the results from one progeny and they suggested that a third locus, *A*, might be involved which modified the expression of resistance. The authors acknowledged some of the difficulties in assessing resistance and accepted that a more complex model could be nearer the truth. It seems likely that this is the case.

It is now known that estimating levels of quantitative resistance is complicated by genotype × environment interactions (Phillips et al., 1979b). Thus the use of a fixed number of cysts or an absolute multiplication rate to delineate categories is not desirable (Phillips, 1984), because absolute values of cyst numbers can vary depending on the conditions under which resistance tests are conducted. It is necessary, therefore, to interpret these results, which were gathered in more than one year, with caution.

De Scurrah et al. (1973b) studied intercrosses between tetraploid clones derived from S. vernei and S. tuberosum subsp. andigena clones with the *H1* gene and then tested the progeny with G. rostochiensis Ro1. The results were interpreted as again indicating the presence of two loci. The results from a progeny of one ex S. vernei line could be explained in terms of one of the genes segregating at the same locus as the *H1* gene. These results are presented as ratios of resistant/susceptible plants, again based on there being a fixed number of cysts defining the division (< 5 cysts per root ball being resistant). There was, however, no indication given of the range of cyst numbers observed in the susceptible category and it is not possible to tell how discrete the segregations were.

Ross (1966) summarizing the sources of major resistance genes, reported that there were likely to be four major dominant genes from S. vernei inherited in a Mendelian way. More recent studies with both G. rostochiensis and G. pallida support the view expressed in earlier reports that resistance from S. vernei is polygenic. Dale and Phillips (1982) tested tetraploid progenies of parents with this source of resistance with a G. pallida population and showed a continuous distribution, as had been observed by Plaisted et al. (1962). Phillips et al. (1979b) examined the general and specific combining abilities of partially resistant parents derived from S. vernei and found that GCA and SCA were equally important. The presence of significant SCA is not compatible with there being a limited number of major dominant resistance genes, but the presence of significant GCA effects implies that there is some consistency in the breeding behaviour of the parents. Phillips and Dale (1982) found that GCA effects dominated when intercrossing ex S. vernei parents with either susceptible clones or material derived from S. tuberosum subsp. andigena CPC 2802.

Other *Solanum* species

Uhrig and Wenzel (1981) reported resistance to G. pallida populations in S. gourlayi. They interpreted their data as indicating a clearer segregation between resistant and susceptible genotypes than is experienced in progenies from crosses with S. vernei and therefore a potentially simpler genetic basis. Their progeny from S. gourlayi had a lower average cyst count than the S. vernei progeny they used as a comparison (i.e. they were more resistant). Data consisting of nematode counts usually require some transformation prior to analysis, as the variance tends to be correlated with the mean. They examined and presented untransformed data and consequently may be mistaken in their conclusions.

Dellaert et al. (1988) studied the inheritance of resistance in several species, including S. vernei, S. spegazzinii, S. leptopytes, S. brevicaule, S. oplocense, S. sparsipilum and S. sucrense, in relation to both species of PCN. They found accessions of these species that gave resistance across a range of *G. pallida* populations. Results with F_1 populations, arising from resistant × susceptible crosses, showed that resistant and susceptible plants could not be clearly separated. The data generally forming a skewed distribution, with a relatively larger number of plants having few cysts. The authors attempted to group their data in relation to the parental phenotypes and show that the distribution varies with the species and the accession of that species. On this basis, they suggested possible explanations of the number of genes that might be involved, which ranged from three or four genes giving incomplete resistance, together with minor genes, to one major gene, with minor genes.

Resistance within *S. tuberosum* and Neotuberosum

Gemmell (1943) was the first person to report reduced susceptibility in cultivated *Solanum* spp. He observed that fewer and smaller cysts were produced on cvs. Doon Star and Epicure than on cvs. Golden Wonder and Majestic. Ellenby (1946) produced similar results, although he more cautiously proposed that, whilst differences between cultivars could be due to resistance, they could also have been the result of differences in seasonal behaviour. Dale and Phillips (1985) tested a number of 'susceptible' cultivars for resistance to *G. pallida* and found significant differences. They also showed that, in crosses between such susceptible cultivars and parents with quantitative resistance, there were significant differences between the former in their contribution to the overall levels of resistance within the individual progenies produced.

Several attempts have been made to produce Neotuberosum populations from primitive *S. tuberosum* Group Andigena cultivars. One such population produced in Britain was tested for resistance to *G. rostochiensis* and *G. pallida* by Glendinning and Phillips (1988). They showed that there were clones within the population that had higher levels of resistance than present-day cultivars, although few had very high levels of resistance. Intercrosses between Neotuberosum clones showed significant GCA and SCA effects, whereas in hybrids with cultivated *S. tuberosum* GCA effects predominated (Dale and Phillips, 1982). Thus this material presents a source of heritable resistance, albeit of a complex nature.

Tolerance of potatoes to cyst nematode attack

Host genotypes that are resistant to nematodes are those that inhibit or reduce nematode multiplication rates. The roots of genotypes that are resistant to PCN are nevertheless invaded by juvenile nematodes and can suffer damage, with large yield losses being sustained when they are grown in infested soil. Genotypes differ, however, in the degree of yield loss sustained in comparable situations (Dale *et al.*, 1988), and these differences are referred to as differences in tolerance. Resistance and tolerance are considered to be independent attributes (Trudgill, 1991), although they can interact. Resistance may contribute to tolerance, especially if it decreases parasitism, and tolerance can affect nematode multiplication, especially at high population densities (Elston *et al.*, 1991). In a resistance breeding programme it is desirable to combine high levels of resistance with high levels of tolerance to ensure both good control of the pest and maximization of the yield. The higher priority has been given to the resistance breeding, and work on tolerance has been limited to identifying methods of screening (Dale *et al.*, 1988; Phillips *et al.*, 1988) or studies of the mechanisms of tolerance, but not investigations into its inheritance. A review of mechanisms of tolerance and factors that affect it is given by Evans and Haydock (1990).

Meloidogyne

Species of root-knot nematode (RKN) that parasitize potatoes are *Meloidogyne arenaria*, *M. incognita*, *M. hapla*, *M. javanica*, *M. thamesi* and *M. chitwoodii*. Like PCN, they are sedentary endoparasites, but they differ in that they are parthenogenic, have a wide host range and can produce several generations in a growing season. Although sources of resistance are known in potato, they have not been widely utilized, perhaps a reflection of their importance relative to PCN.

Nirula *et al.* (1967, 1969) screened a large number of tuber-bearing *Solanum* species for resistance to *M. incognita* and found a high degree of resistance in *S. bulbocastanum*, *S. gandarillasii*, *S. lignicaule*, *S. ajanhuiri*, *S. tuberosum* subsp. *andigena*, *S. spegazzinii* and *S. vernei*. Jatala and Rowe (1976) also screened accessions from 62 tuber-bearing *Solanum* species and found resistance to *M. incognita* in *S. capsicibaccatum*, *S. megistacrolobum*, *S. microdontum*, *S. acroscopicum*, *S. gourlayi* and *S. sparsipilum*. Parallel studies of over 3000 cultivated clones from the International Potato Center (CIP) collection failed to reveal any adequate resistance. Brodie and Plaisted (1976) tested 1473 *S. tuberosum* subsp. *andigena* clones and found *c.* 15% had high levels of resistance.

Gomez *et al.* (1983) investigated the inheritance of resistance to *M. incognita*, *M. javanica* and *M. arenaria* in *S. sparsipilum* by making reciprocal crosses between resistant × resistant and resistant × susceptible diploid clones. In addition, a tetraploid resistant parent was selfed or pollinated with bulk pollen from Neotuberosum clones. Resistance was assessed by counting the numbers of galls or egg masses produced by the nematodes. They concluded that clones with resistance to *M. arenaria* are also resistant to *M. incognita* and *M. javanica* and that reciprocal cross differences were only apparent with *M. arenaria*. They considered that susceptible clones of *S. sparsipilum* possessed recessive or epistatic genes which affected the pattern of segregation for resistance. The results they obtained tended to show continuous ranges of variation. In most crosses there was a predominance of resistant genotypes, whilst in one the majority were intermediate and in another there were few intermediate scores, some resistant and others clearly susceptible. The interpretation of the results in terms of inheritance depends on the definition of the minimum level of gall or egg mass production that constitutes resistance. The authors make this point and conclude that one can only speculate about the number of genes involved. It is evident, though, that more than one gene is involved and they suggest that 'not many loci' are needed to explain the segregation patterns observed.

Mendoza and Jatala (1985) also studied resistance to *M. incognita* in *S. sparsipilum*. Initial experiments indicated that, in crosses with susceptible *S. phureja*, reciprocal differences occurred. From the initial studies, 42 hybrid families were produced between three resistance and four susceptible genotypes in a diallel. Analysis of the data showed that there were significant GCA effects, as well as smaller and significant reciprocal effects. Narrow-sense heritability was calculated as $h^2 = 0.78$. The reciprocal effects indicated either a maternal effect of *S. sparsipilum* or interaction between cytoplasmic elements and nuclear genes. The data from the susceptible × resistant crosses (resistant × susceptible were ignored due to the maternal effects) were used to explore possible genetic models. The authors found that the segregation ratios were not significantly different from a model with three major complementary genes which were present in the heterozygous state in the resistant parent.

Conclusions on the Studies of the Inheritance of Resistance to Nematodes

Ross (1986) summarizes the situation well when he says that finding major genes for resistance is not a problem but that, when found, they tend to be accompanied by minor gene complexes. During several backcrosses, the gene complex is split up and the residual resistance that remains is always lower than in the original wild source. He then adds significantly 'except for the *H1* gene', although the same is probably true for the *H2* gene. This highlights one of the main problems in studying resistance to PCN. Our ability to study the *H1* gene in Europe and the USA is facilitated by the Ro1 pathotype of *G. rostochiensis*. This pathotype is atypical because it is totally avirulent to the *H1* gene and therefore can be used to identify the gene unequivocally. The same is true for pathotype Pa1 of *G. pallida*, in that it is avirulent to the *H2* gene. In South America, where PCN originates, neither of these pathotypes has been detected (Canto-Saenz and de Scurrah, 1977).

In areas where PCN is indigenous it survives on mixed populations of *Solanum* species, with which it has coevolved, and it seems likely that in the evolutionary process the hosts were selected for genes that inhibit the nematodes whilst at the same time the nematodes where selected for genes that overcome the resistance genes. It also seems probable that in host and parasite, which are both outbreeders, not all individual genotypes in a population will necessarily possess all the resistance or virulence genes possible. Thus it should be feasible to select from populations individuals that lack particular resistance or virulence genes. With regard to PCN, it is probable that this has happened, inadvertently, on its introduction into Europe and that Ro1 and Pa1 represent very limited samples of the parent gene pool. Other introductions were possibly larger and contained a broader range of virulence genes, and, indeed, are heterogeneous in this respect, as suggested by Trudgill (1985). This view is endorsed by studies, such as Turner (1990), where it has been shown that populations can be selected for increased virulence, or by investigations of populations which show quantitative rather than qualitative differences (Phillips and Trudgill, 1983; Phillips *et al.*, 1989a, b, c).

Conducting inheritance studies of individual resistance genes is fraught with problems if the test populations consist of unknown mixtures of virulence genes when what is required are avirulent populations. These problems are often confounded by the way that resistance is defined by nematologists. Resistance is not assessed on symptomatology or epidemiology but on the effect of the host on the multiplication rate of the nematode, which in turn can be affected by population density and other environmental factors. Thus, the interpretation of results can be difficult, and the use of qualitative judgements in a quantitative situation must be treated with caution.

References

Anon. (1985) Conclusions of the EPPO Workshop on Cyst Nematodes (Münster, Federal Republic of Germany, 26-28 June 1984). *EPPO Bulletin* 15, 121-122.

Barone, A., Ritter, E., Schachtshabel, U., Debener, T., Salamini, F. and Gebhardt, C. (1990) Localization by restriction fragment length polymorphism mapping in potato of a major dominant gene conferring resistance to the potato cyst nematode *Globodera rostochiensis*. *Molecular and General Genetics* 224, 177-182.

Behrens, E. (1975) *Globodera* Skarbilovich, 1959, eine selbständige Gattung in der Unterfamilie Heteroderinae Skarbilovish, 1947 (Nematoda: Heteroderidae). 1. Vortragstagung zu aktuellen Problemen der Phyto-nematoderidae am 29. 5. 1975 in Rostock, p. 12.

Brodie, B.B. (1984) Nematode parasites of potato. In: Nickle, W.R. (ed.) *Plant and Insect Nematodes*. Marcel Dekker, New York, pp. 167-212.

Brodie, B.B. and Plaisted R.L. (1976) Resistance to root-knot nematodes in *Solanum tuberosum* ssp. *andigena*. *Journal of Nematology* 8, 280-281.

Canto-Saenz, M. and de Scurrah, M.M. (1977) Races of the potato cyst nematode in the Andean region and a new system of classification. *Nematologica* 23, 340-349.

Chavez, R., Jackson, M.T., Schmiedische, P.E. and Franco, J. (1988a) The importance of wild potato species resistant to the potato cyst nematode, *Globodera pallida*, pathotypes P4A and P5A, in potato breeding. 1. Resistance studies. *Euphytica* 37, 9-14.

Chavez, R., Jackson, M.T., Schmiedische, P.E. and Franco, J. (1988b) The importance of wild potato species resistant to the potato cyst nematode, *Globodera pallida*, pathotypes P4A and P5A, in potato breeding. 2. The crossability of resistant species. *Euphytica* 37, 15-22.

Cole, C.S. and Howard, H.W. (1957) The genetics of resistance to potato-root eelworm of *Solanum tuberosum* subsp. *andigena*, clone C.P.C. 1690. *Euphytica* 6, 242-246.

Dale, M.F.B. and Phillips, M.S. (1982) An investigation of resistance to the white potato cyst nematode. *Journal of Agricultural Science* 99, 325-328.

Dale, M.F.B. and Phillips, M.S. (1985) Variation for the degree of susceptibility to the potato cyst-nematode (*Globodera pallida*) within *Solanum tuberosum* ssp. *tuberosum*. *Potato Research* 28, 55–64.

Dale, M.F.B., Phillips, M.S., Ayres, R.M., Hancock, M., Holliday, M., Mackay, G.R. and Tones, S.J. (1988) The assessment of the tolerance of partially resistant potato clones to damage by the potato cyst nematode *Globodera pallida* at different sites and in different years. *Annals of Applied Biology* 113, 79–88.

Dellaert, L.M.W., Vinke, H. and Meyer, K. (1988) The inheritance of resistance to the potato cyst nematode *G. pallida* Pa3 in wild *Solanum* species with broad spectrum resistance. *Euphytica* S, 105–116.

De Scurrah, M.M., Plaisted, R.L. and Harrison, M.L. (1973a) Resistance to the potato nematode *Heterodera rostochiensis* Woll. in clones derived from *Solanum vernei*. *American Potato Journal* 50, 9–18.

De Scurrah, M.M., Mai, W.F. and Plaisted, R.L. (1973b) More about the potato nematode, *Heterodera rostochiensis* Woll. in Peru. *American Potato Journal* 50, 58–61.

Deshmukh, M.G. and Weischer, B. (1970) Resistance in wild species of potato to populations of *Heterodera rostochiensis* Woll. from West Germany. *Potato Research* 13, 129–138.

Dunnett, J.M. (1957) Variation in pathogenicity in the potato root eelworm (*Heterodera rostochiensis* Woll.). *Euphytica* 6, 77–89.

Dunnett, J.M. (1960) Inheritance of resistance to potato root eelworm in a breeding line stemming from *Solanum multidissectum* Hawkes. *Report of the Scottish Plant Breeding Station, 1960*, pp. 39–46.

Dunnett, (1963) *Report of the Scottish Plant Breeding Station, 1962*, pp. 19–21.

Ellenby, C. (1946) The influence of potato variety on the cyst of the potato root eelworm, *Heterodera rostochiensis* Woll. *Annals of Applied Biology* 33, 433–446.

Ellenby, C. (1948) Resistance to the potato root eelworm. *Nature* 162, 704.

Ellenby, C. (1952) Resistance to the potato root eelworm, *Heterodera rostochiensis* Wollenweber. *Nature* 170, 1016.

Ellenby, C. (1954) Tuber forming species and varieties of the genus *Solanum* tested for resistance to the potato root eelworm *Heterodera rostochiensis* Wollenweber. *Euphytica* 3, 195–202.

Elston, D.A., Phillips, M.S. and Trudgill, D.L. (1991) The relationship between initial population density of potato cyst nematode *Globodera pallida* and the yield of partially resistant potatoes. *Revue de Nématologie* 14, 221–229.

Evans, K. and Haydock, P.P.J. (1990) A review of tolerance by potato plants of cyst nematode attack with consideration of what factors may confer tolerance and methods of assaying tolerance and improving it in crops. *Annals of Applied Biology* 117, 702–740.

Evans, K. and Trudgill, D.L. (1992) Pest aspects of potato production. Part 1. The nematode pests of potatoes. In: Harris, P. (ed.) *The Potato Crop,* 2nd edn. Chapman and Hall, London, pp. 438–475.

Franco, J. and Evans, K. (1978) Multiplication of some South American and European populations of potato cyst nematodes on potatoes possessing the resistance genes H1, H2 and H3. *Plant Pathology* 27, 1–6.

Fuller, J.M. and Howard, H.W. (1974) Breeding for resistance to the white potato cyst-nematode *Heterodera pallida*. *Annals of Applied Biology* 77, 121–128.

Gebhardt, C., Ritter, E., Barone, A., Debener, T., Walkemeier, B., Schachtschabel, U., Kaufmann, H., Thompson, R.D., Bonierbale, M.W., Ganal, M.W., Tanksley, S.D. and Salamini, F. (1991) RFLP maps of potato and their alignment with the homoeologous tomato genome. *Theoretical and Applied Genetics* 83, 49–57.

Gebhardt, C., Mugniery, D., Ritter, E., Salamini, F. and Bonnel, E. (1993) Identification of RFLP markers closely linked to the H1 gene conferring resistance to *Globodera rostochiensis* in potato. *Theoretical and Applied Genetics* 85, 541–544.

Gemmell, A.R. (1943) The resistance of potato varieties to *Heterodera schachtii* Schmidt, the potato-root eelworm. *Annals of Applied Biology* 30, 67–70.

Glendinning, D.R. and Phillips, M.S. (1988) Resistance to potato cyst nematodes *Globodera rostochiensis* and *G. pallida* in Neo-tuberosum potatoes. *Potato Research* 31, 477–483.

Goffart, H. and Ross, H. (1954) Untersuchungen zur Frage der Resistenz von Wildarten der Kartoffel gegen der Kartoffelnematoden (*Heterodera rostochiensis* Woll.). *Der Züchter* 24, 193–201.

Gomez, P.L., Plaisted R.L and Brodie, B.B. (1983) Inheritance of the resistance of *Meloidogyne incognita*, *M. javanica* and *M. arenaria* in potatoes. *American Potato Journal* 60, 339–351.

Howard, H.J., Cole, C.S. and Fuller, J.M. (1970) Further sources of resistance to *Heterodera rostochiensis* Woll. in the Andigena potato. *Euphytica* 19, 210–216.

Huijsman, C.A. (1955) Breeding for resistance to the potato root eelworm. *Euphytica* 4, 133–140.

Huijsman, C.A. (1960) Some data on the resistance against the potato root-eelworm (*Heterodera rostochiensis* W.) in *Solanum kurtzianum*. *Euphytica* 9, 185–190.

Janssen, R., Bakker, J. and Gommers, F.J. (1991) Mendelian proof for a gene-for-gene relationship between virulence of *Globodera rostochiensis* and the H1 resistance gene in *Solanum tuberosum* ssp. *andigena* CPC 1673. *Revue de Nématologie* 14, 213–219.

Jatala, P. and Rowe, P.R. (1976) Reaction of 62 tuber-bearing *Solanum* species to the root-knot nematode, *Meloidogyne incognita acrita*. *Journal of Nematology* 8, 290.

Jensen, H.J., Armstrong, J. and Jatala, P. (1979) *Annotated Bibliography of Nematode Pests of Potatoes*. International Potato Center, Lima, Peru.

Kort, J., Ross, H., Rumpenhorst, H.J. and Stone, A.R. (1977) An international scheme to identifying and classifying pathotypes of potato cyst-nematodes *Globodera rostochiensis* and *G. pallida*. *Nematologica* 23, 333–339.

Mai, W.F. and Peterson, L.C. (1952) Resistance of *Solanum ballsi* and *Solanum sucrense* to the Golden Nematode, *Heterodera rostochiensis* Wollenweber. *Science* 116, 224–225.

Mendoza, H.A. and Jatala, P. (1985) Breeding potatoes for resistance to the root-knot nematode *Meloidogyne* species. In: Sasser, J.N. and Carter, C.C. (eds) *An Advanced Treatise on* Meloidogyne. *Volume 1: Biology and Control*. North Carolina State University Graphics, Raleigh, North Carolina, USA.

Momeni D.A., Plaisted, R.L., Peterson, L.C. and Harrison M.B. (1969) The inheritance of resistance to the golden nematode (*Heterodera rostochiensis*) in *Solanum famatinae* and *S. neohawksii*. *American Potato Journal* 46, 128–131.

Nirula, K.K., Nayar, N.K., Bassi, K.K. and Singh, G. (1967) Reaction of tuber-bearing *Solanum* species to root knot nematode, *Meloidogyne incognita*. *American Potato Journal* 44, 66–69.

Nirula, K.K., Khushu, C.L. and Raj, B.T. (1969) Resistance in tuber-bearing *Solanum* species to root knot nematode, *Meloidogyne incognita*. *American Potato Journal* 46, 251–253.

Parrott, D.M. (1981) Evidence for a gene-for-gene relationship between resistance gene H1 from *Solanum tuberosum* ssp. *andigena* and a gene in *Globodera rostochiensis*, and between H2 from *S. multidissectum* and a gene in *G. pallida. Nematologica* 27, 372-382.

Phillips, M.S. (1984) The effect of initial population density on the reproduction of *G. pallida* on partially resistant potato clones derived from *Solanum vernei. Nematologica* 30, 57-65.

Phillips, M.S. and Dale M.F.B. (1982) Assessing potato seedling progenies for resistance to the white potato cyst nematode. *Journal of Agricultural Science* 99, 67-70.

Phillips, M.S. and Trudgill, D.L. (1983) Variations in the ability of *Globodera pallida* to produce females on potato clones bred from *Solanum vernei* or *S. tuberosum* ssp. *andigena* CPC 2802. *Nematologica* 29, 217-226.

Phillips, M.S., Forrest, J.M.S. and Hayter, A.M. (1979a) Genotype × environment interaction for resistance to the white potato cyst nematode (*Globodera pallida*, pathotype E) in *Solanum vernei* × *S. tuberosum* hybrids. *Euphytica* 28, 515-519.

Phillips, M.S., Wilson, L.A. and Forrest, J.M.S. (1979b) General and specific combining ability of potato parents for resistance to the white potato cyst nematode (*Globodera pallida*). *Journal of Agricultural Science* 92, 255-256.

Phillips, M.S., Trudgill, D.L. and Evans, K. (1988) The use of single spaced potato plants to assess their tolerance of damage by potato cyst nematodes. *Potato Research* 31, 469-475.

Phillips, M.S., Rumpenhorst, H.J., Trudgill, D.L., Evans, K., Gurr, G., Heinicke, D., Mackenzie, M. and Turner, S.J. (1989a) Environmental interaction in the assessment of partial resistance to potato cyst nematodes. I. Interactions with centres. *Nematologica* 35, 187-196.

Phillips, M.S., Trudgill, D.L., Rumpenhorst, H.J., Evans, K., Gurr, G., Forrest, J.M.S., Lacey, C.N.D., Mackenzie, M. and Turner, S.J. (1989b) Environmental interactions in the assessment of partial resistance to potato cyst nematodes. II. Interactions with sites and populations. *Nematologica* 35, 197-206.

Phillips, M.S., Rumpenhorst, H.J. and Trudgill, D.L. (1989c) Environmental interactions in the assessment of partial resistance to potato cyst nematodes. III. Interactions with, and virulence differences between populations of *Globodera pallida. Nematologica* 35, 207-215.

Pineda, O., Bonierbale, M.W., Plaisted, R.L., Brodie, B.B. and Tanksley, S.T. (1993) Identification of RFLP markers linked to the H1 gene conferring resistance to the potato cyst nematode (*Globodera rostochiensis*). *Genome* 36, 152-156.

Plaisted, R.I., Harrison, M.B. and Peterson I.C. (1962) A genetic model to describe the inheritance of resistance to the golden nematode *Heterodera rostochiensis* (Wollenweber), found in *Solanum vernei. American Potato Journal* 39, 418-435.

Ross, H. (1962) Über die Vererbung der Resistenz gegen den Kartoffelnematoden (*Heterodera rostochiensis* Wol.) in kreuzungen von *Solanum famatinae* Bitt. et Wittm. mit *Solanum tuberosum* L. und mit *S. chacoense* Bitt. *Der Züchter* 32, 74-80.

Ross, H. (1966) Der Züchtung resistanter Sorten. *Proceedings of the 3rd Triennial Conference of the EAPR.*

Ross, H. (1986) *Potato breeding - problems and perspectives. Advances in Plant Breeding*, 13. P. Parey, Berlin and Hamburg.

Ross, H. and Huijsman, C.A. (1969) Über die Resistenz von *Solanum (Tuberarium)*-Arten gegen europäische Rassen des Kartoffelnematoden (*Heterodera rostochiensis* Woll.) *Theoretical and Applied Genetics* 29, 113-122.

Rothacker, D. (1958) Beitrage zur Resistenzzuchtung gegen den Kartoffelnamatoden (*Heterodera rostochiensis* Wollenweber). *Der Züchter* 28, 133-143.

Rothacker, D., Stelter, H. and Junges, W. (1966) Untersuchungen am Sortiment wilder und kultivierter Kartoffelspecies des Instituttes für Pflanzenzüchtung Gross-Lüsewitz (G-LKS). *Zeitschrift für Pflanzenzüchtung* 56, 101–131.

Stelter, H. (1987) Die Wirtseignung von *Solanum* species für drei *Globodera* Arten. *Nematologica* 33, 310–315.

Stone, A.R. (1973) *Heterodera pallida* n.sp. (Nematoda: Heteroderidae), a second species of potato cyst nematode. *Nematologica* 13, 263–272.

Stone, A.R. (1985) Co-evolution of potato cyst nematodes and their hosts: implications for pathotypes and resistance. *EPPO Bulletin* 15, 131–137.

Toxopeus, H.J. and Huijsman, C.A. (1953) Breeding for resistance to potato root eelworm. *Euphytica* 2, 180–186.

Trudgill, D.L. (1985) Potato cyst nematodes: a critical review of the current pathotype scheme. *EPPO Bulletin* 15, 273–279.

Trudgill, D.L. (1991) Resistance and tolerance of plant parasitic nematodes in plants. *Annual Review of Phytopathology* 29, 167–192.

Turner, S.J. (1989) New sources of resistance to potato cyst-nematodes in the Commonwealth Potato Collection. *Euphytica* 42, 145–153.

Turner, S.J. (1990) The identification and fitness of virulent potato cyst-nematode populations (*Globodera pallida*) selected on resistant *Solanum vernei* hybrids for up to eleven generations. *Annals of Applied Biology* 117, 385–397.

Uhrig, H. and Wenzel, G. (1981) *Solanum gourlayi* Hawkes as a source of resistance against the white potato cyst nematode *Globodera pallida* Stone. *Zeitschrift für Pflanzenzüchtung* 86, 148–157.

Van Soest, L.J.M., Rumpenhorst, H.J. and Huijsman, C.A. (1983) Resistance to potato cyst-nematodes in tuber-bearing *Solanum* species and its geographical distribution. *Euphytica* 32, 65–74.

Wollenweber, H.W. (1923) Krankheiten und Beschadigungen der Kartoffel. *Arbeiten des Forschunginstitutes für Kartoffelbau, Berlin* 7, 1–56.

15 Inheritance of Resistance to Viruses

K.M. SWIEŻYŃSKI

Potato Research Institute, Research Center for Genetics, Breeding and Virology, Młochów, 05-832 Rozalin, Poland.

Introduction

Viruses are important potato pathogens, and introducing resistant cultivars is one of the most efficient ways of reducing the losses which they cause.

The virus–potato interactions which result in resistance are still poorly understood, although they are being extensively investigated. Research progress in this area has been recently summarized by Ross (1986), Zadina (1986) and Tavantzis (1990); virus resistance in other species has been referred to by Ponz and Bruening (1986), Evered and Harnett (1987), Kegler and Kleinhempel (1987), Kegler and Schenk (1990) and Fraser (1990). Most of the information on inheritance of resistance to viruses refers to segregation in the progeny, with few attempts to understand the physiological or biochemical nature of resistance.

No official names for individual plant viruses have yet been approved by the International Committee for Taxonomy of Viruses. In the literature double names are being increasingly used, consisting of a vernacular name followed by the family, genus or group name (Matthews, 1992). This terminology will be adopted.

Types of Resistance – Terminology

In this chapter only genetically controlled resistance is considered. Let us adopt the definition (Anon., 1950) that resistant means: 'possessing qualities that hinder the development of a given disease'. Putting an agronomic interpretation on this definition describes resistant genotypes as those which produce crops less affected by a given pathogen.

Genetic variation in both the virus and its host plant may influence the level of resistance if it affects the virus-plant interaction. The level of resistance may also depend on the conditions in which the virus and the plant interact, e.g. temperature and the stage of growth at which the plant is exposed to infection. The situation is often complex and it may happen that a particular gene may or may not confer resistance depending on the activity of other genes in the host, on genes present in the virus or on environmental conditions. Examples will be given later in this chapter.

Virus strains differ in pathogenicity (ability to cause disease). Pathogenicity can be separated into two elements: aggressiveness (unspecific pathogenicity) and virulence (pathogenicity specifically dependent on plant genotype). Strains of many potato viruses differ both in aggressiveness (inducing mild or severe symptoms without detectable virus strain-potato genotype interaction) and virulence (disease symptoms dependent on virus strain-plant genotype interaction).

The plant may be either immune to or infectible by the virus. If it is infectible, it may be either resistant or susceptible (Cooper and Jones, 1983), with all intermediate situations possible.

Various types and various levels of resistance to viruses occur in the potato. The terminology used to define them is not always precise. The following types of resistance will be considered.

Resistance associated with hypersensitivity

According to a definition cited by Russell (1978), hypersensitivity is 'rapid death of infected cells which restricts the spread of obligate pathogens'. It is switched on by the very specific recognition between the pathogen's avirulence gene product and the host's resistance gene product (Fritig et al., 1987). Very little is known about these gene products, but many dominant genes have been found in plants which trigger hypersensitivity, and many virulent strains of viruses are known to overcome this resistance (a virulent strain does not possess the avirulence genes necessary for the initiation of hypersensitivity).

Hypersensitivity, determined by dominant major genes and usually resulting in no virus spread under field conditions, is sometimes called field immunity. Field-immune plants may be artificially infected, and often show top necrosis after graft inoculation.

The expression of hypersensitivity may be modified by factors such as the genetic background, external conditions (temperature) or the age of the plant. In some situations the restriction of the pathogen by hypersensitivity is so ineffective that plants with this reaction are not resistant. Thus hypersensitivity to potato M carlavirus could be associated with considerable virus spread under field conditions (Dziewońska and Ostrowska, 1977). Some potato cultivars hypersensitive to the potato leafroll luteovirus became almost 100% infected within 1 year. The infected plants produced no sprouting tubers, which resulted in considerable yield loss (M. Chrzanowska, pers. comm.).

Resistance associated with tolerance

Tolerance to pathogens has been defined in various ways and many types of tolerance have been recognized (Russell, 1978). In this chapter tolerance will be defined as a reaction not associated with hypersensitivity and resulting in little damage to the host. Depending on circumstances, tolerance may or may not provide resistance. Thus tolerance to tobacco rattle virus in potato cultivars results in no losses due to virus-induced necroses. The virus does not become permanently established in either tolerant or sensitive cultivars. In contrast, cultivars tolerant to potato Y potyvirus may carry a high concentration of virus, which spreads easily in such plants, affecting their yield. In the first example tolerance does provide resistance, whereas in the second it does not. There are several examples of the successful breeding of virus-resistant cultivars in which the resistance is associated with tolerance (Russell, 1978).

Many different biological phenomena may result in resistance depending on tolerance. If a virus enters a plant but is ill-adapted to it and unable to cause disease symptoms, e.g. its multiplication in infected cells or its cell-to-cell movement is restricted for reasons other than hypersensitivity, the likely effect is resistance associated with tolerance.

Evidence is accumulating that the movement of virus particles between cells is an active process, is genetically determined (Hull, 1989) and may depend on light quality (Thomas *et al.*, 1988), as well as temperature.

Detailed information on the effects of some genes determining tolerance is already available. Thus gene *Tm1* regulates both tolerance and resistance to tobacco mosaic tomovirus in the tomato. Its presence is associated with considerable reduction of virus concentration in infected plants, and protoplasts with this gene are unable to support virus multiplication. The resistance is temperature-dependent, for it is not expressed at 33°C, and virulent strains of the virus can overcome it (Fraser and Loughlin, 1980).

Extreme resistance

This term has been used to describe a high level of resistance controlled by dominant major genes (Ross, 1986). The virus does not become established even after graft inoculation in plants carrying such genes. Inoculated plants usually produce no disease symptoms, and virus multiplication is reduced.

It is possible that extreme resistance is an extreme expression of hypersensitivity, as necrotic lesions are occasionally found on inoculated plants.

Resistance to infection

This term is often used in a general sense to indicate that a plant does not easily become infected, with no consideration as to what might be the reason.

This resistance may be associated with hypersensitivity or tolerance; it may be also associated with other factors, e.g. those interfering with the initiation of direct contact between the virus particle and the host cell.

Resistance to virus vectors

This type of resistance is being utilized in potato breeding as well as in other crops (Russell, 1978; Beekman, 1987). An interesting type of resistance is associated with the presence of glandular hairs trapping insects (Tingey, 1991).

Resistance to Potato Leafroll Luteovirus (PLRV)

Strains of the virus

Strains of PLRV differ in the severity of the symptoms they produce on potato and on test plants (Beemster and de Bokx, 1987). There are indications of interaction between virus strains and plant species (Tamada *et al.*, 1984) and between virus strains and potato cultivars (Syller, 1985).

Inheritance of resistance

Evaluating PLRV resistance on the basis of external disease symptoms is laborious and unreliable. However, in recent years the evaluation of virus concentration in inoculated plants by enzyme-linked immunosorbent assay (ELISA) has opened up possibilities for more rapid progress to be made in the study of the inheritance.

A high level of resistance to PLRV exists in old potato cultivars and in several species widely used in potato breeding. According to Ross (1986), much of the resistance in present-day cultivars can be traced back to *S. demissum* and *S. tuberosum* subsp. *andigena* hybrids (W races) bred by K.O. Mueller, as well as to clone MPI 19268, which is an ancestor of many leafroll-resistant Dutch cultivars. A third source of resistance is the MPI clone 44.1016-10, which originates from *S. acaule*.

Two types of resistance may be identified.

Resistance associated with hypersensitivity

Hypersensitivity is found in potato cultivars and in wild species, e.g. *S. raphanifolium* (Ross, 1986). It is controlled by the dominant gene N_L (Zadina and Novak, 1983). The expression of hypersensitivity depends on modifying genes and external conditions (Butkiewicz, 1978; Zadina and Novak, 1983).

Not much is known about the modifying genes, but it is generally accepted that the presence of the N_L gene without a suitable genetic background does not sufficiently protect a plant from PLRV. A high level of resistance associated with a hypersensitive reaction occurs in the German cv. Apta and in Polish cvs Dryf and Irga. In the latter, which is highly resistant, hypersensitivity was detected at 27°C, but not 15°C (Syller, 1991). In the hypersensitive breeding line PW 129, plants were difficult to infect, even by graft inoculation, but, when infected, reacted with an absence of sprouting in the tubers or with the occurrence of dwarf plants in the daughter tubers (Chrzanowska, 1988).

Resistance associated with tolerance

Swieżyński et al. (1989) found a high level of resistance in diploid breeding lines, which was apparently not associated with hypersensitivity. Among ancestors of the lines are PLRV-resistant genotypes of hybrid origin, widely utilized in potato breeding: MPI 44.1016-10, MPI 44.335-130 and cvs Apta and Schwalbe. The resistant breeding lines were difficult to infect, even after graft inoculation, and infected plants had a low concentration of virus, which was unevenly distributed. The diploid clone DW 84-1457, of similar origin, was outstanding in resistance and could not be infected by either aphids or graft inoculation (Dziewońska, 1990). However, some virus could occasionally be detected in the daughter tubers of graft-inoculated plants of this clone (M. Was, pers. comm.). Swieżyński et al. (1990) found that the progeny could exceed the parents in the level of this type of resistance. They suggested that at least two different dominant genes are responsible for the high level of resistance.

A similar type of resistance was apparently detected in Scotland by Barker and Harrison (1986), who found resistance associated with low virus concentration in infected plants. The low concentration was the result of impaired virus spread within the plant, for the virus titre was not reduced in infected cells. Monogenic segregation for resistance (restricted virus accumulation) was observed in the progeny of a resistant clone (Barker and Solomon, 1990).

This type of resistance was possibly observed by Baerecke (1956). Using resistant parents of different origin, she found a considerable transgression of level of resistance in the progeny.

In recent years two new sources of resistance have been found and are being utilized in breeding work (Ross, 1986). The first was detected in *S. etuberosum* and *S. brevidens*. Preliminary results indicate an oligogenic control and low virus concentration in resistant genotypes (Chavez et al., 1988). The second was found in *S. acaule* (Brown et al., 1984b). The relation between the resistance originating from the new sources and the high level of resistance identified in existing breeding lines is not known.

Resistance to Potato Y Potyvirus (PVY)

Strains of the virus

There are three main groups of strains (Beemster and de Bokx, 1987):

1. PVY^O *group (common strains).* These induce generally severe systemic symptoms of crinkle, rugosity or leaf-drop streak in potato, systemic necrosis in *Physalis floridana* and systemic mottling in tobacco.
2. PVY^N *group (tobacco veinal necrosis strains).* These strains produce systemic veinal necrosis in tobacco and mild mottling in almost all potato cultivars.
3. PVY^C *group (stipple streak strains).* Previously they were separately recognized as virus C, which, in contrast to other groups of strains, was not aphid-transmitted.

Occasionally PVY strains are found which do not fit the above separation criteria (Davidson, 1980; Jones, 1990). A strain of PVY^N causes potato tuber ring necroses in Slovenia (Kus, 1990) and several other mid-European countries (Schiessendoppler, 1990).

Chrzanowska (1991) found PVY^N isolates which produced exceptionally mild symptoms on most potato cultivars, although they quickly reached a high concentration in inoculated plants.

Inheritance of resistance

Types of genes controlling resistance

Genes triggering hypersensitivity were found to control resistance to some or all groups of PVY strains. Genes controlling extreme resistance to PVY have also been described.

Resistance associated with hypersensitivity

This has been found both in cultivars and in potato species.

Clone 11-79 was bred in Australia from cvs Katahdin and Snowflake, both of which are reputed to show a necrotic response when inoculated with PVY (Hutton, 1951). Cultivars tracing their resistance from 11-79 are not completely resistant to the PVY^O group of strains but show a high degree of resistance to them in the field (Davidson, 1980). Solomon (1978) reported that *Ny* was the gene (or complex) responsible for this resistance. She found that sap inoculating plants carrying the *Ny* gene produced only local lesions. The gene *Nc* controls resistance to PVY^C strains, and is carried by most potato cultivars (Ross, 1986).

Cockerham (1970) described the genes Ny_{chc}, Ny_{dms}, Ry_{sto}^{n1} and Ry_{sto}^{n2} from wild species, and found that they gave hypersensitivity to all tested strains of PVY. There are indications that the genes Ny_{chc} and Ny_{dms} provide a satisfactory level of resistance only in the presence of a suitable genetic background (Davidson, 1980). Cockerham (1970) found Ny_{chc} in 32 of 48 samples of *S. chacoense* and in 11 of 27 samples of *S. microdontum*. He also reported that gene Ny_{dms} in *S. demissum* controls necrotic reaction to both PVY and PVA. *S. demissum* CPC 2103 is homozygous for this gene. The genes with subscript *sto* were identified by Cockerham (1970) in *S. stoloniferum*. He found that the genes Ry_{sto}^{n1} and Ry_{sto}^{n2} control lethal necrosis to PVY with susceptibility to PVA.

A type of resistance, believed to be polygenically inherited, was found in *S. phureja* CPC 979. Davidson (1980) found that, when the gene Ny_{chc} was placed in a background of resistance derived from CPC 979, PVY inoculation resulted in infection restricted to the production of local lesions only at the site of inoculation. Solomon (1978) found that clones derived from *S. phureja*, with apparently polygenic inheritance, carried no *Ny* genes and reacted as susceptibles in glasshouse tests, and their resistance was detectable only in field trials.

Extreme resistance

Genes controlling extreme resistance are known in *S. stoloniferum*, *S. hougasii* and *S. tuberosum* subsp. *andigena*.

Extreme resistance to all tested strains of PVY and PVA was found to be determined by a dominant gene present in *S. stoloniferum*. The gene was named by Ross (1961) *Ry* and by Cockerham (1970) Ry_{sto}. Cockerham found this gene in *S. stoloniferum* CPC 9, CPC 28.4 and PI 160226, the last two being homozygous for the gene. After graft-inoculating plants carrying the gene, the virus produced symptoms ranging from no visible response to a variety of localised necrotic flecks in leaves and/or stems, accompanied by stunting of axillary shoots and also of the scions. No virus was recoverable either from the grafted plant itself or from its tuber progeny (Cockerham, 1970). In graft-inoculated breeding lines carrying this gene, Delhey (1975) found small chlorotic–necrotic lesions and tubers with necrotic symptoms, but no virus could be recovered. Barker and Harrison (1984) evaluated the reaction of potato cvs Corine and Pirola carrying this gene. Cv. Corine produced no symptoms, even after graft inoculation. In cv. Pirola, manually inoculated leaves developed a few necrotic streaks on the underside of the veins, but the virus could not be recovered. Protoplasts of both cultivars were resistant to infection. They were much less affected in comparison with protoplasts originating from susceptible genotypes or from resistant genotypes with a hypersensitive reaction.

In *S. hougasii* the gene Ry_{hou} controlled extreme resistance to all tested strains of PVY and to PVA (Cockerham, 1970). It was present in a sample of PI 161726 × PI 161740 (from Wisconsin) and in samples GLKS 66-3 and 66-4 (from Gross Lüsewitz).

In *S. stoloniferum* and *S. tuberosum* subsp. *andigena*, dominant genes controlling extreme resistance to PVY, but not to PVA, are known. In the former species, gene Ry^{na}_{sto} controls extreme resistance to PVY and lethal necrosis to PVA, whereas gene Ry^{rna}_{sto} controls extreme resistance to PVY and non-lethal rusty necrosis to PVA. Both were found by Cockerham (1970), the first in PI 161172 and the other in CPC 12.

A dominant gene controlling resistance to PVY was found in some accessions of S. tuberosum subsp. andigena obtained from the John Innes Institute (Munoz et al., 1975). Inability to recover the virus from graft-inoculated plants indicated the presence of extreme resistance. Later work demonstrated that clones originating from this source were hypersensitive to PVA (Anon., 1984). Ross (1986) refers to this gene as Ry_{adg}.

Fernandez-Northcote (1990) found that potato genotypes resistant to PVY^O show various levels of resistance to other strains of PVY and to related viruses (Table 15.1).

According to Ross (1986), the segregation for extreme resistance usually deviates little from expected values, but deviations were found to occur due to minor genes. He lists the following cultivars with extreme resistance originating from S. stoloniferum: Corine, Sante (The Netherlands), Bóbr, Brda, Bzura, Pilica, San (Poland), Magyar Rosa, Szignal (Hungary), Barbara, Bison, Cordia, Esta, Fanal, Forelle, Franzi, Heidrun, Pirola, Wega (Germany).

Linkage

According to Cockerham (1970), the gene Ny_{chc} is linked to genes Nx_{chc} and Nx^{spl}_{tbr} determining the reaction to potato X potexvirus; Ry_{hou} is independent of Ny_{dms}; Ry_{sto} is an allele of Ry^{n1}_{sto} and probably an allele of Ry^{rna}_{sto} and Ry^{na}_{sto}.

Resistance to Potato A Potyvirus (PVA)

Mild, moderate and severe strains are known (Smith, 1972; Beemster and de Bokx, 1987). They differ in severity of symptoms produced in potato cultivars. A severe strain, called PVA^N, has been described by Kowalska et al. (1981).

Inheritance of resistance

The gene Na, present in many cultivars, protects the plant from infection under natural pressure from PVA by means of a hypersensitive response (Ross, 1986). It was named by Cockerham (1970) Na_{tbr}, but this new name seems not to have been accepted. It gives no protection from PVY. Cockerham (1970) also found genes controlling hypersensitivity to PVA, but not to PVY, in both S. tuberosum subsp. andigena and S. demissum.

As gene Na is present in many potato cultivars and an increasing number

Table 15.1 Reaction to PVY and related viruses of extremely resistant potato genotypes of various origins (Fernandez-Northcote, 1990).

Virus	Origin of the strain	Extreme resistance to PVY from				Controls	
		S. stoloniferum		S. tuberosum subsp. andigena			
		cv. Bzura	V 3	V 2	XY 13.14	cv. Conchita	cv. Molinera
PVY^O	Peru	–	–	–	–	h	S
PVY^N	Peru	–	–	–	–	h	S
PVY^C	Holland	–	–	–	–	h	S
PVA	Holland	–	–	H	–	R	S
PVV^1	Holland	–	–	Ha	Ha	S	S
PVV	Ireland	–	–	Ha	H	h	S
PVV	Peru	–	–	–	–	S	S
PVT^2	Peru	–	–	–	–	–	–

[1] PVV = potato V potyvirus (Fribourg and Nakashima, 1984); [2] PVT = potato T closterovirus (?) (Salazar and Harrison, 1978).
I - extreme resistance; H - hypersensitivity; Ha - hypersensitivity only after grafting; h - low hypersensitivity; R - no infection after mechanical inoculation, but after grafting; S - susceptibility.

of cultivars carry the gene Ry, a high level of resistance to PVA often results. Among 1080 potato cultivars described by Joosten and van der Woude (1985), 193 are highly resistant to PVA.

The gene Na is linked to gene Nx_{tbr}, which controls resistance to potato virus X (Cockerham, 1970).

Resistance to Potato X Potexvirus (PVX)

Strains of the virus

Various criteria have been used to separate strains of PVX (Smith, 1972; Beemster and de Bokx, 1987). Torrance *et al.* (1986) found serological differences between strains, using monoclonal antibodies.

Strains differ both in aggressiveness and virulence, and mild, moderate and severe strains are known. Unusually severe strains also occur, to which many potato cultivars react with top necrosis (Smith, 1972). Strains differing in virulence to genotypes carrying the genes Nx and Nb were separated by Cockerham (1955) into four groups. He called the genes Nx_{tbr} and Nb_{tbr} and the groups 1, 2, 3 and 4. Strains of group 1 were not pathogenic to potato genotypes carrying the gene Nx_{tbr} or Nb_{tbr}, strains of group 2 were able to overcome the resistance provided by Nx_{tbr}, strains of group 3 were able to overcome the resistance provided by Nb_{tbr}, and strains of group 4 were able to overcome the resistance provided by both these genes. Later a separation was made according to the presence of the gene $Rx_{(scr)}$, providing resistance to strain PVX_{HB} (Tavantzis, 1990). Strain PVX_{HB} has been reported from South America; it overcomes the extreme resistance that protects the plant from other strains (Moreira *et al.*, 1980). Fernandez-Northcote (1990) recently suggested separating PVX strains according to their serological reaction into two main pathotypes: 1 and HB, the four groups of Cockerham (1955) belonging to pathotype 1. The strain of PVX to which resistance is controlled by the gene Nb was called B in some early papers.

The danger of spread of virulent strains of the virus is difficult to evaluate. Tavantzis (1990) isolated PVX from a naturally infected plant of cv. Atlantic which showed extreme resistance to PVX, and demonstrated that this strain deviated in ways which might hamper its spread. On the other hand, Jones (1985) found no evidence of low viability in the strain PVX_{HB}.

Inheritance of resistance

Resistance associated with hypersensitivity

Cockerham (1970) identified several genes conferring hypersensitivity:

1. Nx_{tbr}. This gene, originally named Nx, is present in many old potato cultivars. It controls hypersensitivity to strains of groups 1 and 3 of Cockerham (1955). Cv. Cardinal is duplex for this gene. Its allele, Nx_{tbr}^{spl}, originating from *S. sparsipilum* CPC 71, gives a similar reaction to PVX strains and is probably identical to Nx_{tbr} (Cockerham, 1970).

2. Nb_{tbr}. This gene, present in many old potato cultivars, controls hypersensitivity to PVX strains of groups 1 and 2 of Cockerham (1955). Cv. Catriona is duplex for this gene and Cockerham (1970) obtained from it dihaploids homozygous for the gene.

3. Nx_{chc}. This gene controls hypersensitivity to all the four groups of strains recognized by Cockerham (1955). It is frequently found in *S. chacoense* and *S. microdontum*. In a survey of the two species, a similar reaction to PVX was found in all 43 samples of *S. chacoense* and in three of 27 samples of *S. microdontum* (Cockerham, 1970).

4. Rx^n_{acl}. The gene occurs in both *S. andigena* and *S. acaule*. It controls hypersensitivity to all strains of PVX tested so far.

Extreme resistance

Genes conferring extreme resistance are known to occur in several species.

Rx_{adg}, found in *S. tuberosum* subsp. *andigena*, controls extreme resistance to all PVX strains except PVX_{HB}. According to Cockerham (1970), the gene is present in USDA seedling 41956 and in *S. tuberosum* subsp. *andigena* genotypes CPC 141, CPC 189, CPC 244, CPC 1673 and cv. Collajera. Ross (1986) has pointed out that USDA seedling 41956 originates from the Chilean cv. Villaroela, not from subsp *andigena*. The gene present in USDA 41956 is referred to by Ross as Rx, by Tavantzis (1990) as Rx_{adj} and by Fernandez-Northcote (1990) as Rx_{tbr}. Cockerham (1970) identified some genotypes which were triplex for the gene.

Rx_{acl}, found in *S. acaule*, controls extreme resistance to all strains of PVX except PVX_{HB}. Ross (1954) found *S. acaule* lines which were homozygous for the gene.

Brown *et al.* (1984a) found two dominant genes in *S. sucrense* that controlled extreme resistance to PVX. One provided extreme resistance to PVX_{HB}, but not to a PVX strain of group 2 of Cockerham (1955). The other provided extreme resistance to the strain of group 2, but not to PVX_{HB}. The two genes were inherited independently. Tavantzis (1990) refers to the gene providing extreme resistance to PVX_{HB} as $Rx_{(scr)}$.

Ross (1986) recognized another gene for extreme resistance (or perhaps hypersensitivity). Its origin is unknown, but it is carried by the nematode-resistant *S. vernei* backcross hybrid 62-33-3 and has been introduced into several Dutch cultivars.

In a study of the expression of genes controlling extreme resistance, Delhey (1974) inoculated genotypes carrying the gene Rx_{adg} with PVX. He found that the reaction depended on the temperature, and that, at or below 20°C, the resistance mechanism was hampered. At 20°C, systemic infection occurred in plants inoculated mechanically. Freshly harvested tubers from graft-inoculated plants grown at 8°C showed necrotic symptoms and PVX could be isolated from such tubers. He concluded that extreme resistance is related to hypersensitivity.

Tavantzis (1990) reported that in PVX-inoculated protoplasts of genotypes carrying the genes Rx_{adg} or Rx_{acl} virus accumulation is considerably reduced. He also evaluated the incidence of PVX in naturally infected plants of cv. Atlantic carrying the gene Rx_{adg}. In field-grown plants in Maine, where temperatures of 15–20°C are not unusual, the percentage of tubers carrying low levels of PVX depended on the origin of the sample. A very small proportion (less than 1% of the tubers examined) contained levels of PVX comparable to those found in tubers of PVX-susceptible cultivars. The virus isolate recovered from such plants appeared to reach a lower concentration in tobacco, and to induce local lesions in *Gomphrena globosa*. He concluded that such deviating strains are unlikely to attain epidemiological significance.

A considerable number of potato cultivars carry genes for extreme resistance. According to Ross (1986), the gene Rx_{acl} is present in German cvs Aguti, Assia, Barbara, Moni, Natalie, Roeslau and Saphir and the Argentine cv. Serrana Inta. The gene Rx_{adg} (Rx), originating from USDA seedling 41956, is present in the US cvs Atlantic, Carlton, Jemsec, Reliance, Saco, Shoshoni and Tawa. Gene Rx_{adg}, originating from the nematode-resistant accession CPC 1673, is present in many nematode-resistant cultivars. The gene from the *S. vernei* hybrid 62-33-3 is present in Dutch cvs Atrela, Darwina, Pansta, Produzent, Promesse, Proton and Sante. In Poland, cvs Bóbr, Brda, Bzura, Dryf, Fregeta, Olza, Pilica and San are extremely resistant to PVX (M. Chrzanowska, pers. comm.).

Linkage

According to Cockerham (1970), Nx_{tbr}^{spl} is linked to Nx_{chc} and to Na, which controls a hypersensitive reaction to PVA. It is not linked to Nb_{tbr}. Rx_{acl}^{n} is an allele of Rx_{acl}. Rx_{acl} is not linked to Nb_{tbr} (Solomon, 1985). Little is known about the relationship between Rx_{adg} and Rx_{acl}, but they are apparently located on different chromosomes (Ritter *et al.*, 1991).

Resistance to Potato M Carlavirus (PVM)

Strains of the virus

Strains differ in the type and severity of symptoms they produce, as well as in their transmissibility by aphids (Beemster and de Bokx, 1987). Symptoms varying from mild to very severe were found on cv. Uran, depending on the virus isolate (Kowalska, 1978). PVM strains also differ in their rate of multiplication, in the concentration reached and in their ability to induce hypersensitivity in plants carrying the gene *Rm*. In plants graft-inoculated with strain M55a, 80% of necrotic reactions were found within 2 weeks and 100% within 3 weeks, but in plants graft-inoculated with strain M3 no plants were necrotic after 2 weeks and 30% of plants did not react with necroses (Kowalska, 1981).

Inheritance of resistance

Resistance associated with hypersensitivity

The gene *Rm* was found in *S. megistacrolobum* EBS 1787 (Ross and Jacobsen, 1976). Plants carrying this gene may remain healthy under high natural virus pressure, the level of resistance depending on the genetic background (Dziewońska *et al.*, 1978). At present, breeding lines are available in which gene *Rm* is associated with a genetic background such that plants did not become infected after 4–5 years of growth under high natural virus pressure. They could not be mechanically inoculated and after graft inoculation they reacted with hypersensitivity (Chrzanowska, 1988). The inheritance of this genetic background is not understood.

Hypersensitivity to PVM was also found in *S. cardiophyllum* and *S. microdontum* (Oertel *et al.*, 1978). It is not known whether it is due to the presence of the gene *Rm* and whether it may be utilized to develop resistant genotypes.

Resistance associated with tolerance

In two progenies of *S. gourlayi*, INTA 7330 and INTA 7356, a high level of resistance was found to be associated with the gene *Gm*, which was expressed in reduced virus spread in graft-inoculated plants (Swieżyński *et al.*, 1981). The high level of resistance is expressed only in genotypes carrying two dominant genes, *Gm* and *Ma*, the second gene being found both in *S. gourlayi* hybrids and in various breeding lines not originating from this species. Even in the presence of both genes, some effect of genetic background is detectable (Pochitonow, 1989). Breeding lines carrying the genes remained virus-free for 6 years under high natural PVM pressure (Dziewońska *et al.*, 1988). They were not infected after mechanical inoculation, and reacted with delayed virus accumulation after graft inoculation (Waś *et al.*, 1980).

There are other reports on resistance not associated with hypersensitivity. Oertel *et al.* (1978) found reduced infection after mechanical inoculation in some accessions of *S. chacoense*, *S. megistacrolobum*, *S. microdontum* and *S. tuberosum* subsp. *andigena*. Less susceptible were cvs Ada, Karsa, Kardula, Tunika and Elgina. Dziewońska (1975) found a comparatively high level of resistance in the breeding line S 5592 and in potato cvs Schwalbe and Aquila. In matings, the mean progeny level was slightly above the midparent value, which may indicate dominance in the inheritance of the resistance.

Resistance to Potato S Carlavirus (PVS)

Strains of the virus

Kowalska (1981) observed considerable differences between PVS isolates in their ability to induce necrotic symptoms in graft-inoculated plants carrying the *Ns* gene. Isolate S 66 induced necrotic reaction in most plants after 3 weeks and in all of them after 6 weeks. In contrast, isolate S 28 induced a necrotic reaction in only one of eight graft-inoculated plants. In The Netherlands a virulent strain has occasionally been found to evoke a bronze-coloured necrosis in many cultivars, later accompanied by withering and dropping of the lower and middle leaves (Beemster and de Bokx, 1987).

Inheritance of resistance

Gene *Ns*, controlling a resistance associated with hypersensitivity, was found by Baerecke (1967a) in *S. tuberosum* subsp. *andigena* cv. Huaca Nahui, PI 258907. The dominance of *Ns* was confirmed by Makarov (1975). Plants carrying the gene were not infected after mechanical inoculation. Some variation was found among genotypes carrying *Ns* in their reaction to graft inoculation (Pietrak, 1985). Plants carrying the gene are seldom found to be infected in field conditions (M.A. Dziewońska, pers. comm.). The gene is expressed also in the presence of the gene *Gm* controlling PVM resistance (Waś and Dziewońska, 1984).

A similar type of reaction has also been found in accessions of *S. chacoense*, *S. megistacrolobum* and *S. microdontum* (Kegler and Kleinhempel, 1987). Cvs Szignal in Hungary and Fantasia in Germany were the first potato cultivars to carry the gene *Ns* (Ross, 1986).

A recessive gene *s*, providing a high level of resistance to PVS, was found in cv. Saco by Baerecke (1967b). This cultivar did not become infected either naturally or after mechanical inoculation, and graft-inoculated plants usually remained healthy. Plants occasionally reacted with slight necrosis, but never with top necrosis. Tubers of plants which were graft-inoculated usually produced healthy progeny. Occasionally, deviating plants (small with less dissected leaves) were obtained, which were infected with PVS (Baerecke, 1967b). Although Bagnall and Young (1972) found no deviations from expectation in the progeny segregation of various crosses, Baerecke (1967b) found indications that other genes modify the reaction, as, after selfing cv. Saco, a small proportion of the progeny appeared to be susceptible. Baerecke (1967a) also found that mating cv. Saco with a genotype carrying the gene *Ns* resulted in an increased frequency of progeny with a necrotic reaction. Dziewońska and Pochitonow (1971) confirmed the resistance of cv. Saco to mechanical inoculation, but obtained 100% infection after graft inoculation.

Many authors have noted differences between potato cultivars in their level of resistance to PVS in the apparent absence of genes *Ns* or *s*. The German cv. Adretta is resistant (Ross, 1986), as are the Polish cvs Narew and Uran (Kapsa *et al.*, 1983).

Resistance to Tobacco Rattle Tobravirus (TRV)

Strains of the virus

TRV isolates fall into two main classes: those producing long or short nucleoprotein particles containing parts 1 and 2 of the RNA genome, and those producing RNA 1 but no virus nucleoprotein in infected plants. Both types of isolate are found in nature (Robinson, 1989). The second type tends to produce more pronounced symptoms on infected potato plants (Waś, 1978). TRV does not spread easily in potato plants and most tubers with spraing symptoms, when grown the following year, give rise to healthy plants (Dale and Solomon, 1988). Robinson (1990) recognized six serotypes among TRV isolates. Different vector species were found to be associated with the different serotypes. TRV isolates differ in ability to become systemic in potato plants, and a strain that becomes systemic in one cultivar may not do so in another (Harrison, 1968).

Reaction in the potato

A typical symptom is tuber spraing. Some tolerant cultivars do not produce spraing symptoms: cvs Bintje, Record, Arran Pilot and Stormont Enterprise. Jellis and Gray (1980) evaluated the progeny of the last cultivar and concluded that tolerance was controlled by a single dominant gene with modifiers.

Harrison (1968) evaluated 14 cultivars with various levels of tolerance to spraing. He found that TRV more often passed from affected tubers to progeny plants and tubers in cultivars showing slight spraing. Infected plants of cv. Arran Pilot showed neither spraing nor stem-mottle; their roots resisted infection by vector-borne TRV, but leaves could be infected by manual inoculation with infective sap.

Resistance to Other Viruses

Some genes controlling extreme resistance to PVY also control extreme resistance to potato V potyvirus or potato T closterovirus (Fernandez-Northcote, 1990).

Resistance to potato mop-top furovirus (PMTV) has been found in cvs Pentland Dell (Harrison, 1968), Gloria, Ostara and Resy (Nohejl and Rasocha, 1989), and Pito, Hertha and Record (Kurppa, 1989).

Resistance to *Mycoplasma*-like Organisms

Vulchev (1969) found some resistance to stolbur in cvs Civa, Home Guard, Jara, Red Warba and Saskia. Among wild species, resistance to stolbur occurs in some accessions of *S. acaule, S. chacoense, S. cardiophyllum,*

S. demissum, *S. pinnatisectum*, *S. stoloniferum* and *S. tarijense* (Bukasov and Kameraz, 1972).

Resistance to the Potato Spindle Tuber Viroid (PSTV)

Mild and severe strains are known. Beemster and de Bokx (1987) report that a special strain called potato unmottled curly dwarf virus has been described.

No cultivar shows hypersensitivity or extreme resistance (Ross, 1986). Kegler and Kleinhempel (1987) reported resistance in cvs Amsel, Norland and Ukrainskij Rannij, as well as in several wild species, including *S. berthaultii* PI 473340 and *S. acaule* OCH 11603.

Resistance to Vectors of Viruses

The most promising form of resistance appears to be the production of glandular hairs in *S. berthaultii*. Progress in breeding potatoes with this character was reported by Tingey (1991). New sources of aphid resistance have recently been found at the International Potato Center (Anon., 1990). A high level of antixenosis was found in some CIP clones, and antibiosis was confirmed in *S. neocardenassii*.

Breeding Virus-resistant Cultivars

Objectives

Resistant cultivars can reduce the direct losses in yield and occasionally also in tuber quality; they can lower the cost of seed, for the grower may not need to buy certified seed if he can use his own, and the cost of seed production, because less elaborate production methods are sufficient and the use of insecticides is unnecessary. Virus-resistant cultivars are of special importance in areas where viruses cause serious losses and healthy seed is expensive.

Ideally, multiple resistance to all economically important viruses is required. PLRV and PVY cause the greatest losses in most potato-growing areas. In Central and Eastern Europe, PVM is also important. TRV and PMTV may be of great economic importance locally, and there is often interest in resistance to PVA, PVX and PVS, although they are seldom responsible for serious losses.

Sources of resistance

Durable resistance provided by dominant genes, with no negative side-effects, is always desirable. Sources of resistance to all the important viruses exist and many of them have already been introduced into cultivars. Extreme resistance to PVY and PVA is controlled by a single dominant gene Ry_{sto}, providing comprehensive resistance to all known strains of both viruses. Cultivars with this gene occupied 5797 ha of certified seed planted in Poland in 1990, and there are as yet no indications that this resistance is breaking down.

Some dominant genes (N_L for PLRV and Rm for PVM) control hypersensitivity. They provide a high level of resistance if in a suitable genetic background. Highly resistant cultivars having the gene N_L are available, but gene Rm has not yet been introduced into cultivars.

Alternative sources of high-level resistance not associated with hypersensitivity have been found for both PLRV and PVM. There is as yet insufficient information to decide which is the better type of resistance. Both may result in no or negligible virus spread in natural conditions, and both are likely to be determined by a few dominant genes. There is insufficient experience of their durability to make an informed choice between them.

Because genes controlling resistance often originate from wild species, undesirable genes may be closely linked to the resistance genes and may be difficult to eliminate. In fact, a considerable introgression of germplasm from wild species has been noted in potato hybrids originating from them (Debener et al., 1991) and there are also indications that it may result in distortions of segregation ratios (Ritter et al., 1991). The introduction of foreign cytoplasm may also be harmful, and the possibility of negative pleiotropic effects of the genes themselves should also be considered.

These dangers are probably not very serious, as several such genes have already been successfully introduced into cultivars, and Swieżyński et al. (1988) found no correlation between tuber yield and resistance to PLRV in segregating potato families. Nevertheless, some difficulties may arise. Most breeding lines with the gene Ry_{sto} are male-sterile, probably due to cytoplasmic incompatibility. Ross (1986) reported the occurrence of a gene restoring male fertility, and the Młochów Research Station has received male-fertile hybrids which are extremely resistant to PVY, from Professor Bukasov of the Institute for Plant Production in St Petersburg.

Methods of selection

Virus resistance is often expressed so early that young seedlings may be selected without much effort (Wiersema, 1961; Solomon, 1978). Aggressive virus isolates should be used in order to produce disease symptoms that are easy to recognize without further testing. Young seedlings can be screened by mechanical inoculation with four viruses: PVY, PVX, PVM and PVS. At the Młochów Research Station, young seedlings are often selected for multiple resistance to the first three. For this purpose five successive inoculations are applied at 2–3-day intervals. The first two inoculations are done only with PVM, the third with PVY and the last two with PVY and PVX.

For the selection of genotypes with high levels of resistance to PLRV, aphid inoculation of young seedlings may be used (Chuquillanqui and Jones, 1980). At Młochów the evaluated genotypes are usually graft-inoculated and the virus concentration is determined by ELISA in scions developing on inoculated stocks and in their tuber progeny. For the evaluation of TRV resistance, a glasshouse test has been devised (Dale and Solomon, 1988).

To avoid escapes, tests must be repeated. Graft inoculation is usually used, but Waś (1988) found that sap-inoculating plants and keeping them at high temperature, followed by a thermal shock, is more efficient and less laborious when testing for the presence of extreme resistance to PVY.

Various resistance genes may be unconsciously accumulated in breeding material, thus complicating the breeding procedure. Periodic checking is therefore desirable to ascertain whether the segregation obtained is consistent with expectations. Galvez and Brown (1980) checked 19 clones identified as extremely resistant to PVY by graft testing. The clones were crossed with susceptible genotypes and the progeny evaluated for resistance. Two progenies segregated 1 : 3, twelve 1 : 1 and one 5 : 1 (resistant : susceptible). Four failed to show a significant number of resistant individuals.

It follows that not only clones simplex and duplex for the gene Ry were present (1 : 1 and 5 : 1 segregation), as expected, but also clones with a much higher proportion of susceptible progeny. In such clones, either Ry was absent (possibly being replaced by other genes providing resistance) or its expression was prevented.

The possibility that other genes may complicate the evaluation is indicated by the fact that a parallel segregation of genes controlling extreme resistance and hypersensitivity was found in the diploid progeny of extremely resistant parents (Anon., 1989).

Breeding cultivars with resistance to several viruses

There are still no cultivars with a high level of resistance to PVM and there are no reports of cultivars in which extreme resistance to PVY is associated with high levels of resistance to PLRV. The development of cultivars highly resistant to two or three of these viruses remains a challenge for the breeder. There are probably two main reasons for the lack of such cultivars: a high level of resistance to PVM and the application of ELISA for efficient screening for PLRV resistance are of so recent origin that there has not been time to develop new cultivars with their use; secondly, since the breeder must select for many important characters, introducing even a few genes for resistance to viruses is a difficult task. A solution may be provided by parental line breeding (Swieżyński, 1983). If the breeder receives a parent whose progeny consists of a large proportion of genotypes with multiple resistance to viruses, the chance to introduce a cultivar with multiple resistance increases considerably.

After the last war, parental line breeding was developed in the Max Planck Institute (MPI) für Züchtungsforschung, Cologne, Germany. Ross (1977)

Table 15.2 Potato genotypes with resistance to several viruses at the Młochów Research Station in 1991.

Ploidy level	Number of genotypes	First seedling year	High resistance to viruses					
			PLRV	PVY	PVX	PVA	PVM	PVS
Diploid	3	1984	+	+^{1}	+		+	
	1	1984	+	+^{1}	+			+
	24	1988	+	+^{1}	+		+	
Tetraploid	5	1986	+	+	+	+	+	
	9	1987	+	+	+	+	+	
	2	1987	+	+	+	+	+	+
	26	1988	+	+	+	+	+	
	13	1988	+	+	+	+	+	+

^{1}Field immunity only.

reports on several clones developed at that institute which were extremely resistant to PVY and PVX and resistant to PLRV. In addition, some were nematode-resistant. Clones from the MPI have been used by breeders in many countries.

In Poland, parental genotypes resistant to viruses have been developed over the last 35 years (Dziewońska, 1986). The breeding lines which resulted are shown in Table 15.2. Some of them are also resistant to nematodes or to *Phytophthora infestans*. Polish breeders received the first parental genotypes with multiple resistance to PLRV, PVY, PVA, PVX, PVM and PVS in 1990.

In Scotland, cultivars carrying four genes for virus resistance were soon obtained (Davidson, 1980). At present, multiplex breeding lines are being developed for genes providing extreme resistance to PVY and PVX (Solomon *et al.*, 1989).

At the International Potato Center in Lima, breeding lines with multiple resistance to PLRV, PVY and PVX are being developed (Anon., 1990).

Development of transgenic potatoes

As an alternative to traditional breeding, foreign DNA providing resistance to viruses can be introduced into cultivars. There are many reports of research progress in this area (e.g. Kaniewski *et al.*, 1990) and new procedures are being developed. The results are difficult to predict. However, transgenic cultivars with virus resistance have not yet been registered, and there are many reports that creation of transgenic plants may be associated with negative side-effects. At the same time, progress in breeding resistant cultivars with traditional methods is also considerable. Therefore, it seems safe to assume that at least in the next two decades the main progress in virus resistance will still be associated with traditional breeding methods.

Acknowledgements

The author is highly indebted to Drs M. Chrzanowska, M. Dziewońska, J. Syller and M. Waś for comments on the manuscript, to Dr R. Wastie for correcting the English of the manuscript, and to M. Regulska for help in its technical preparation.

References

Anon. (1950) Definition of some terms used in plant pathology. *Transactions of the Mycological Society* 33, 154-160.
Anon. (1984) Potato virus research. *Annual Report for 1983, International Potato Center*, Lima, pp. 63-74.
Anon. (1989) Control of viruses and virus-like diseases. *Annual Report, International Potato Center*, Lima, pp. 51-64.
Anon. (1990) Control of viruses and virus-like diseases. *Annual Report, International Potato Center*, Lima, pp. 51-62.
Baerecke, M.-L. (1956) Ergebnisse der Resistenzzüchtung gegen Blattrollvirus der Kartoffel. *Zeitschrift für Pflanzenzüchtung* 36, 395-412.
Baerecke, M.-L. (1967a) Überempflindlichkeit gegen das S-Virus der Kartoffel in einem bolivianischen Andigena-Klon. *Züchter* 37, 281-286.
Baerecke, M.-L. (1967b) Prüfung von Saco und Saco-Kreuzungen auf Resistenz gegen das S-Virus der Kartoffel. *European Potato Journal* 10, 206-220.
Bagnall, R.H. and Young, D.A. (1972) Resistance to virus S in the potato. *American Potato Journal* 49, 196-201.
Barker, H. and Harrison, B.D. (1984) Expression of genes for resistance to potato virus Y in potato plants and protoplasts. *Annals of Applied Biology* 105, 539-545.
Barker, H. and Harrison, B.D. (1986) Restricted distribution of potato leafroll antigen in resistant potato genotypes and its effect on transmission of the virus by aphids. *Annals of Applied Biology* 109, 595-604.
Barker, H. and Solomon, R.M. (1990) Evidence of simple genetic control in potato of ability to restrict potato leafroll virus concentration in leaves. *Theoretical and Applied Genetics* 80, 188-192.
Beekman, A.G.B. (1987) Breeding for resistance. In: de Bokx, J.A. and van der Want, J.P.H. (eds) *Viruses of Potatoes and Seed-potato Production*. PUDOC, Wageningen, pp. 162-170.
Beemster, A.B.R. and de Bokx, J.A. (1987) Survey of properties and symptoms. In: de Bokx, J.A. and van der Want, J.P.H. (eds) *Viruses of Potatoes and Seed-potato Production*. PUDOC, Wageningen, pp. 84-113.
Brown, C.R., Salazar, L., Ochoa, C. and Chuquillanqui, C. (1984a) Strain-specific immunity to PVX_{HB} is controlled by a single dominant gene. *Abstracts of the Ninth Triennial Conference of EAPR*, Interlaken, pp. 249-250.
Brown, C.R., Salazar, L., Chavez, R., Schilde-Rentschler, L. and Lizarraga, C. (1984b) Ploidy manipulation of a new source of resistance to PLRV from *Solanum acaule*. *Abstracts of the Ninth Triennial Conference of EAPR*, Interlaken, pp. 288-289.
Bukasov, S.M. and Kameraz, A.J. (1972) Selekcia i semenowodstwo kartofelia. Izdatielstwo 'kołos', Leningrad.

Butkiewicz, H. (1978) Intolerance to potato leafroll virus (PLRV) occurring in potato plants. *Ziemniak - the Potato*. Bonin, pp. 5-37. (In Polish with English summary.)

Chavez, R., Brown, C.R. and Iwanaga, M. (1988) Transfer of resistance to PLRV titre buildup from *Solanum etuberosum* to a tuber-bearing *Solanum* gene pool. *Theoretical and Applied Genetics* 76, 129-135.

Chrzanowska, M. (1988) Reaction of potato clones, field resistant to leafroll virus (PLRV) and resistant to virus Y (PVY) and M (PVM) after artificial infection by graft method. *Biuletyn Instytutu Ziemniaka* 38, 9-20. (In Polish with English summary.)

Chrzanowska, M. (1991) New isolates of the necrotic strain of potato virus Y (PVY^N) found recently in Poland. *Potato Research* 34, 179-182.

Chuquillanqui, C. and Jones, R.A.C. (1980) A rapid technique for assessing the resistance of families of potato seedlings to potato leafroll virus. *Potato Research* 23, 121-128.

Cockerham, G. (1955) Strains of potato virus X. *Proceedings of the 2nd Conference on Potato Virus Diseases*, Lisse-Wageningen, 1954, pp. 89-92.

Cockerham, G. (1970) Genetical studies on resistance to potato viruses X and Y. *Heredity* 25, 309-348.

Cooper, J.I. and Jones, A.T. (1983) Responses of plants to viruses: proposals for the use of terms. *Phytopathology* 73, 127-128.

Dale, M.F.B. and Solomon, R.M. (1988) A glasshouse test to assess the sensitivity of potato cultivars to tobacco rattle virus. *Annals of Applied Biology* 112, 225-229.

Davidson, T.M.V. (1980) Breeding for resistance to virus disease of the potato (*Solanum tuberosum*) at the Scottish Plant Breeding Station. *Fifty-ninth Annual Report 1979-80*, pp. 100-108.

Debener, T., Salamini, F. and Gebhardt, C. (1991) The use of RFLPs (restriction fragment length polymorphism) detects germplasm introgressions from wild species into potato (*Solanum tuberosum* ssp. *tuberosum*) breeding lines. *Plant Breeding* 106, 173-181.

Delhey, R. (1974) Zur Natur der extremen Virusresistenz bei der Kartoffel I. Das X-Virus. *Phytopathologische Zeitschrift* 80, 97-119.

Delhey, R. (1975) Zur Natur der extremen Virusresistenz bei der Kartoffel II. Das Y-Virus. *Phytopathologische Zeitschrift* 82, 163-168.

Dziewońska, M.A. (1975) The differences in resistance to potato virus M. Part 1 and 2. *Ziemniak - the Potato*, Bonin, Vol. 2, pp. 195-234. (In Polish with English summary.)

Dziewońska, M.A. (1986) Development of parental lines for breeding of potatoes resistant to viruses and associated research. In: Beekman, A.G.B., Louwes, K.M., Dellaert, L.M.W. and Neele, A.E.F. (eds) *Potato Research of Tomorrow*. PUDOC, Wageningen, pp. 96-100.

Dziewońska, M.A. (1990) A very high level of resistance to PLRV in a diploid potato clone. *Potato Research* 33, 145 (Abstract).

Dziewońska, M.A. and Ostrowska, K. (1977) Necrotic reaction to potato virus M in *Solanum stoloniferum* and *S. megistacrolobum*. *Phytopathologische Zeitschrift* 88, 172-179.

Dziewońska, M.A. and Pochitonow, Z. (1971) Development of potato clones resistant to viruses. *Zeszyty Problemowe Postępów Nauk Rolniczych* Z. 118, 97-118. (In Polish with English summary.)

Dziewońska, M.A., Czech, B., Ostrowska, K. and Waś, M. (1978) Reaction to potato virus M (PVM) of hybrids with gene *Rm* derived from *Solanum megistacrolobum*. *Abstracts of the Seventh Triennial Conference of EAPR*, Warsaw, pp. 155-156.

Dziewońska, M.A., Butkiewicz, H. and Ostrowska, K. (1988) Advances in the years 1980-1984 in the development of diploid potatoes resistant to viruses. In: Swieżyński, K., Zimnoch-Guzowska, E. and Czech, B. (eds) *Genetic Principles of Potato Breeding*. Bonin, pp. 58-69. (In Polish with English summary.)

Evered, D. and Harnett, S. (eds) (1987) *Plant Resistance to Viruses*. CIBA Symposium 133, J. Wiley and Sons, Chichester.

Fernandez-Northcote, E.N. (1990) Variability of PVX and PVY and its relationship to genetic resistance. In: *Control of Virus and Virus-like Diseases of Potato and Sweet Potato*, 3rd Planning Conference. International Potato Center, Lima, pp. 131-139.

Fraser, R.S.S. (1990) The genetics of resistance to plant viruses. *Annual Review of Phytopathology* 28, 179-200.

Fraser, R.S.S. and Loughlin, S.A.R. (1980) Resistance to tobacco mosaic virus in tomato: effects of the *Tm-1* gene on the virus multiplication. *Journal of General Virology* 48, 87-96.

Fribourg, C.E. and Nakashima, J. (1984) Characterization of a new potyvirus from potato. *Phytopathology* 74, 1363-1369.

Fritig, B., Kaufmann, S., Dumas, B., Feoffroy, P., Kopp, M. and Legrand, M. (1987) Mechanisms of the hypersensitive reaction in plants. In: Evered, D. and Harnett, S. (eds) *Plant Resistance to Viruses*. CIBA Symposium 133, J. Wiley and Sons, Chichester, pp. 92-107.

Galvez, R. and Brown, C.R. (1980) Inheritance of extreme resistance to PVY derived from *Solanum tuberosum* ssp. *andigena*. *American Potato Journal* 57, 476-477.

Harrison, B.D. (1968) Reactions of some old and new British potato cultivars to tobacco rattle virus. *European Potato Journal* 11, 165-176.

Hull, R. (1989) The movement of viruses in plants. *Annual Review of Phytopathology* 27, 213-240.

Hutton, E.M. (1951) Possible genotypes conditioning virus resistance in the potato and the tomato. *Journal of the Institute of Agricultural Science* 17(3), 132-138.

Jellis, G.J. and Gray, L.E.M. (1980) Potatoes: pathology. *Annual Report of the Plant Breeding Institute, Cambridge, 1979*, pp. 45-46.

Jones, R.A.C. (1985) Further studies on resistance breaking strains of potato virus X. *Plant Pathology* 34, 182-189.

Jones, R.A.C. (1990) Strain group specific and virus specific hypersensitive reactions to infection with potyviruses in potato cultivars. *Annals of Applied Biology* 117, 93-105.

Joosten, A. and van der Woude, K. (1985) Geniteurslijst voor aardappel rassen. Commissie ter bevordering van het kweken en het onder zoek van nieuwe aardappel rassen (COA), Wageningen.

Kaniewski, W., Lawson, C., Sammons, B., Haley, L., Hart, J., Delannay, X. and Tumer, N.E. (1990) Field resistance of transgenic Russet Burbank potato to effects of infection by potato virus X and potato virus Y. *Biotechnology* 8, 750-754.

Kapsa, E., Gabriel, W. and Iskrzycka, T. (1983) Resistance to leafroll virus and to PVY, PVM and PVS of 19 potato cultivars included into national list in the years 1967-1975. *Biuletyn Instytutu Ziemniaka* 29, 7-15. (In Polish with English summary.)

Kegler, H. and Kleinhempel, H. (1987) *Virusresistenz bei Pflanzen*. Akademie Verlag, Berlin.

Kegler, H. and Schenk, G. (1990) Connections and correlations of traits and influence factors in quantitative virus resistance of plants (review). *Archiv für Phytopathologie und Pflanzenschutz* 26, 427-439.

Kowalska, A. (1978) Differences among isolates of potato virus M and potato virus S. *Phytopathologische Zeitschrift* 93, 227-240.

Kowalska, A. (1981) Zmienność wirusów M i S ziemniaka. Instytut Ziemniaka, Oddział Naukowo-Badawczy Młochów.

Kowalska, A., Skrzeczkowska, S., Syller, J. and Rudzińska-Langwald, A. (1981) New strain of potato virus A isolated from hybrids of *S. megistacrolobum*. In: Kochman, J., Błaszczak, W. and Książek, D. (eds), *Virus Diseases of Plants. Proceedings of the XIII Conference of Polish Plant Virologists*, Jabłonna, October 1978, pp. 153-165.

Kurppa, A.H.J. (1989) Reaction of potato cultivars to primary and secondary infection by potato mop-top furovirus and strategies for virus detection. *Bulletin OEPP - EPPO Bulletin* 19, 593-598.

Kus, M. (1990) Potato tuber ring necrotic disease (PTRN) in Slovenia. *Abstracts of the Eleventh Triennial Conference of EAPR*, Edinburgh, p. 196.

Makarov, P.P. (1975) Inheritance of resistance to potato virus S. *Potato Research* 18, 326-329.

Matthews, R.E.F. (1992) *Fundamentals of Plant Virology*. Academic Press, San Diego, California.

Moreira, A., Jones, R.A.C. and Fribourg, C.E. (1980) Properties of the resistance breaking strain of potato virus X. *Annals of Applied Biology* 95, 93-103.

Munoz, F.J., Plaisted, R.L. and Thurston, H.D. (1975) Resistance to potato virus Y in *Solanum tuberosum* ssp. *andigena*. *American Potato Journal* 52, 107-115.

Nohejl, J. and Rasocha, V. (1989) Field provoking test for the establishment of degree of potato resistance to mop-top virus. *Proceedings of the 10th Conference of Czechoslovak Plant Virologists*, p. 140 (Abstract).

Oertel, H., Hamann, U. and Runge, M. (1978) Zum Stand der M-Virus-Resistenzzüchtung in der DDR. *Abstracts of the Seventh Triennial Conference of EAPR*, Warsaw, pp. 153-154.

Pietrak, J. (1985) Reaction of potato clones with the gene *Ns* to inoculation with PVS by grafting. *Biuletyn Instytutu Ziemniaka* 32, 17-24. (In Polish with English summary.)

Pochitonow, Z.M. (1989) Dziedziczenie odporności na wirus M ziemniaka (PVM) w diploidalnych mięszańcach od Solanum gourlayi Haw. Unpublished PhD thesis, Institute for Potato Research, Bonin.

Ponz, F. and Bruening, G. (1986) Mechanisms of resistance to plant viruses. *Annual Review of Phytopathology* 24, 355-381.

Ritter, T., Debener, T., Barone, A., Salamini, F. and Gebhardt, C. (1991) RLFP mapping on potato chromosomes of two genes controlling extreme resistance to potato virus X (PVX). *Molecular and General Genetics* 227, 81-85.

Robinson, D.J. (1989) Tobacco rattle tobravirus: variation among strains and detection by c-DNA probes. *Bulletin OEPP - EPPO Bulletin* 19, 619-623.

Robinson, D.J. (1990) Serological variation among tobacco rattle tobravirus (TRV) isolates. *Scottish Crop Research Institute Annual Report for 1989*, p. 80.

Ross, H. (1954) Die Vererbung der 'Immunität' gegen das X-Virus in tetraploidem *Solanum acaule*. *Caryologia* 6 (Suppl.), 1128-1132.

Ross, H. (1961) Über die Vererbung von Eigenschaften für Resistenz gegen das Y-und A-Virus in *Solanum stoloniferum* und die mögliche Bedeutung für eine allgemeine Genetik der Virusresistenz in *Solanum* sect. Tuberarium. *Proceedings of the Fourth Conference on Potato Virus Diseases*, Braunschweig, 1960, pp. 40-49.

Ross, H. (1977) Methods for breeding virus resistant potatoes. *Report of the Planning Conference on Developments in the Control of Potato Virus Diseases.* International Potato Center, Lima, pp. 93-114.

Ross, H. (1986) Potato breeding - problems and perspectives. *Advances in Plant Breeding: Supplement 13. Journal of Plant Breeding.* Paul Parey, Berlin and Hamburg.

Ross, H. and Jacobsen, E. (1976) Beobachtungen an Nachkommenschaften aus Kreuzungen zwischen dihaploiden und tetraploiden kartoffelformen: Samenzahl, Ploidiestufen sowie Spaltungsverhaeltnisse des Gens für extreme Resistenz gegen das X-Virus (Rx_{acl}). *Zeitschrift für Pflanzenzüchtung* 76, 265-280.

Russell, G.E. (1978) *Plant Breeding for Pest and Disease Resistance.* Butterworths, London, Boston.

Salazar, L.F. and Harrison, B.D. (1978) *Potato Virus T.* Commonwealth Mycological Institute - Association of Applied Biologists Descriptions of Plant Viruses, No. 187. Commonwealth Agricultural Bureaux, Farnham Royal, UK.

Schiessendoppler, E. (1990) PVY as causal agent of tuber necrotic ring disease. *Abstracts of the Eleventh Triennial Conference of EAPR*, Edinburgh, pp. 194-195.

Smith, K.M. (1972) *A Textbook of Plant Virus Diseases*, 3rd edn. Longman, London.

Solomon, R.M. (1978) Methods of screening for resistance to potato viruses X and Y. *Abstracts of the Seventh Triennial Conference of EAPR*, Warsaw, pp. 159-160.

Solomon, R.M. (1985) The relationship of genes X^i and *Nb* for resistance to potato virus X. *Heredity* 55, 135-138.

Solomon, R.M., Mackay, G.R. and Muir, J.S. (1989) Virus diseases. *Scottish Crop Research Institute, Annual Report 1988*, pp. 77-78.

Swieżyński, K.M. (1983) Parental line breeding in potatoes. *Genetica (Yugoslavia)* 15, 243-256.

Swieżyński, K.M., Dziewońska, M.A. and Ostrowska, K. (1981) Inheritance of resistance to potato virus M found in *Solanum gourlayi. Genetica Polonica* 22, 1-8.

Swieżyński, K.M., Dziewońska, M.A. and Ostrowska, K. (1988) Reaction to the potato leafroll virus (PLRV) in diploid potatoes. *Potato Research* 31, 289-296.

Swieżyński, K.M., Dziewońska, M.A. and Ostrowska, K. (1989) Resistance to the potato leafroll virus (PLRV) in diploid potatoes. *Plant Breeding* 103, 221-227.

Swieżyński, K.M., Dziewońska, M.A. and Ostrowska, K. (1990) Inheritance of the resistance to the potato leafroll virus (PLRV) in the potato. *Abstracts of the Eleventh Triennial Conference of EAPR*, Edinburgh, pp. 538-539.

Syller, J. (1985) Comparison of some isolates of potato leafroll virus in Poland. *Journal of Phytopathology* 113, 17-23.

Syller, J. (1991) The effects of temperature on the susceptibility of potato plants to infection and accumulation of potato leafroll virus. *Journal of Phytopathology* 133, 216-224.

Tamada, T., Harrison, B.D. and Roberts, I.M. (1984) Variation among British isolates of potato leafroll virus. *Annals of Applied Biology* 104, 107-116.

Tavantzis, S.M. (1990) Outlook on the molecular biology of virus resistance in the potato. In: Vayda, M.E. and Park, W.D. (eds) *The Molecular and Cellular Biology of the Potato*, CAB International, Wallingford, pp. 113-135.

Thomas, P.E., Hassan, S. and Mink, G.I. (1988) Influence of light quality on translocation of tomato yellow top virus and potato leafroll virus in *Lycopersicon peruvianum* and some of its tomato hybrids. *Phytopathology* 78, 1160-1164.

Tingey, M.W. (1991) Potato glandular trichomes. In: Hedin, P.A. (ed.) *Naturally*

Occurring Pest Regulators. American Chemical Society Symposium Series 449, pp. 126-135.

Torrance, L.A., Larkins, P. and Butcher, G.W. (1986) Characterization of monoclonal antibodies against potato virus X and comparison of serotypes with resistance groups. *Journal of General Virology* 67, 57-67.

Vulchev, P. (1969) Studies of the field resistance of a collection of potato varieties to stolbur. *Proceedings of the Fourth Triennial Conference of EAPR*, Brest, pp. 140-141.

Waś, M. (1978) Stable and unstable isolates of TRV in potato plants. *Biuletyn Instytutu Ziemniaka* 21, 47-61. (In Polish with English summary.)

Waś, M. (1988) The influence of temperature and inoculation methods on the reaction of potato clones to potato virus Y (PVY). *Ziemniak - the Potato*, Bonin, pp. 79-89.

Waś, M. and Dziewońska, M.A. (1984) Reaction to PVM and PVS in potato clones with the genes Gm and Ns. *Abstracts of the Ninth Triennial Conference of EAPR*, Interlaken, pp. 245-246.

Waś, M., Dziewońska, M.A., Ostrowska, K. and Kowalska, A. (1980) Reaction of *Solanum gourlayi* and its hybrids with *Solanum tuberosum* to potato virus M (PVM). *Phytopathologische Zeitschrift* 97, 186-191.

Wiersema, H.T. (1961) Methods and means used in breeding potatoes with extreme resistance to viruses X and Y. *Proceedings of the Fourth Conference on Potato Virus Diseases*, Braunschweig, 1960, pp. 30-36.

Zadina, J. (1986) Genetics of the resistance of potatoes to the main and economically most significant viruses in CSRR and its application in potato breeding. Vedecke prace. *Vyzkumny a Slechtitelsky Ustav Bramborarsky v Havlickove Brode* 10, 15-30. (In Czech with English summary.)

Zadina, J. and Novak, F. (1983) The inheritance of extreme intolerence to potato leafroll virus (PLRV). *Genetika a Slechteni* 19, 189-194. (In Czech with English summary.)

16 Inheritance of Resistance to Late Blight

V. UMAERUS[1] AND M. UMAERUS[2]

[1]*Department of Plant and Forest Protection, Swedish University of Agricultural Sciences, PO Box 7044, S-750 07 Uppsala, Sweden and* [2]*Department of Plant Breeding, Swedish University of Agricultural Sciences, PO Box 7003, S-750 07, Uppsala, Sweden.*

Introduction

New results have accumulated in recent years on the cytology, genetics and epidemiology of the fungus, *Phytophthora infestans* (Mont.) de Bary, causing late blight in potatoes, which may call for a reappraisal of resistance breeding (Figure 16.1). A wealth of results is also accumulating, although not always easy to interpret, on interactions between the fungus and *Solanum* species, and on new techniques for incorporating and breeding for resistance. This chapter is an attempt to survey recent publications, using the large number of survey papers and books published during the last few years as a background (Erwin *et al.*, 1983; Bajaj, 1987; Jellis and Richardson, 1987; Vayda and Park, 1990; Ingram and Williams, 1991; Lucas *et al.*, 1991).

Variability in *Phytophthora infestans*

P. infestans was classified as a specialized biotroph when it was first described in the middle of the 19th century. It was later considered a bionecrotroph, but is now considered a biotroph (Robertson, 1991). It reproduces by asexual zoospores or, when its two mating types are present, by sexual oospores.

The fact that until recently the two compatibility groups of *P. infestans* had been identified only in Mexico and the very great variability of the fungus in this area are considered to be indications of its origin in central Mexico (Niederhauser, 1991; further references below).

Sexual reproduction

A substantial amount of knowledge has accumulated in the last 10 years related to the population genetics and the ploidy levels of *P. infestans*

Fig. 16.1. (a) A devastating attack by late blight. (b) Attack by late blight in cv. Bintje, sprayed and unsprayed plots. (Photograph by Karl Arne Hedene.)

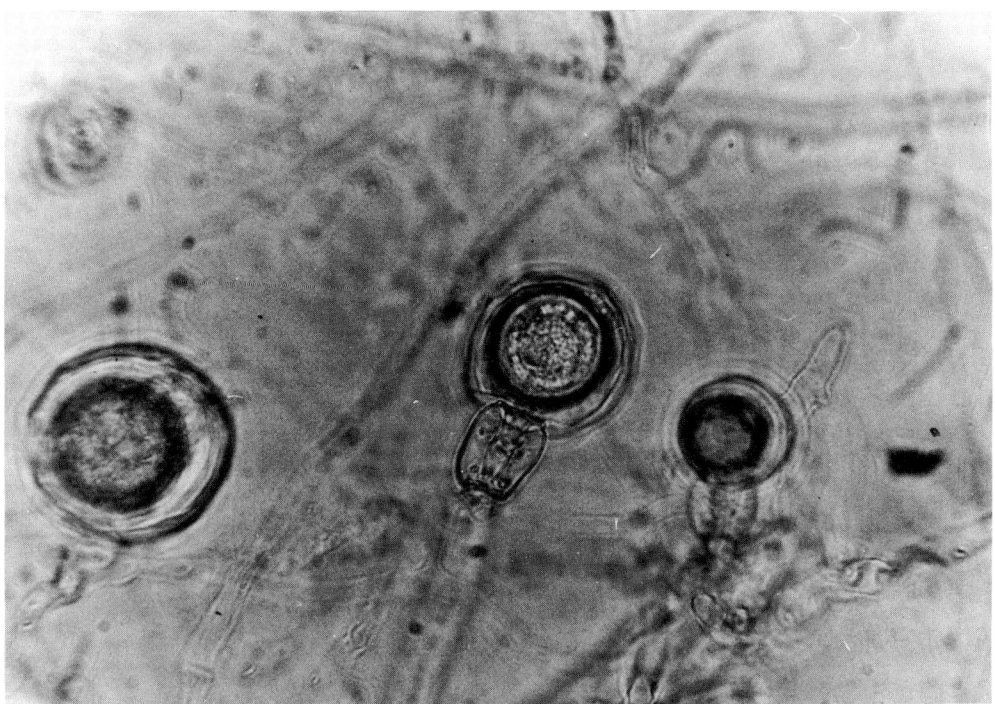

Fig. 16.2. Oospores formed on agar in the presence of the two mating types, A1 and A2 (×1200). (Photograph by Sardar Kadir.)

and has been reviewed recently by Shaw (1991) and Fry and Spielman (1991).

Mating types

The most dramatic discovery of the 1980s was the presence of the A2 mating type in several countries in Europe, the Middle East, Asia and South America (Spielman et al., 1991) and also in the USA and Canada according to a preliminary report (Deahl et al., 1991). Until the 1980s all strains collected outside Mexico were believed to be of the A1 mating type and this still seems to be the case in the USA, Canada and Peru. It has also become apparent that major changes in genotype frequencies are occurring or have recently occurred in Europe, not only in mating types but also in allozyme patterns, virulence frequencies and fungicide resistance.

From observations of mating types and allozyme polymorphisms, Spielman et al. (1991) postulate that two different types of P. infestans populations occur in Europe: pre-A2 or 'old' populations (with only the A1 mating type), and post-A2 or 'new' populations (with both A1 and A2 mating types) (Figure 16.2). Their data indicate that collections from different countries containing new genotypes are closely related to each other, and are distantly related to collections containing old genotypes. They also report evidence that the patterns of genotypic distribution in Europe are not typical of samples from central Mexico (which have contained both mating types and in almost equal proportions). Diversity is much higher in central Mexico, and the distribution of genotypes and mating types appears to be the result of random mating. This is in contrast with the European collections, where observations on genotypes are consistent with clonal reproduction, as shown by Fry et al. (1991) for isolates from The Netherlands collected in 1989.

Ploidy

A surprising result of recent investigations of the ploidy of *P. infestans* isolates from different locations is that central Mexico almost exclusively harbours a diploid population, but isolates from other locations have higher ploidy levels. This work has been summarized by Fry and Spielman (1991) and Shaw (1991). Thus many isolates from the USA, Canada, Peru and Europe have about twice the amount of DNA per zoospore nucleus compared with isolates from central Mexico.

Tooley and Therrien (1991) present a detailed report of the variation in ploidy of *P. infestans* and its implications, particularly with respect to the recent spread of the A2 mating type and possibilities of producing sexual progenies with a presumably tetraploid A1 mating type. The measured DNA contents suggested an increase in tetraploid A2 isolates in The Netherlands through the 1980s. A2 isolates from Poland showed a very wide range of DNA contents, some of them might even be pentaploids or hexaploids. In contrast, the situation in Japan seemed quite different. The A1 isolates had a significantly higher DNA content than the A2 isolates. Also the isozyme composition was different between the two mating types. The discrete and uniform ploidy groups thus found indicate that sexual reproduction is less probable in this population.

Already Sansome and Brasier in their pioneering work (see Sansome, 1977) could demonstrate that British isolates were probably tetraploid while Mexican isolates were diploid. Sansome (1977) suggested two reasons why polyploidy might be advantageous to *P. infestans*. Firstly, tetraploids may be better adapted to the cooler conditions found in temperate zones compared with their related diploids. Secondly, polyploids can store a wider array of virulence allele combinations.

Spielman et al. (1991) and Fry and Spielman (1991) also present data which suggest that individuals in the 'new' population have more virulence factors and in this respect are close to or exceed those found in central Mexico. They also find it likely that the new genotypes are fitter than the old in Europe, because of their rapid increase in frequency after introduction.

Tooley and Therrien (1991) find it premature to draw final conclusions on the impact of polyploidy in *P. infestans*. *In vitro* studies by Spielman and associates (cited by Tooley and Therrien, 1991) did not produce consistent results. Hybrid progenies have been obtained from several presumed triploid × diploid crosses, but in other cases, where a viable progeny was not obtained, the parental isolates were found to differ in DNA content.

Whittaker et al. (1991) were able to show that sexual reproduction can occur in the laboratory between diploid and hyperploid isolates. Viable progenies with a range of DNA contents similar to what has been observed in the field in the UK and in The Netherlands were generated. The percentage germination of oospores was generally lower in the diploid × tetraploid and tetraploid × tetraploid crosses than in the diploid × diploid cross.

Spielman et al. (1991) and Fry and Spielman (1991) interpret the ploidy situation revealed during the 1980s as most probably due to a recent migration of strains from Mexico and population displacement of the fungus. The new and old population types are characterized by unique allozyme alleles and genotypes. A second explanation would be that the European A2 strains arose by mutation from an A1 strain or strains, but they find that the sudden widespread occurrence of the A2s over a large area is more consistent with the hypothesis of migration. A third theory assumes that the A2 isolates could have been present at very low levels and merely increased to detectable frequencies in the late 1970s. Shaw (1987, 1991) and Shattock et al. (1990) do not exclude the possibility that low frequencies of A2 could have been present in Europe for a very long time.

Oospore production

Although no reports have been published which demonstrate the presence of oospores in nature outside Mexico, the two mating types are now spread almost worldwide, major changes in genotype frequencies are found in many locations (in the 'new' population) and *in vitro* experiments show the potential of sexual propagation. In central Mexico oospores were found in blighted leaves from field crops (Niederhauser, 1956; Smoot et al., 1958) and in soils where a potato crop had been grown in the previous season, as observed by A.J. Galindo (Fry et al., 1989; J.S. Niederhauser, pers. comm.). In Europe there are very fragmentary reports. Rullich and Schöber (1988) examined under the microscope 158 samples of leaves and stems of potatoes from 59 locations in the former West Germany but were unable to prove a natural occurrence of oospores.

Oospore production as a result of selfing is not uncommon. A number of external agents have been reported to induce selfing in normally self-sterile A2 isolates (for references see Brasier, 1983; Shaw, 1988). The germination capacity of selfed oospores is very low and colony establishment still lower, as reported by Shattock et al. (1986).

Survival studies have shown oospores to survive freezing and thawing, long-term storage, and even passage through the digestive system of snails. Therefore, they may have the potential of survival in soils not immediately associated with potato tubers, and over long periods, probably more than 10 years if their behaviour is similar to other *Phytophthora* species (Fry et al., 1989). Schöber (1990) reports from one field experiment that oospores may survive in soil and serve as primary source of infection.

The prospects of oospores in the soil as a source of inoculum will drastically change present disease management strategies.

Parasexuality?

Many observations indicate that some sort of somatic hybridization (two hyphae of different genotypes fuse and the nuclei of the two hyphae are mixed) can take place in *Phytophthora*. The perspectives of somatic compatibility and incompatibility are discussed by Shaw and Shattock (1991). The phenomenon of parasexuality – the production and characterization of heterokaryons and their formation – has recently been reviewed by Kuhn (1991). Although the majority of studies were made under laboratory conditions, Layton and Kuhn (1990) demonstrated that heterokaryons formed *in planta*, using *P. megasperma* f. sp. *glycinea*. Kuhn (1991) suggests that the larger karyotypes (polyploids and aneuploids) found in isolates of *P. infestans* and *P. megasperma* could have their bases in parasexual events.

Genetic markers

P. infestans is at present the subject of intensive genetic studies (probably more than any *Phytophthora* species) and many genetic markers have been identified. Three systems of isozymes have been studied (Spielman, 1991) and the genetics of the mating type locus (Shaw and Shattock, 1991). Polymorphisms in ribosomal, mitochondrial and nuclear DNA are valuable tools for population studies, and RFLP markers in both mitochondrial and nuclear DNA have proved useful for studies of populations of *P. infestans* (Carter *et al.*, 1991; Goodwin, 1991).

Goodwin (1991) gives one example: the distribution pattern of *P. infestans* mtDNA forms in different regions. Goodwin and his associates at Cornell have described four variant forms of the mitochondrial genome. The most common has been designated form A and is spread worldwide. A variant of this form is called form B, and also has a worldwide distribution. Carter *et al.* (1991) have designated A and B forms as types I and II respectively. The two forms are strongly associated with other genetic markers (isozymes) in some locations. The B form has not been found in central Mexico but is more common than the A form in samples from both northeastern and northwestern Mexico. Goodwin concludes that the strong association between mtDNA forms and isozyme genotypes suggests that sexual reproduction is probably not occurring within these populations, although both mating types are present. The distribution pattern of the B form within Mexico and its prevalence outside Mexico indicate that the B form mtDNA and the A2 mating type were introduced to Europe concurrently. Also Carter *et al.* (1991) report evidence of conservation of genotypes of type I, indicating that 'members of this group have not readily exchanged genetic material with *P. infestans* belonging to other groups'.

Finding and mapping nuclear DNA markers is generally accomplished by the production of a genetic map of restriction fragment length polymorphisms (RFLPs). Although time-consuming, RFLP mapping of *P. infestans* is in progress, as reported by Goodwin (1991) and Carter *et al.* (1991).

The use of 'DNA fingerprints' to identify individual isolates has only recently been demonstrated for fungi (for references see Goodwin, 1991). Several laboratories have none the less demonstrated the potential of this technique, both for RFLP mapping and for understanding population biology with reference to *P. infestans* (Goodwin, 1991; Carter *et al.*, 1991; A. Drenth, pers. comm.).

The presence of viruses or virus-like particles has been observed in nuclei of certain strains of *P. drechsleri* (Roos and Shaw, 1985) and *P. infestans* (Styer, 1978) and is probably the explanation for the presence of double-stranded RNA in certain isolates. Tooley *et al.* (1989) proposed that the dsRNA could be valuable as a genetic marker to analyse segregation of cytoplasmic traits in *P. infestans* and as an epidemiological marker to monitor worldwide migration of Mexican isolates. Newhouse *et al.* (1992) verified this potential and demonstrated the similarities between isolates from Mexico and The Netherlands in this respect.

Transformation

The first description of a gene transfer procedure involving *P. infestans* was made by Judelson and Michelmore (1991) and Judelson *et al.* (1991). Drug-resistant transformants were obtained without loss of pathogenicity. The technique of gene transfer will be essential for the characterization of genes controlling specificity and pathogenicity. To date, very little is known about the molecular basis of pathogenicity. Pieterse *et al.* (1991) have made some progress in characterizing *P. infestans* genes, which show induced expression during a compatible interaction with the host. Three putative *in planta* induced genes from *P. infestans* have been isolated. One of them codes for ubiquitin, which plays a key role in several cellular processes.

Adaptation

The concept of 'fitness' of *P. infestans* has always been of major concern to potato breeders and plant protectionists. A general treatment of this subject related to fungal pathogen systems has been made by Antonovics and Alexander (1989). Fry and Spielman (1991) and Turkensteen (1992) discuss 'adaptation' of certain strains of *P. infestans* (aggressiveness) to certain cultivars with polygenic partial resistance. Fry *et al.* (1989) address the question specifically in relation to the importance of the perfect stage. Evidence exists in the literature that some isolates of the fungus are more aggressive on certain resistant cultivars. The effects may be small though significant and also quite variable between years. Bjor and Mulelid (1991) reported on such a case in tubers, where there was a significant isolate/cultivar interaction. Fry and Spielman (1991) express their opinion that, if adaptation occurs, 'its influence is likely to be very limited because the degree of adaptation must be very small'.

Increased fitness of metalaxyl-resistant isolates has been demonstrated in Israel (Kadish and Cohen, 1988), but a differential effect was not found when late-blight-susceptible and resistant cultivars were tested (Kadish et al., 1990).

There is at present no convincing proof that isolates from sexual populations would be more aggressive than from asexual populations. Preliminary growth-chamber studies at Cornell University compared a sexual population from central Mexico with a mixture of asexual populations from outside Mexico and measured infection frequency, lesion size and sporulation capacity on leaves. Although some of the components differed, there was no difference between the two populations in the overall effect as expressed by an index combining the three componental effects. Field studies emphasized a bigger role of environment. Fry et al. (1989) concluded that the controversy remains regarding adaptation. Hypothetically, sexual recombination might prevent or enhance adaptation.

Host–Pathogen Interaction

The ability of a plant to stop an invasion of pathogenic fungi depends partly on the presence of preformed barriers (passive defence) and partly on pathogen-induced active responses. Plants respond to pathogen attack with the more or less rapid activation of a complex pattern of defence reactions, which appear to be similar in many diverse plant species. Similar responses have been found to be induced by different pathogens in non-host and host-incompatible interactions.

Types of resistance

In the potato a working concept has been to recognize two main types of resistance, race-specific and general. This is necessarily an oversimplification, but has long served as a framework for resistance breeding. In the concept of general resistance, all aspects of resistance have been included that cannot be attributed to monogenically inherited, dominant R genes, expressed as a hypersensitive reaction. The terms race-specific and general resistance have several approximate synonyms (see, for example, Umaerus et al., 1983).

Race-specific resistance

Breeding for resistance has resulted in the definition of R genes which correspond to virulence genes of the respective pathotypes (Black et al., 1953).

Potato genotypes with race-specific resistance respond with a hypersensitive reaction to incompatible races of *P. infestans*. In compatible interactions the degree of infection depends on the level of underlying general resistance.

General resistance

General resistance may be defined as the ability of the host plant to limit the development of any isolate of the fungus. No implication of type of inheritance is inherent in the term, although it is generally thought that it has a multigenic background and gives greater hopes of durable resistance than the R genes. It may be expressed at different stages of the infection process, and attempts have been made to separate a series of components, e.g. infection efficiency, lesion growth and sporulation, that influence individual steps of the disease cycle.

EXPRESSION OF RESISTANCE IN TUBERS VS. LEAVES

A number of earlier investigations have pointed out that only in some populations is there a correlation between resistance in the foliage and in the tubers. Recent publications have given further evidence of this. Indeed, mechanisms of resistance may be different, since phytoalexins have been found to accumulate in tubers but not in leaves (Rohwer *et al.*, 1987).

INFLUENCE OF R GENES ON LEVEL OF GENERAL RESISTANCE

Darsow (1987) compared resistance in a large number of cultivars and breeding lines with varying numbers of R genes. The trend was that resistance to late blight in the tuber increased with increasing numbers of R genes. These relations were confirmed by evaluation of data reported by other authors, and are in agreement with a number of earlier reports. Turkensteen (1989) found a certain discrepancy compared with expected results in general resistance levels of foliage of subpopulations of four families with and without R genes. However, the presence of R genes was not a prerequisite for increased resistance, since clones with the highest level of resistance were found in both types of subpopulations.

Non-host resistance

The limited host range of most fungal pathogens means that all plants are resistant to most potential pathogens, and that such resistance is very durable. The great similarity in many responses to pathogen attack between host and non-host plant species (e.g. Cuypers and Hahlbrock, 1988) and the possibility of durability have given hope that genes could be transferred between species.

Induced resistance

Induced systemic resistance has been observed in potato cultivars without R-genes after local inoculation of leaves with zoospores of a race of

P. infestans or of *P. cryptogea* isolated from wheat or a solution of K_2HPO_4 (Strömberg and Brishammar, 1991), or after treatment of leaves with hyphal wall components of *P. infestans* (Doke et al., 1987).

Pathogenicity-related proteins (PR proteins) have been discussed in connection with induced resistance. A number of PR proteins have been identified in potatoes, but their role is not clear. Also, it is not clear whether genetic differences exist between potato genotypes in their ability to become induced.

Histology and cytology

Histological and cytological events in the interaction between potatoes and *P. infestans* have been reviewed by Coffey and Gees (1991). The sequence of events in the interaction between *P. infestans* and *Solanum* was described in detail on the basis of available results in the review by Coffey and Wilson (1983a).

The fungus produces both sporangia and zoospores, but only zoospores are believed to cause significant infection of the leaves. Fungal zoospores encyst on the leaf surface and cysts germinate and form appressoria. There have been indications that prepenetration effects on resistance exist (Nandris et al., 1979), but most authors have not been able to prove effects of resistance in this phase (e.g. Berggren et al., 1988; Gees and Hohl, 1988).

Penetration takes place through the periclinal wall of an epidermal cell. Some authors have found preferential penetration of stomata. The wounding destroys the membranes of the ruptured cells and induces wound responses, which may have an effect in delimiting the fungal growth. This is clear, for example, in the interaction in tubers, where a period of wound healing influences the outcome of resistance screening (Schöber and Schiessendoppler, 1983).

In a susceptible leaf the hyphae have been seen to grow intracellularly through the tightly packed palisade layer, but they continue their growth intercellularly, producing intracellular haustoria that end in the host cells. The haustoria are formed in advance of host cell death. In leaf tissue with either specific or general resistance, fungal growth is restricted and more rapid host cell necrosis takes place, generally before observable death of the fungal parasite.

Gees and Hohl (1988) compared several cultivars with different levels of general resistance, which additionally differed by the presence or absence of R3. Penetration frequency did not differ among the cultivar–pathogen combinations. The main cytological feature of incompatibility was that all cells originally penetrated turned brown and prevented the pathogen from invading adjacent cells. Exceptions to this were the guard cells, from which the pathogen frequently colonized a neighbouring subsidiary cell. During compatible interactions this browning reaction was variable (31–98% of the cells). The time required for lesions to start expanding tended to increase in cultivars containing the R3 gene, but speed of tissue colonization following this latent period was fairly uniform in all cultivars where growth beyond the inoculation site occurred, and sporulation, expressed as numbers of sporangia formed per unit area of infected tissue, was similar in all cultivars. Gees and Hohl (1988) concluded that high general resistance may have a phenotype indistinguishable from a hypersensitive reaction evoked by specific resistance genes.

Doke *et al.* (1982) observed varying proportions of hypersensitively reacting and non-reacting cells in tuber tissue challenged by an incompatible race of *P. infestans*.

Histological events were evaluated by Berggren *et al.* (1988), using leaves of potato cultivars with different levels of general resistance. Host genotype did not influence the preinfectional processes of encysted zoospores, for example, frequency of germination, length of germ tubes or frequency of formation of appressoria. There were significant differences between cultivars in the extension of intercellular hyphal growth and branching. In contrast to cv. Bintje, the hyphae in the most resistant cv. Matilda were mostly confined to the epidermal and palisade cells. Even 48 h after infection, the number of hyphae per section was much lower, and sporulation capacity was reduced in the resistant cultivar.

Cuypers and Hahlbrock (1988) found in cv. Datura that spores of two *P. infestans* isolates, one compatible and one incompatible, and one of *P. megasperma* all germinated rapidly and the developing hyphae penetrated leaf tissue within 1–2 h. Epidermal cells turned necrotic in all three cases within a few hours after inoculation. Early and rapid formation of callose-like material was noted in all three cases. The growth of *P. megasperma* hyphae reached a boundary between spongy and palisade parenchyma. *P. infestans* hyphae penetrated the palisade parenchyma in both compatible and incompatible interactions. Rapid and strong necrosis was observed in the incompatible reaction and in the non-host response, and was observed, although rarely, in the compatible interaction. Subsequent sporulation appeared only in the compatible interaction. Cuypers and Hahlbrock (1988) concluded that mesophyll cells have an important role in disease resistance.

Histological evaluation of induced resistance was made by Strömberg and Brishammar (1993). They found that, when systemic disease resistance was induced in the susceptible cv. Bintje by preinfection with the same fungus, the germination of cysts on the leaf surface was enhanced. In the epidermis, papilla deposition increased, penetration decreased and the spread of hyphae in the mesophyll decreased. The reduction of disease severity was thus the result of the combined action of several successive defence reactions.

Biochemistry

The biochemistry of host–pathogen interaction in *Solanum/P. infestans*, and particularly the hypersensitive resistance response in tubers, has been the subject of several recent reviews (Preisig and Kuc, 1987; Tomiyama, 1989; Friend, 1991a; Hohl, 1991a, b; Parker *et al.*, 1991), and potato/late blight is used as an example in most general reviews of host–pathogen interactions (e.g. Dixon and Harrison, 1990).

The following is an attempt to summarize, very briefly and mainly from reviews, the conclusions from the many investigations that have been made on the biochemical defence mechanisms in the potato/*P. infestans* interaction.

Recognition

It has been postulated that recognition is the key event in host–pathogen interaction. However, in spite of enormous efforts and many investigations there is no clear identification of the mechanism of recognition, or knowledge of how recognition leads to either compatibility or incompatibility and resistance. Parker *et al.* (1991) used plant cell cultures in conjunction with fungal elicitors to examine recognition and early signalling events leading to assayable defence responses. They concluded: 'We envisage the presence of many different target sites for pathogen recognition on the plant cell surface funnelling the signal through a transduction chain, which in turn switches on multiple defence reactions.' Friend (1991b) sums up the state of knowledge of biochemistry in the host–pathogen interaction, and particularly recognition, in the following way: 'I would suggest that we have insufficient knowledge at this stage to ask the questions in the correct manner, let alone to answer them correctly.'

Adhesion

Surface-related host–pathogen interactions in *Phytophthora* have been reviewed by Hohl (1991a). Physical contact triggers production of adhesive material in the fungus and callose production in the host. Different types of molecules, produced by the pathogen, appear to be involved in adhesion of the pathogen to different kinds of surfaces, such as the plasma membrane, the host cell wall or a Petri dish. Definite information on the identity of the molecules is not available, but a lectin–ligand-type interaction to host surfaces is implied. No race specificity has been found, and it has not been established that lectin–ligand interactions are involved in recognition, or that adhesion is a prerequisite for infection.

Penetration

Galactanase and galacturonase enzymes produced by the pathogen are important in penetration of the cell wall, with the greater importance attached to galactanase enzymes (Keenan and Friend, 1989). In spite of a diversity of enzymes produced by *P. infestans in vitro*, the late blight disease is not characterized by a massive disintegration of tissue and cell walls. The reason does not seem to have been clarified.

Callose

Callose formation at the point of entry of the fungus occurs in the *P. infestans*/potato relationship as well as in many other host/pathogen interactions. In the early stages of penetration of leaf epidermal cells, host cell wall appositions of callose-like material are often seen. In susceptible reactions with *P. infestans*, the formation of such material appears to be suppressed (Coffey and Wilson, 1983b). In resistant potato cultivars, the haustoria are often small and typically encased by wall-like material composed of callose and cellulose (Hohl, 1991b), while in susceptible cultivars they tend to be larger and usually not encased. The formation of callose-like material in both compatible and incompatible interactions has been noted in several investigations, and it is enhanced in induced resistance. Its importance in the infection process is not clear, however. It is seen by some authors as a general wound response rather than being specifically related to pathogen defence (Cuypers and Hahlbrock, 1988).

Phenolic compounds

In the 1950s and first half of the 1960s, the physiological studies on disease resistance centred around the physiology related to the oxidation of phenolic compounds. However, the toxicities of phenolic compounds were seen as insufficient to explain the inhibition of hyphal growth, and decisive conclusions could not be drawn. Research on phytoalexins instead came into focus in the 1960s. Tomiyama (1989) discusses the hypersensitive defence response in terms of a reaction where polyphenolic and sesquiterpenoid compounds cooperate to protect the tissue from the hyphal development of *P. infestans*, and where the importance of their respective roles varies with the cultivar's type of resistance. Alteration in the host cell wall is considered to be caused by intermediately produced oxidation products caused by the hypersensitive cell death.

Phytoalexins

A very large number of papers on phytoalexins have been published by research groups in the USA and Japan. Only some information from reviews (Keenan *et al.*, 1986; Stössel, 1988; Tomiyama, 1989) and a few selected papers is cited here.

Phytoalexins are defined as 'low molecular weight compounds that are both synthesized by and accumulated in plant cells after exposure to microorganisms' (Paxton, 1981). The accumulation of antibiotically active compounds in potato tubers infected with *P. infestans* served as the basis for the phytoalexin hypothesis proposed by Müller and Börger (1940). Since then, tuber tissue of R-gene resistant cultivars has been used in many investigations. The cultivars Kennebec in the USA and Rishiri in Japan were used in many investigations, and generalizations about resistance mechanisms have been based on the results and taken to confirm the importance of the sesquiterpenoid phytoalexins in the resistance reaction. Rohwer et al. (1987), however, compared the reaction of tubers, leaves and cell suspension cultures from leaves. In tubers, sesquiterpenoid phytoalexins accumulated, although with different speed, in all cases. In leaves, phytoalexins were not detectable. Cell cultures of some cultivars transiently accumulated phytoalexins. Based on this and other results, they expressed doubt as to the importance of the phytoalexins in disease resistance. Brishammar (1987) is also very critical of the concept of phytoalexins in the potato and questions several of the assumptions made.

The phytoalexins of potato are sesquiterpenoid compounds. Those considered to be of importance in disease resistance are rishitin, lubimin and phytuberin. Their biosynthetic pathways have been elucidated. Solavetivone is a key metabolic intermediate for all. Steroid glycoalkaloids and sesquiterpenoid phytoalexins share in part the same biosynthetic pathway, but fungal infection diverts this pathway to the production of sesquiterpenoids.

It has been suggested that non-specific elicitors and specific suppressors produced only by virulent races of the fungus lead to specificity expressed *in planta*. Various types of molecules act as elicitors of phytoalexins. In potato tubers, unsaturated fatty acids – arachidonic and eicosapentaenoic acids – released from the cell walls of *P. infestans* elicit sesquiterpenoid accumulation. Water-soluble glucans from the fungus can suppress the hypersensitive reaction elicited by fungal elicitors and have been proposed as race-specific suppressors of horizontal resistance and phytoalexin response believed to confer gene-for-gene specificity.

Doke and Tomiyama (1980) studied the interaction with the host cell membrane of hyphal wall components of various races of *P. infestans* on protoplasts prepared from tuber tissues of potato cultivars having various resistance genes. There was no significant difference in the physiological activities of the components from seven races of the fungus. Significant differences in the reactivity of protoplasts from different cultivars used were found regardless of the R genes of the cultivars. The higher the field resistance of the cultivars used, the more reactive to hyphal wall components were the protoplasts prepared from them.

The results clearly showed that hyphal components are non-specific elicitors of hypersensitive responses in potato plants rather than determinants of host–parasite specificity. The results also indicate a relationship to general resistance. However, the continued work of the group centred on hypersensitivity and the R-gene response, and as far as we know no further investigations on general resistance have been published.

PR proteins

Pierpoint et al. (1990) reported the presence of inducible pathogenesis-related (PR) proteins in potato, a thaumatin-like protein, chitinase and β-1,3-glucanase. Chitinase and β-1,3-glucanase might be elicitors of other resistance responses, but would not be expected to act directly on *P. infestans*, since oomycetes contain no chitin, and the *Phytophthora* species tested were insensitive to a mixture of chitinase and β-1,3-glucanase (Mauch et al., 1988).

Woloshuk et al. (1991) found that proteins isolated from virus-infected tobacco plants or from tomato plants inoculated with *P. infestans* acted directly on *P. infestans*, causing lysis of sporangia and inhibition of growth *in vitro*. Summing up very recent plant pathological research, Moffat (1992) attaches great importance to the finding that the onset of the plant response has been found to correlate with the production of a group of messenger RNAs, believed to direct the synthesis of PR proteins. Salicylic acid induces expression of the same messenger RNAs.

Genetic basis of resistance

Inheritance of R-gene resistance

In potato there are 11 major genes for hypersensitive reaction resistance to *P. infestans* (Black et al., 1953; Malcolmson and Black, 1966), derived from *S. demissum*, which have been seen to interact with corresponding virulence genes in *P. infestans* in a gene-for-gene relationship, with resistance genes in the host dominant and virulence genes in the pathogen recessive, similar to several other gene-for-gene interactions. A further number of R genes have been investigated or postulated in other *Solanum* species, but their identity to the genes derived from *S. demissum* does not seem to be established. The reaction of some of the R genes has been called 'weak', since they may not under all conditions express a clear hypersensitive necrotic reaction, and may be mistaken for general resistance.

Inheritance of general resistance

Field resistance is believed to be a quantitative character resulting from the joint action of polygenes (Black, 1970). In a number of studies, data have been examined from breeding programmes with introgression of *S. demissum*, where large numbers of crosses and seedlings have been available, and have mainly verified significant GCA effects (Tai and Hodgson, 1975; Malcolmson and Killick, 1980), although an earlier publication by Killick and Malcolmson (1973) indicated considerable SCA effects.

COMPONENTS OF GENERAL RESISTANCE

Attempts have been made to divide the infection process into components which might be separably evaluated and which could be combined in new cultivars. In some cases, indications have been found of the possibility that different components are under separate genetic control, in others not.

Colon and Budding (1988, 1990) tested populations of ten wild *Solanum* species in the field and followed up the field observations with analysis of the type and components of resistance in selected clones of six South American species in comparison with susceptible and field-resistant (Pimpernel) cultivars. A large variation in resistance was found. General resistance was present in all the species. Infection efficiency was found to be an important component of general resistance in all six species and in Pimpernel. This was combined with different rates of lesion growth in different clones.

Beekman (1987) tested 80 genotypes from 12 species for penetration, number of sporangia produced and the latent period. Correlations between these parameters varied from $r = +0.37$ to $r = +0.95$.

The examples indicate that there is at least a certain degree of independence between the components of general resistance, which could be made use of in breeding. It is well known, however, that environmental conditions, for example, day length and mineral nutrition, are important in the expression of general resistance, and there will always be a degree of uncertainty of how this type of resistance is expressed in a new environment.

Pathogen avirulence genes, host resistance genes and host defence response genes

On the basis of the gene-for-gene model, three classes of genes are central to the expression of resistance in plant–pathogen interactions (Dixon *et al.*, 1990): pathogen avirulence genes, which elicit a defensive response by the host, the corresponding host resistance genes, and host defence response genes, which are activated as a result of the interaction between avirulence genes and host resistance genes.

Molecular techniques and potato genetics

Gabriel and Rolfe (1990) state that the cloning of avirulence genes in pathogens has proved that the gene-for-gene relationship is correct – single avirulence genes encoding single protein products are the genetic elements that interact with disease resistance genes. The step from this general statement to detailed knowledge of gene products and mechanisms, and to conclusions about their applicability in the *Solanum/P. infestans* system and in resistance breeding has not yet been taken. Great strides have been made, however, in the use of molecular techniques for different purposes in the potato. Identification of potato chromosomes through Giemsa C-banding (Pijnacker and Ferwerda, 1984) or *in situ* hybridization (Visser *et al.*, 1988) is possible. RFLP techniques have been used for the construction of linkage maps and analysis of genetic variability (Bonierbale *et al.*, 1988; Gebhardt *et al.*, 1989a, b), and markers for all 12 linkage groups in potato are available (Bonierbale *et al.*, 1988). Genetic distance based on RFLPs has been successfully measured in several plant species, including *Solanum* (Debener *et al.*, 1990). Several genes have been characterized, and the process is under way for R genes. The R1 gene has been localized to chromosome 5 (Leonards-Schippers *et al.*, 1991). Molecular techniques are a prerequisite for the characterization of interspecific hybrids, which as we have seen are of great interest for resistance breeding in potato (Pehu *et al.*, 1989, 1990). RFLP analysis can be used to detect chromosomal regions descended from more distantly related *Solanum* species and may be used for mapping traits for which the wild species were introgressed (Debener *et al.*, 1991).

The great efforts to identify general resistance mechanisms may become of very great interest for future potato breeding, and molecular techniques to simplify screening in a breeding programme may become important.

Identity of hypersensitive and general resistance?

Evidence is accumulating that the symptoms of R-gene resistance (hypersensitive reaction) and general resistance are often similar or overlapping. This is in fact an old observation among breeders, who have had difficulty in some cases separating screened material into the expected groups, and was also often reported on in earlier investigations, but it has been accentuated in recent publications. The similarity exists both in the macroscopically observed lesion spread (Colon and Budding, 1988; Turkensteen, 1989; Rivera-Peña, 1990c; Swieżyński, 1990; Swieżyński *et al.*, 1991) and in histological observations (Gees and Hohl, 1988).

The possibility that general and specific resistance could be expressions of the same gene in different environments has been discussed but not found plausible (references in Gees and Hohl, 1988). A possibility to resolve the question might be expected when R genes have been isolated and cloned and their primary products identified. If specific and general resistance genes are different, the early biochemical events in general and specific resistance should differ.

The origin of resistance

It is repeated as axiomatic in virtually all textbooks and reviews from the time of Vavilov (1935) that the coexistence and mutual selection between a local population of host plants and an indigenous pathogen leads to a balanced coexistence. Resistant wild relatives of cultivated species should therefore be looked for in the centres of origin/variability, where the centre of origin/ variation of the pathogen is also to be found. A heavy infection pressure selects for high levels of resistance, implying that the pathogen kills off or prevents from multiplication the susceptible part of the host population, which in turn selects for aggressiveness and virulence in the pathogen. Corroboration has been found in the high incidence of resistance to *P. infestans* in the Mexican gene pool, considered to be the original home and centre of variation of *P. infestans* (Niederhauser, 1991).

Very complex and strong resistance can be found in Mexican wild species (Rivera-Peña, 1990c), which is expected to be created through intense continuous infection pressure. However, no plant could be found during several years of observation that was killed or prevented from reproducing in the wild habitat (Rivera-Peña, 1990a). According to Niederhauser (1991, and references), Mexican wild species are protected by a high level of general (horizontal) resistance. Unnecessary resistance genes with no selective advantage would then be expected to be selected against. Symptoms of hypersensitive and general resistance were difficult to separate in highly resistant wild species (Rivera-Peña, 1990c), but indications were that very complex race-specific resistance was present.

Many races of *P. infestans*, both very virulent and very aggressive, exist in Mexico, both in wild habitats and in cultivated fields (Turkensteen, 1989; Rivera-Peña, 1990b). The most complex races in Mexico have all 11 known virulence factors (Rivera-Peña, 1990b). A rapid change in the race spectrum is taking place in Europe (Schöber and Turkensteen, 1992), and races with nine virulence factors are repeatedly found. In 1991 all virulence factors were found to be present in the pathogen population, although most cultivars do not carry corresponding resistance genes, and some of the genes have never been used in European breeding. Most of the complex races are very aggressive and most virulence genes are superfluous. These observations are not in agreement with the hypothesis of stabilizing selection (Van der Plank, 1963), and support the criticism of Parlevliet (1981), as pointed out by Turkensteen (1992).

Resistance exists in the South American gene pool, although indications are (Niederhauser, 1991) that the existence of *P. infestans* may not have been very long in evolutionary terms. The Neotuberosum project (Glendinning, 1989; Plaisted and Hoopes, 1989) has shown, as did the existence of a degree of resistance in the European cultivars remaining after the heavy attacks in the 1840s, that potential for resistance exists in very susceptible material.

Breeding for Resistance

Probably breeding for resistance to late blight in potatoes (Salaman, 1911) was the first attempt at scientifically based resistance breeding. A single entry in the *Solanum* collection in the Edinburgh Botanical Garden was found to be resistant during a year of heavy late blight attack. It was known to be of Mexican origin. It was called *S. edinense*, and was later determined to be a hybrid with *S. demissum*.

It has often been found difficult, especially under long-day North European and North American conditions, to combine a high level of general resistance with early maturity. Toxopeus (1958) reported that, in cultivars produced before ones with wild-species ancestry were grown, resistance in the foliage and in the tubers tended to be associated and to be related to lateness of maturity. Swieżyński (1990; see this 1990 paper for earlier references) discussed this relationship on the basis of the distribution of potato cultivars from some European countries according to maturity and resistance. According to varietal assessment data, there are no first early cultivars with highly resistant foliage and few late-maturing cultivars with susceptible foliage or tubers. Swieżyński found it plausible that the negative correlation is physiologically determined and therefore difficult to break, rather than an independent trait introduced together with resistance from, for example, *S. demissum*. There are several examples, however, that show that extreme lateness is not a prerequisite for resistance, for example, Beekman (1987). In Neotuberosum (Glendinning, 1989) a similar relationship also exists, but less close than in Tuberosum and some early-maturing Neotuberosum clones have good levels of resistance.

Sources of resistance

S. tuberosum subspp. *tuberosum* and *andigena*

There are several ways of using the available cultivars of (mainly) subsp. *tuberosum* origin, or their ancestral subsp. *andigena* material, in breeding for increased resistance to *P. infestans* without introducing (new) exotic germplasm: conventional breeding or population breeding, somaclonal selection in selected cultivars, and selection in subsp. *andigena* to produce Neotuberosum material.

CULTIVARS AND BREEDING CLONES

Present-day cultivars from a number of countries possess a considerably improved level of general resistance compared with the original material introduced to Europe, even if the results of resistance breeding are sometimes said to be discouraging (e.g. Swieżyński, 1990). Wastie (1991) has recently, in a comprehensive review, described the development of resistance breeding.

The improved general resistance is partly the result of the discarding of the most susceptible clones due to the heavy late blight attacks starting in the 1840s, and partly a result of many decennia of conscious breeding effort. Introgression has taken place of a number of wild *Solanum* species into many existing cultivars. The crosses with exotic germplasm had to be followed by several cycles of backcrossing to eliminate undesirable traits from the exotic species: short-day dependency for tuberization, long stolons, small tuber size, low yield, deep eyes, etc. The result is that the genomes of present-day cultivars and breeding clones are a mosaic of subsp. *tuberosum* with varying numbers and probably sizes of fragments from several other species. Continued breeding efforts based on this material are probably the approach which will pay off most quickly for the production of cultivars combining general resistance with acceptable earliness, tuber morphology, quality and yield. Breeding may take place at the tetraploid level or alternating between the diploid and tetraploid levels, using conventional crosses between selected parents and/or population breeding in wide populations.

SOMACLONAL VARIATION

The possibility of retaining the desirable characteristics of a cultivar with regard to yield, quality and resistance traits, but adding late blight resistance to cultivars lacking in this respect, became an attractive prospect when Shepard *et al.* (1980) observed resistance to *P. infestans* in eight lines regenerated from protoplasts.

In order for the technique to be attractive for practical breeding, it must be shown to be more efficient and/or cheaper than conventional breeding and produce cultivars which are stable. The possibility of screening for resistance *in vitro* is very important for the efficiency of the technique. There has been some reluctance to use the method on a large scale in breeding. Yield ability was questioned and stability over tuber generations was not proved. The suggestion was put forward that higher-yielding somaclones with improved resistance may be bolters (Sanford *et al.*, 1984), which are often met with in seed production programmes, and usually discarded, although a few have become cultivars attractive for different ecological zones from those of the original cultivar. It has been repeatedly shown that many variant somaclones show detrimental changes, e.g. are polyploids, aneuploids, etc. (Karp *et al.*, 1989).

Reports are becoming available from long-term field observation of somaclones, indicating that stability may not be a large issue. Cassels *et al.* (1991) found clones (adventitious regenerants of cv. Bintje) with improved field resistance that were stable in successive years' trials. Most of such clones were of a later-maturity type, however. Clones with greater susceptibility than the original clone were also found. Cassels *et al.* (1987, 1991) applied a screening method in which adventitious regenerants were screened for normal phenotype and vigour, then passed through one cycle of vegetative replication to break down chimeras, and finally screened for late blight resistance under natural infection in Ireland and retention of economically important traits in several years of field trials. Rietveld *et al.* (1991) reported results of observations over five vegetative generations in the field of a large number of clones regenerated *in vitro* from meristems adventitiously initiated on tuber disc explants. Variability was reported, but with only low frequencies of plants with gross phenotypic aberrations, and in some somaclonal populations a shift in the population mean occurred in the desirable direction with stability over tuber generations.

Neotuberosum

N.W. Simmonds began his programme of selection for adaptation of *S. tuberosum* subsp. *andigena* to the environment of the UK in 1959 (Simmonds, 1966; Glendinning, 1975). Neotuberosum derives from a mass selection programme, which included also selection for resistance to late blight through natural late blight occurrence in the field, although improvement of resistance traits was not a primary concern. Tests of the resistance level have been made in the foliage and in the tubers (Glendinning, 1989). A similar programme was initiated in New York in 1963 (Plaisted, 1987), where artificial inoculations were made to select for an increased level of late blight resistance.

Wild and cultivated exotic species

Incorporation of resistance from wild species was attempted very early (Salaman, 1911), and *S. demissum* was used extensively in the 1940s for its hypersensitive monogenic resistance (Ross, 1986; Plaisted and Hoopes, 1989). This species continued to be used because of its general resistance, even after race-specific late blight resistance was considered a failure, and R genes from *S. demissum* are present in most cultivars. Resistance has long been looked for, and has been identified in many species. Many surveys were published earlier and also within the last decade (e.g. Colon *et al.*, 1987; Schmiediche, 1988; Hawkes, 1990; Darsow and Hinze, 1991). Resistance to *P. infestans* is an important part of the information from gene bank collections (Hanneman and Bamberg, 1986; Hoekstra and Seidewitz, 1987; Huaman, 1987). Investigations on the existence and types of resistance of additional species are continually being published.

A very large number of *Solanum* species are potential sources of resistance to *P. infestans*, and the choice of species and accessions to be used in breeding can also take into consideration possible combinations of resistance components and other traits of the species, as well as the ease of incorporating desirable traits in an agronomically usable end-product. For example, Darsow and Hinze (1991) list 60 tuber-forming species in which late blight resistance has been found, and also a number of clones with a satisfactory resistance in the tubers. In some genotypes of 27 species, a combination of resistance in the leaves and in the tubers was found.

The coevolution of the fungus and the wild species in Mexico also indicated the possibility of finding resistance in other Mexican species. However, Mexican species belong to crossability groups that are distinct from most South American taxa, and this has made sexual crosses difficult, although embryo culture has been successful in many cases (Schmiediche, 1988). The occurrence of general resistance was tested in a large number of accessions in the German–Netherlands potato collection by van Soest *et al.* (1984). They confirmed the widespread occurrence of resistance in Mexican accessions, but also found high levels of resistance in several species from the Andes, with a concentration in the Bolivian–Argentinian parts of the centres of origin of the potato.

Methods of incorporating resistance

Sexual hybridization

Not long ago the prerequisite which had to be be fulfilled before a wild species could be considered to be useful in the improvement of cultivated varieties was that it should be crossable with the cultivated ones. Abdalla and Hermsen (1973) identified the diploid *S. verrucosum* as the 'least troublesome' Mexican diploid species for use in potato breeding because of good crossability and regular meiosis.

Possibilities to incorporate traits from other species via sexual crosses have increased through knowledge of sexual mechanisms, such as the endosperm balance number (EBN) (Ortiz and Ehlenfeldt, 1992) and the use of dihaploids of *S. tuberosum*. Peloquin and coworkers, summarized in Jansky *et al.* (1990), stress the advantages of incorporating the germplasm of $2x$ *Solanum* species into the cultivated potato through hybridization with (di)haploids of $4x$ *S. tuberosum*. In products of such crosses, tuber formation and tuber traits are often much improved in comparison with the exotic parent.

Also, it has been shown that additional chromosomes, probably from other species, may be more or less stably incorporated into the potato genome and are not necessarily detrimental (Wilkinson, 1992).

Somatic hybridization, asymmetric fusion and cybrid production

Somatic fusion may provide means by which traits from sexually incompatible wild species can be incorporated into crop plants. Fusion and regeneration have been carried out in a large number of both intra- and interspecific combinations. Refinement of techniques is still part of the investigations on fusion. Somatic fusion in plant breeding has been reviewed by Deimling *et al.* (1988). In *Solanum*, a number of experiments have been made in order to transfer disease resistance from wild species, but fusion experiments have so far been made only in a few cases with the explicit purpose of incorporating resistance to *P. infestans*. In fusion experiments, as with sexual hybridization, the use of dihaploids of cultivars is advantageous in combination with the wild species, which are often diploid. The difficulty with male sterility often encountered in dihaploids should not be important in fusion.

Both somatic fusion and sexual hybridization have been tried in order to incorporate late blight resistance from the Andean diploid species *S. circaeifolium* subsp. *circaeifolium* in combination with dihaploid *S. tuberosum* (Louwes *et al.*, 1992; Mattheij *et al.*, 1992). At least some fertile progenies were produced, which will allow further breeding to be carried out.

Since late-blight-resistant wild species contain not only the trait of interest but also many unfavourable genes, in most cases the transfer of only a small part of the donor genome to the recipient species is intended. In order to limit the need for backcrossing, it may be possible to produce an asymmetrical hybrid. One way to achieve partial genome transfer is to irradiate the donor species with a high dose of X- or gamma rays. Several investigators have used this strategy in other species. In order to be efficient in resistance breeding, the technique needs to be combined with an efficient *in vitro* screening procedure.

Cybrid production in *Solanum* has been reported by Perl *et al.* (1990). The immediate interest is for production of male-sterile lines to be used in hybrid, true potato seed production. The existence of extranuclear inheritance of resistance to *P. infestans* has not been clearly documented.

Gene transfer

So far, knowledge of the genetic and biochemical background in the potato/ *P. infestans* interaction has not been sufficient for gene transfer to have a real impact in cultivar improvement. According to Dixon *et al.* (1990), the host defence response genes, which are activated as a result of the interaction between avirulence genes and host resistance genes, are the type of genes that would be best suited for transfer to different species. A number of defence response genes have been cloned, some from potato (cited by Dixon *et al.*, 1990). The technique may also open possibilities for the transfer of genes influencing resistance from non-*Solanum* species when general mechanisms for resistance reactions are better understood. Experiments are under way to transfer genes from insects encoding antimicrobial proteins, which have been shown to also have effects against fungi (Destefano-Beltran *et al.*, 1990).

Selection for resistance

A survey of techniques used for evaluation of resistance in field and greenhouse experiments was published by Schöber and Umaerus (1987). In breeding programmes, a stepwise selection is often carried out, starting with mass screening in the greenhouse, sometimes followed by a more detailed evaluation of resistance components (Umaerus and Lihnell, 1976) and testing under the very severe conditions in Mexico, and finishing with an evaluation in the field under local conditions and testing of tuber resistance.

Screening for specific and general resistance in the foliage

Drop inoculation of detached leaflets has been used to evaluate both specific (Schick and Hopfe, 1962) and general (Killick and Malcolmson, 1973) resistance. Swieżyński *et al.* (1991) applied the method in an evaluation of a large number of seedlings from resistant parents of *S. tuberosum* introgressed with species known to carry genes for specific resistance: *S. demissum*, *S. stoloniferum*, *S. verrucosum* and *S. microdontum*. In the progenies evaluated, seedling populations which were exclusively resistant (no spreading lesions) or exclusively susceptible (large spreading lesions) were more frequent than populations with a variable reaction, and the frequency of genotypes with moderately resistant leaflets was low. This caused the authors to suspect that the resistance of the parents was due mainly to the presence of genes for specific resistance, and they cautioned that a similar situation may be found in many present-day resistant cultivars.

Tuber resistance tests

Although tuber testing is included in most breeding programmes, it is not always regarded as the most important component of resistance. However, according to Darsow (1987), tuber slice testing started in 1967 in the DDR and has been applied every year since 1970 on a large scale. Several different tests are used in different countries and breeding programmes (Bjor, 1987; Wastie et al., 1987), some of which use tuber slices, with or without a period of wound healing, and some whole tubers. Testing is sometimes done at harvest, sometimes during storage. Variable results are often reported, which makes testing over several years necessary. This in turn makes it difficult to screen for tuber resistance early in a breeding programme. Part of the difficulty in evaluating tuber resistance may depend on the influence of the physiological age of the tubers. This has long been known, but not always taken into account. Immature tubers are usually susceptible. Bhatia and Young (1985) found an interaction between physiological age and cultivar; most of the tested clones became increasingly susceptible over storage time, sometimes with dramatic changes, while some remained resistant.

Screening in vitro

In vitro selection for disease resistance was reviewed by Jones (1990).

A number of attempts have been made to obtain resistance by use of the fungal toxic metabolite of the pathogen (Stolle and Schöber, 1984) as selective agent. Treatment of plants with the fungal filtrate showed no relationship with degree or type of cultivar resistance. Culture filtrates were used as selective agents *in vitro* using callus (Behnke, 1980; Foroughi-Wehr and Stolle, 1985), micronodes on mutagen-treated axillary buds (Crinò et al., 1990), or plantlets (Meulemans et al., 1987). A correlation between *in vitro* response and field performance has not always been found, and results have been disappointing with regard to effect on other plant traits when calli have been used. Field testing over a 5-year period showed changes in an undesired direction to have been induced, particularly in other resistance traits, in the material selected by Behnke (1980), as reported by Wenzel et al. (1987) and Wenzel (1988). However, this was seen as due to mutations during the callus phase, and modification of the screening procedure might lead to less detrimental effects.

Meulemans and Fouarge (1986) screened plantlets regenerated from protoplasts, from internode calli, from internodes or from leaflets, which were sprayed under axenic conditions with a sporangia suspension of *P. infestans*. Resistant regenerants (slow or no parasite extension when inoculated *in vitro* with race 0) were obtained from all sources except true-to-type micropropagation, and cocultivation *in vitro* of adventitious shoots and *P. infestans* was seen as a possible method by Meulemans et al. (1987). Götz (1991) tested *in vitro*-grown plants for horizontal resistance and compared the results with field tests. Her conclusion was that the *in vitro* test would be useful for preselection, but did not always show good agreement with field results.

Conclusions

A change in the *P. infestans* population outside Mexico is evident from the many reports published during the 1980s. Irrespective of its origin, whether due to a recent migration of strains from Mexico or a slow shift which is now noticeable, the 'new' population has certain characteristics which are of concern to those engaged in breeding and plant protection. These are the increasing frequency of the A2 mating type and of virulence genes, the selection towards higher DNA content and a probable better fitness than the 'old' population.

The presence of the two mating types in the same population most probably indicates that the new population has or will have the potential of sexual recombination besides vegetative propagation (although somatic hybridization has most probably played a role, as well as mutation). At present, there are no convincing proofs that viable oospores or hybrid progenies exist in nature outside central Mexico. Only *in vitro* studies have demonstrated the potential of sexual reproduction. An explanation might be the variation in ploidy (unbalanced ploidy situation) between isolates within the new as well as between the new and old population. This could be a transition phase in an evolutionary process. It also remains to be seen if sexual reproduction has an advantage over clonal reproduction.

The presence of many virulence genes in the new population emphasizes the shortcomings of breeding for race-specific resistance, as already elaborated by Turkensteen (1989, 1992).

An increased fitness, in terms of ecological fitness or adaptation to different environments, could lead to increased disease problems in marginal areas of the present distribution of the disease. If oospores become a means of survival in the soil and a primary source of inoculum, a dramatic change in the epidemiology of the disease will take place. Another question is raised by the variability in ploidy levels of the fungus. Are dosage effects for virulence probable?

In general, the information on variability of the fungus which has accumulated during the last decade points only in one direction and that is an increased need of resistant cultivars and thus a higher priority of breeding for resistance to *P. infestans*.

The interactions of hosts and pathogens appear more complex as our understanding of them increases. The potato appears to have a multicomponent mechanism for disease resistance. From a breeding point of view, the difference between race-specific and general resistance is still important. The durability of resistance cannot be proved until it has been 'broken down'. Ample evidence exists, through many years of practical experience, that R-gene resistance against *P. infestans* in the potato is not durable. General resistance has been equated with durable resistance, since it is expected to depend on many genes and possibly several mechanisms. The durability has not so far been disproved. The difference between race-specific and general resistance resides in the race specificity, i.e. the reaction of a host genotype as resistant towards some fungal pathotypes and susceptible to others. It cannot be defined by the type of reaction, which may be similar both histologically and biochemically, and the gene-for-gene relationship is not as clear as it appeared.

The last few years have not clarified the exact events in the infection process. According to Coffey and Gees (1991):

> It appears to us that the results of many years of study of resistance mechanisms have revealed little useful information. The one clear conclusion is that cytological and biochemical events in resistant interactions appear to be decidedly non-specific in nature. We would like to propose that specificity resides in the susceptible event.

This has also been argued by Van der Plank (1989).

Although the R-gene gene-for-gene concept is well established, recent results indicate a more complex picture, not yet clearly interpreted, and raise some doubt about the relationship of recessive virulence genes to dominant resistance genes. In the cases investigated more thoroughly, i.e. virulence against host resistance genes R1, R2, R3 and R4, virulence in diploid *P. infestans* was under single-locus control, with avirulence dominant (Al-Kherb, 1988). Spielman *et al.* (1989) could not verify these results in all aspects. Virulence appeared to be dominant and controlled by one locus against potato resistance genes R2 and R4. Avirulence against R3 appears to be dominant and possibly controlled by two loci in the fungus. The segregation for virulence against R1 did not indicate which phenotype was dominant, but did suggest single-locus control.

There seems to be a necessity to continuously revise priorities and test methods both for resistance in the foliage and in the tubers, and their relative role. The role of tuber resistance may have been underestimated in many breeding programmes.

Although no breeding programme relies solely on hypersensitive resistance, there is no unanimity in the conclusions of different authors as to the desirability of complementing general resistance with R genes. Darsow (1987) and Turkensteen (1989) draw somewhat different conclusions for breeding from results showing a certain degree of influence of the existence and dosage of R genes in cultivars with general resistance. Darsow's conclusion is as follows:

> Relative resistance was found to occur in conjunction with R-genes necessarily causatively related to them. The existence of only a loose connection between those two forms of resistance does not support a need for preselection for R-genes, but equally it would not justify any discarding of forms with R-genes.

Turkensteen, advising CIP, sees no necessity to include R-genes, since they make screening for general resistance more difficult, especially under the conditions in developing countries. The difficulty of separating symptoms of race-specific and general resistance causes Swieżyński *et al.* (1991) to conclude that progeny evaluation is necessary for evaluation of the type of resistance present in a clone.

Variation in the *Solanum* gene pool is tremendous, and the difficulty for a breeder is not in identifying sources of resistance, but in deciding which source is to be preferred from the points of view of expected durability of resistance, ease of incorporation into the cultivated material, and expected efficiency in breeding, due to the necessity of combining the many desirable traits necessary to make a cultivar both acceptable to the farmer and attractive on the market. For both these goals of cultivar production, environmental aspects have become prominent during the last few years.

Classical breeding is time-consuming, and any method that is more efficient is of advantage. *In vitro* methods offer freedom from the effects of a variable natural environment, thus making it easier to measure slight quantitative differences in polygenically inherited characters. Large numbers of individuals can be handled in a very small space. Working with microspores and dihaploids promises ease in the discovery of recessive traits and additive characters and a reduction in the population size required. Various *in vitro* techniques are being included in practical breeding programmes. However, all techniques do not work in all genotypes, which limits their usefulness, and undesirable side-effects have hampered the use of *in vitro* screening.

Durability of resistance is the first concern in late blight resistance breeding. The hopes were great, when R gene resistance was introduced, that a lasting solution was possible. The hope now is that information on resistance mechanisms may give rise to more accurate predictions which allow the choice of the best type(s) of resistance as well as the most efficient breeding method. Non-host resistance (Niks, 1988) has received increased attention, since techniques are beginning to be available for the transfer of genes between distantly or non-related species. Not enough is known to predict its potential and durability. The results surveyed by Moffat (1992) indicate possibilities of strengthening the overall resistance to pathogens, not only *P. infestans*. With the very great resources allotted to biotechnology, results will be expected very soon – indeed, many more publications can be expected before this book is published, and conclusions made now may need to be revised.

References

Abdalla, M.M.F. and Hermsen, J.G.Th. (1973) An evaluation of *Solanum verrucosum* Schlechtd. for its possible use in potato breeding. *Euphytica* 22, 19–27.

Al Kherb, S.M. (1988) The inheritance of host-specific pathogenicity in *Phytophthora infestans*. Unpublished PhD thesis, University College of North Wales, Bangor (cited by Spielman *et al.*, 1989).

Antonovics J. and Alexander, H.M. (1989) The concept of fitness in plant–fungal pathogen systems. In: Leonard, K.J. and Fry, W.E. (eds) *Plant Disease Epidemiology, Genetics, Resistance and Management*. McGraw-Hill, New York, Toronto, London, Vol. 2, Chapter 7, pp. 185–214.

Bajaj, Y.P.S. (ed.) (1987) *Biotechnology in Agriculture and Forestry, 3. Potato*. Springer Verlag, Berlin, Heidelberg, New York.

Beekman, A.G.B. (1987) *Phytophthora* research at the Foundation for Agricultural

Plant Breeding (SVP), The Netherlands. In: Jellis, G.J. and Richardson, D.E. (eds) *The Production of New Potato Varieties: Technological Advances.* Cambridge University Press, Cambridge, pp. 99-101.

Behnke, M. (1980) General resistance to late blight of *Solanum tuberosum* plants regenerated from callus resistant to culture filtrates of *Phytophthora infestans. Theoretical and Applied Genetics* 56, 151-152.

Berggren, B., Widmark, A.-K. and Umaerus, V. (1988) The expression of general resistance to late blight (*Phytophthora infestans*) in potato leaves. *Potato Research* 31, 611-616.

Bhatia, S.K. and Young, R.J. (1985) Reaction of potato tuber slices to *Phytophthora infestans* in relation to physiological age. *American Potato Journal* 62, 471-478.

Bjor, T. (1987) Testing the resistance of potato genotypes to tuber blight. *Potato Research* 30, 525-532.

Bjor, T. and Mulelid, K. (1991) Differential resistance to tuber late blight in potato cultivars without R-genes. *Potato Research* 34, 3-8.

Black, W. (1970) The nature of inheritance of field resistance to late blight (*Phytophthora infestans*) in potatoes. *American Potato Journal* 47, 279-288.

Black, W., Mastenbroek, C., Millsand, W.R. and Peterson, L.C. (1953) A proposal for an international nomenclature of races of *Phytophthora infestans* and of genes controlling immunity in *Solanum demissum* derivatives. *Euphytica* 2, 173-178.

Bonierbale, M.W., Plaisted, R.L. and Tanksley, S.D. (1988) RFLP maps based on a common set of clones reveal modes of chromosomal evolution in potato and tomato. *Genetics* 120, 1095-1103.

Brasier, C.M. (1983) Problems and prospects in *Phytophthora* research. In: Erwin, D.C., Bartnicki-Garcia, S. and Tsao, P.H. (eds) *Phytophthora, its Biology, Taxonomy, Ecology and Pathology.* American Phytopathological Society, St Paul, Minnesota, pp. 351-364.

Brishammar, S. (1987) Critical aspects of phytoalexins in potato. *Journal of Agricultural Sciences in Finland* 59, 217-230.

Carter D.A., Archer, S.A., Buck, K.W., Shaw, D.S. and Shattock, R.C. (1991) DNA polymorphism in *Phytophthora infestans*: the UK experience. In: Lucas, J.A., Shattock, R.C., Shaw, D.S. and Cooke, L.R. (eds) *Phytophthora.* Symposium of the British Mycological Society, the British Society for Plant Pathology and the Society of Irish Plant Pathologists held at Trinity College, Dublin, September 1989. Cambridge University Press, Cambridge, pp. 272-294.

Cassels, A.C., Austin, S. and Götz, E.M. (1987) Variation in tubers in single cell-derived clones of potato in Ireland. In: Bajaj, Y.P.S. (ed.) *Biotechnology in Agriculture and Forestry. 3. Potato.* Springer-Verlag, Berlin, pp. 375-391.

Cassels, A.C., Deadman, M.L., Brown, C.A. and Griffin, E. (1991) Field resistance to late blight (*Phytophthora infestans* (Mont.) de Bary) in potato (*Solanum tuberosum* L.) somaclones associated with instability and pleiotropic effects. *Euphytica* 56, 75-80.

Coffey, M.D. and Gees, R. (1991) The cytology of development. In: Ingram, D.S. and Williams, P.H. (eds) *Advances in Plant Pathology, Volume 7, Phytophthora infestans, the Cause of Late Blight of Potato.* Academic Press, London, pp. 31-52.

Coffey, M.D. and Wilson, U.E. (1983a) Histology and cytology of infection and disease caused by *Phytophthora.* In: Erwin, D.C., Bartnicki-Garcia, S. and Tsao, P.H. (eds) *Phytophthora: Its Biology, Taxonomy, Ecology and Pathology.* American Phytopathological Society, St Paul, Minnesota, pp. 289-301.

Coffey, M.D. and Wilson, U.E. (1983b) An ultrastructural study of the late blight fungus *Phytophthora infestans* and its relation with the foliage of two potato cultivars possessing different levels of general (field) resistance. *Canadian Journal of Botany* 61, 2669-2685.

Colon, L.T. and Budding, D.J. (1988) Resistance to late blight (*Phytophthora infestans*) in ten wild *Solanum* species. *Euphytica Supplement* (December 1988), 77-86.

Colon, L.T. and Budding, D.J. (1990) Components of field resistance to *Phytophthora infestans* in *Solanum* ssp. *Abstracts of the Eleventh Triennial Conference of EAPR*, Edinburgh, UK, 8-13 July 1990, pp. 81-82.

Colon, L.T., Sijpkes, L. and Budding, D.J. (1987) *Aardappel. Phytophthora infestans. Resistentie in wilde Solanumsorten.* Jaarverslag, SVP, Wageningen, pp. 44-45.

Crinò, P., Lai, A., Penuela, R. Martino, L., Papacchioli, V. and Sonnino, A. (1990) In vitro mutation breeding for resistance to *Phytophthora infestans*. *Abstracts of the Eleventh Triennial Conference of EAPR*, Edinburgh, pp. 8-9.

Cuypers, B. and Hahlbrock, K. (1988) Immunohistochemical studies of compatible and incompatible interactions of potato leaves with *Phytophthora infestans* and the nonhost response to *Phytophthora megasperma*. *Canadian Journal of Botany* 66, 700-705.

Darsow, U. (1987) Long-term results of a tuber slice test for relative resistance to late blight. *Potato Research* 30, 9-22.

Darsow, U. and Hinze, E. (1991) Nutzung von Wildarten zur Verbesserung der relativen *Phytophthora*-Resistenz in der Kartoffelzüchtung. *Kartoffelforschung aktuell, Gross Lüsewitz*, pp. 43-51.

Deahl, K.L., Goth, R.W., Young, R., Sinden, S.L. and Gallegly, M.E. (1991) Occurrence of the A2 mating type of *Phytophthora infestans* in potato fields in the United States and Canada. *American Potato Journal* 68, 717-725.

Debener, T., Salamini, F. and Gebhardt, C. (1990) Phylogeny of wild and cultivated *Solanum* species based on nuclear restriction fragment length polymorphisms (RFLPs). *Theoretical and Applied Genetics* 79, 360-368.

Debener, T., Salamini, F. and Gebhardt, C. (1991) The use of RFLPs (restriction fragment length polymorphisms) detects germplasm introgressions from wild species into potato (*Solanum tuberosum* ssp. *tuberosum*) breeding lines. *Plant Breeding* 106, 173-181.

Deimling, S., Zitzlsperger, J. and Wenzel, G. (1988) Somatic fusion for breeding of tetraploid potatoes. *Plant Breeding* 101, 181-189.

Destefano-Beltran, L., Nagpala, P.G., Cetiner, M.S., Dodds, J.H. and Jaynes, J.M. (1990) Enhancing bacterial and fungal disease resistance in plants: application to potato. In: Vayda, M.E. and Park, W.D. (eds) *The Molecular and Cellular Biology of the Potato*. CAB International, Wallingford, pp. 205-221.

Dixon, R.A. and Harrison, M.J. (1990) Activation, structure, and organization of genes involved in microbial defense in plants. *Advances in Genetics* 28, 165-234.

Dixon, R.A., Blyden, E.R. and Ellis, J.A. (1990) Biochemistry and molecular genetics of plant-pathogen systems. In: Khanna, K.R. (ed.) *Biochemical Aspects of Crop Improvement*. CRC Press, Boca Raton, Ann Arbor, Boston, Chapter 7, pp. 179-222.

Doke, N. and Tomiyama, K. (1980) Effect of hyphal wall components from *Phytophthora infestans* on protoplasts of potato tuber tissues. *Physiological Plant Pathology* 16, 179-186.

Doke, N., Tomiyama, K. and Furuichi, N. (1982) Elicitation and suppression of the hypersensitive response in host-parasite specificity. In: Asada, Y., Bushnell,

W.R., Ouchi, S. and Vance, C.P. (eds) *Plant Infection: the Physiological and Biochemical Basis*, Springer Verlag, Berlin, pp. 79-96.

Doke, N., Ramirez, A.V. and Tomiyama, K. (1987) Systemic induction of resistance in potato plants against *Phytophthora infestans* by local treatment with hyphal wall components of the fungus. *Journal of Phytopathology* 119, 232-239.

Erwin, D.C., Bartnicki-Garcia, S. and Tsao, P.H. (eds) (1983) *Phytophthora, its Biology, Taxonomy, Ecology and Pathology*. American Phytopathological Society, St Paul, Minnesota.

Foroughi-Wehr, B. and Stolle, K. (1985) In vitro selection for resistance exemplified by the system potato/*Phytophthora infestans* (Mont.) de Bary. *Nachrittenblatt des Deutschen Pflanzenschutzdienstes* 37, 170-173.

Friend, J. (1991a) The biochemistry and cell biology of interaction. In: Ingram, D.S. and Williams, P.H. (eds) *Advances in Plant Pathology, Volume 7, Phytophthora infestans, the Cause of Late Blight of Potato*. Academic Press, London, pp. 85-129.

Friend, J. (1991b) Host-pathogen interactions: current questions. In: Lucas, J.A., Shattock, R.C., Shaw, D.S. and Cooke, L.R. (eds) *Phytophthora*. Symposium of the British Mycological Society, the British Society for Plant Pathology and the Society of Irish Plant Pathologists held at Trinity College, Dublin, September 1989. Cambridge University Press, Cambridge, pp. 46-49.

Fry, W.E. and Spielman, L.J. (1991) Population biology. In: Ingram, D.S. and Williams, P.H. (eds) *Advances in Plant Pathology, Volume 7, Phytophthora infestans, the Cause of Late Blight of Potato*. Academic Press, London, pp. 171-192.

Fry, W.E., Tooley, P.W. and Spielman, L.J. (1989) The importance of the perfect stage of *Phytophthora infestans* from the standpoint of epidemiology and adaptation. *Report of the Planning Conference on Fungal Diseases of the Potato*, CIP, Lima (Peru) 21-25 September 1987, pp. 17-30.

Fry, W.E., Drenth, A., Spielman, L.J., Mantel, B.C., Davidse, L.C. and Goodwin, S.B. (1991) Population genetic structure of *Phytophthora infestans* in the Netherlands. *Phytopathology* 81, 1330-1336.

Gabriel, D.W. and Rolfe, B.G. (1990) Working models of specific recognition in plant-microbe interactions. *Annual Review of Phytopathology* 28, 365-391.

Gebhardt, C., Blomendahl, U., Schachtschnabel, U., Debener, T., Salamini, F. and Ritter, E. (1989a) Identification of 2n breeding lines and 4n varieties of potato (*Solanum tuberosum*, ssp. *tuberosum*) with RFLP-fingerprints. *Theoretical and Applied Genetics* 78, 16-22.

Gebhardt, C., Ritter, E., Debener, T., Schachtschnabel, U., Walkemeier, B., Uhrig, H. and Salamini, F. (1989b) RFLP analysis and linkage mapping in *Solanum tuberosum*. *Theoretical and Applied Genetics* 78, 65-75.

Gees, R. and Hohl, H.R. (1988) Cytological comparison of specific (R3) and general resistance to late blight in potato leaf tissue. *Phytopathology* 78, 350-357.

Glendinning, D.R. (1975) Neo-tuberosum: new potato breeding material. 1. The origin, composition and development of the Tuberosum and Neo-tuberosum gene pools. *Potato Research* 18, 256-261.

Glendinning, R. (1989) Late blight resistance in the British Neo-Tuberosum potato population. *Potato Research* 32, 321-330.

Goodwin (1991) DNA polymorhism in *Phytophthora infestans*: the Cornell experience. In: Lucas, J.A., Shattock, R.C., Shaw, D.S. and Cooke, L.R. (eds) *Phytophthora*. Symposium of the British Mycological Society, the British Society for Plant Pathology and the Society of Irish Plant Pathologists held at Trinity College, Dublin, September 1989. Cambridge University Press, pp. 256–271.

Götz, E. (1991) Untersuchungen zur vertikalen und horizontalen *Phytophthora*-Resistenz an *in-vitro*-Pflanzen der Kartoffel. *Kartoffelforschung Aktuell, Groß Lüsewitz*, pp. 19–24.

Hanneman, R.E. and Bamberg, J.B. (1986) Inventory of tuber bearing Solanum species. *Wisconsin Agricultural Experiment Station Bulletin* 533.

Hawkes, J.G. (1990) *The Potato: Evolution, Biodiversity and Genetic Resources*. Belhaven Press, London.

Hoekstra, R. and Seidewitz, L. (1987) *Evaluation Data on Tuber-bearing* Solanum *Species*, 2nd edn. German–Dutch Curatorium for Plant Genetic Resources. FAL, Braunschweig, and SvP, Wageningen.

Hohl, H.R. (1991a) Surface-related host–pathogen interactions in *Phytophthora*. In: Lucas, J.A., Shattock, R.C., Shaw, D.S. and Cooke, L.R. (eds) *Phytophthora*. Symposium of the British Mycological Society, the British Society for Plant Pathology and the Society of Irish Plant Pathologists held at Trinity College, Dublin, September 1989. Cambridge University Press, Cambridge, pp. 70–89.

Hohl, H.R. (1991b) Nutrition. In: Ingram, D.S. and Williams, P.H. (eds) *Advances in Plant Pathology, Volume 7, Phytophthora infestans, the Cause of Late Blight of Potato*. Academic Press, London, pp. 53–83.

Huaman, Z. (1987) *Inventory of Andean Potato Cultivars with Resistance to Some Pests and Diseases and Other Desirable Traits*. CIP, Lima, Peru.

Ingram, D.S. and Williams, P.H. (eds) (1991) *Advances in Plant Pathology, Volume 7, Phytophthora infestans, the Cause of Late Blight of Potato*. Academic Press, London.

Jansky, S.H., Yerk, G.L. and Peloquin, S.J. (1990) The use of potato haploids to put $2x$ wild species germplasm into a usable form. *Plant Breeding* 14, 290–294.

Jellis, G.J. and Richardson, D.E. (eds) (1987) *The Production of New Potato Varieties: Technological Advances*. Cambridge University Press, Cambridge.

Jones, P.W. (1990) Disease resistance. In: Dix, P.J. (ed.) *Plant Cell Line Selection*. VCH Verlagsgesellschaft, Weinheim, pp. 113–149.

Judelson, H.S. and Michelmore, R.W. (1991) Transient expression of genes in the oomycete *Phytophthora infestans* using *Bremia lactucae*. *Current Genetics* 19, 453–459.

Judelson, H.S., Tyler, B.M. and Michelmore, R.W. (1991) Transformation of the oomycete pathogen, *Phytophthora infestans*. *Molecular Plant–Microbe Interactions* 4, 602–607.

Kadish, D. and Cohen, Y. (1988) Fitness of *Phytophthora infestans* isolates from metalaxyl-sensitive and -resistant populations. *Phytopathology* 78, 912–915.

Kadish, D., Grinberger, M. and Coohen, Y. (1990) Fitness of metalaxyl-sensitive and metalaxyl-resistant isolates of *Phytophthora infestans* on susceptible and resistant potato cultivars. *Phytopathology* 80, 200–205.

Karp, A., Jones, M.G.K., Foulger, D., Fish, N. and Bright, S.W.J. (1989) Variability in potato tissue culture. *American Potato Journal* 66, 669–684.

Keenan, P.J. and Friend, J. (1989) The degradation of potato cell walls by pathogens. In: Osborne, D.J. and Jackson, M.B. (eds) *Cell Separation in Plants*. Springer-Verlag, Berlin, pp. 179–187.

Keenan, P.J., Ellis, J.S., Rathmell, W.G. and Friend, J. (1986) Carbohydrate and lipid-containing elicitors from *Phytophthora infestans*. Do they have a common mechanism of action? In: Bailey, J. (ed.) *Biology and Molecular Biology of Plant–Pathogen Interactions*. Springer-Verlag, Berlin, pp. 185–189.

Killick, R.J. and Malcolmson, J.F. (1973) Inheritance in potatoes of field resistance to late blight (*Phytophthora infestans* (Mont.) de Bary). *Physiological Plant Pathology* 3, 121–131.

Kuhn, D.N. (1991) Parasexuality in *Phytophthora*? In: Lucas, J.A., Shattock, R.C., Shaw, D.S. and Cook, L.R. (eds) *Phytophthora*. Symposium of the British Mycological Society, the British Society for Plant Pathology and the Society of Irish Plant Pathologists held at Trinity College, Dublin, September 1989. Cambridge University Press, Cambridge, pp. 242–255.

Layton, A. and Kuhn, D.N. (1990) *In planta* formation of heterokaryons of *Phytophthora megasperma* f. sp. *glycenea*. *Phytopathology* 80, 602–606.

Leonards-Schippers, C., Gieffers, W., Ritter, E., Salamini, F. and Gebhardt, C. (1991) RFLP-marker based localization of the gene R1 conferring vertical resistance to *Phytophthora infestans* in potato. *The 2nd International Potato Molecular Biology Symposium*, St Andrews, Scotland, August 1991 (Abstract).

Louwes, K.M., Hoekstra, R. and Mattheij, W.M. (1992) Interspecific hybridization between the cultivated potato (*Solanum tuberosum*) subspecies *tuberosum* L. and the wild species *S. circaeifolium* subsp. *circaeifolium* Bitter exhibiting resistance to *Phytophthora infestans* (Mont.) de Bary and *Globodera pallida* (Stone) Behrens. 2. Sexual hybrids. *Theoretical and Applied Genetics* 84, 362–370.

Lucas, J.A., Shattock, R.C., Shaw, D.S. and Cooke, L.R. (eds) (1991) *Phytophthora*. Symposium of the British Mycological Society, the British Society for Plant Pathology and the Society of Irish Plant Pathologists held at Trinity College, Dublin, September 1989. Cambridge University Press, Cambridge.

Malcolmson, J.F. and Black, W. (1966) New R-genes in *Solanum demissum* Lindl. and their complementary races of *Phytophthora infestans* (Mont.) de Bary. *Euphytica* 15, 199–203.

Malcolmson, J.F. and Killick, R.J. (1980) The breeding value of potato parents for field resistance to late blight measured by whole seedlings. *Euphytica* 29, 489–495.

Mattheij, W.M., Eijlander, R., de Koning, J.R.A. and Louwes, K.M. (1992) Interspecific hybridization between the cultivated potato *Solanum tuberosum* subspecies *tuberosum* L. and the wild species *S. circaeifolium* subsp. *circaeifolium* Bitter exhibiting resistance to *Phytophthora infestans* (Mont.) de Bary and *Globodera pallida* (Stone) Behrens. 1. Somatic hybrids. *Theoretical and Applied Genetics* 83, 459–466.

Mauch, F., Mauch-Mani, B. and Boller, T. (1988) Antifungal hydrolases in pea tissue. II. Inhibition of fungal growth by combinations of chitinase and β-1,3-glucanase. *Plant Physiology* 88, 936–942.

Meulemans, M. and Fouarge, G. (1986) Regeneration of potato somaclones and *in vitro* selection for resistance to *Phytophthora infestans* (Mont.) de Bary. *International Symposium on Crop Protection*, Gent, Belgium, 6 May 1986. Mededelingen van de Faculteit Landbouwwetenschappen, Rijksuniversiteit, Gent, Vol. 38, pp. 533–545.

Meulemans, M., Duchene, D. and Fouarge, G. (1987) Selection of variants by dual culture of potato and *Phytophthora infestans*. In: Bajaj, Y.P.S. (ed.) *Biotechnology in Agriculture and Forestry, 3. Potato*. Springer Verlag, Berlin, Heidelberg, New York, pp. 318–331.

Moffat, A.S. (1992) Improving plant disease resistance. *Science* 257, 482–483.

Müller, K.O. and Börger, H. (1940) Experimentelle Untersuchungen über die *Phytophthora*-Resistenz der Kartoffel. *Arbeiten aus der biologischen Bundesanstalt für Land- und Forstwirtschaft (Berlin-Dahlem)* 23, 189–231.

Nandris, D., de Vallavielle, C. and Bouvier, J. (1979) Studies of some interactions between potatoes and *Phytophthora infestans*. *Physiological Plant Pathology* 15, 1–12.

Newhouse, J.R., Tooley, P.W., Smith, O.P. and Fishel, R.A. (1992) Characterization of double-stranded RNA in isolates of *Phytophthora infestans* from Mexico, the Netherlands, and Peru. *Phytopathology* 82, 164–169.

Niederhauser, J.S. (1956) The blight, the blighter, and the blighted. *Transactions of the New York Academy of Sciences* 19, 55–63.

Niederhauser, J.S. (1991) *Phytophthora infestans*: the Mexican connection. In: Lucas, J.A., Shattock, R.C., Shaw, D.S. and Cooke, L.R. (eds) *Phytophthora*. Symposium of the British Mycological Society, the British Society for Plant Pathology and the Society of Irish Plant Pathologists held at Trinity College, Dublin, September 1989. Cambridge University Press, Cambridge, pp. 25–45.

Niks, R.E. (1988) Nonhost plant species as donors for resistance to pathogens with narrow host range. II. Concepts and evidence on the genetic basis of nonhost resistance. *Euphytica* 37, 89–99.

Ortiz, R. and Ehlenfeldt, M.K. (1992) The importance of endosperm balance number in potato breeding and the evolution of tuber-bearing *Solanum* species. *Euphytica* 60, 105–113.

Parker, J.E., Knogge, W. and Scheel, D. (1991) Molecular aspects of host–pathogen interactions in *Phytophthora*. In: Lucas, J.A., Shattock, R.C., Shaw, D.S. and Cooke, L.R. (eds) *Phytophthora*. Symposium of the British Mycological Society, the British Society for Plant Pathology and the Society of Irish Plant Pathologists held at Trinity College, Dublin, September 1989. Cambridge University Press, Cambridge, pp. 90–103.

Parlevliet, J.E. (1981) Stabilizing selection in crop pathosystems: an empty concept or a reality? *Euphytica* 30, 259–269.

Paxton, J.D. (1981) Phytoalexins – a working definition. *Phytopathologische Zeitschrift* 101, 106–109.

Pehu, E., Karp, A., Moore, K., Steele, S., Dunckley, R. and Jones, M.G.K. (1989) Molecular, cytogenetic and morphological characterization of somatic hybrids of dihaploid *Solanum tuberosum* and diploid *S. brevidens*. *Theoretical and Applied Genetics* 78, 696–704.

Pehu, E., Thomas, M., Poutala, T., Karp, A. and Jones, M.G.K. (1990) Species specific sequences of *Solanum brevidens* and *Solanum tuberosum* and their application to study somatic hybrids. *Theoretical and Applied Genetics* 80, 693–698.

Perl, A., Aviv, D. and Galun, E. (1990) Protoplast-fusion-derived *Solanum* cybrids: application and phylogenetic limitations. *Theoretical and Applied Genetics* 79, 632–640.

Pierpoint, W.S., Jackson, P.J. and Evans R.M. (1990) The presence of a thaumatin-like protein, a chitinase and a glucanase among the pathogenesis-related proteins of potato (*Solanum tuberosum*). *Physiological and Molecular Plant Pathology* 36, 325–338.

Pieterse, C.M.J., Risseeuw, E.P. and Davidse, L.C. (1991) An *in planta* induced gene of *Phytophthora infestans* codes for ubiquitin. *Plant Molecular Biology* 17, 799–811.

Pijnacker, L.P. and Ferwerda, M.A. (1984) Giemsa C-banding of potato chromosomes. *Canadian Journal of Genetics and Cytology* 26, 415–419.

Plaisted, R.L. (1987) Advances and limitations in the utilization of Neotuberosum in potato breeding. In: Jellis, G.J. and Richardson, D.E. (eds) *The Production of New Potato Varieties: Technological Advances*, Cambridge University Press, Cambridge, pp. 186–196.

Plaisted, R.L. and Hoopes, R.W. (1989) The past record and future prospects for the use of exotic germplasm. *American Potato Journal* 66, 603-628.

Preisig, C.L. and Kuć, J.A. (1987) Phytoalexins, elicitors, enhancers, suppressors and other considerations in the regulation of R-gene resistance to *Phytophthora infestans* in potato. In: Nishimura, S., Vance, C.A. and Doke, N. (eds) *Molecular Determinants of Plant Diseases*. Japan Science Press, Tokyo, Springer Verlag, Berlin, pp. 203-221.

Rietveld, R.C., Hasegawa, P.M. and Bressan, R.A. (1991) Somaclonal variation in tuber disc-derived populations of potato I. Evidence of genetic stability across tuber generations and diverse locations. *Theoretical and Applied Genetics* 82, 430-440.

Rivera-Peña, A. (1990a) Wild tuber-bearing species of *Solanum* and incidence of *Phytophthora infestans* (Mont.) de Bary on the western slopes of the volcano Nevado de Toluca. 2. Distribution of *P. infestans. Potato Research* 33, 341-347.

Rivera-Peña, A. (1990b) Wild tuber-bearing species of *Solanum* and incidence of *Phytophthora infestans* (Mont.) de Bary on the western slopes of the volcano Nevado de Toluca. 3. Physiological races of *Phytophthora infestans. Potato Research* 33, 349-355.

Rivera-Peña, A. (1990c) Wild tuber-bearing species of *Solanum* and incidence of *Phytophthora infestans* (Mont.) de Bary on the western slopes of the volcano Nevado de Toluca. 5. Type of resistance to *P. infestans. Potato Research* 33, 479-486.

Robertson, N.F. (1991) The challenge of *Phytophthora infestans*. In: Ingram, D.S. and Williams, P.H. (eds) *Advances in Plant Pathology, Volume 7, Phytophthora infestans, the Cause of Late Blight of Potato*. Academic Press, London, pp. 1-30.

Rohwer, F., Fritzemeier, K.-H., Scheel, D. and Hahlbrock, K. (1987) Biochemical reactions of different tissues of potato (*Solanum tuberosum*) to zoospores and elicitors from *Phytophthora infestans. Planta* 170, 556-561.

Roos, U.-P. and Shaw, D.S. (1985) Intranuclear virus-like particles in a laboratory strain of *Phytophthora drechsleri. Transactions of the British Mycological Society* 84, 340-344.

Ross, H. (1986) *Potato Breeding - Problems and Perspectives. Advances in Plant Breeding* 13, P. Parey, Berlin and Hamburg.

Rullich, G. and Schöber, B. (1988) Untersuchungen über das natürliche Vorkommen von Oosporen des Pilzes *Phytophthora infestans* im Feld. *Jahresbericht, BBA, Braunschweig*.

Salaman, R.N. (1911) *Studies in Potato Breeding*. IV Conférence Internationale de Génétique, Paris, Masson, pp. 573-575.

Sanford, J.C., Weeden, N.F. and Chyi, Y.S. (1984) Regarding the novelty and breeding value of protoplast-derived variants of 'Russet Burbank' (*Solanum tuberosum* L.). *Euphytica* 33, 709-715.

Sansome, E. (1977) Polyploidy and induced gametangial formation in British isolates of *Phytophthora infestans. Journal of General Mycology* 99, 311-316.

Schick, R. and Hopfe, A. (1962) Die Züchtung der Kartoffel. In: Schick, R. and Klinkowski, M. (eds) *Die Kartoffel*. VEB Verlag, Berlin, Vol. 2, pp. 1462-1583.

Schmiediche, P. (1988) The utilization of wild potato species in breeding. *29th Planning Conference*, 9-13 February 1987, CIP, Lima, Peru, pp. 135-149.

Schöber, B. (1990) Untersuchungen zur Keimung der Oosporen von *Phytophthora infestans* (Mont) de Bary im Freiland. *Jahresbericht, BBA, Braunschweig*.

Schöber, B. and Schiessendoppler, E. (1983) Vergleichende Resistenzprüfung von

Kartoffelknollen gegen der Erreger der Braunfäule *Phytophthora infestans* (Mont.) de Bary. *Potato Research* 26, 179-181.

Schöber, B. and Turkensteen, L.J. (1992) Recent and future developments in potato fungal pathology. Proceedings of Advances in Potato Crop Protection. International Symposium of Netherlands Society of Plant Pathology and Pathology and Virology Sections of the EAPR, 1-6 September 1990, Wageningen, The Netherlands. *Netherlands Journal of Plant Pathology*, Supplement (in press).

Schöber, B. and Umaerus, V. (1987) Late blight of potato - Kraut- und Braunfäule der Kartoffel (*Phytophthora infestans* (Mont) de Bary). *Potato Disease Assessment Keys*. Pathology, Committee for Disease Assessment, Ås-Trykk, Norway, pp. 19-35.

Shattock, R.C., Tooley, P.W. and Fry, W.E. (1986) Genetics of *Phytophthora infestans*: characterization of single-oospore cultures from A1 isolates induced to self by intraspecific stimulation. *Phytopathology* 76, 407-410.

Shattock, R.C., Shaw, D.S., Fyfe, A.M., Dunn, J.R., Loney, K.H. and Shattock, J.A. (1990) Phenotypes of *Phytophthora infestans* collected in England and Wales from 1985 to 1988: mating type, response to metalaxyl and isoenzyme analysis. *Plant Pathology* 39, 242-248.

Shaw, D.S. (1987) The breeding system of *Phytophthora infestans*: the role of the A2 mating type. In: Day, P.R. and Jellis, G.J. (eds) *Genetics and Plant Pathogenesis*. Blackwell Scientific Publications, Oxford, pp. 161-174.

Shaw, D.S. (1988) The *Phytophthora* species. In: Ingram, D.S. and Williams, P.H. (eds) *Advances in Plant Pathology*, Volume 6, Sidhu, G.S. (ed.) *Genetics of Plant Pathogenic Fungi*. Academic Press, London, San Diego, pp. 27-51.

Shaw, D.S. (1991) Genetics. In: Ingram, D.S. and Williams, P.H. (eds) *Advances in Plant Pathology, Volume 7, Phytophthora infestans, the Cause of Late Blight of Potato*. Academic Press, London, San Diego, pp. 131-170.

Shaw, D.S. and Shattock, R.C. (1991) Genetics of *Phytophthora infestans*: the Mendelian approach. In: Lucas, J.A., Shattock, R.C., Shaw, D.S. and Cooke, L.R. (eds) *Phytophthora*. Symposium of the British Mycological Society, the British Society for Plant Pathology and the Society of Irish Plant Pathologists held at Trinity College, Dublin, September 1989. Cambridge University Press, Cambridge, pp. 218-230.

Shepard, J.F., Bidney, D. and Shahin, E. (1980) Potato protoplasts in crop improvement. *Science* 208, 17-24.

Simmonds, N.W. (1966) Studies of the tetraploid potatoes. III. Progress in the experimental re-creation of the Tuberosum group. *Journal of the Linnean Society (Botany)* 59, 269-288.

Smoot, J.J., Gough, F.J., Lamey, H.A., Eichenmuller, J.J. and Gallegly, M.E. (1958) Production and germination of oospores of *Phytophthora infestans*. *Phytopathology* 48, 165-171.

Spielman, L.J. (1991) Isozymes and the population genetics of *Phytophthora infestans*. In: Lucas, J.A., Shattock, R.C., Shaw, D.S. and Cooke, L.R. (eds) *Phytophthora*. Symposium of the British Mycological Society, the British Society for Plant Pathology and the Society of Irish Plant Pathologists held at Trinity College, Dublin, September 1989. Cambridge University Press, Cambridge, pp. 231-241.

Spielman, L.J., McMaster, B.J. and Fry, W.E. (1989) Dominance and recessiveness at loci for virulence against potato and tomato in *Phytophthora infestans*. *Theoretical and Applied Genetics* 77, 832-838.

Spielman, L.J., Drenth, A., Davidse, L.C., Sujkovski, L.J., Gu, W., Tooley, P.W.

and Fry, W.E. (1991) A second world-wide migration and population displacement of *Phytophthora infestans*? *Plant Pathology* 40, 422-430.

Stolle, K. and Schöber, B. (1984) Wirkung eines Toxins von *Phytophthora infestans* (Mont.) de Bary auf Kartoffelknollengewebe. *Potato Research* 27, 173-184.

Stössel, P. (1988) Pathogenesis and host-parasite specificity in *Phytophthora*. In: Hess, W.M., Singh, R.S., Singh, U.S. and Weber, D.J. (eds) *Experimental and Conceptual Plant Pathology*. Gordon and Breach Science Publishers, New York, pp. 273-300.

Strömberg, A. and Brishammar, S. (1991) Induction of systemic resistance in potato (*Solanum tuberosum* L.) plants to late blight by local treatment with *Phytophthora infestans* (Mont.) de Bary, *Phytophthora cryptogea* Pethyb. & Laff., or dipotassium phosphate. *Potato Research* 34, 219-225.

Strömberg, A. and Brishammar, S. (1993) A histological evaluation of induced resistance to *Phytophthora infestans* (Mont.) de Bary in potato leaves. *Journal of Phytopathology* 137, 15-25.

Styer, E.L. (1978) Electron microscopy of intranuclear viruslike particles in *Phytophthora infestans*. Unpublished PhD thesis, University of Maryland, College Park (cited by Tooley *et al.*, 1989).

Swieżyński, K.M. (1990) *Resistance to Phytophthora infestans in the Potato*. Institute for Potato Research, Bonin.

Swieżyński, K.M., Sieczka, M.T., Sujkowski, L.S., Zarzycka, H. and Zimnoch-Guzowska, E. (1991) Resistance to *Phytophthora infestans* in potato genotypes originating from wild species. *Plant Breeding* 107, 28-38.

Tai, G.C.C. and Hodgson, W.A. (1975) Estimating general combining ability of potato parents for field resistance to late blight. *Euphytica* 24, 285-289.

Tomiyama, K. (1989) Biochemical aspects of potato late blight with respect to compatibility and incompatibility reactions. *Fungal Diseases of Potato*. Report of the Planning Conference on Fungal Diseases of the Potato, CIP, Lima, Peru, 21-25 September 1987, pp. 21-25.

Tooley, P.W. and Therrien, R.C. (1991) Variation in ploidy in *Phytophthora infestans*. In: Lucas, J.A., Shattock, R.C., Shaw, D.S. and Cooke, L.R. (eds) *Phytophthora*. Symposium of the British Mycological Society, the British Society for Plant Pathology and the Society of Irish Plant Pathologists held at Trinity College, Dublin, September 1989. Cambridge University Press, Cambridge, pp. 204-217.

Tooley, P.W., Hewings, A.D. and Falkenstein, K.F. (1989) Detection of double-stranded RNA in *Phytophthora infestans*. *Phytopathology* 79, 470-474.

Toxopeus, H.J. (1958) Some notes on the relations between field resistance to *Phytophthora infestans* in leaves and tubers and ripening time in *Solanum tuberosum* subsp. *tuberosum*. *Euphytica* 7, 123-130.

Turkensteen, L.J. (1989) Interaction of R-genes in breeding for resistance of potatoes against *Phytophthora infestans*. *Fungal Diseases of Potato*. Report of the Planning Conference on Fungal Diseases of the Potato, CIP, Lima, Peru, 21-25 September 1987, pp. 85-96.

Turkensteen, L.J. (1993) Durable resistance of potatoes against *Phytophthora infestans*. In: Parlevliet, J.E. and Jacobs, Th. (eds) *Durability of Disease Resistance*. Kluwer Academic, Dordrecht, pp. 115-124.

Umaerus, V. and Lihnell, D. (1976) A laboratory method for measuring the degree of attack by *Phytophthora infestans*. *Potato Research* 19, 91-107.

Umaerus, V., Umaerus, M., Erjefält, L. and Nilsson, B.-A. (1983) In: Erwin, D.C., Bartnicki-Garcia, S. and Tsao, P.H. (eds) *Phytophthora, its Biology, Taxonomy,*

Ecology and Pathology. American Phytopathological Society, St Paul, Minnesota, pp. 315-326.

Van der Plank, J.E. (1963) *Plant Disease, Epidemics and Control.* Academic Press, New York.

Van der Plank J.E. (1989) A paradox as an aid to understanding host-pathogen specificity. *Plant Pathology* 38, 144-145.

van Soest, L.J.M., Schöber, B. and Tazelaar, M.F. (1984) Resistance to *Phytophthora infestans* in tuber-bearing species of *Solanum* and its geographical distribution. *Potato Research* 27, 393-411.

Vavilov, N.I. (1935) The origin, variation, immunity and breeding of cultivated plants. *Chronica Botanica* 1, 1-366.

Vayda, M.E. and Park, W.D. (eds) (1990) *The Molecular and Cellular Biology of the Potato.* CAB International, Wallingford.

Visser, R.G.F., Hoekstra, R., van der Leij, F.R., Pijnacker, L.P., Witholt, B. and Feenstra, W.J. (1988) *In situ* hybridization to somatic metaphase chromosomes of potato. *Theoretical and Applied Genetics* 76, 420-424.

Wastie, R.L. (1991) Breeding for resistance. In: Ingram, D.S. and Williams, P.H. (eds) *Advances in Plant Pathology, Volume 7, Phytophthora infestans, the Cause of Late Blight of Potato.* Academic Press, London, pp. 193-224.

Wastie, R.L., Caligari, P.D.S., Stewart, H.E. and Mackay, G.R. (1987) A glasshouse progeny test for resistance to tuber blight (*Phytophthora infestans*). *Potato Research* 30, 533-538.

Wenzel, G. (1988) Biotechnology in agriculture - an overview. In: Rehm, H.J. and Reed, G. (eds) *Biotechnology.* VCH Publishers, Germany, Chapter 21, pp. 771-796.

Wenzel, G., Debnath, S.C., Schuchmann, R. and Foroughi-Wehr, B. (1987) In: Jellis, G.J. and Richardson, D.E. (eds) *The Production of New Potato Varieties.* Cambridge University Press, Cambridge, pp. 277-288.

Whittaker, S.L., Shattock, R.C. and Shaw, D.S. (1991) Inheritance of DNA contents in sexual progenies of *Phytophthora infestans. Mycological Research* 95(9), 1094-1100.

Wilkinson, M.J. (1992) The partial stability of additional chromosomes in *Solanum tuberosum* cv. Torridon. *Euphytica* 60, 115-122.

Woloshuk, C.P., Meulenhoff, J.S., Seela-Buurlage, M., Elzen, P.J.M. van den and Cornelissen, B.J.C. (1991) Pathogen-induced proteins with inhibitory activity toward *Phytophthora infestans. Plant Cell* 3, 619-628.

17 Inheritance of Resistance to Warm-growing-season Fungal Diseases

J.J. PAVEK AND D.L. CORSINI

United States Department of Agriculture, Agricultural Research Service, Pacific West Area, PO Box AA, Aberdeen, Idaho 83210, USA.

Verticillium wilt (*Verticillium dahliae*) and early blight (*Alternaria solani*) are the most common diseases of potato plants in the warm-growing-season areas of potato production. The diseases often occur concurrently, depending on the presence of inoculum and on environmental conditions. Symptoms of both diseases are often seen at about the same time during the development of the crop, during the later stages of tuber bulking (Nielsen, 1948; McLean, 1955). The result of the diseases is a hastening of maturity, i.e. early dying, which results in smaller tubers and lower dry-matter content.

McLean and others (McLean, 1955; Busch, 1966; Krikun and Orion, 1979) have reported yield losses of 30% or more from Verticillium wilt alone, and early blight can be equally devastating (Rotem and Feldman, 1965; Douglas and Pavek, 1972). Combined, they can result in serious crop losses.

Warm-growing-season potato production areas occur around the world. In the USA these include western Nebraska, southern Idaho and the Columbia Basin of Oregon and Washington, all with an arid, desert climate. In these areas, summer daytime high temperatures average 28–36°C with night-time lows of 10–16°C. July soil temperatures under the plant canopy are in the 16–22°C range. These temperatures are very favourable for the development of Verticillium wilt and early blight.

Earlier publications (Raeder, 1944; Edmundson *et al.*, 1951) indicated that Fusarium wilt was important in these production areas, but later studies showed that what appeared to be Fusarium wilt was in fact Verticillium wilt (Nielson, 1948). Fusarium wilt, however, can be a problem in lower-latitude areas where soil temperatures during the growing season are near 30°C (Goss, 1936).

Verticillium Wilt

Verticillium wilt (caused by *Verticillium dahliae* Kelb.; also referred to as microsclerotial strain of *V. albo-atrum* Reinke and Berthold) has received the most attention by breeding programmes in areas where it occurs, because of the drastic yield reductions it causes in very susceptible varieties and because it is difficult to control by chemical means. Although infection occurs early, wilt symptoms do not develop until the later part of the growing season, when rapid tuber bulking occurs. Juvenile plants in the rapid vegetative growth stage are resistant to colonization by the fungus, but once tuber bulking occurs considerable differences in resistance to colonization and symptom development can be observed (McLean, 1955). Varieties that are late in maturity appear resistant because they remain in the juvenile stage longer, but if they truly lack resistance they will also succumb once the plant enters the rapid, tuber-bulking stage (Busch and Edgington, 1967). Resistant varieties, however, continue to restrict colonization during this later stage of crop development (Davis *et al.*, 1983). Some varieties, such as Kennebec, show tolerance; they are susceptible to infection and subsequent colonization of the vascular system of the stem, and yet they normally yield well in spite of the presence of the fungus (Mohan *et al.*, 1990).

The first documented evaluations for Verticillium wilt resistance were performed by Nielsen (1948) in an arid, irrigated area of southern Idaho (USA). He demonstrated that what had earlier been called Fusarium wilt by Raeder (1944) was Verticillium wilt. More recently, the importance of this disease in limiting potato production in many irrigated areas (Krikun and Orion, 1979) has been realized, and extensive testing for resistance or tolerance has begun (Susnoschi *et al.*, 1976; Corsini *et al.*, 1985, 1988, 1990).

In 1949, the breeding programme at Aberdeen, Idaho (USA), was initiated, with resistance to Verticillium wilt a major goal. The breeder, John McLean (1955), tested numerous varieties and breeding selections and used the most resistant ones as parents to increase the general level of resistance in his breeding population. One of these, USDA 41956, was a particularly important parent, used in both the Aberdeen and the Beltsville, Maryland (USA), programmes. The Aberdeen breeding programme has continued to emphasize Verticillium wilt resistance in the development of new varieties (Corsini *et al.*, 1990; Mohan *et al.*, 1990).

Immunity to infection by *Verticillium dahliae* has not been identified in *S. tuberosum* or in wild tuber-bearing *Solanum* species (Webb and Hougas, 1959; Corsini *et al.*, 1988). When working with unadapted species and with late-maturing *S. tuberosum* selections, one must be aware that day-length may have an important influence in determining resistance (Busch and Edgington, 1967).

Hunter *et al.* (1968), working in Wisconsin, conducted a genetic study of Verticillium wilt resistance and concluded that resistance was heritable and that one of their apparently susceptible parents, Katahdin, possessed recessive genes for resistance. But their scale delineating resistance and susceptibility may have incorrectly classified this variety. In 4 years of testing, McLean

Table 17.1. Verticillium wilt on seedlings of selfs (1961) and on seedlings of crosses (1963) of parents with a range of reactions (from Hunter et al., 1968).

Parent[1]	Wilt index means[2]		Progeny in wilt index class[2] (%)			Number scored
	Parent	Progeny	1	2 and 3	4 and 5	
Selfed						
1	1.1	1.4	78	18	5	257
2	1.1	1.3	83	14	3	263
3	1.3	1.7	68	19	13	237
4	2.3	2.0	42	46	12	256
5	2.5	2.3	38	46	16	214
6	2.6	2.5	32	45	23	199
7	2.8	1.7	65	25	10	229
8	3.3	2.4	23	58	19	235
Crossed	Mid-parent					
3 × 1	1.2	1.5	71	24	5	280
2 × 1	1.1	1.5	75	17	8	287
3 × 2	1.2	1.6	77	8	15	289
8 × 2	2.2	1.6	65	28	7	295
5 × 1	1.8	1.7	57	36	7	214
4 × 2	1.7	1.7	63	25	12	291
6 × 1	1.8	1.8	46	51	4	264
2 × 5	1.8	1.9	62	24	14	289
8 × 6	3.0	2.0	56	29	15	95
7 × 5	2.7	2.1	51	34	15	261
4 × 5	2.4	2.3	29	58	13	286

[1] Parents: 1 = RD31 − 31, 2 = RD6 − 15, 3 = B3641 − 1, 4 = B929 − 23, 5 = B3620 − 1, 6 = B3627 − 18, 7 = Katahdin, 8 = B1172 − 14.
[2] Wilt class: 1 = 0 wilt/escape, 2 and 3 = slight to moderate wilt, 4 and 5 = severe wilt/dead.

(1955) found that Katahdin had a wilt index of 5 (where 0 is none and 100 is maximum wilt), while Russet Burbank had 26 and Irish Cobbler 47. Clearly, Katahdin shows resistance, which we have verified recently (Mohan et al., 1990), and the case for recessive genes is not very strong.

The results of Hunter et al. (1968) in testing 4741 seedlings from 11 crosses and eight selfs indicate that resistance is due mainly to additive genes (Table 17.1). They cited McLean (1955) and also stated that Katahdin transmitted recessive genes for resistance to its progeny, but they failed to point out that in three of the crosses resistance was dominant to susceptibility (Table 17.1). However, no information is given on replication or randomization, and careful scrutiny of their scoring reveals that their highest category of wilt resistance, no symptoms, could also include escapes, which may occur with the method of inoculation used at their location in northern Wisconsin. Furthermore, because maturity of the seedlings was ignored, valid conclusions regarding any particular gene action cannot be made.

Malamud (1975) bypassed the root system, a likely site for resistance, by using stem inoculation in his study of Verticillium wilt resistance in *S. tuberosum*, wild species and cultivated diploids. His attempt to fit the progeny reaction into discrete classes and to fit proposed segregation ratios appears unjustified. However, it is clear that the degree of parental resistance determined the proportion of progeny in resistant versus susceptible categories.

Results at Aberdeen with progeny from numerous crosses of various parents (Corsini *et al.*, 1990; Mohan *et al.*, 1990), ranging from resistant to susceptible, support the additive gene hypothesis (J.J. Pavek and D.L. Corsini, unpublished). Differences in maturity are hard to eliminate in evaluation procedures, and this complicates conclusions drawn from any genetic analysis. Nevertheless, progress can be made by selecting for resistance within maturity classes.

Early Blight

Early blight (*Alternaria solani* (Ellis and Martin) Jones and Grout) occurs throughout most of the world's potato-growing regions (Rotem and Reichert, 1964; Potter and Hooker, 1967; Prasad and Dutt, 1980). The disease is adapted to a wide range of environments. It commonly occurs, along with late blight, in cooler, more humid climates, but it is a particularly serious problem in arid, warm-growing-season areas under sprinkler irrigation (Douglas and Pavek, 1972; Holley *et al.*, 1985). It can be especially devastating to susceptible, early-maturing varieties. Like Verticillium wilt, early blight is a disease of the maturing plant. Juvenile plants in a rapid, vegetative growth phase are usually resistant to infection. Maintaining plants in rapid vegetative growth will delay disease development (Rotem and Feldman, 1965; Johanson and Thurston, 1990). As with Verticillium wilt, the breeder has the problem of distinguishing true resistance from late maturity.

Testing germplasm of *S. tuberosum* and various other species for resistance to early blight has been carried out in various studies (LeClerg, 1946; O'Brien and Akeley, 1971; Douglas and Pavek, 1972; Bussey and Stevenson, 1991; Christ, 1991). Ranges of reactions to the pathogen were usually obtained. But, with a long growing season and with favourable conditions for disease development, all plants eventually showed defoliation. Immunity was not found in *S. tuberosum* or other tuber-bearing species and the inherent resistance, i.e. field resistance, identified in these studies affects the rate of development, not the presence or absence of disease (Holley *et al.*, 1993; Pelletier and Fry, 1989, 1990). There has been little use of this germplasm to specifically develop early-blight-resistant varieties.

Herriott *et al.* (1986) reported that early blight resistance was highly heritable in a population derived from the cultivated diploids *S. phureja* and *S. stenotomum*. Narrow-sense heritability of resistance was 0.83, with no allowance made for maturity differences. The authors concluded that relatively few genes were involved, because of the rapid progress they made in identifying resistant plants. However, our experience has been that, when attempting to sort out resistance from late maturity, progress is very slow.

Breeders in warm-growing-season areas should be concerned about the tuber-rot phase of early blight, which is rarely mentioned in the literature (Folsom and Bonde, 1925). Highly susceptible varieties, such as BelRus and Onaway, can be rendered unusable in storage if they are exposed to inoculum during harvest (Potter and Hooker, 1967). No inheritance studies have been conducted because of the sporadic nature of this phase of early blight and because of difficulties with artificial inoculation. However, we have observed differing degrees of susceptibility among clones, which suggest that there may be one or more genes, with dosage effects, for susceptibility (J.J. Pavek and D.L. Corsini, unpublished).

Fusarium Wilt

Wilt caused by *Fusarium* species (*F. solani* (Mart) App. and Wr. var. *eumartii* (Carp.) Wr. and *F. oxysporum* Schlecht) has been reported in many potato-growing areas (Goss, 1936; McLean and Walker, 1941). Fusarium wilt causes little loss in yield and appears to have been mainly a localized problem in the past. This may be due to the fact that the pathogens occurred in virgin soil (Goss, 1936) and disappeared after cultivation, or because of other unknown factors. These *Fusarium* species are relatively weak pathogens on the potato plant, needing a favourable environment for infection (Goss, 1936; McLean and Walker, 1941).

Little effort has been devoted to evaluating resistance to *F. solani* or *F. oxysporum* wilts in potato germplasm. Goss (1928) and McLean and Walker (1941) identified differences in the degree of susceptibility in a limited number of varieties. Of these only Katahdin, which appeared somewhat resistant to both species, is still of commercial importance. Malamud (1975) also studied Fusarium wilt resistance in *S. tuberosum*, cultivated diploids and wild species. He identified resistance in accessions of some cultivated diploids and wild species in greenhouse tests. Although, after some manipulation, his data from $2x$ and $4x$ R × S crosses fit a 1 : 2 : 1 (resistant/tolerant/susceptible) ratio, they would apparently also fit a normal distribution, which may be expected with additive genes and alleles. We believe that further research is needed. However, it is clear from this work that resistance to Fusarium wilt is not controlled by a dominant gene, as in tomato (see also Goth and Webb, 1981). We found no other reports of research on the inheritance of resistance to Fusarium wilt in potato.

References

Busch, L.V. (1966) Susceptibility of potato varieties to Ontario isolates of *Verticillium albo-atrum*. *American Potato Journal* 43, 439–442.

Busch, L.V. and Edgington, L.V. (1967) Correlation of photoperiod with tuberization and susceptibility of potato to *Verticillium albo-atrum*. *Canadian Journal of Botany* 45, 691–693.

Bussey, M. and Stevenson, W.R. (1991) A leaf disk assay for detecting resistance to early blight caused by *Alternaria solani* in juvenile potato plants. *Plant Disease* 75, 385–390.

Christ, B.J. (1991) Effect of disease assessment method on ranking potato cultivars for resistance to early blight. *Plant Disease* 75, 353–356.

Corsini, D.L., Davis, J.R. and Pavek, J.J. (1985) Stability of resistance of potato to strains of *Verticillium dahliae* from different vegetative compatibility groups. *Plant Disease* 69, 980–982.

Corsini, D.L., Pavek, J.J. and Davis, J.R. (1988) Verticillium wilt resistance in non-cultivated tuber-bearing *Solanum* species. *Plant Disease* 72, 148–151.

Corsini, D.L., Pavek, J.J. and Davis, J.R. (1990) *Verticillium* wilt resistant potato germplasm: A66107-51 and A68113-4. *American Potato Journal* 67, 517–525.

Davis, J.R., Pavek, J.J. and Corsini, D.L. (1983) A sensitive method for quantifying *Verticillium dahliae* colonization in plant tissue and evaluating resistance among potato genotypes. *Phytopathology* 73, 1009–1014.

Douglas, D.R. and Pavek, J.J. (1972) Screening potatoes for field resistance to early blight. *American Potato Journal* 49, 1–6.

Edmundson, W.C., Schaal, L.A. and Landis, B.J. (1951) Potato growing in the Western States. *US Department of Agriculture Farmers' Bulletin* No. 2034.

Folsom, D. and Bonde, R. (1925) *Alternaria solani* as a cause of tuber rot in potatoes. *Phytopathology* 15, 282–286.

Goss, R.W. (1928) Varietal susceptibility of potatoes to *Fusarium* wilt and stem-end rot. *Phytopathology* 18, 307–309.

Goss, R.W. (1936) *Fusarium* wilts of potato: their differentiation and the effect of environment upon their occurrence. *American Potato Journal* 13, 172–180.

Goth, R.W. and Webb, R.E. (1981) Sources and genetics of host resistance in vegetable crops. In: Mace, M.E., Bell, A.A. and Beckman, C.H. (eds) *Fungal Wilt Diseases of Plants*. Academic Press, New York, pp. 377–409.

Herriott, A.B., Haynes, F.L., Jr, and Shoemaker, P.B. (1986) The heritability of resistance to early blight in diploid potatoes (*Solanum tuberosum* subspp. *phureja* and *stenotomum*). *American Potato Journal* 63, 229–232.

Holley, J.D., Hall, R. and Hofstra, G. (1983) Identification of rate-reducing resistance to early blight in potato. *Canadian Journal of Plant Pathology* 5, 111–114.

Holley, J.D., Hall, R. and Hofstra, G. (1985) Effects of cultivar resistance, leaf wetness duration and temperature on rate of development of potato early blight. *Canadian Journal of Plant Science* 65, 179–184.

Hunter, D.E., Darling, H.M., Stevenson, F.J. and Cunningham, C.E. (1968) Inheritance of resistance to *Verticillium* wilt in Wisconsin. *American Potato Journal* 45, 72–78.

Johanson, A. and Thurston, H.D. (1990) The effect of cultivar maturity on the resistance of potatoes to early blight caused by *Alternaria solani*. *American Potato Journal* 67, 615–623.

Krikun, J. and Orion, D. (1979) *Verticillium* wilt of potato: importance and control. *Phytoparasitica* 7, 107–116.

LeClerg, E.L. (1946) Breeding for resistance to early blight in the Irish potato. *Phytopathology* 36, 1011–1015.

McLean, J.G. (1955) Selecting and breeding potatoes for field resistance to *Verticillium* wilt in Idaho. *University of Idaho Research Bulletin* No. 30.

McLean, J.G. and Walker, J.C. (1941) A comparison of *Fusarium avenaceum*,

F. oxysporum, and *F. solani* var. *eumartii* in relation to potato wilt in Wisconsin. *Journal of Agricultural Research* 63, 495-525.

Malamud, O.S. (1975) Inheritance of the reaction to *Verticillium* and *Fusarium* wilts in tuber-bearing *Solanum* species and species-hybrids. Dissertation, University of Nebraska, Lincoln. Available from: University Microfilm, Ann Arbor, MI 48104; Order No. 76-02036. *Dissertation Abstracts* Abs. 36, 3706B.

Mohan, S.K., Davis, J.R., Corsini, D.L., Sorensen, L.H. and Pavek, J.J. (1990) Reaction of potato clones and accessions of *Solanum* spp. to *Verticillium dahliae* Kleb. and its toxin. *Potato Research* 33, 449-458.

Nielsen, L.W. (1948) *Verticillium* wilt of potatoes in Idaho. *University of Idaho Research Bulletin* No. 13.

O'Brien, M.J. and Akeley, R.V. (1971) Evaluation of some potato varieties and breeding lines for resistance to early blight. *US Department of Agriculture, ARS Production Research Report* No. 140.

Pelletier, J.R. and Fry, W.E. (1989) Characterization of resistance to early blight in three potato cultivars: incubation period, lesion expansion rate, and spore production. *Phytopathology* 79, 511-517.

Pelletier, J.R. and Fry, E. (1990) Characterization of resistance to early blight in three potato cultivars: receptivity. *Phytopathology* 80, 361-366.

Potter, H.S. and Hooker, W.J. (1967) Early blight of potato. *Michigan State University Extension Bulletin* No. 572.

Prasad, B. and Dutt, B.L. (1980) Reaction of *Solanum* spp. against early blight of potatoes. *Journal of Indian Phytopathological Association* 7, 68-72.

Raeder, J.M. (1944) Disease of potatoes in Idaho. *Idaho Agricultural Experiment Station Bulletin* No. 254.

Rotem, J. and Feldman, S. (1965) The relation between the ratio of yield to foliage and the incidence of early blight in potato and tomato. *Israel Journal Agricultural Research* 15, 115-122.

Rotem, J. and Reichert, I (1964) Dew – a principal moisture factor enabling early blight epidemics in a semi-arid region of Israel. *Plant Disease Reporter* 48, 211-215.

Susnoschi, M., Krikun, J. and Zuta, Z. (1976) Trial of common potato varieties in relation to their susceptibility to *Verticillium* wilt. *Potato Research* 19, 323-334.

Webb, R.E. and Hougas, R.W. (1959) Preliminary evaluation of *Solanum* species and species hybrids for resistance to disease. *Plant Disease Reporter* 43, 144-151.

18 Inheritance of Resistance to Fungal Diseases of Tubers

R.L. WASTIE

Crop Genetics Department, Scottish Crop Research Institute, Invergowrie, Dundee DD2 5DA, UK.

Introduction

Any consideration of the genetics of resistance to fungal diseases other than late blight is likely to be noteworthy for what is not known about such diseases rather than for what is. This is more a reflection of the varied nature of resistance mechanisms and a lack of close understanding of host–parasite relationships than of the importance of the diseases themselves. Indeed, breeding for resistance, which has had a greater or lesser degree of success depending on the assiduousness with which it has been pursued, has by and large proceeded empirically without any great understanding of the underlying mechanisms or of their genetic control. Where major gene resistance has been identified (or even suspected), as with resistance to wart (*Synchytrium endobioticum*) and late blight, it has been readily and relatively quickly utilized; resistance mechanisms which are under the combined control of several genes, each acting on one or more components of resistance, are more intractable to handle and much more difficult to identify.

The diseases considered in this chapter are primarily diseases of tubers, and range from those caused by persistent soil-borne fungi, such as wart and common and powdery scab (*Streptomyces scabies* and *Spongospora subterranea* respectively), to those which are tuber-borne, at least in part: black scurf (*Thanatephorus cucumeris*), gangrene (*Phoma foveata*), silver scurf (*Helminthosporium solani*), dry rot (*Fusarium* spp.), skinspot (*Polyscytalum pustulans*) and black dot (*Colletotrichum coccodes*). Fusarium wilt (caused by *Fusarium* spp.) and powdery mildew (*Erysiphe cichoracearum* DC ex Merat), which are warm-climate diseases, have not been considered, nor have tuber rots caused by *Pythium* and *Phytophthora* spp., Sclerotinia mould (*S. sclerotiorum* (Lib.) de Bary) and the range of mainly leaf-infecting diseases listed by Hooker (1981). All are of local or relatively minor interest, and very little information is available on aspects of resistance, other than the names of a few resistant or particularly susceptible cultivars.

The diseases which follow are described in an order that reflects the extent to which the genetics of resistance have been explored and explained.

Wart

Synchytrium endobioticum (Schilb.) Perc.

Wart disease is one of the few potato diseases for which resistance breeding has provided a continuing and satisfactory means of control, albeit allied to restriction orders that prevent the cultivation of susceptible varieties on land known to be infested. No chemical or rotational measures are practicable for this persistent soil-borne disease, and this, together with its disastrous effect on crop yield and quality, has focused the minds of pathologists and breeders alike on devising screening techniques for detecting resistance to wart and in exploring the inheritance of resistance. For these reasons, the genetics of resistance to wart is better understood than that of most other fungal or bacterial diseases of potatoes.

For a general account of the epidemiology, distribution, identification and infection techniques of the fungus, as well as a full bibliography, the reader is referred to the comprehensive account by Langerfeld (1984b); general information on resistance breeding and genetics will be found in Howard (1970), Langerfeld (1984b) and Ross (1986), as well as in other chapters in this book.

From the time of its appearance in Europe towards the end of the 19th century, and until the beginning of the fifth decade of the 20th, only one form of *S. endobioticum* was known. Following the discovery in 1909 (Gough, 1920) that certain varieties, notably cv. Snowball, were resistant, several cultivars with effective resistance to this common race 1 were produced: in Germany, for example, some 75% of the potato cultivars used in the former Federal Republic are resistant to pathotype 1 (Langerfeld, 1984a). With the help of legislation which required outbreaks to be notified and the growing of susceptible varieties thereafter forbidden, the disease had to all intents and purposes been apparently routed and contained. However, since 1942 the situation has changed, in that many more races of the fungus have been identified, and the disease is now known to occur in all continents except Australia (Langerfeld, 1984b). The progress of the disease in Europe since 1940, in terms of the occurrence of new pathotypes (now totalling 17, with an additional race in Newfoundland) is given by Bojnansky (1984). Most of the newer races, which are defined in terms of formerly resistant cultivars which they are able to infect, are localized in montane and submontane regions of Czechoslovakia, Germany and the former USSR, and are not of general occurrence; other new outbreaks have often been local and temporary in nature, resulting from isolated importations of infected material or from weather conditions particularly favourable to the disease (Bojnansky, 1984). Thus, although race 1 (originally reported in 1888 from what is now the Czech and Slovak Republics) is of negligible importance in Germany, some seven additional races are now known (Stachewicz, 1989); a survey in the former FRG indicated race 6 to be the commonest (Langerfeld, 1984a). The precise identification of such pathotypes is complicated by the lack of a comprehensive and generally available differential series of host genotypes. One such set has been suggested (Anon., 1983). It is significant that some of the new races apparently appeared on what were pathotype 1 foci (Ullrich, 1959), and that many German potato cultivars with the highest resistance to pathotype 1 are highly susceptible to other pathotypes.

If the scenario against which resistance breeding has been conducted is complicated by the pathotype situation, it is complicated further by the nature of disease expression and the means by which it is assessed. Resistance (expressed as immunity) can be distinguished from susceptibility in a field test, but the existence of several pathotypes makes field testing too confusing, cumbersome and slow to be acceptable for breeding (as distinct from statutory testing) purposes.

Laboratory test methods are well established; four are outlined by Ross (1986) and details of the operation and interpretation of standard tests are given by Langerfeld (1987b); but they tend to be more sensitive than field tests and present a gradation between resistance and susceptibility, rather than a clear expression of one or the other. Both resistant and susceptible cultivars vary in their degree of susceptibility, and in the scheme of Hille (1965), interpreted by Pratt (1976), six infection categories are recognized. Three of them involve various degrees of necrosis of infected epidermal cells, which subsequently slough off, taking the fungus with them and allowing the underlying tissues to remain wart-free; and three involve degrees of intensity of infection and thus represent a susceptible reaction. The resistant reaction is subdivided by Pratt (1976) into two groups: one (RG1) in which the fungus dies out before it has had time to become established, and the other (RG2) in which the fungus becomes established but is mostly (but not always completely) sloughed off. In so far as RG2 types are field-resistant, they can legitimately be regarded as immune as far as their field behaviour is concerned, but there are dangers in attempting to interpret on a major gene basis a resistance/susceptibility pattern which seems on closer scrutiny to be more similar to a polygenically controlled system.

Such interpretations of the genetic control of resistance as have been made have generally been based on a monogenic dominant mode of inheritance, at least as far as the control of the necrotic response is concerned. However, modifying genes may also be present which condition the nature and extent of the response, especially where more than one race of the fungus is present.

Early workers published extensive data from field tests for resistance. Salaman and Lesley (1923) in England found that selfing immune varieties produced mostly immune progeny, and that selfing susceptibles likewise produced mainly susceptible offspring. They deduced that immunity was dominant and susceptibility recessive, and suggested that two factors, which they named X and Y, could induce immunity independently provided 'factor

Z' was present. In the absence of factor Z, immunity was only induced by both X and Y acting together. Black (1935) in Scotland was aware that both immunity and susceptibility had different degrees of intensity, and to explain this he postulated the existence of three interacting resistance factors, which he designated A, B and C. Lunden and Jørstad (1934) in Norway found that different ratios of immune : susceptible progeny resulted from intercrossing immune parents. They postulated the existence of dominant immunity factors X' and X", and also complementary factors Y and Z which acted independently of X and, if present together, could also confer immunity. Lunden (1950) gave additional examples of the complex ratios obtained in crosses involving immune varieties where the immunity factor was present in simplex or duplex form and where Y and Z were also implicated to further complicate the ratios obtained.

A comparison of resistance to more than one pathotype was undertaken by Maris (1961) in The Netherlands, who noticed, when testing the progeny of parents susceptible to race 6 against this same race, that a surprisingly high number of resistant seedlings was obtained – far more than were found in similar screening for race 1 resistance. He concluded that the inheritance of resistance to race 6 is different from that to race 1, and involves both dominant and recessive resistance genes. From further investigations, using *S. tuberosum* dihaploids, Maris (1973) suggested that the presence of resistant clones in progenies from certain selfed susceptible varieties could be explained by resistance being determined by a single dominant gene A, the action of which was inhibited by three complementary dominant genes B, C and D.

Scheidt and Hunnius (1981) in Germany interpreted the inheritance of resistance on the basis of one dominant gene, with resistance to pathotypes 2 and 6 being inherited independently. They studied the progenies of five crosses with the resistant Max Planck Institute clone 50.247/2, and recorded 51% of 794 clones resistant to pathotype 2, 52% of 1089 clones resistant to pathotype 6, but only 34% of 612 clones resistant to both pathotypes. Of the remaining 66%, one-half was susceptible to both pathotypes and one-quarter resistant either to one or the other.

Rothacker *et al.* (1974) in Germany also compared resistance to wart pathotypes, using hybrids of South American *S. andigena* with domestic cultivars resistant to pathotype 1. Resistance to this pathotype among the 1621 seedlings of 112 progenies ranged from 4 to 100%; about 30% of progenies contained plants resistant to pathotype 2, and eight forms susceptible to pathotype 1 were resistant to some of the other eight pathotypes used. Langerfeld and Bätz (1990) screened 411 breeding lines in Germany and found 36% to be susceptible to races 1, 2, 6 and 8, 56% resistant to race 1, and 7% resistant to one or more of the races 2, 6 and 8. Nine lines (2%) were resistant to all four races.

Ross (1986) invoked the action of polygenes to explain the fact that resistance to more than one race is not comprehensive. Polygenes are also implicated by Sedova (1980) in work in Russia with diploid species, and in Czechoslovakia by Zadina (1983), who in part attributed the unexpected deviations observed in resistance patterns from crosses involving parents with simplex or duplex resistance to race 1 ($R_1r_1r_1r_1$ and $R_1R_1r_1r_1$ respectively) to non-race-specific polygenic resistance.

The conclusion from a rather comprehensive survey of the inheritance of wart resistance in Czechoslovakia (Zadina and Findejs, 1988) indicates that resistance to the common race 1 is conditioned by the dominant gene R_1, for which 71 (mostly continental European) cultivars are listed as simplex and 27 cultivars and hybrids as duplex. Resistance to pathotype 2, which was first recorded in 1941, and to the Czech pathotypes 15, 16 and 17 (using the nomenclature of Langerfeld, 1984b), recorded in 1965, 1967 and 1975 respectively, is assumed to be similarly based on major genes at a single locus, but, apart from three cultivars which had a duplex basis of resistance to pathotype 2, only the simplex type of resistance was found (Table 18.1).

The breeding line Lü 58.526/22 possessed simplex resistance to four races and duplex resistance to the common race (Table 18.1).

In a recent study on resistance to pathotype 1 in Germany (Lellbach and Effmert, 1990), the progeny of a diallel cross involving nine resistant and one susceptible clone were screened in the laboratory. On the basis of the overall percentage of resistant types in all nine combinations, and selfs, the parental genotypes could be separated into five significantly different groups (Table 18.2). Intercrossing clones in three of the groups indicated that the progeny inherited their resistance as if it were controlled by a single major dominant gene in the simplex or duplex condition in the autotetraploid, the susceptible parent being nulliplex. Two of the parental clones did not fit into the three groups. The authors infer that greater genetic variation is present than is theoretically possible in the single-factor model of autotetraploidy; certainly the observations of other authors quoted above support this conclusion.

Major gene resistance thus only accounts for a part of the resistance to wart present in resistant cultivars, and further studies along the lines of that described above are needed to elucidate the nature of this resistance and the

Table 18.1. Genetic basis of major gene resistance to pathotypes of *Synchytrium endobioticum* (data from Zadina and Findejs, 1988). R^2r^2 = RRrr etc.

Genotype	Pathotype[1] for which the appropriate resistance gene is present				
	1	2	15	16	17
Appollo (= Argo)	R^2r^2	R^2r^2	Rr^3	—	Rr^3
Apta	Rr^3	Rr^3	Rr^3	—	—
Bürs 58.80/15	R^2r^2	Rr^3	Rr^3	Rr^3	—
Fortuna	Rr^3	Rr^3	—	—	—
Hilla	Rr^3	R^2r^2	—	—	—
Hochprozentige	R^2r^2	R^2r^2	—	—	—
Li 963/57	R^2r^2	Rr^3	Rr^3	—	—
Lü 58.526/22	R^2r^2	Rr^3	Rr^3	Rr^3	Rr^3
Ora (= Mira)	—	Rr^3	—	—	—
Saphir	Rr^3	Rr^3	r^4	Rr^3	r^4

[1] Nomenclature according to Langerfelt (1984b).

number of major genes involved. To this end an agreed set of differential genotypes is needed. Some of the resistance observed is likely to be conditioned, at least in part, by polygenes, some of which may operate by modifying the background in which the major genes are expressed.

No information is available as to the mechanism by which major resistance genes (or polygenes) might operate. Lipsits (1964) in the former USSR reported the existence of specific immunological differences between proteins of varieties resistant and susceptible to wart, but to what extent they are related to resistance mechanisms was not determined.

Dry Rot

1. *Fusarium coeruleum* (Lib.) ex Sacc. (syn. *F. solani* (Mart.) Sacc. var. *coeruleum* (Sacc.) Booth).
2. *Fusarium sulphureum* Schlecht. (syn. *F. sambucinum* Fuckel f.6 Wollenw.) Teleomorph: *Gibberella cyanogena* (Desm.) Sacc.
3. *Fusarium sambucinum* Fuckel. Teleomorph: *Gibberella pulicaris* (Fr.) Sacc.
4. *Fusarium avenaceum* (Corda ex Fr.) Sacc. Teleomorph: *Gibberella avenacea* R.J. Cook.

The pathology of dry rot is complicated by the fact that several species of *Fusarium* can cause the disease. Their relative importance differs from one

Table 18.2. Overall proportion of resistant individuals among progenies of a diallel crossing scheme involving ten clones resistant or susceptible to race 1 of *Synchytrium endobioticum* (data from Lellbach and Effmert, 1990).

Parental genotype[1]	Overall percentage of resistant types in nine crosses and selfing[2]	Group	Basis of resistance
541/9	91.9 a	I	Duplex
Lisera	90.4 a		
751/132	89.8 a		
562/3	81.3 b	II	Not known
125/15	74.4 c	III	Simplex
Adretta	74.2 c		
7074/314	73.4 c		
17/6	73.4 cd		
539/6	70.2 d	IV	Not known
Swerdlowski	49.3 e	V	Nulliplex

[1] All resistant except the last.
[2] Figures followed by the same letter are not significantly different ($P \leq 0.01$).

region to another, but their epidemiology, with the disease they cause, is similar, although, because of their somewhat complicated taxonomy, it is often unclear as to exactly which species is being referred to. The above list, based on the nomenclature of Gerlach and Nirenberg (1982), lists the four most important species, three of which have teleomorphs in the genus *Gibberella*, which are therefore the preferred names.

Although resistance to dry rot has been a stated objective of several potato-breeding schemes, there is very little information on how this objective is being achieved. Sources of resistance (e.g. to *G. pulicaris*) have been identified in several wild species (Podgaetski and Koval, 1989), and resistance is readily transmissible to the progeny (Corsini and Pavek, 1986). That this resistance is highly heritable is suggested by the high percentage of resistant progeny obtained by intercrossing resistant parents (Corsini and Pavek, 1986; Bebre, 1987). One complication lies in the fact that despite their similarity in source (soil) and mode of attack (wound pathogens), resistance/susceptibility to each dry rot pathogen appears to be independently inherited. Several authors in widely separated locations have commented on this (see Wastie *et al.*, 1989), and recent data from progeny tests confirm this (Wastie and Bradshaw, 1993). A consideration of the nature of resistance affords no clue as to why this should be so, although the different responses to the several pathogens suggest that several mechanisms of resistance are involved. O'Brien and Leach (1983) showed that the formation of suberin and wound periderm was an important mechanism of resistance to invasion by *G. pulicaris*, although it is probably one of a general nature since it is basically a wound response. Seppanen (1983) claimed to have identified both horizontal and vertical resistance to the six *Fusarium* species, which he compared on tuber discs, but he equated 'horizontal' with resistance to several species and 'vertical' with resistance to only one.

A role for chemical inhibitors in pathogenesis has been attributed by Desjardins and Gardner (1991) to the phytoalexin rishitin, which they suggest is important in conditioning the resistance of potato tubers to dry rot (caused in their studies by *G. pulicaris*). However, rishitin is not likely to be the only chemical involved in resistance, for some strains of *G. pulicaris* were highly tolerant of rishitin and yet of low aggressiveness. It seems likely that other chemicals, and indeed other mechanisms, including those involved in wound healing, are also involved in the determination of resistance to these *Fusarium* spp. It is thus probable that the genetic control of resistance to each species is independent and certainly not straightforward. Suska (1989) confirmed the polygenic nature of the inheritance of resistance to *G. cyanogena*, and in a study of resistance in the progeny of crosses between resistant clones from wild forms of the Commersoniana group and susceptible *S. tuberosum* dihaploids, found that broad- and narrow-sense heritability were high. In some cases the inheritance of resistance was accompanied by high starch content.

The fact that resistance to the various dry-rot-causing species of *Fusarium* is inherited separately, but is expressed similarly and is apparently readily transmissible, augurs well for the future of resistance breeding. Test techniques are well established (see Langerfeld, 1987a) and tuber discs afford

Fig. 18.1. Necrosis of stolons caused by *Rhizoctonia solani*.

a convenient way of assessing tissue penetration. If combined with tests for mechanical injury, the development of varieties with both biochemical and morphological/mechanical resistance and maybe even immunity should be possible, as Leach and Webb (1981) have suggested. Nevertheless, a greater understanding of the genetic control of such resistance mechanisms is desirable, and probably not as difficult to achieve as with those pathogens that can enter the tuber directly without the help of damage or wounding, and for which resistance to both entry and spread within the tissue is required.

Black Scurf and Stem Canker

Rhizoctonia solani Kühn. Teleomorph: *Thanatephorus cucumeris* (Frank) Donk

Stolons, stems, roots and tubers are all susceptible to this soil- and tuber-borne fungus, isolates of which vary in virulence to the various parts of the plant being attacked (Figure 18.1). Furthermore, the different parts of the plant differ in relative susceptibility between genotypes (Frank et al., 1976). It is thus likely that different resistance factors operate in the various parts of the plant, and it is not surprising, therefore, that, with its multiple entry sites and inherent variability, little progress has been made in breeding for resistance and none in understanding the genetics of resistance.

A method of testing stems, stolons and tubers for resistance to *R. solani* has been described by Frank et al. (1976), and methods of testing for resistance to stem canker by Tolstrup et al. (1990). Sources of resistance to stem canker have been identified in several wild species (Gadhziev and Lebedeva, 1983). Nazarova and Turuleva (1987) found four wild species with resistance to both stem and tuber infection; Banville (1978) observed resistance to stem canker in North American cultivars, but none to the black scurf expression of the disease. Gadzhiev (1986) crossed cultivars of *S. tuberosum* with forms of *S. phureja* and *S. vernei* which had resistance to the stem canker phase of the disease, and screened more than 7000 seedlings for resistance. The very small number of resistant individuals in the F_2 generation led to the conclusion that resistance was under polygenic control. However, it is possible that this observation could also be explained by the presence of a few major genes, with resistance being recessive. The widespread occurrence of this disease and the ease with which resistance screening can be carried out, at least for its stem canker form, appear to make it a fertile topic for further investigation, although, as Ross (1986) has pointed out, the different components of resistance will have to be identified, and their genetics investigated, before effective resistance breeding can begin.

Common Scab

Streptomyces scabies (Thaxt.) Waksman & Henrici

Common scab is a major and common cause of disfigurement of potato tubers (Figure 18.2), and as such is an important target for resistance breeding. As with other diseases, resistance has been identified and handled with little understanding of underlying mechanisms of resistance, and practically none of their mode of inheritance. Many sources of resistance have been identified, in both breeding material (Leach et al., 1938; Krantz and Eide, 1941, 1948; Bedin, 1985; Ross, 1986) and wild species (Pfeffer and Effmert, 1985; Hawkes, 1990). Resistance appears to be associated, at least in part, with the effectiveness of the periderm at isolating an infection and preventing it from spreading more deeply into the cortex. Whether any chemical inhibitors are also involved is not known.

Breeding and selecting for scab resistance have been widely practised in many potato improvement programmes, and some attempts have been made to base this on a degree of understanding of the underlying genetics. Thus Krantz and Eide (1941) in the USA distinguished five breeding types among F_1, F_2 and F_3 generations of a cross between 'Accession 123' and cv. Lookout Mountain, based on the assumption that the former was triplex and the latter

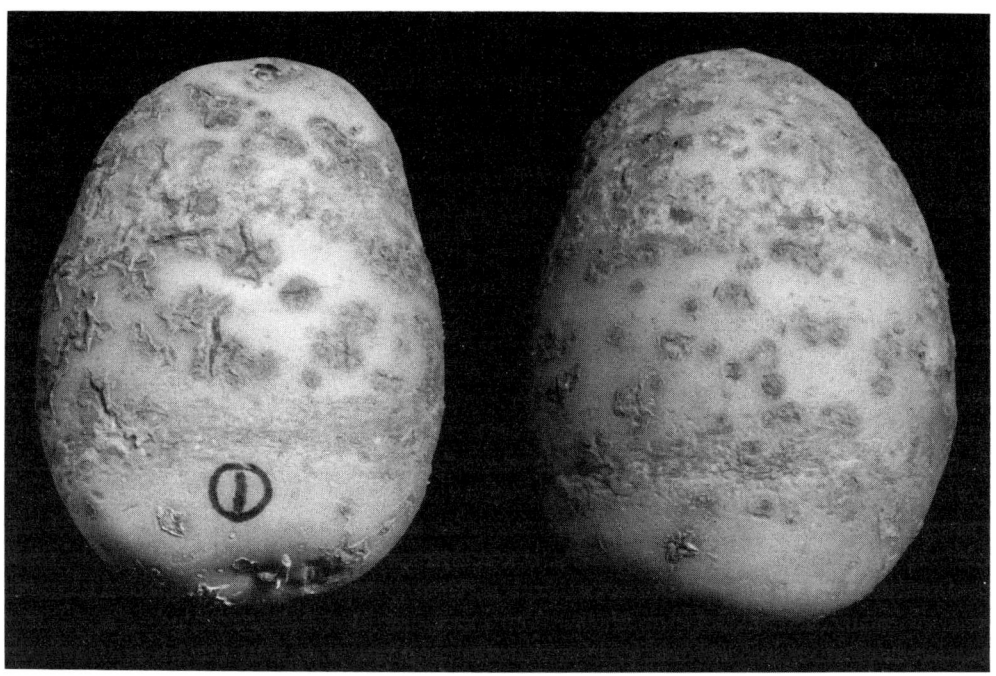

Fig. 18.2. Common scab: the left tuber is 73% covered with lesions caused by *Streptomyces scabies*.

simplex for a gene influencing the reaction to scab. The five types were thus quadruplex, triplex, duplex, simplex and nulliplex for this particular gene. The cvs Hindenburg and Jubel, considered to be quadruplex and triplex respectively, are certainly 'good' parents for scab resistance, and have been used in breeding work in Germany also (Pfeffer and Effmert, 1967). Lauer and Eide (1963) considered that the inheritance of resistance could be explained on a single-gene basis, with the duplex level necessary for effective resistance. However, there is likely to be more control of the resistance mechanism(s) than that conferred by a single major gene. Thus Zadina (1958), working with the progeny of Czech varieties whose resistance/susceptibility was known, obtained a higher proportion of resistant types when a resistant cultivar was used as the female parent rather than as the pollen parent, hence indicating a maternal (cytoplasmic) influence. Cipar and Lawrence (1972) and Howard (1978) were of the opinion that more than one gene pair is implicated in the resistance reaction; the former authors, working with dihaploids of cvs Avon and Hindenburg, concluded that resistance was controlled at more than one locus, and yet the mode of inheritance was relatively simple. They obtained hybrid diploid test progenies from crossing scab-susceptible *S. phureja* clones and cv. Hindenburg (resistant) dihaploids, and suggested that their evaluation could rapidly clarify the mode of inheritance of resistance. Pfeffer and Effmert (1985) recognized that resistance is inherited polygenically, and identified genotypes with high GCA for intercrossing to produce progenies with a high proportion of resistant individuals. Blomquist (1963) attempted a quantitative genetic analysis of scab resistance; from crosses involving 13 clones he calculated the additive, dominance and epistatic components of genetic variance at 65%, 33% and 2% respectively of the total genetic variance.

In conclusion, it might appear that genetic studies into the inheritance of scab resistance would be relatively straightforward, given the simplicity of the host response (periderm proliferation), the ease of assessment and the relative abundance of resistant and susceptible genotypes.

Powdery Scab

Spongospora subterranea (Wallr.) Lagerh.

Powdery or corky scab, as its name implies, is another disfiguring blemish-forming disease of tubers; like common scab and wart it is caused by a persistent soil-inhabiting fungus, but it has a more sophisticated life cycle and attacks its host primarily in two morphologically distinct regions – root hairs and tubers. Being an obligate parasite, the causal organism does not lend itself easily to being manipulated, and, although a considerable amount of resistance screening has been carried out, the nature and control of resistance have not been elucidated. It is likely that this control, exercised as it must be at several locations on the plant (roots, stolons, tubers), is under the control of several genes. The continuity of the scale between resistance and susceptibility, at least as far as tubers are concerned, suggests that resistance is inherited polygenically. In tests carried out in Scotland on 19 progenies from 11 parental cultivars, Wastie (1991) concluded that the resistance of a progeny could be predicted from the phenotypic resistance of its parents. Such resistance is highly heritable, and likely to be amenable to manipulation.

Gangrene

Phoma foveata Foister (syn. *Phoma exigua* Desm. var. *foveata* (Foister) Boerema)

Gangrene is an important cause of spoilage of tubers in store, particularly in northern Europe. Tubers carrying latent infection with *P. foveata*, originating either from soil or from pycnospores produced on an infected mother tuber or washed down from the haulm, will exhibit symptoms of gangrene (Figure 18.3) after storage if the pathogen is able to penetrate the tuber via eyes, lenticels or wounds, and then spread within the tissue. Thus resistance to gangrene is dependent upon several interacting factors, including resistance to damage as well as the reaction of the host tissue to penetration and invasion.

Although cultivars differ widely in susceptibility to the disease, little information is available as to the precise mechanisms of resistance. That it is highly heritable has been demonstrated by Wastie *et al.* (1988), who estimated the broad-sense heritability for 36 progenies from ten parents at 49%. They found

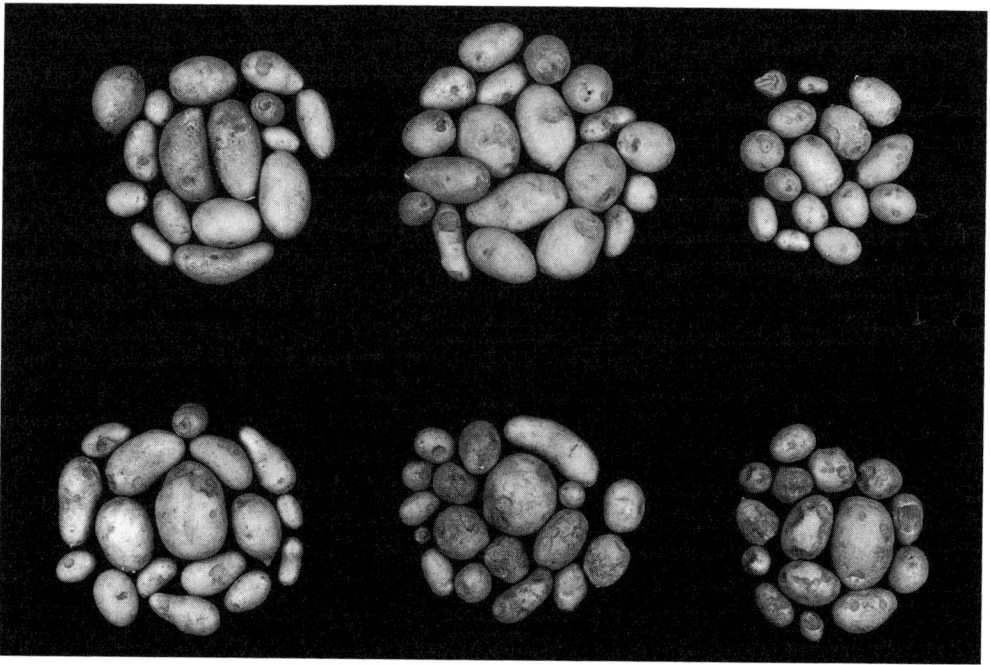

Fig. 18.3. Progenies exhibiting different reactions to gangrene, from resistant (top left) to susceptible (bottom right).

a slight tendency for dominance to favour susceptibility, and concluded that the mean phenotypic resistance of the parents would give an indication of the likely resistance of the progeny. Apart from this comment, no other information appears to be forthcoming concerning the genetical control of resistance.

Skinspot

Polyscytalum pustulans (Owen & Wakef.) M.B. Ellis

Silver Scurf

Helminthosporium solani Dur. & Mont.

and Black Dot

Colletotrichum coccodes (Wallr.) Hughes (syn. *C. atramentarium* (Berk. & Br.) Taub.)

Fig. 18.4. Tubers infected by *Polyscytalum pustulans*, the cause of skinspot.

These three tuber-blemishing diseases are considered together because they have a number of similarities, not least in that nothing is known of resistance mechanisms and rather little concerning genotypic differences in susceptibility. All three pathogens cause a deterioration of tubers in store; all are mainly or partly tuber-borne as well as soil-borne, and all are relatively common. No resistance breeding appears to have been done, and, although several cultivars are known to be susceptible to one or other of the diseases, few cultivars with high levels of resistance have been identified. Manturova (1960, reported by Boyd, 1972) observed that some Russian genotypes with thick, rough periderms were highly resistant to skinspot (Fig. 18.4), and Nagdy and Boyd (1965) made a similar observation in relation to cvs Golden Wonder and Dunbar Rover in Scotland. Bannon (1975) considered resistance to stem base infection to be an important component of resistance to skinspot. However, until resistance mechanisms are better understood and progress is made with progeny testing for resistance, no understanding of the inheritance of underlying resistance mechanisms is possible.

Conclusions

The basis of resistance to most of the foregoing diseases appears to be primarily polygenic. In no case have the individual genes responsible for resistance been identified. Unfortunately, this statement is as true now as it was 20 years ago when Howard (1970) wrote his treatise on potato genetics. Such progress as has been made in improving the resistance of potatoes to diseases has come from the plant breeder rather than from any efforts by geneticists. However, when the genetics of resistances are better understood and resistance genes identified, modern manipulative techniques may enable them to be quickly exploited. Hitherto we have looked to major genes for this exploitation; field resistance, which is more widely available but less well understood, must be a better choice and demands the joint attention of the plant pathologist and breeder/geneticist.

References

Anon. (1983) Panel on potato wart disease, second meeting, Prague, 1982. European and Mediterranean Plant Protection Organization, Paris.

Bannon, E. (1975) Susceptibility of potato cultivars to skinspot disease. *Potato Research* 18, 531-538.

Banville, G.J. (1978) Studies on the *Rhizoctonia* disease of potatoes. *American Potato Journal* 55, 56. Abstract.

Bebre, G.T. (1987) Producing *Fusarium* resistant potato hybrids. In: Rashal, I.D. (ed.) *Genetika i selektsiya v Latviiskoi SSR*. Zinatne, Riga, Latvian SSR, pp. 66-67.

Bedin, P. (1985) Le point sur la gale commune. *Pomme de Terre Française* 47, 13-16.

Black, W. (1935) Studies on the inheritance of resistance to wart disease [*Synchytrium endobioticum* (Schilb.) Perc.] in potatoes. *Journal of Genetics* 30, 127-146.

Blomquist, A.W. (1963) A quantitative genetic analysis of resistance to common scab (*Streptomyces scabies* Thaxt.) in the Irish potato (*Solanum tuberosum* L.). *Dissertation Abstracts* 24, 1800.

Bojnansky, V. (1984) Potato wart pathotypes in Europe from the ecological point of view. *European and Mediterranean Plant Protection Organization Bulletin* 14, 141-146.

Boyd, A.E.W. (1972) Potato storage diseases. *Review of Plant Pathology* 51, 297-321.

Cipar, M.S. and Lawrence, C.H. (1972) Scab resistance of haploids from two *Solanum tuberosum* cultivars. *American Potato Journal* 49, 117-120.

Corsini, D. and Pavek, J.J. (1986) *Fusarium* dry rot resistant potato germplasm. *American Potato Journal* 63, 629-638.

Desjardins, A.E. and Gardner, H.W. (1991) Virulence of *Gibberella pulicaris* on potato tubers and its relationship to a gene for rishitin metabolism. *Phytopathology* 81, 429-435.

Frank, J.A., Leach, S.S. and Webb, R.E. (1976) Evaluation of potato clone reaction to *Rhizoctonia solani*. *Plant Disease Reporter* 60, 910-912.

Gadzhiev, N.M. (1986) The nature of the inheritance of black scurf resistance in interspecific potato hybrids. *Selektsiya i Semenovodstvo, USSR* 1, 18-19.

Gadzhiev, N.M. and Lebedeva, V.A. (1983) Some aspects of resistance to *Rhizoctonia* in potato varieties. *Selektsiya i Semenovodstvo, USSR* 1, 23-24.

Gerlach, W. and Nirenberg, H. (1982) The genus *Fusarium* - a pictorial atlas. *Mitteilungen aus der Biologischen Bundesanstalt für Land- und Forstwirtschaft, Berlin-Dahlem* 209.

Gough, G.C. (1920) Wart disease of potatoes (*Synchytrium endobioticum* Perc.): a study of its history, distribution, and the discovery of immunity. *Journal of the Royal Horticultural Society* 45, 301-312.

Hawkes, J.G. (1990) *The Potato: Evolution, Biodiversity and Genetic Resources*. Belhaven Press, London.

Hille, M. (1965) Die Beurteilung von Kartoffelsorten hinsichtlich ihres Verhaltens gegenüber *Synchytrium endobioticum* (Schilb.) Perc., dem Erreger des Kartoffelkrebses. *Nachrichtenblatt des Deutsche Pflanzenschutzdienst* 17, 137-142.

Hooker, W.J. (ed.) (1981) *Compendium of Potato Diseases*. American Phytopathological Society, St Paul, Minnesota.

Howard, H.W. (1970) *Genetics of the Potato, Solanum tuberosum*. Logos Press, London.

Howard, H.W. (1978) The production of new varieties. In: Harris, P.M. (ed.) *The Potato Crop*. Chapman & Hall, London, pp. 607-646.

Krantz, F.A. and Eide, C.J. (1941) Inheritance of reaction to common scab in the potato. *Journal of Agricultural Research* 63, 219-231.

Krantz, F.A. and Eide, C.J. (1948) Resistance to common scab of potatoes in parental clones and in their hybrid progenies. *American Potato Journal* 25, 294-300.

Langerfeld, E. (1984a) Potato wart in the Federal Republic of Germany. *European and Mediterranean Plant Protection Organization Bulletin* 14, 135-139.

Langerfeld, E. (1984b) Zusammenfassende Darstellung des Erregers des Kartoffelkrebses anhand von Literaturberichten. *Mitteilungen aus der Biologischen Bundesanstalt für Land- und Forstwirtschaft, Berlin-Dahlem* 219.

Langerfeld, E. (1987a) Methoden bei der Prüfung der Reaktion von Kartoffelsorten gegen Lagerfäule erreger der Gattung Fusarium. In: Førsund, E. (ed.) *Potato Disease Assessment Keys*. European Association for Potato Research, Ås, Norway, pp. 101-110.

Langerfeld, E. (1987b) Methoden zur Prüfung der Reaktion von Kartoffelsorten gegen den Erreger des Kartoffelkrebses [*Synchytrium endobioticum* (Schilb.) Perc.]. In: Førsund, E. (ed.) *Potato Disease Assessment Keys*. European Association for Potato Research, Ås, Norway, pp. 112-124.

Langerfeld, E. and Bätz, W. (1990) Verhalten von Kartoffel-Neuzüchtungen gegenüber verschiedenen Pathotypen von *Synchytrium endobioticum* (Schilb.) Perc., dem Erreger des Kartoffelkrebses. *Nachrichtenblatt des Deutschen Pflanzenschutzdienstes (Braunschweig)* 42, 44-45.

Lauer, F.I. and Eide, C.J. (1963) Evaluation of parent clones of potato for resistance to common scab by the 'highest scab' method. *European Potato Journal* 6, 35-44.

Leach, J.G., Krantz, F.A., Decker, P. and Mattson, H. (1938) The measurement and inheritance of scab resistance in selfed and hybrid progenies of potatoes. *Journal of Agricultural Research* 56, 843-853.

Leach, S.S. and Webb, R.E. (1981) Resistance of selected potato cultivars and clones to *Fusarium* dry rot. *Phytopathology* 71, 623-629.

Lellbach, H. and Effmert, M. (1990) Ergebnisse einer Diallelanalyse zur Vererbung der Resistenz gegen *Synchytrium endobioticum* (Schilb.) Perc., pathotyp 1 (D1) bei Kartoffeln (*Solanum tuberosum* L.). *Potato Research* 33, 251-256.

Lipsits, D.V. (1964) Immunological investigations on proteins of potato varieties resistant and susceptible to potato wart. *Doklady Academii Nauk SSSR* 158, 1443-1446.

Lunden, A.P. (1950) Investigations on the reaction of potato to wart disease (*Synchytrium endobioticum*). *Melding Norges Landbrukshøgskole* 30, 1-48.

Lunden, A.P. and Jørstad, I. (1934) Investigations on the inheritance of immunity to wart disease [*Synchytrium endobioticum* (Schilb.) Perc.] in the potato. *Journal of Genetics* 29, 375-385.

Maris, B. (1961) Races of the potato wart causing fungus *Synchytrium endobioticum* (Schilb.) Perc. and some data on the inheritance of resistance to race 6. *Euphytica* 10, 269-276.

Maris, B. (1973) Studies with potato dihaploids on the inheritance of resistance to wart disease. *Potato Research* 16, 324. (Abstract).

Nagdy, G.A. and Boyd, A.E.W. (1965) Susceptibility of potato varieties to skin spot (*Oospora pustulans*) in relation to the structure of the skin and eye. *European Potato Journal* 8, 200-214.

Nazarova, L.P. and Turuleva, L.M. (1987) Resistance of North American wild species of potato to *Rhizoctonia*. *Sbornik Nauchnykh Trudov po Prikladnoi Botanike, Genetike i Selektsii* 115, 103-106.

O'Brien, V.J. and Leach, S.S. (1983) Investigations into the mode of resistance of potato tubers to *Fusarium roseum* 'sambucinum'. *American Potato Journal* 60, 227-233.

Pfeffer, C. and Effmert, M. (1967) Die auslese geeigneter Eltern und Kombinationem für die Schorfresistenz-Züchtung. *European Potato Journal* 10. 286-295.

Pfeffer, C. and Effmert, M. (1985) Die Züchtung homozygoter Eltern für Resistenz gegen Kartoffelschorf, verursacht durch *Streptomyces scabies* (Thaxt.) Waksman & Henrici. *Archiv für Züchtungsforschung* 15, 325-333.

Podgaetski, A.A. and Koval, N.D. (1989) Sources of resistance to *Fusarium* dry rot in potato. *Selektsiya i Semenovodstvo (Moskva)* 4, 33-34.

Pratt, M.A. (1976) The relation between field and laboratory susceptibility of potato cultivars to wart disease. *European and Mediterranean Plant Protection Organization Bulletin* 6, 111-117.

Ross, H. (1986) Potato breeding - problems and perspectives. *Advances in Plant Breeding*, Supplement 13 to *Journal of Plant Breeding*.

Rothacker, D., Effmert, M., Gottschling, W., Jakowlewa, W.I. and Lechnowicz, W.S. (1974) Möglichkeiten der Züchtung krebsbiotypen-resistenter Kartoffeln auf der Grundlage von *S. andigenum* × *S. tuberosum*-Bastarden. *Archiv für Züchtungsforschung* 4, 45-55.

Salaman, R.N. and Lesley, J.W. (1923) Genetic studies in potatoes: the inheritance of immunity to wart disease. *Journal of Genetics* 13, 177-186.

Scheidt, M. and Hunnius, W. (1981) Vererbung der Resistenz gegen die Pathotypen 2 und 6 des Kartoffelkrebses (*Synchytrium endobioticum*). *Zeitschift für Pflanzenzüchtung* 86, 158-173.

Sedova, T.S. (1980) Inheritance of wart resistance by interspecies potato hybrids. *Selektsiya i Semenovodstvo* 8, 18-19.

Seppanen, E. (1983) Fusariums of the potato in Finland. 6. Varietal tuber resistance to *Fusarium* species. *Annales Agriculturae Fenniae* 22, 8-17.

Stachewicz, H. (1989) 100 Jahre Kartoffelkrebs - seine Verbreitung und derzeitige Bedeutung. *Nachrichtenblatt für den Pflanzenschutz in der DDR* 43, 109-111.

Suska, M. (1985) (published 1989) Inheritance of resistance of potato tubers to dry rot caused by *Fusarium sulphureum* Schl. *Hodowla Roslin, Aklimatyzacja i Nasiennictwo* 29, 23-36.

Tolstrup, K., Bjor, T., Böös, U., Erjefält, L., Kirk, H.G., Kivinen, M., Oloffson, B., Pietilä, L., Sundheim, L. and Löschenkohl, L. (1990) Methods for testing resistance to *Rhizoctonia solani*. *Abstracts of the 11th Triennial Conference of the European Association for Potato Research*, 255-256.

Ullrich, J. (1959) Physiological specialization in *Synchytrium endobioticum* (Schilb.) Perc. in FRG. *Rostlinna Vyroba* 5, 111-116.

Wastie, R.L. (1991) Resistance to powdery scab of seedling progenies of *Solanum tuberosum*. *Potato Research* 34, 249-252.

Wastie, R.L. and Bradshaw, J.E. (1993) Inheritance of resistance to *Fusarium* spp. in tuber progenies of potato. *Potato Research* 36, (in press).

Wastie, R.L., Caligari, P.D.S., Stewart, Helen E. and Mackay, G.R. (1988) Assessing the resistance to gangrene of progenies of potato (*Solanum tuberosum* L.) from parents differing in susceptibility. *Potato Research* 31, 355-365.

Wastie, R.L., Stewart, Helen E. and Brown, J. (1989) Comparative susceptibility of some potato cultivars to dry rot caused by *Fusarium sulphureum* and *F. solani* var. *coeruleum*. *Potato Research* 32, 49-55.

Zadina, J. (1958) A genetical evaluation of some potato varieties resistant to scab [*Actinomyces* scabies (Thaxter) Güssow]. *Vědecké Práce Výzkumného Ustavu Bramborǎrského Csazv v Havlíčkové Brodě 1958*, 59-79.

Zadina, J. (1983) Genetics of the resistance of potatoes to the D pathotype of wart disease [*Synchytrium endobioticum* (Schilb.) Perc.]. *Sbornik UVTIZ, Genetika a Slechtění* 19, 285-296.

Zadina, J. and Findejs, R. (1988) Genetics of potato resistance to wart disease [*Synchytrium endobioticum* (Schilb.) Perc.] and its use in potato breeding. *Vědecké Práce - Výzkumný a Slechtitelský Ustav Bramborářský v Havlíčové Brodě* 11, 41-56.

19 Inheritance of Resistance to Bacterial Diseases

J.G. ELPHINSTONE

Department of Plant Pathology, Rothamsted Experimental Station, ARFC Institute of Arable Crops Research (IACR), Harpenden, Herts AL5 2JQ, UK.

Introduction

Bacterial diseases severely hinder production, storage, transport and marketing of potatoes worldwide. The most important diseases and their causal agents are:

1. Blackleg and bacterial soft rot (Perombelon and Kelman, 1980), caused by *Erwinia carotovora* subsp. *atroseptica* (van Hall) Dye, *E. carotovora* subsp. *carotovora* (Jones) Dye and *E. chrysanthemi* (Burkh. *et al.*).
2. Bacterial wilt or brown rot (Kelman, 1953; Hayward, 1991), caused by *Pseudomonas solanacearum* E.F. Smith.
3. Bacterial ring rot (Slack, 1987), caused by *Clavibacter michiganensis* subsp. *sepedonicus* (Spieck. and Kotth.) Davis *et al.* (syn. *Corynebacterium sepedonicum* (Spieck. and Kotth.) Skapt. and Burkh.).

There is no practical means available for the chemical control of these diseases. Current control strategies rely on the integrated use of resistant or tolerant varieties, seed certification, improved crop husbandry and general on-farm hygiene, but they are usually only partially effective (French, 1988; De Boer, 1990; Murphy, 1990). Improved cultivar resistance therefore remains the most desirable target for sustainable and economically viable control in the future. Evidence of genetic variation exists amongst modern potato cultivars, with respect to their levels of susceptibility to blackleg and soft rot (Wastie and Mackay, 1985), bacterial wilt (Eden-Green and Elphinstone, 1993) and bacterial ring rot (De Boer and McCann, 1990). However, a significant improvement in the resistance of commercially acceptable cultivars to these diseases has yet to be achieved through traditional plant breeding approaches. In fact, the use of currently available tolerant cultivars has been discouraged in some cases (Granada, 1988; De Boer and McMann, 1990) due to fears that they can act as symptomless carriers of inoculum.

There are several reasons for the apparent slow rate of achievement in the development of acceptable potato cultivars with resistance to bacterial diseases:

1. The distinction between resistance and tolerance to the bacterial diseases has not been clearly defined in most selection programmes. In most cases genotypes have been selected with resistance to pathogen multiplication *in planta* and/or symptom expression, but not to infection. These apparently resistant genotypes subsequently behave as symptomless carriers of latent populations of virulent bacteria which can eventually cause disease when specific environmental conditions become conducive.

2. No sources of immunity to the bacterial diseases have been identified. In fact, the level of resistance in tetraploid cultivars is generally low and is readily overcome as pathogen populations increase.

3. Accurate identification and quantification of resistance or susceptibility within a given population has not been possible because results (from the numerous screening methods employed) have been unpredictably influenced by interactions with the environment or other pathogens and pests. Consequently, the mechanisms of resistance and the nature of their inheritance have been difficult to determine and little precise information is currently available.

4. The utilization of higher levels of resistance, found in wild *Solanum* spp. and primitive cultivars, has been hindered by sexual imcompatibility between tetraploid cultivars and these species. Even when hybridization has been possible, the accompanying transfer of unwanted genes and associated undesirable phenotypic characteristics from tne wild species has hindered breeding efforts.

5. The genetic control of resistance to bacterial diseases in modern potato cultivars seems to be complex. The additional complexity associated with the tetrasomic inheritance patterns of tetraploids has meant that accurate information on modes of inheritance is scarce.

6. Resistance to a particular bacterial disease will not guarantee the acceptance of a particular potato cultivar unless it is combined with numerous other desirable traits such as high yield, agronomic and cooking quality, adaptation to specific agroecological conditions, and resistance to late blight, viruses, nematodes and insects. Intensive breeding with recurrent selection and an associated segregation and loss of target genes has therefore been unavoidable.

The aim of this chapter is to pool the rather limited and often contradictory information on the genetic control and inheritance of resistance to the important bacterial diseases of potato. This will include an account of the inheritance of resistance genes from modern and primitive cultivars and wild *Solanum* spp. with respect to their relevance for use in plant breeding. In addition, various techniques for the transmission of resistance genes to modern cultivars will be compared. Factors which determine pathogenicity in the bacteria will also be discussed in terms of their role in the characterization of interacting host resistance factors.

Inheritance of Resistance in Modern Cultivars

Resistance to Erwinia spp.

Although tetraploid cultivars vary with respect to the severity of blackleg and tuber soft rot symptom development (Hahn, 1974; Pietkiewicz, 1980; Bourne et al., 1981; Krause et al., 1982; Bushkova, 1983; Ciampi and Andrade-Soto, 1984; Lapwood et al., 1984; Lewosz, 1984; McGuire and Kelman, 1984; Vlasov et al., 1987), the degree to which these represent genetically stable differences in resistance to infection or disease tolerance has been difficult to determine (Wastie and Mackay, 1985). No cultivar has been considered highly resistant, although variation in relative susceptibility is apparent. Nevertheless, some attempts have been made to breed for resistance to the pectolytic Erwinia spp. amongst modern cultivars.

The existence of heritable resistance to both blackleg and soft rot amongst modern tetraploid cultivars has been demonstrated on the basis that progenies from resistant × resistant or resistant × susceptible parents contain a higher frequency of resistant genotypes than those resulting from two susceptible parents (Zadina and Dobias, 1976; Dobias, 1977). Significant differences were also found between progenies from self-pollinated resistant and susceptible cultivars. Lellbach (1976) estimated that a high percentage of the variation in susceptibility between cultivars (around 85%) was due to stable genetic differences, although the response to the environment was not clearly determined. Results from a simple diallel set of crosses, involving only three cultivars, showed that resistance was probably controlled by minor genes whose additive effects could increase resistance levels (Lellbach, 1978). No evidence for the existence of vertical resistance to Erwinia spp. in potato cultivars has been presented and all variation has therefore been attributed to horizontal genes (Van der Plank, 1978). Munzert (1984) determined a heritability coefficient of $h^2 = 0.12$–0.45 for blackleg resistance in a study of 25 European cultivars over 5 years. Environmental interactions were again unpredictable, affecting the reactions of different groups of cultivars to different extents in different years, and in some cases accounting for more of the variance than that due to differences in genotype.

Both Dobias (1975) and Munzert (1984) found positive correlations between the levels of resistance to blackleg and tuber soft rot in tetraploid cultivars ($r = 0.65$ and 0.57 respectively). This could be expected since blackleg, by definition (Perombelon and Kelman, 1987), is seed-borne and develops from the rotting seed tuber. Nevertheless, significant correlations have not consistently been reported (Dopke and Heitefuss, 1987) and it has been proposed that some resistance mechanisms in the tuber and stem base may be under independent genetic control (Lyon, 1989).

Resistance to **Pseudomonas solanacearum**

European and American tetraploid cultivars which exhibit unexpected levels of variation in field resistance to *P. solanacearum* have been catalogued recently (Eden-Green and Elphinstone, 1993). In general, results vary greatly with environment and pathogen strain, and the level of resistance encountered has been generally low. Their apparent disease tolerance in certain locations may reflect their degree of adaptation to particular environments (Tung *et al.*, 1990a). One of these cultivars (Cruza-148) has been used in breeding (Schmiediche, 1986; Tung *et al.*, 1990a) but the source of resistance in this case is unknown and its mode of inheritance was not discussed specifically.

Resistance to **Clavibacter michiganensis** *subsp.* **sepedonicus**

European and North American cultivars also vary in field response to infection by *Clavibacter michiganensis* subsp. *sepedonicus* (Zielke, 1983; Olsson, 1987; De Boer and McCann, 1990). Most of these cultivars were bred from a small number of tolerant sources identified by Bonde *et al.* (1942). Little was determined of the mode of inheritance except that resistance from the cultivar President could be transmitted to almost half of the progeny in a cross with the susceptible cultivar Katahdin. Multiplication of the pathogen in the stem is slower in the more tolerant varieties (De Boer and McCann, 1990), which show resistance to symptom expression rather than infection. The use of such tolerant cultivars has been discouraged in North America since they support viable populations of the bacterium and thus act as sources of inoculum.

Inheritance of Resistance from Wild Species and Primitive Cultivars

Resistance to **Erwinia** *spp.*

Numerous sources of resistance to *Erwinia* spp. have been detected amongst wild diploid *Solanum* spp. and primitive cultivars, as well as amongst cultivated tetraploids of *S. tuberosum* subsp. *andigena* (Koromyslova, 1972; Dobias, 1978; Hidalgo and Echandi, 1982; Bushkova and Vlasov, 1983; Van Soest, 1983; French and de Lindo, 1985; Kogut and Polozhenets, 1985; Morozova, 1987; Pawlak *et al.*, 1987; Vlasov *et al.*, 1987; Austin *et al.*, 1988; Huaman *et al.*, 1988; Lojkowska and Kelman, 1989). Significantly higher levels of resistance and greater variability in disease response have been recorded amongst these materials than amongst cultivars of *S. tuberosum* subsp. *tuberosum* and their progeny (Bushkova and Vlasov, 1983; Pawlak *et al.*, 1987; Huaman *et al.*, 1988; Elphinstone, 1990). As in the tetraploids, sexual transmission of resistance genes resulted in higher frequencies of resistant genotypes amongst hybrids of resistant rather than susceptible diploid parents (Bushkova and Vlasov, 1983).

At the International Potato Center, relatively high levels of resistance to *E. chrysanthemi* from selected clones of *S. tuberosum* subsp. *andigena* have been combined with adaptation to the warm tropics by hybridization with advanced breeding lines of *S. tuberosum* subsp. *tuberosum*. Inheritance of resistance in the progeny, and in hybrids of the clonal selections from subsp. *andigena*, is currently under investigation (Huaman et al., 1988, 1990). Combined resistance to *Fusarium oxysporum*, *Fusarium solani* (the causal agents of dry rot) and *Erwinia chrysanthemi* has been detected amongst the subsp. *andigena* selections (Huaman et al., 1989) – an important finding, considering the significant interaction which occurs between these fungal and bacterial pathogens worldwide, and the fact that resistance to dry rot is rare and is usually not correlated with soft rot resistance (Ratuszniak, 1981; Pawlak et al., 1987). Reports that resistance to combined infection by *Fusarium* and *Erwinia* spp. does not correlate with resistance to either pathogen alone (Krause et al., 1982; Dorozhkin et al., 1985) stress the need for further investigation on this complex interaction between pathogens and host.

Unlike the case for modern cultivars, variation in phenotypic resistance levels between diploid species seems to be affected relatively more by genotype than by environmental influences, and inheritance patterns are simpler. Thus, new opportunities have arisen to more accurately investigate the mechanisms of resistance and their genetic control. Studies at North Carolina State University (NCSU), on the heritability of resistance to soft rot in diploid populations involving *S. stenotomum* and *S. phureja*, showed that both level and frequency of resistance increased with recurrent selection. Furthermore, resistance to each of the pectolytic *Erwinia* spp. (Eca, Ecc and Echr) was highly correlated. Broad-sense heritability estimates were fairly high for this resistance ($h^2 = 0.43$–0.82), whereas narrow-sense estimates from half-sib analysis were variable but averaged 0.33 (P. Wolters, unpublished). A significant correlation ($r = 0.719$) between resistances to blackleg and soft rot has been detected amongst primitive cultivars of this type (Elphinstone, 1990).

Resistance to Pseudomonas solanacearum

Since 1967, *Solanum phureja* has been used as a major source of resistance to *P. solanacearum* in potato breeding. This resistance was thought to be governed by a series of independent dominant genes under the possible control of other modifying genes (Rowe and Sequeira, 1970; Rowe et al., 1972). The expression of resistance to a particular biovar 1 strain (K-60) suggested that three dominant genes were required. Similarly, the inheritance of resistance to a second biovar 1 strain (S-123) fitted a three-gene model. However, their results indicated that only one gene was common to both models. Other reports suggested that four major genes could be involved (Sequeira, 1979).

Since clones of *S. phureja* vary in resistance reactions to different strains of *P. solanacearum* (Rowe and Sequeira, 1970; French and de Lindo, 1982; Huang *et al.*, 1989), it is possible that numerous resistance genes could influence strain specificity. Resistance to specific strains, however, appears to be simply inherited in this *Solanum* sp. It was assumed that resistance to latent infection is also under genetic control because symptomless tuber and stem infection was restricted to certain genotypes of *S. phureja*, and strains of *P. solanacearum* differed in their ability to cause latent infection (Ciampi and Sequeira, 1980).

Newly developed breeding populations contain resistance genes from several diploid *Solanum* spp. (Schmiediche, 1986, 1988), including *S. phureja*, *S. sparsipilum*, *S. chacoense*, *S. microdontum* and *S. raphanifolium*. Since the genetic basis for resistance may be different in each species, it is thought that its inheritance in the new lines could be polygenically controlled (Schmiediche, 1987). However, no information on the patterns of inheritance of resistance from the individual species is available in the literature.

Tung *et al.* (1990a) have demonstrated that the stable expression of resistance, under warm tropical conditions, is enhanced when resistance genes are combined with genes conferring adaptation to that environment (e.g. earliness and heat tolerance). They also indicated that the frequency of resistant genotypes to a particular strain of the pathogen was highest when different sources of resistance were combined in a single population. It was therefore concluded that widening of the genetic base for both resistance and adaptation is of great significance to the breeding of bacterial wilt-resistant potatoes. Parental combining ability influenced the inheritance of resistance; some resistant genotypes expressed high levels of either specific or general combining ability. Further exploration showed that, although differential interactions between resistant cultivars and pathogen strains were significant, the more important interaction was between temperature and pathogen aggressiveness (Tung *et al.*, 1990b). Indirect evidence from these studies supports the theory that resistance in populations with multiple sources of resistance is controlled by minor genes.

Resistance to **Clavibacter michiganensis** *subsp.* **sepedonicus**

High levels of resistance to bacterial ring rot (Kogut and Polozhenets, 1985; Slack and Gedye, 1988) have been detected in wild *Solanum* spp. and primitive cultivars. However, attempts to exploit this resistance have yet to appear in the literature.

The Use of Ploidy Level Manipulation (4x–2x Crossing) in the Transfer of Resistance from Primitive to Modern Cultivars

Ploidy level manipulation has been used to exploit a variety of useful genes from diploids and transfer them to tetraploids (Iwanaga, 1985). Advantage may be taken of diploid mutants which produce $2n$ pollen, formed by first-division restitution (FDR), following a mutation in a single gene which causes the formation of parallel spindles during meiosis (Mok and Peloquin, 1975). It has been estimated that 80% of the diploid genotype can be transferred to the tetraploid progeny which result from $4x-2x$ and $2x-2x$ crosses. Inheritance patterns are far simpler at the diploid (disomic) than the tetraploid (tetrasomic) level. This breeding strategy therefore facilitates the combination of genes from diploid *Solanum* spp. governing resistance to the bacterial diseases as well as other necessary characteristics (e.g. late blight, virus, nematode and insect resistances, tolerances to environmental stresses and desirable agronomic traits).

Diploids with soft rot and blackleg resistance from *S. stenotomum* and *S. phureja*, which formed unreduced $2n$ gametes, were selected at NCSU (Iwanaga, 1985). Transfer of this resistance to fertile tetraploid progenies has already been demonstrated by fertilization of tetraploid female parents with $2n$ pollen from the diploids. Similarly, breeding at the diploid level has enabled the selection of genotypes with improved agronomic traits, high starch content and resistance to bacterial soft rot at the Potato Research Institute, Bonin, Poland (E. Zimnoch-Guzowska and E. Lojkowska, pers. comm.). Several of the advanced hybrids produced $2n$ pollen and will be used to transfer the resistance to tetraploid cultivars. During the course of this research, *S. chacoense* was identified as a useful source of resistance.

The introduction of bacterial wilt resistance from *S. phureja* into tetraploid cultivars was first accomplished by breeding for improved agronomic characteristics at the diploid level. This involved the crossing of selected *S. phureja* clones with *S. tuberosum* dihaploids ($2n = 2x = 24$), followed by $4x-2x$ crossing with a $2n$ pollen-producing diploid clone (Rowe and Sequeira, 1970; Schmiediche, 1986). A similar strategy was used at the International Potato Center to produce tetraploid populations with broad genetic bases and bacterial wilt resistance genes from numerous diploid *Solanum* spp., including *S. phureja*, *S. chacoense*, *S. microdontum*, *S. sparsipilum* and *S. raphanifolium* (Schmiediche, 1986, 1987; Mendoza, 1988). Breeding at the diploid level was facilitated by simple inheritance patterns, as well as the high levels of fertility amongst the primitive diploid cultivars used. Dihaploids of Tuberosum cultivars were not used in this case due to their high levels of male and also female, sterility. Instead, advanced diploid selections of *S. phureja* and *S. stenotomum*, from NCSU, were used to breed improved quality and agronomic traits into the resistant diploid populations. Resistance to root-knot nematodes (*Meloidogyne* spp.) and to potato tuber moth (*Phthorimaea operculella*) were also detected amongst the improved diploid populations. Resistant tetraploids were then selected following fertilization of advanced ($4x$) cultivars and breeding lines with $2n$ pollen from selected diploids.

Interspecific Somatic Hybridization

Tubers from hexaploid somatic hybrids between *Solanum brevidens*, a non-tuber-forming diploid wild species, and a tetraploid potato cultivar (PI 203900) have been shown to demonstrate high levels of resistance to soft rot caused by Eca, Ecc and Echr (Austin *et al.*, 1988). Furthermore, some of the hybrids (formed by fusion between isolated tetraploid and diploid protoplasts, followed by regeneration) were fertile, despite the fact that interspecific somatic hybrids are often sterile. Some of the pentaploid progeny from crosses between these somatic hybrids and the tetraploid cultivar, Katahdin, also demonstrated high levels of resistance, showing that the resistance could be sexually transferred. The resistance could also be stably transferred following a second cross between resistant pentaploids and Katahdin.

The transmission genetics of somatic hybrids is not clearly understood. In some cases pairing between chromosomes derived from highly divergent parental species is inhibited. Since the pentaploid progeny from the first and second backcrosses with Katahdin were phenotypically variable with respect to their soft rot resistance, RFLP analysis was used to analyse chromosome transmission from diploid and tetraploid parents during somatic hybridization and subsequent sexual crossing (Williams *et al.*, 1990). The results indicated that the somatic hybrids themselves contained at least one complete copy of each chromosome from each parent. Furthermore, the loss of some *S. brevidens*-specific DNA in the pentaploid sexual progeny was explained by intergenomic pairing and recombination rather than loss of whole chromosomes during meiosis. It therefore follows that somatic fusion may be a useful way to bypass problems of sexual incompatibility and introduce stable resistance genes from wild species to new potato cultivars.

Molecular Aspects of Host–Pathogen–Environment Interactions: the Key to Understanding the Mechanisms and Genetics of Resistance

Significant advances have been made in recent years in the study of the molecular aspects of phytopathogenicity in bacteria and their genetic control. Since the bacterial genome is relatively simple compared with that of the host plant, most research has concentrated on the identification of pathogenicity factors which can be used to trace the plant's reaction to them. Hence, the environment *in planta*, under which pathogenicity is expressed or suppressed, may be determined in terms of the biochemical pathways involved and the genes which regulate them. Of the bacterial plant pathogens, *Erwinia* spp. and *P. solanacearum* have been amongst the most intensively studied in this way, because extensive information was already available on the closely related organisms *Escherichia coli* and *Pseudomonas aeruginosa*. Knowledge of the factors involved in pathogenesis will greatly assist in the improvement of techniques for the accurate identification and selection of resistant genotypes in potato breeding programmes, as well as in the ultimate goal of engineered resistance.

Interactions with Erwinia spp.

The molecular aspects of pathogenicity (Kotoujansky, 1987) and the biochemical basis of resistance to the pectolytic *Erwinia* spp. (Lyon, 1989) have both been reviewed recently. In addition, the molecular responses of the potato tuber to environmental stresses brought about by mechanical wounding and hypoxia and their effect on soft rot susceptibility have been discussed in detail (Davis *et al.*, 1990; see also chapter 11 by Vayda). In summary, pathogenicity in *Erwinia* is mainly due to the production of cell wall degrading enzymes, which primarily consist of a number of different pectinases and some cellulases and proteases. However, the impairment of natural host resistance, by wounding and the induction of anaerobic conditions, is also required for infection and disease initiation (Perombelon and Kelman, 1980).

Several possible resistance mechanisms have been described (Table 19.1) but most are active only under aerobic conditions. Under natural field and storage conditions, however, disease occurs when hypoxia is induced after tubers become covered with a film of moisture (Maher and Kelman, 1983). Furthermore, hypoxia overrides wound-healing responses and therefore facilitates infection (Davis *et al.*, 1990). The most effective resistance mechanisms should therefore be active following infection under anaerobic conditions. A recent study of the high levels of resistance found in somatic hybrids between *S. brevidens* and *S. tuberosum* has identified such a mechanism. Resistance, in this case, was due to the reduced ability of a pectinase (pectate lyase), produced by Eca, to degrade cell wall pectic components because of their high degree of methyl esterification (Hedley and Perombelon, 1990). Similar resistance mechanisms in tetraploid cultivars were proposed by Weber (1983).

A system has recently been described which enables the simultaneous study of bacterial and plant gene expression in potatoes affected by soft rot (Yang *et al.*, 1989). The biochemical changes taking place in pathogen and host tissue, which were separated by an inert membrane, were studied as the disease progressed. In tuber slices inoculated with *Erwinia carotovora*, induction of phenylalanine ammonia lyase (PAL) activity and mRNA levels were greater and more rapid than in non-inoculated controls. PAL is involved in the synthesis of phenolics and lignin, two oxygen-dependent resistance factors. Further analysis of other resistance factors using this kind of system would be useful.

Interactions with Pseudomonas solanacearum

Pathogenicity factors in *P. solanacearum* include the production of an extracellular polysaccharide (EPS), polygalacturonases (PG), an endoglucanase (a

Table 19.1. Possible mechanisms of resistance to the pectolytic *Erwinia* spp. in potato (Lyon, 1989).

Resistance mechanism	Effect	Oxygen-dependent
Rapid wound periderm formation	Prevents infection through wounds	Yes
Lignin, cutin and suberin formation	Present chemical barrier in cell walls	Yes
Esterification of cell wall pectin	Inhibits cell wall degradation by *Erwinia* pectic enzymes	No
Production of oxidized phenolic compounds	Inhibits bacterial growth and possibly inhibits polygalacturonase	Yes
Phytoalexin elicitation (e.g. rishitin) by cell wall fragments	Inhibits bacterial growth	Yes
Production of inhibitory proteins	Inhibits bacterial growth; Inhibits pectic enzyme activity; Possibly agglutinates bacterial cells (e.g. lectin)	Yes
High calcium content	Increased resistance of cell wall to polygalacturonase.	No

type of cellulase) and plant growth hormones, such as auxin, cytokinin and ethylene (Hayward, 1991). It has been demonstrated that genes governing the production of EPS are activated when the bacteria are transferred from a nutrient-rich to a nutrient-poor medium (Denny and Baek, 1991). A similar phenomenon has also been observed for those genes which regulate PG production (Allen *et al.*, 1991). In other studies, genes controlling the aggressiveness of the bacterium have been shown to be affected by pH (Arlat and Boucher, 1991). Further investigation will hopefully determine how these findings relate to the environments in susceptible and resistant plant tissues and vessels, and the way in which they influence bacterial pathogenicity.

Most resistance mechanisms have been studied for the interaction between *P. solanacearum* and *S. phureja*. Zalewsky and Sequeira (1975) isolated a bacterial inhibitor which was much more potent when isolated from resistant than susceptible *S. phureja* clones. However, when progenies from crosses between resistant and susceptible clones were studied, resistance and inhibition were found to be inherited independently. Non-compatibility between virulent strains of *P. solanacearum* and resistant clones of *S. phureja* and non-host plants has been shown to be the result of a hypersensitive response (HR) in the plant, the initiation of which is regulated by specific (*hrp*) genes in the pathogen (Boucher *et al.*, 1988; Huang *et al.*, 1989). Current investigation is focused on the determination of molecular pathways involved in the induction of the HR and the possible role of soluble plant products in the regulation of the pathogenicity genes involved (Allen *et al.*, 1991).

A procedure which can be used to monitor the expression of pathogenicity factors *in planta* has been described by Boucher and Boistard (1988). They fused the structural gene encoding β-galactosidase production with the regula-

Table 19.2. Somaclonal variation in susceptibility to bacterial potato diseases.

Pathogen	Cultivars tested	References
E. carotovora	Crystal	Taylor et al., 1988
		Taylor and Secor, 1990
	Zarevo	Marunenko et al., 1988
C. michiganensis subsp. sepedonicus	Zarevo	Marunenko et al., 1988
		Rassadina et al., 1988

tory sequences of the *hrp* region of the bacterial megaplasmid. Activity of β-galactosidase could therefore be used as a measure of the expression of the pathogenicity genes also present in the *hrp* region. This technique will be used to investigate the role of plant factors in the regulation and expression of pathogenicity genes in susceptible and resistant plants.

Apart from the HR, other resistance mechanisms are evident which do not prevent infection or pathogen multiplication at the infection site, but seem to restrict the movement of the pathogen within the plant (Ciampi and Sequeira, 1980; Bowman and Sequeira, 1982; Prior et al., 1990). This type of resistance mechanism is responsible for symptomless carriers of latent infection.

Somaclonal Variation

Somaclonal variation, a common occurrence during potato regeneration from tissue culture, often results in detectable phenotypic changes caused by chromosomal deletions and rearrangements (Scowcroft et al., 1983). Variation between somaclones has been reported with respect to susceptibility to bacterial potato diseases (Table 19.2).

Differences in tuber susceptibility to *Erwinia* soft rot have been noted between protoplast-derived regenerants (Taylor et al., 1988). Moreover, differential responses have been observed, under aerobic conditions, between the survival rates of *in vitro* cell cultures challenged directly with bacterial cultures or culture-free bacterial enzyme/metabolite complexes (Rehbein, 1983; Marunenko et al., 1988, Taylor and Secor, 1990). Similarly, cell cultures showed differential survival of repeated doses of a toxin from *C. michiganensis* subsp. *sepedonicum*, and could be selected and regenerated for determination of resistance under field conditions (Marunenko et al., 1988; Rassadina et al., 1988).

The nature of somaclonal variation was reviewed by Scowcroft et al. (1983) and Ross (1986), but no specific information on the nature of resistance to bacterial diseases has been presented. The level and stability of increased resistance obtained through somaclonal selection remain to be compared with those inherited through interspecific hybridization. Nevertheless, providing yield and other agronomic quality characters remain unaffected, this phenomenon represents a potentially valuable means through which the field tolerance of a well-established cultivar may be improved. The further development of procedures for the *in vitro* selection of bacterial resistant potato lines will depend on the degree of correlation which is found between the responses of cell cultures *in vitro* and mature plants under field conditions.

Genetic Engineering

Advances in molecular biology have enabled the various limitations of conventional breeding approaches to be overcome through the introduction of genes encoding the production of potent antibacterial peptides into the potato genome (Montanelli and Nascari, 1989; Destefano-Beltran *et al.*, 1990). Artificially synthesized gene constructs which control the production of lytic peptides, modelled on those found in the giant silk moth (*Hyalophora cecropia*), have been transferred into potato via *Agrobacterium rhizogenes* and *A. tumefaciens*. Low doses of the various peptides used (cecropins, attacins and lysozyme) are highly bactericidal *in vitro* and act synergistically against both *P. solanacearum* and *Erwinia* spp., as well as other important fungal potato pathogens. Moreover, by combining the relevant structural DNA constructs with additional DNA involved in proteinase regulation in response to wounding (Davis *et al.*, 1990), it was possible to cause the peptide genes to be expressed in the same way *in planta*. Trials are now under way to compare the levels of resistance of transformed and wild-type plants to *P. solanacearum* and *Erwinia* spp. (see chapter 11 by Vayda). Many aspects of this research remain to be examined before this type of engineered resistance will be available at the commercial level. However, it has been clearly demonstrated that the procedures used are operative and will be in place for the introduction and evaluation of newly identified resistance genes as they become available.

Future Prospects

Although significant improvements in the understanding of the various biochemical and physiological mechanisms of pathogenicity in *Erwinia* spp. and *P. solanacearum* have taken place in recent years, the capacity of the host to respond to the pathogens has received relatively little attention. However, as the numerous pathogenicity factors become increasingly better understood, it is anticipated that attention will shift to focus more on the factors affecting the expression of pathogenicity within the host plant. Identification of the biochemical pathways involved in host–pathogen interactions should shed light on the host genes which play a role in conferring resistance. Indeed, experimental systems have already been described which can monitor the gene products of both pathogen and host during the development of these bacterial diseases (Boucher and Boistard, 1988; Yang *et al.*, 1989). Simultaneous analysis of the differences between resistant and susceptible plants at the DNA level will also be needed if the elusive resistance genes are to be characterized and their modes of inheritance properly evaluated.

Compared with the pectolytic *Erwinia* spp. and *P. solanacearum*, virtually nothing is known of pathogenicity factors of *Clavibacter michiganensis* subsp. *sepedonicus* or the mechanisms and modes of inheritance of resistance to it in the host plant. Advances in the investigation of the interactions between the host and this less-studied bacterium should be greatly facilitated by the application of research findings on the other pathogens.

The wild and primitive relatives of modern potato varieties should play a major role in the identification of resistance mechanisms and the genes which control them. The higher levels of resistance detected in these materials, together with the relative simplicity of their genomes, should make studies in these materials easier than in modern tetraploid cultivars. Prolonged breeding programmes and the introduction of new breeding methods have enabled the production of highly acceptable diploid potato clones which can carry useful genes from diploid to tetraploid progenies. Furthermore, advances in the field of genetic engineering offer a realistic means of introducing characterized genes from wild species directly into improved cultivars.

This chapter has reviewed research findings over several decades and has highlighted the scarcity of accurate information on the inheritance of bacterial disease resistance in potato. However, it seems that the rapidly increasing amount of information which has begun to accumulate on the physiological, biochemical and genetic interaction between bacteria and potato will almost certainly lead to substantial advances in the near future.

References

Allen, C., Huang, Y. and Sequeira, L. (1991) Cloning of genes affecting polygalacturonase production in *Pseudomonas solanacearum*. *Molecular Plant-Microbe Interactions* 4, 147-154.

Arlat, M. and Boucher, C. (1991) Identification of a *dsp* DNA region controlling aggressiveness of *Pseudomonas solanacearum*. *Molecular Plant-Microbe Interactions* 4, 211-213.

Austin, S., Lojkowska, E., Ehlenfeldt, M.K., Kelman, A. and Helgeson, J.P. (1988) Fertile interspecific somatic hybrids of *Solanum*: a novel source of resistance to *Erwinia* soft rot. *Phytopathology* 78, 1216-1220.

Bonde, R., Stevenson, F.J., Clark, C.F. and Akeley, R.V. (1942) Resistance of certain potato varieties and seedling progenies to ring rot. *Phytopathology* 32, 813-819.

Boucher, C. and Boistard, P. (1988) Genetic approach to pathogenicity determinants of *Pseudomonas solanacearum*. *Bacterial Wilt Newsletter* 3, 1-2. ACIAR, Canberra.

Boucher, C., Arlat, M., Zischek, C. and Boistard, P. (1988) Genetic organization of pathogenicity determinants of *Pseudomonas solanacearum*. In: Keen, N., Kosage, T. and Walling, L. (eds) *Physiology of Plant-Microbial Interactions*. American Society of Plant Physiologists, Bethesda, Maryland.

Bourne, W.F., McCalmont, D.C. and Wastie, R.L. (1981) Assessing potato tubers for susceptibility to bacterial soft rot (*Erwinia carotovora* var. *atroseptica*). *Potato Research* 24, 409-415.

Bowman, J.E. and Sequeira, L. (1982) Resistance to *Pseudomonas solanacearum* in potato: infectivity titrations in relation to multiplication and spread of the pathogen. *American Potato Journal* 59, 155-164.

Bushkova, L.N. (1983) [Resistance of potato varieties to the blackleg pathogen.] *Potato Abstracts* (1986) 11, 70.

Bushkova, L.N. and Vlasov, N.M. (1983) [Resistance of potato to *Erwinia atroseptica.*] *Kartofel i Ovoshchi* 11, 17.

Ciampi, L. and Andrade-Soto, N. (1984) Preliminary evaluation of bacterial soft rot resistance in native Chilean potato clones. *American Potato Journal* 61, 109-112.

Ciampi, L. and Sequeira, L. (1980) Multiplication of *Pseudomonas solanacearum* in resistant potato plants and the establishment of latent infections. *American Potato Journal* 57, 319-329.

Davis, M.C., Butler, W. and Vayda, M.E. (1990) Molecular responses to environmental stresses and their relationship to soft rot. In: Vayda, M.E. and Park, W.D. (eds) *The Molecular and Cellular Biology of the Potato*. CAB International, Wallingford, UK, pp. 71-87.

De Boer, S.H. (1990) Control of bacterial ring rot. In: Boiteau, G., Singh, R.P. and Parry, R.H. (eds.) *Potato Pest Management in Canada*. Fredrickton, New Brunswick, pp. 242-253.

De Boer, S.H. and McCann, M. (1990) Detection of *Corynebacterium sepedonicum* in potato cultivars with different propensities to express ring rot symptoms. *American Potato Journal* 67, 685-694.

Denny, T.P. and Baek, S.R. (1991) Genetic evidence that extracellular polysaccharide is a virulence factor of *Pseudomonas solanacearum*. *Molecular Plant-Microbe Interactions* 4, 198-206.

Destefano-Beltran, L., Nagpala, P.G., Cetiner, M.S., Dodds, J.H. and Jaynes, J.M. (1990) Enhancing bacterial and fungal disease resistance in plants: application to potato. In: Vayda, M.E. and Park, W.D. (eds) *The Molecular and Cellular Biology of the Potato*. CAB International, Wallingford, UK, pp. 205-221.

Dobias, K. (1975) Relation between resistance to bacterial soft rot of tubers and resistance to blackleg of potatoes. *Rostlinna Vyroba* 21, 145-150.

Dobias, K. (1977) Possibilities of breeding for resistance to bacterial soft rot (*Erwinia carotovora* (Jones) Holland). *Rostlinna Vyroba* 23, 255-260.

Dobias, K. (1978) Resistance of some potato species to blackleg. *Potato Abstracts* (1981) 6, 66.

Dopke, F. and Heitefuss, R. (1987) Blackleg of potato cultivars in relation to stem and tuber resistance to *Erwinia* spp. *Abstracts of Conference Papers and Posters of the 10th Triennial Conference of the EAPR*, Albourg, Denmark, pp. 138-139.

Dorozhkin, N.A., Bel'skaya, S.I. and Novikova, L.M. (1985) Resistance of potato varieties to combined *Fusarium*-bacterial rot. *Kartofelevodstvo* 6, 28-31.

Eden-Green, S.J. and Elphinstone, J.G. (1993) Bacterial wilt disease caused by *Pseudomonas solanacearum*. *Natural Resources Institute (NRI) Bulletin*. Chatham, UK.

Elphinstone, J.G. (1990) Selection of cultivars, from the world potato germplasm collection, with potential resistance to blackleg and soft rot caused by *Erwinia chrysanthemi*. In: *Abstracts of Conference Papers and Posters of the 11th Triennial Conference of the EAPR*, pp. 515-516.

French, E.R. (1988) Strategies for bacterial wilt control. In: French, E.R. (ed.) *Report of the Planning Conference on Bacterial Diseases of the Potato 1987*. International Potato Center, Lima, Peru, pp. 133-142.

French, E.R. and de Lindo, L. (1982) Resistance to *Pseudomonas solanacearum* in potato: specificity and temperature sensitivity. *Phytopathology* 72, 1408-1412.

French, E.R. and de Lindo, L. (1985) Sources of resistance in tuberiferous *Solanum* spp. to soft rot by *Erwinia*. In: Graham, D.C. and Harrison, M.D. (eds) *Report of the International Conference on Potato Blackleg Disease*. Potato Marketing Board, Oxford, pp. 79-80.

Granada, G.A. (1988) Latent infections induced by *Pseudomonas solanacearum* in potato and symptomless plants. In: French, E.R. (ed.) *Report of the Planning Conference on Bacterial Diseases of the Potato 1987*. International Potato Center, Lima, Peru, pp. 93-108.

Hahn, W. (1974) Resistenzprufung der Kartoffelknolle gegen den Erreger der Nassfaule, *Pectobacterium carotovorum* (Jones) Waldee. *Archin Für Zuchtungsforschung* 4, 133-140.

Hayward, A.C. (1991) Biology and epidemiology of bacterial wilt caused by *Pseudomonas solanacearum*. *Annual Review of Phytopathology* 29, 65-87.

Hedley, D. and Perombelon, M.C.M. (1990) *Scottish Crop Research Institute Annual Report for 1989*, SCRI, Invergowrie, Dundee, p. 58.

Hidalgo, O.A. and Echandi, E. (1982) Evaluation of potato clones for resistance to tuber and stem rot induced by *Erwinia chrysanthemi*. *American Potato Journal* 59, 585-592.

Huaman, Z., de Lindo, L. and Elphinstone, J.G. (1988) Resistance to blackleg and soft rot and its potential use in breeding. In: French, E.R. (ed.) *Report of the Planning Conference on Bacterial Diseases of the Potato 1987*. International Potato Center, Lima, Peru, pp. 215-228.

Huaman, Z., Tivoli, B. and de Lindo, L. (1989) Screening for resistance to *Fusarium* dry rot in progenies of cultivars of *S. tuberosum* ssp. *andigena* with resistance to *Erwinia chrysanthemi*. *American Potato Journal* 66, 357-364.

Huaman, Z., de Lindo, L. and Elphinstone, J.G. (1990) Breeding for resistance to *Erwinia chrysanthemi* in potatoes for the tropics. In: *Abstracts of Conference Papers and Posters of the 11th Triennial Conference of the EAPR*, pp. 6-7.

Huang, Y., Helgeson, J.P. and Sequeira, L. (1989) Isolation and purification of a factor from *Pseudomonas solanacearum* that induces a hypersensitive-like response in potato cells. *Molecular Plant-Microbe Interactions* 2, 132-138.

Iwanaga, M. (1985) Ploidy level manipulation approach: development of diploid populations with specific resistance and FDR 2*n* pollen production. In: *Present and Future Strategies for Potato Breeding and Improvement*. International Potato Center, Lima, Peru, pp. 57-70.

Kelman, A. (1953) The bacterial wilt caused by *Pseudomonas solanacearum*. *North Carolina Agricultural Experimental Station Technical Bulletin* 99.

Kogut, I.D. and Polozhenets, V.M. (1985) Testing cultivated potato species and their hybrids for resistance to bacterial diseases. *Selektsiya i semenovodstvo kartofelya* 85, 65-74.

Koromyslova, M.I. (1972) Breeding potatoes for resistance to blackleg. *Byulleten Vsesoyuznogo Ordena Lenina Institut Rastenievodstva Imeni N.I. Vavilova* 28, 73-80.

Kotoujansky, A. (1987) Molecular genetics of pathogenesis by soft rot erwinias. *Annual Review of Phytopathology* 25, 405-430.

Krause, B., Koczy, T., Komorowska-Jedrys, J. and Ratuszniak, E. (1982) Laboratory determinations of tuber resistance to the chief storage rots in a world collection of potato varieties. *Biuletyn Instytutu Ziemniaka* 27, 111-134.

Lapwood, D.H., Read, P.J. and Spokes, J. (1984) Methods for assessing the susceptibility of potato tubers of different cultivars to rotting by *Erwinia carotovora* subspecies *carotovora* and *atroseptica*. *Plant Pathology* 33, 13-20.

Lellbach, H. (1976) Anfalligkeit des Kulturkartoffelsortiments gegenuber der Knollennassfaule. *Tagungsberichte Deutsche Akademie der Landwirtschaftswissenschaften zu Deutschen Demokratischen Republik* 140, 199-206.

Lellbach, H. (1978) Estimating genetic parameters derived from diallel crosses in cases of susceptibility to soft rot in the potato. *Archiv fur Zuchtungsforschung* 8, 193-199.

Lewosz, W. (1984) Blackleg of potatoes in the light of the literature and personal research. *Biuletyn Instytutu Ziemniaka* 31, 103-123.

Lojkowska, E. and Kelman, A. (1989) Screening of seedlings of wild *Solanum* species for resistance to bacterial stem rot caused by soft rot erwinias. *American Potato Journal* 66, 379-390.

Lyon, G.D. (1989) The biochemical basis of resistance of potatoes to soft rot *Erwinia* spp. - a review. *Plant Pathology* 38, 313-339.

McGuire, R.G. and Kelman, A. (1984) Reduced severity of *Erwinia* soft rot in potato tubers with increased calcium content. *Phytopathology* 74, 1250-1256.

Maher, E.A. and Kelman, A. (1983) Oxygen status effects on maceration of potato tissue by pectic enzymes produced by *Erwinia carotovora*. *Phytopathology* 73, 536-539.

Marunenko, I.M., Kuchko, A.A., Oleinik, T.N. and Predko, M.N. (1988) Cell selection in potato for resistance to pathogens. *Biologiya Kul'tiviruemykh Kletok i Biotekhnologiya* 1, 169. (Abstract).

Mendoza, H.A. (1988) Progress in resistance breeding in potatoes as a function of efficiency of screening procedures. In: French, E.R. (ed.) *Report of the Planning Conference on Bacterial Diseases of the Potato 1987*. International Potato Center, Lima, Peru, pp. 39-64.

Mok, D.W.S. and Peloquin, S.J. (1975) Three mechanisms of $2n$ pollen formation in diploid potatoes. *Canadian Journal of Geneties and Cytology* 17, 217-225.

Montanelli, C. and Nascari, G. (1989) Transformation of potato for bacterial disease resistance using antibacterial genes from insects. *Rivista di Agricoltura Subtropicale e Tropicale* 83, 375-385.

Morozova, E.V. (1987) Sources of resistance among forms of *Solanum andigenum* Juz. et Buk. *Sbornik Nauchnykh Trudov po Prikladnoi Botanike Genetike i Selektsii* 115, 45-48.

Munzert, M. (1984) The breeding possibilities for improving resistance to blackleg (*Erwinia atroseptica*). *Abstracts of Conference Papers of the 9th Triennial Conference of the EAPR*, Interlaken, Switzerland, p. 385.

Murphy, A.M. (1990) Control of other bacterial diseases. In: Boiteau, G., Singh, R.P. and Parry, R.H. (eds) *Potato Pest Management in Canada*. Fredrickton, New Brunswick, pp. 254-270.

Olsson, K. (1987) Screening for resistance to ring rot (*Corynebacterium michiganensis* pv. *sepedonicum* in potatoes. *Vaxtskyddsnotiser* 51, 47-55.

Pawlak, A., Pavek, J.J. and Corsini, D.L. (1987) Resistance to storage diseases in breeding stocks. In: Jellis, G.J. and Richardson, D.E. (eds) *The Production of New Potato Varieties*. Cambridge University Press, Cambridge, pp. 96-98.

Perombelon, M.C.M. and Kelman, A. (1980) Ecology of the soft rot erwinias. *Annual Review of Phytopathology* 18, 361-387.

Perombelon, M.C.M. and Kelman, A. (1987) Blackleg and other potato diseases caused by soft rot erwinias: a proposal for a revision of terminology. *Plant Disease* 71, 283-285.

Pietkiewicz, J.B. (1980) Variation in the reaction of potato tubers to diseases. *Potato Research* 23, 473. (Abstract).

Prior, P., Beramis, M., Chillet, M. and Schmit, J. (1990) Preliminary studies for tomato bacterial wilt (*Pseudomonas solanacearum*) resistance mechanisms. *Symbiosis* 9, 393–400.

Rassadina, G.V., Khromova, L.M. and Butenko, R.G. (1988) *Biologiya Kul'tiviruemykh Kletok i Biotekhnologiya* 1, 167–168.

Ratuszniak, E. (1981) Variability in the resistance of potato tubers to *Phytophthora infestans*, *Erwinia carotovora* var. *atroseptica*, *Fusarium sulphureum* and mechanical damage in laboratory assessment. *Ziemniak*. 1981, 45–58.

Rehbein, E. (1983) Selektion auf *Erwinia*-Resistenz bei Kartoffeln mittels Gewebekultur. Unpublished PhD thesis, University of Bonn.

Ross, H. (1986) *Potato Breeding – Problems and Perspectives*. Advances in Plant Breeding 13, Paul Parey, Berlin and Hamburg.

Rowe, P.R. and Sequeira, L. (1970) Inheritance of resistance to *Pseudomonas solanacearum* in *Solanum phureja*. *Phytopathology* 60, 1499–1501.

Rowe, P.R., Sequeira, L. and Gonzales, L.C. (1972) Additional genes for resistance to *Pseudomonas solanacearum* in *Solanum phureja*. *Phytopathology* 62, 1093–1094.

Schmiediche, P. (1986) Breeding potatoes for resistance to bacterial wilt caused by *Pseudomonas solanacearum*. In: Persley, G.J. (ed.) *Bacterial Wilt Disease in Asia and the South Pacific*, proceedings of an international workshop held at PCARRD, Los Banos, Philippines, 8–10 October 1985. *ACIAR Proceedings* 13, pp. 105–111.

Schmiediche, P. (1987) The utilisation of wild potato species in breeding. In: *Strategies for the Conservation of Potato Genetic Resources IV. XXIX Planning Conference*. International Potato Center, Lima, Peru, pp. 135–150.

Schmiediche, P. (1988) Breeding for resistance to *Pseudomonas solanacearum*. In: French, E.R. (ed.) *Report of the Planning Conference on Bacterial Diseases of the Potato 1987*. International Potato Center, Lima, Peru, pp. 19–28.

Scowcroft, W.R., Larkin, P.J. and Bretter, R.I.S. (1983) Genetic variation from tissue culture. In: Helgeson, J.P. and Deverall, B.J. (eds) *Use of Tissue Culture and Protoplasts in Plant Pathology*. Academic Press, Sydney, Australia, pp. 139–162.

Sequeira, L. (1979) Development of resistance to bacterial wilt derived from *Solanum phureja*. In: *Report of a Planning Conference on Developments in Control of Potato Bacterial Diseases*. International Potato Center, Lima, Peru, pp. 55–62.

Slack, S.A. (1987) Proceedings of the symposium on the eradication of bacterial ring rot: biology and ecology of *Corynebacterium sepedonicum*. *American Potato Journal* 64, 665–670.

Slack, S.A. and Gedye, C. (1988) Screening for bacterial ring rot resistance in the inter-regional potato (IR-1) germplasm collection. *American Potato Journal* 65, 500–501.

Taylor, R.J. and Secor, G.A. (1990) Potato protoplast-derived callus tissue challenged with *Erwinia carotovora* subsp. *carotovora*: survival, growth and identification of resistant callus lines. *Journal of Phytopathology* 129, 228–236.

Taylor, R.J., Ruby, C.L. and Secor, G.A. (1988) Assessment of field performance and soft rot resistance in a population of protoplast-derived potato clones. *Phytopathology* 78, 1595. (Abstract).

Tung, P.X., Rasco, E.T., Vander Zaag, P. and Schmiediche, P. (1990a) Resistance to *Pseudomonas solanacearum* in the potato: I. Effects of sources of resistance and adaptation. *Euphytica* 45, 203–210.

Tung, P.X., Rasco, E.T., Vander Zaag, P. and Schmiediche, P. (1990b) Resistance to *Pseudomonas solanacearum* in the potato: II. Aspects of host–pathogen–environment interaction. *Euphytica* 45, 211–215.

Van der Plank, J.E. (1978) *Genetic and Molecular Basis of Plant Pathogenesis*. Springer-Verlag, Berlin.

Van Soest, L.J.M. (1983) Evaluation and distribution of important properties in the German–Netherlands potato collection. *Potato Research* 26, 109–121.

Vlasov, N.M., Bushkova, L.N. and Pereverzev, D.S. (1987) Genetic sources for breeding potato varieties resistant to blackleg. *Sbornik Nauchnykh Trudov po Prikladnoi Botanike, Genetike i Selektsii* 115, 54–59.

Wastie, R.L. and Mackay, G.R. (1985) Breeding for resistance to blackleg: the present and the future. In: Graham, D.C. and Harrison, M.D. (eds) *Report of the International Conference on Potato Blackleg Disease*. Potato Marketing Board, Oxford, pp. 75–76.

Weber, J. (1983) The role of pectin in the expression of varietal and seasonal differences in soft rot susceptibility of potato tubers. *Phytopathologische Zeitschrift* 108, 135–142.

Williams, C.E., Hunt, G.J. and Helgeson, J.P. (1990) Fertile somatic hybrids of *Solanum* species: RFLP analysis of a hybrid and its sexual progeny from crosses with potato. *Theoretical and Applied Genetics* 80, 545–551.

Yang, Z., Cramer, C.L. and Lacy, G.H. (1989) System for simultaneous study of bacterial and plant gene expression in soft rot of potato. *Molecular Plant–Microbe Interactions* 2, 195–201.

Zadina, J. and Dobias, K. (1976) [Possibilities of breeding for resistance to tuber rot in potatoes.] *Potato Abstracts* (1977) 2(8), 120.

Zalewsky, J.C. and Sequeira, L. (1975) An antibacterial compound from *Solanum phureja* and its role in resistance to bacterial wilt (*Pseudomonas solanacearum*). *Phytopathology* 65, 1336–1341.

Zielke, R. (1983) Possibilities of testing potatoes for resistance to bacterial ring rot, *Corynebacterium sepedonicum* (Spiek. et Kotth.) Skapt. et Burkh. *Tagungsbericht Akademie der Landwirtschaftswissenschaften der DDR* 216, 441–450.

20 Inheritance of Resistance to Insects and Mites

K.V. RAMAN, A.M. GOLMIRZAIE, M. PALACIOS AND J. TENORIO

International Potato Center (CIP), Apartado Postal 5969, Lima, Peru.

Introduction

The greatest diversity of insects associated with potatoes occurs in South and Central America, where the crop and its many related wild species originated. The use of chemical pesticides to control pests and diseases is now a common practice in many countries. The potato is the heaviest user of chemical pesticides of all the major food crops. Costs often total up to 20% of production. A conservative estimate of insecticide costs for potato production in developing countries exceeds $300 million per year (Horton, 1987; CIP, 1992; Raman, 1992).

The adverse consequences of the use of insecticides in potato production have been summarized by Cisneros (1984) and Raman (1988). These include the development of insects resistant to insecticides, persistence of residues in tubers for consumption, emergence of new pests, destruction of beneficial organisms, human intoxication and contamination of the environment. Recognizing that sustainable potato production cannot be achieved as long as pest and disease control depends on chemical pesticides, the International Potato Center (CIP), located in Lima, Peru, has focused its activities on the development of integrated pest management (IPM) strategies to help developing country potato producers to substantially reduce their use of insecticides and fungicides. IPM is based on the concept that pesticide use can be reduced to minimum levels through a combination of control practices. The use of varietal resistance remains the first defence in the control of insect pests and is accorded high priority in IPM. The development of potatoes with multiple resistance to biotic and abiotic stresses represents a major research effort at CIP, involving many scientists from different disciplines (Mendoza, 1992). Our intent in this chapter is to report the advances made at CIP in the development of insect-resistant germplasm. For reviews of work conducted elsewhere, the reader is referred to the following papers: Gibson (1978); Radcliffe *et al.* (1981); Radcliffe (1982); Flanders *et al.* (1992); Raman and Radcliffe (1992) and Tingey and Yencho (1992).

Potato Insect and Mite Complex

The potato ecosystem is inhabited by numerous insect and mite species. These pests can damage the potato plant by feeding on leaves, reducing photosynthetic area and efficiency; by attacking stems, weakening the plant and inhibiting nutrient transport; and by attacking potato tubers destined for consumption or use as seed.

Seventeen general groups of pests are reported to attack the potatoes in the developing world (Raman, 1991). Our present knowledge indicates that the potato tuber moth (PTM), *Phthorimaea operculella*, Andean potato weevil (APW), *Premnotrypes* spp., green peach aphid (GPA), *Myzus persicae*, and leafminer flies (LMF), *Liriomyza huidobrensis*, are major pests in many developing countries (Raman, 1992). When pest pressure is severe, many of these individual pest species can routinely cause tuber yield reductions of 30–70%. Losses of this magnitude have been shown in untreated plots for aphid-transmitted viruses, PTM, APW and LMF (Raman and Radcliffe, 1992). These pests attack the plant over a prolonged period and reduce yield more than those that defoliate or damage plant parts for a brief period, such as leaf-feeding cutworms and leaf beetles. The potato plant recuperates from this latter type of damage under favourable environmental conditions. Midmore (1986) has discussed the compensatory mechanisms in potato in relation to insect damage. The potato has considerable ability to compensate for early-season loss of foliage, but little for adverse effects on plant health, as caused by virus infection, disruption of nutrient transport or reduced photosynthetic efficiency. For more information on the economic importance, distribution, biology, nature of damage and yield losses by major pests on a global basis, the reader is referred to Raman and Radcliffe (1992).

Criteria for Potato Insect Resistance Programme

Concentrated team-orientated research for developing insect-resistant germplasm for developing countries is recent. At CIP, several criteria were considered before deciding to establish an insect resistance programme for specific potato pests:

1. The level of economic damage being caused by a particular pest should be significant. Aphid-transmitted viruses, PTM, APW, LMF, thrips, and mites reduce yield in the 30–70% range in susceptible cultivars (Raman and Radcliffe, 1992).

2. Genetic resistance should be sought to those pests for which it is considered feasible to find such resistance. For example, it is highly unlikely that resistance will be found to such pests as cutworms, leafcutter ants and white grubs; limited resources should not be utilized in this direction.

3. The level of resistance needed to reduce pest populations should be considered. Some potato varieties have high economic thresholds to pests and can withstand considerable early-season damage by leaf-feeding pests (30% or more for certain varieties) without reducing yields (Raman and Midmore, 1983). Therefore, high levels of resistance to some pests may not be necessary.

4. Low levels of resistance may be combined with other methods of control, such as biological control or cultural practices, to maintain insect populations below economic damage levels. For example, only moderate to low levels of resistance have been found for APW. By combining this resistance with a strong biological and cultural control programme, it is possible to adequately control this pest (Raman, 1992). The planting system in which potatoes are grown can dictate the level of resistance that might be needed. In many areas of the tropics, potatoes are grown in multicropping systems or in association with other crops, such as corn, wheat or garlic, and pest populations in these systems are reduced (Potts, 1988). The level of resistance needed in such systems may be lower.

Status of Resistance to Insects

Resistance to insects or mites damaging potatoes in developing countries is not extensively reported in the literature. Many of the reports deal only with field observations, and until recently there was little systematic evaluation of germplasm. Systematic evaluation of the world potato germplasm collection began in 1979. The collection maintained at CIP consists of two major components of genetic resources, which are wild species and primitive cultivated species (Huaman, 1986). The wild gene pool consists of more than 200 tuber-bearing *Solanum* species (Hawkes, 1992). The cultivated germplasm collection originally consisted of more than 15,000 samples. This has now been reduced to about 5000 because of duplicate identification. This collection is now considered to be the largest, most diverse collection in existence. A computerized data bank using all available data from collecting expeditions and evaluations has been organized by Huaman (1986). A number of these accessions have been evaluated and data have been recorded following a standardized list of descriptors. A total of 65 descriptors are used for each accession, of which 17 are on disease and pest reaction. Many of the 3000 introductions of wild and cultivated species maintained in the IR-1 collection at Wisconsin, USA, have also been screened for resistance to major pests and diseases at CIP (Hanneman and Bamberg, 1986; Hanneman, 1989). CIP's germplasm bank is evaluated on a continuing basis for resistance to PTM, GPA, APW, LMF and mites (Raman, 1988).

Potato tuber moth

Four species of PTM are reported to attack potatoes. Of these, *Phthorimaea operculella* (Zeller) is the most damaging and generally of greatest importance in warmer climates (Raman and Radcliffe, 1992). CIP's host plant resistance research for this pest was initiated by screening the world collection. A total of 3747 primitive native cultivars, comprising *Solanum tuberosum, S. andigena, S. chaucha, S. curtilobum, S. juzepczukii* and *S. stenotomum*, were evaluated. Of the wild species maintained in the IR-1 collection, 452 were tested by Raman and Palacios (1982), including some interspecific hybrid populations. In Colombia, Estrada and Valencia (1987) screened several clones obtained from CIP, IR-1 and a Colombian collection. The advanced clones screened were crosses involving *S. tuberosum* × *S. andigena*. In other studies, conducted in Peru, Chavez (1984) screened 61 selected genotypes of *S. sparsipilum* and 2062 hybrid clones from the crosses between resistant wild and susceptible cultivated germplasm. Several sources of resistances were identified. In the study by Raman and Palacios (1982), 22 primitive and 21 wild potato accessions were identified as highly resistant and are now being used in breeding resistance into high-yielding potato varieties adapted to warm climates. Resistance in these clones has been attributed to antibiosis (Raman and Palacios, 1982). No correlation existed of resistance with glycoalkaloid levels in tubers (Raman and Palacios, 1982). Of the wild species tested by Chavez (1984), 12 highly resistant clones were selected and used as breeding stock. These clones belonged to five species, *S. pinnatisectum, S. commersonii, S. tarijense, S. sparsipilum* and *S. sucrense*. Interspecific crosses between them and susceptible progeny of *S. sucrense* were successful. The hybrids were further screened and 171 resistant hybrids obtained. These hybrids have now been hybridized with Tuberosum dihaploids for transferring resistance into the cultivated tetraploid potato. Work with diploid populations is still continuing.

Attempts have been made to study the inheritance of resistance in three diploid clones isolated from an interspecific hybrid population involving *S. chacoense, S. phureja* and *S. sparsipilum* (Raman et al., 1981). The results indicated that resistance was mostly dominant and controlled by a few major genes. In one reciprocal cross, clear cytoplasmic effects on resistance were observed. Although the three resistant diploids did not produce $2n$ pollen, one of them produced $2n$ eggs. Thus, the resistance of the diploid was transferred to the $4x$ level through $2x$–$4x$ crosses. Present research concentrates on elucidating the inheritance of resistance, increasing the frequency of resistance genes and obtaining diploids with $2n$ gametes in order to incorporate the resistance into the tetraploid breeding population. Scurrah et al. (1987) and Scurrah and Raman (1984) have screened several tetraploid populations obtained by crossing $2x$ clones with $4x$ clones adapted to warm climates. The resulting populations showed a low resistance frequency (1.5%) compared with diploid populations (36%), indicating slow progress at the tetraploid level. However, after several cycles of selection and screening, $4x$ clones with

Fig. 20.1. Resistance to potato tuber moth, *Phthorimaea operculella*. Tubers on left are from an advanced resistant clone.

a high level of resistance have been created (Scurrah and Raman, 1984; CIP, 1991). These clones act as good transmitters of PTM resistance and can be used as progenitors in the breeding programmes of NARS to transfer resistance to PTM into the commonly grown cultivars of the tropics. These clones exhibit high levels of antibiosis under both laboratory and storage conditions. It is now essential that detailed inheritance studies at the $4x$ level be conducted to determine the most appropriate method of breeding.

Screening has been expanded to obtain new sources of resistance to widen the genetic base. Raman (1992), after screening several wild species, encountered high levels of resistance in *S. berthaultii*. This species has high densities of glandular trichomes, which are effective in reducing oviposition and larval feeding. Scurrah *et al.* (1987) have developed several trichome families with resistance to this pest. After several cycles of selection and screening, a base population with a high level of resistance has been created (Raman, 1992). Clones from this population act as good transmitters of PTM resistance and can be used as progenitors in breeding by national programmes to transfer PTM resistance into commonly grown tropical cultivars (Figure 20.1). Mendoza (1992) reported heritability in the broad sense (h^2) for resistance to this pest as varying widely from high to low. CIP also collaborates with several institutions in industrialized countries to develop transgenic potatoes with resistance to PTM. The goal is to insert the gene for production of *Bacillus thuringiensis* (*Bt*) endotoxin into potato plants. Transformed potato plants have now been tested for resistance to PTM in Belgium and at Michigan State University, USA (CIP, 1992). Highly resistant clones have now been selected. These clones now need to be tested under field conditions to confirm resistance. Protocols related to biosafety and testing methods for use of these clones in developing countries need to be developed.

Leafminer fly

Several species of leafminer flies are reported to attack potatoes, the most important being *Liriomyza huidobrensis* (Raman and Radcliffe, 1992). A total of 323 clones comprising varieties and advanced clones from CIP and from the world collection have been screened, together with 408 from the true potato seed (TPS) breeding programme at CIP, and 33 clones have been selected. Both resistance and yield were significantly higher in resistant clones when compared with the commonly grown Peruvian cultivar 'Revolucion'. These selected clones are being used as progenitors for the genetic study of leafminer fly resistance and for obtaining populations with high resistance levels. Improved resistant clones are now being tested by several national programmes of developing countries (Raman, 1992). Heritability in the broad sense (h^2) for resistance appears high and good screening procedures are available (Mendoza, 1992). Glandular trichomes are effective in reducing oviposition and longevity (Figure 20.2). These trichomes have now been incorporated into improved germplasm (Raman, 1992).

Fig. 20.2. Resistance to leafminer fly, *Liriomyza huidobrensis*, in clone K 419-8, with high densities of glandular trichomes.

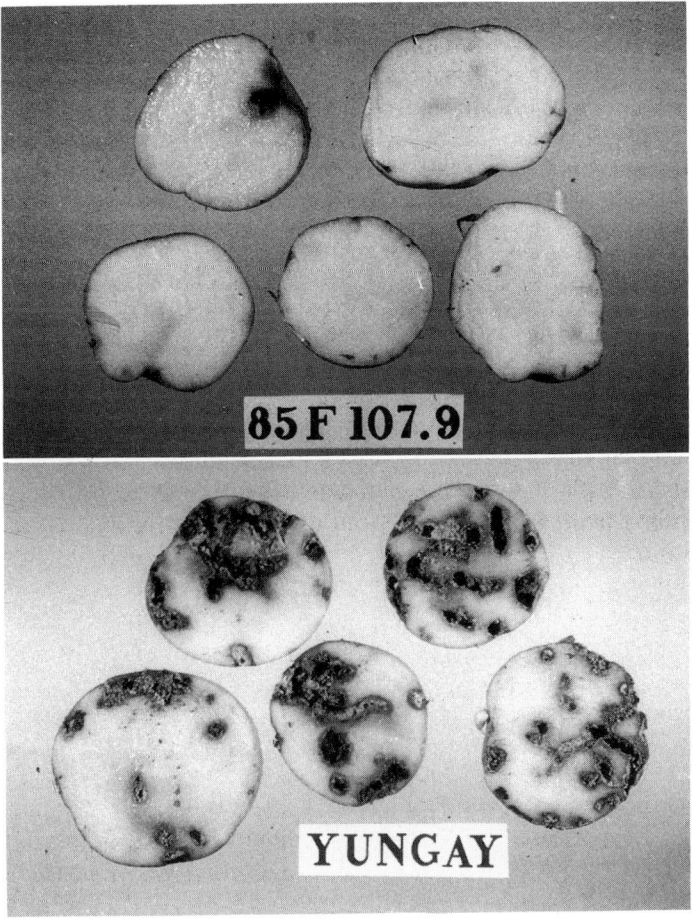

Fig. 20.3. Resistance to Andean potato weevil, *Premnotrypes* spp. Tubers on top are from an advanced resistant clone.

Andean potato weevil (APW)

Several genera of APW are reported to occur in the Andes, the most important being *Premnotrypes* spp. (Raman, 1992; Raman and Radcliffe, 1992). Field screening for this pest is difficult due to a lack of uniform distribution. Several clones from the highland-developed population at CIP have been field-screened. Highly variable data were reported from one year to another. These discrepancies in resistance data were mostly due to varying infestation levels (densities) from year to year (Raman, 1991). Laboratory screening methods seem to be more reliable. With this method, a total of 1308 clones from the world collection and 136 advanced clones have been screened. Two cultivars, HJT 5523 (*S. andigena*) and CUA 74 (*S. goniocalyx* × *S. stenotomum*), and 30 advanced clones were selected (Figure 20.3). Three of these clones yielded an average of 2 kg per plant (Raman, 1991). Screening work has mostly concentrated on identifying sources of resistance. At present, no breeding effort is being made to incorporate the identified resistance into other cultivars, and the genetics are not known.

Aphids, PTM, LMF and mites

The most advanced population selected involves clones with glandular trichomes from the K series developed by Cornell University. The biological activity, chemistry and genetic exploitation of glandular trichomes have been under intensive study at Cornell for several years (Tingey and Yencho, 1992). Two types of glandular trichomes have been identified. The first (type A) is a short type with a tetralobulate gland at the tip. The second is a longer multicellular hair with an ovoid gland at its apex (type B). Both type A and type B are needed in high densities to confer resistance. The presence of these hairs has been reported for relatively few tuber-bearing species, notably *S. polyadenium*, *S. berthaultii*, *S. tarijense*, *S. lignicaule*, *S. wittmackii* and, most recently, *S. flavoviridens* and *S. neocardenasii* (Ochoa, 1990). As for the non-tuber-bearing species, the best known example is *S. penelli*. Tests at CIP have confirmed that resistance to GPA is directly related to trichome density and to the amount of exudate produced. The trichome exudate accumulating on the bodies and proboscis of GPA reduces survival by 70%. Mortality of GPA nymphs can reach up to 60%, with the highest mortality occurring in the first instar stages. For PTM, three K clones, K421.2, K432.5 and K434.1, exhibited an antixenosis type of resistance to egg laying, and larval development was reduced. For LMF, the oviposition capacity and adult longevity were significantly lower. In the case of the red spider mite, *Tetranychus urticae*, egg laying and larval and adult populations were reduced (Figure 20.4) (CIP, 1991, 1992).

Using RFLP markers, Bonierbale (1991) has identified two genomic regions, one on chromosome 6 and the other on chromosome 10, which coincide very tightly with loci influencing the density of type A trichomes. Interaction between the genotypes at these two loci indicates epistatic effects on the type A trichome phenotype. The organization of genes influencing the type B trichome properties is more complex. In a backcross to *S. berthaultii*, Bonierbale (1991) identified loci on chromosomes 5 and 11 whose genotypes are correlated with the density of type B trichomes.

Procedures for Screening Potato Germplasm for Resistance

Techniques for evaluating large amounts of genetic material differ for each insect pest. However, there are several standard procedures that must be considered in potato germplasm evaluation and these include:

Fig. 20.4. Resistance to red spider mite, *Tetranychus urticae*. Clone on left is resistant, with high densities of glandular trichomes.

1. A screening programme must ensure large and uniform insect populations to guarantee adequate selective pressure. Planting date and application of selective insecticides to boost target pest populations have been effective for increasing populations of PTM, LMF and GPA (Raman and Booth, 1983).
2. When field infestations are not uniform, screening should be done under controlled conditions. In Peru, APW populations in the field are highly variable. The initial screening, therefore, is done under controlled greenhouse conditions.
3. A host reaction scale needs to be developed to describe accurately the damage levels. In potato research at CIP, a 0 to 5 scale is usually recommended. A rating of 0 to 2 indicates some resistance and suggests further testing; 4 to 5 is used for highly susceptible for discard; and 3 suggests an intermediate range. Examples of these scales are given in the following section. These visual scales are usually subjective and, where possible, should be complemented by collecting pest population data.
4. Methods should be designed to evaluate rapidly large numbers of clones. This offers a useful time-saving tool. Storage evaluation for PTM resistance and visual evaluation of glandular trichomes are useful for discarding susceptible genotypes at the early stages. Recent research on the use of RFLP/PCR techniques to identify resistance genes should help the rapid selection of segregating resistant progeny (Bonierbale, 1991).

Potato tuber moth

A procedure for evaluating resistance to this pest has been developed by Raman and Palacios (1982) and Raman and Booth (1983). Two tests are employed.

Antibiosis

This test is conducted under controlled laboratory conditions. Susceptible cultivars are included as checks. Pupation is evaluated at 3 weeks after infestation with first-instar larvae. Based on the mean number of pupae recovered per tuber, accessions that have up to 0.5 are considered highly resistant (HR), more than 0.5 to 1.0 resistant (R), more than 1 to 2 susceptible (S) and more than 2 highly susceptible (HS).

Antixenosis or non-preference

Small quantities of different clones are screened in stores with a high PTM population. Six tubers of each clone, as well as tubers of known susceptible varieties, are placed individually in paper bags and exposed to PTM oviposition and after 120 days the damage is evaluated, using a 1-4 scale: class 1 = 0-1 holes per tuber (very slight damage, resistant) (R); class 2 = >1-2 holes per tuber (slightly damaged, moderately resistant) (MR); class 3 = >2-4 holes per tuber (damaged, susceptible) (S); and class 4 = >4 holes per tuber (highly damaged, highly susceptible) (HS). When clone numbers are high (>1000 clones), tubers are cut into half and scored on the basis of total area mined by PTM larvae within the cut tuber. For more details on evaluation procedures, the reader is referred to Raman and Booth (1983).

Leafminer fly

Screening for this pest is conducted in the field using natural populations. In the initial screening, the emphasis is on discarding susceptibles. Susceptible cultivars are included as checks for comparison. The larvae of this pest mine through the foliage. The mining usually starts from the basal half and progresses upwards. Severe damage results in complete drying up of foliage. A 1-5 scale based on foliar damage is used: class 1 = no damage, highly resistant (HR); class 2 = slight damage, mostly restricted to basal half of plant, approximately 1-25% damage, resistant (R); class 3 = moderate damage, mostly restricted to middle and basal half of plant, approximately 26-50% damage, moderately resistant (MR); class 4 = except terminal leaves, entire plant damaged, approximately 51-75% damage, susceptible (S); class 5 = most of the entire plant damaged, approximately 76-100% damage, highly susceptible (HS). For more details, the reader is referred to Yabar (1988).

Andean potato weevil

Screening is conducted under field and laboratory conditions. In the field test, ten tubers per clone are grown in a weevil-infested field along with a known susceptible check. At harvest, tubers are evaluated for damage, using a visual scale of 5 grades, where grade $1 = 0$ damage and grade $5 = 75$ to 100% damage. A laboratory test to identify escapes under field conditions has also been used (CIP, 1985). Weevils collected from the field readily lay eggs when maintained at 10°C. First-instar larvae hatching from these eggs are used to infest tubers placed in plastic containers. At 60 days after infestation, tubers are evaluated for pupation. Those with less than 20% pupation are selected as resistant.

Glandular trichomes

Gregory *et al.* (1986) and Tingey (1991) have developed biochemical and visual methods for selecting clones with glandular trichomes. At CIP, three methods are used. The first involves visual assessment. Fully expanded leaflets from the terminal portion of plants are examined using a portable field microscope with $30\times$ magnification and selected for type A and B glandular trichomes. Clones are classified as having low, medium and high densities of glandular trichomes. An enzymic browning assay developed by Ave *et al.* (1986) is used on clones selected from the visual test to further discard those that are not chemically active. In the final test, the selected clones are infested with mites, aphids, PTM and LMF. Clones least damaged are selected.

Breeding Approaches

The work on host plant resistance to major pests of the tropics is quite recent. The scheme for developing germplasm with resistance to pests is outlined in Figure 20.5. High levels of resistance have not been encountered in any single cultivar, but crosses between cultivars have been useful in increasing the level of resistance. Breeding for pest resistance at the tetraploid level presents some problems, due to tetrasomic inheritance. Changes in genotypic frequencies in a $4x$ population under selection are slower than in a $2x$ one. The genetic progress in one generation of selection in a $2x$ population is five times greater than in a $4x$ population (Mendoza and Sawyer, 1985). Breeding at the diploid level through ploidy level manipulation has been effective for developing populations with specific pest resistances.

Iwanaga (1983) and Watanabe (1991) have outlined strategies for germplasm enhancement for pest and disease resistance through the use of diploids and $2n$ gametes. The main scheme for germplasm enhancement employed at CIP for pest resistance is shown in Figure 20.6. Tetraploid populations ($4x$) with resistance to PTM and LMF, clones with high densities of glandular

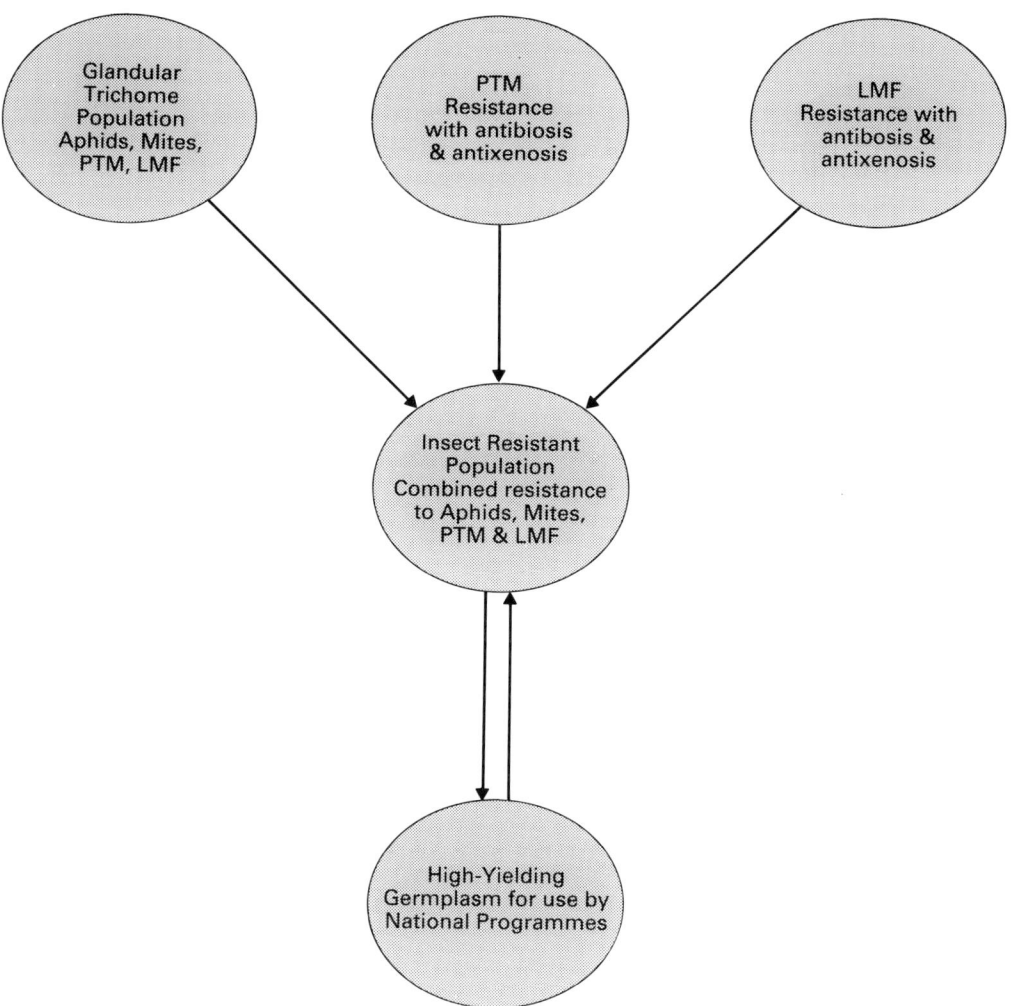

Fig. 20.5. Germplasm improvement scheme for developing resistance to potato pests.

trichomes, a TPS population and CIP pathogen-tested clones are hybridized and crossed to diploid (2x) clones that produce the 2n gametes and have resistance to a wide range of pests; 4x–2x progenies are generated. These progenies are then tested for pest resistance. Clones selected as resistant are then crossed with tetraploids to obtain clones with multiple pest resistance and good agronomic traits, such as high yield and good tuber quality. Dihaploids induced from advanced tetraploid breeding lines have been useful in capturing valuable genes from wild species and maintaining cultivated characters in hybrids with wild species. Élite dihaploids with resistance to pests and diseases have been extracted from selected tetraploids (Watanabe, 1991). As a tool to transfer pest resistance from the improved hybrids to tetraploid potatoes, 2n pollen has been employed.

Inheritance of Resistance to Insects and Mites

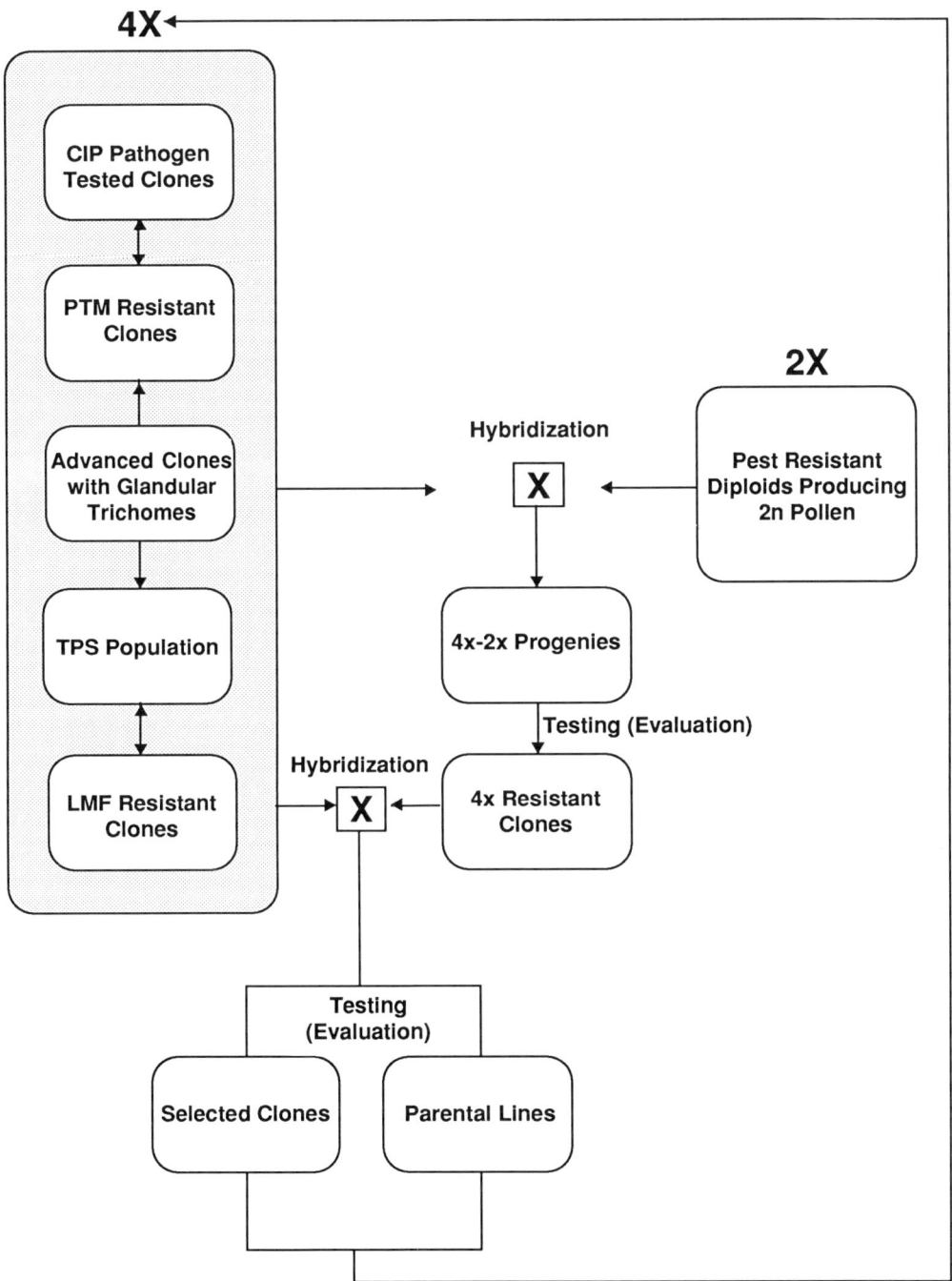

Fig. 20.6. Breeding scheme for resistance to potato pests.

Progress has also been made in solving cross-incompatibility problems between tetraploid potatoes and disomic tetraploid wild species. Methods such as embryo rescue and counterfeit pollinations help in obtaining tetraploid hybrids with workable levels of male and female fertility and intergenomic recombination (Iwanaga et al., 1991). These approaches have aided in the development of potato lines with multiple resistance to insect and mite pests.

Summary

There is substantial potential for breeding potato plants resistant to insect and mite pests. Work in developing countries has barely begun, and there is little information available about the genetics of insect and mite resistance. Breeding for pest resistance has a number of advantages: no significant costs to the farmer are involved, it does not impair the quality of the environment and it is generally compatible with other methods of pest control. Since the potato is the largest user of pesticides worldwide and costs of these chemicals are increasing in the developing countries, potato varieties resistant to insect pests offer a great potential, because they imply a certain amount of independence from chemically dependent pest control systems. Genetic resistance in potato plants, wherever available, should be combined with other desirable plant characters, such as high yields and good quality, and should provide the basic foundation on which to build integrated pest management systems. Adapted resistant germplasm demonstrated to be of value in the suppression of major pests should be increased and tested under conditions representative of actual production areas. Particular attention should be paid to demonstration/evaluation trials of this germplasm for performance against the complex of pest species and of the cumulative benefit of resistance in reducing losses in yield and quality. These demonstration trials should be structured so as to incorporate as many other promising control tactics as possible. For example, the integration of resistant germplasm with biological control may provide visible and dramatic evidence that both tactics together result in greater levels of control and greater economic advantage than the use of either tactic alone. Resistant germplasm found useful in these trials should be integrated into breeding programmes for improved adaptation, further testing, and distribution to national programmes for evaluation, refinement and consideration for use in pest management programmes. More work should be done on genetics to facilitate breeding.

References

Ave, D.A., Eannetta, N.T. and Tingey, W.M. (1986) A modified enzymic browning assay for potato glandular trichomes. *American Potato Journal* 63, 553-558.
Bonierbale, M.W. (1991) RFLP mapping of trichome traits from *S. berthaultii* and continued map development. In: *Annual Report to the International Potato*

Center for Research on the Development of Pest Resistance Populations. New York State College of Agriculture and Life Sciences at Cornell University, Ithaca, New York, pp. 11–18.

Chavez, R. (1984) The use of wide crosses in potato breeding. Unpublished PhD thesis, University of Birmingham.

CIP (1985) *Annual Report.* International Potato Center, Lima, Peru, pp. 38–39, 80–81.

CIP (1991) *Annual Report.* International Potato Center, Lima, Peru, pp. 63–77.

CIP (1992) *Annual Report.* International Potato Center, Lima, Peru, pp. 83–87.

Cisneros, F.H. (1984) The need for integrated pest management in developing countries. In: *Report of the XXII Planning Conference on Integrated Pest Management*, 4–8 June 1984. International Potato Center, Lima, Peru, pp. 19–30.

Estrada, N. and Valencia, L. (1987) Mejoramiento de la papa para resistencia a la palomilla *Phthorimaea operculella* (Zeller) en Colombia. In: *Memoria de la XII Reunión de la Asociación Latinoamericana de la Papa*, 9–13 March 1987, Panama, pp. 212–219.

Flanders, K.L., Hawkes, J.G., Radcliffe, E.B. and Lauer, F.I. (1992) Insect resistance in potato: sources, evolutionary relationships, morphological and chemical defenses, and ecological defenses. *Euphytica* 61, 83–111.

Gibson, R.W. (1978) Pest aspects of potato production. In: Harris, P.M. (ed.) *The Potato Crop: the Scientific Basis for Improvement.* Chapman and Hall, London, pp. 470–503.

Gregory, P., Tingey, W.M., Ave, D.A. and Bouthyette, P.Y. (1986) Insect defensive chemistry of potato glandular trichomes. In: Juniper, B.E. and Southwood, T.R.E. (eds) *The Plant Surface and Insects.* Blackwell Scientific Publications, Oxford, pp. 173–183.

Hanneman, R.E., Jr (1989) The potato germplasm resource. *American Potato Journal* 66, 665–667.

Hanneman, R.E., Jr and Bamberg, J.B. (1986) Inventory of tuber bearing *Solanum* species. *Wisconsin Agricultural Experiment Station Bulletin* 533, Revised in November 1986: 216pp.

Hawkes, J.G. (1992) Biosystematics of the potato. In: Harris, P.M (ed.). *The Potato Crop: the Scientific Basis for Improvement*, 2nd edn. Chapman and Hall, London, pp. 13–64.

Horton, D. (1987) Potatoes in the Third World. *Courier* 101, 82–84.

Huaman, Z. (1986) Conservation of potato genetic resources at CIP. *CIP Circular* 14(2), 1–7.

Iwanaga, M. (1983) Ploidy level manipulation approach: development of diploid populations with specific resistance and FDR 2*n* pollen production. In: *Report of the XXVI Planning Conference 'Present and Future Strategies for Potato Breeding and Improvement'*, 12–14 December 1983, Lima, Peru, pp. 57–70.

Iwanaga, M., Freyre, R. and Watanabe, K. (1991) Breaking the crossability barriers between disomic tetraploid *Solanum acaule* and tetrasomic *S. tuberosum.* *Euphytica* 52(3), 183–191.

Mendoza, H. (1992) Development of potatoes with multiple resistance to biotic and abiotic stresses. In: Zehnder, G., Jansson, R.K., Powelson, M. and Raman, K.V. (eds.) *Advances in Potato Pest Biology and Management.* American Phytopathological Society, New York.

Mendoza, H.A. and Sawyer, R.L. (1985) The breeding program at the International Potato Center. In: Russell, G.E. (ed.) *Progress in Plant Breeding.* Butterworths, London, pp. 117–137.

Midmore, D.J. (1986) Respuesta de la planta, de papa (*Solanum* spp.) al daño de insectos: Algunos efectos de compensación. In: Valencia, L. (ed.) *Memoria del Curso sobre Control Integrado de Plagas de Papa*. Centro Internacional de la Papa – Instituto Colombiano Agropecuario, Bogotá, Colombia, pp. 176–200.

Ochoa, C.M. (1990) *The Potatoes of South America: Bolivia*, 1st edn. Cambridge University Press, Cambridge, UK.

Potts, M.J. (1988) The influence of intercropping on pests and diseases of potato with special reference to their control. In: *Asian Potato Association Proceedings*, 12–26 June 1988, Kunming, China, pp. 118–119.

Radcliffe, E.B. (1982) Insect pests of potato. *Annual Review of Entomology* 27, 173–204.

Radcliffe, E.B., Lee, F.I., Lee, M. and Robinson, D.P. (1981) Evaluation of the United States potato collection to green peach aphid and potato aphid. *University of Minnesota Agriculture Experimental Station Technical Bulletin* 331, 1–41.

Raman, K.V. (1988) Integrated insect pest management for potatoes in developing countries. International Potato Center (CIP), Lima, Peru. *CIP Circular* 16(1), 1–8.

Raman, K.V. (1991) Varietal resistance of potato to arthropod pests. In: *Proceedings of the 11th International Congress of Plant Protection*, 5–9 October 1987, Manila, Philippines, Vol. 1, pp. 556–568.

Raman, K.V. (1992) Potato pest management in developing countries. In: Zehnder, G., Jansson, R.K., Powelson, M. and Raman, K.V. (eds) *Advances in Potato Pest Biology and Management*. American Phytopathological Society, New York.

Raman, K.V. and Booth, R.H. (1983) *Evaluation for Technology for Integrated Control of Potato Tuber Moth in Field and Storage*. Technical Evaluation Series No. 10, International Potato Center, Lima, Peru.

Raman, K.V. and Midmore, D.J. (1983) Efficacy of insecticides against major insect pests of potatoes in hot climates of Peru. *Crop Protection* 2, 483–489.

Raman, K.V. and Palacios, M. (1982) Screening potato for resistance to potato tuber worm. *Journal of Economic Entomology* 75(1), 47–49.

Raman, K.V. and Radcliffe, E.B. (1992) Pest aspects of potato production, Part 2. Insect pests. In: Harris, P.M. (ed.) *The Potato Crop: the Scientific Basis for Improvement*, 2nd edn. Chapman and Hall, London, pp. 476–506.

Raman, K.V., Iwanaga, M., Palacios, M. and Egusquiza, R. (1981) Breeding for resistance to potato tuber worm *Phthorimaea operculella* (Zeller). *American Potato Journal* 58, 516.

Scurrah, M. and Raman, K.V. (1984) Breeding and screening for resistance to major potato pests. In: *Report of XXII Planning Conference on Integrated Pest Management*, 4–8 June 1984. International Potato Center, Lima, Peru, pp. 103–113.

Scurrah, M., Raman, K.V., Manrique, L. and Palacios, M. (1987) Estudios de resistencia en la papa a *Phthorimaea operculella* (Zeller) a nivel diploide y tetraploide. In: *Memoria de la XII Reunión de la Asociación Latinoamericana de la Papa*, 9–13 March 1987, Panama, pp. 184–190.

Tingey, W.M. (1991) Potato glandular trichome defensive activity against insect attack. In: Hedin, P.A. (ed.) *Naturally Occurring Pest Bioregulators*. ACS Symposium Series 449, ACS, Washington, DC, pp. 126–135.

Tingey, W.M. and Yencho, G.C. (1992) Insect resistance in potato: a decade of progress. In: Zehnder, Z., Jansson, R.K., Powelson, M. and Raman, K.V. (eds) *Advances in Potato Pest Biology and Management*. American Phytopathological Society, New York.

Watanabe, K. (1991) Present status of germplasm enhancement at CIP and application of biotechnology. In: *Report of Planning Conference on Application of Molecular Techniques to Potato Germplasm Enhancement*, 5-9 March 1990, Lima, Peru, pp. 135-140.

Yabar, E.L. (1988) La mosca minadora de la papa en el Perú. Sector Agrario, Instituto Nacional de Investigación Agraria y Agroindustrial, Lima, Peru. *Informe Especial* 1(2), 1-37.

VI POTATO BREEDING

21 Breeding Strategies for Clonally Propagated Potatoes

J.E. BRADSHAW AND G.R. MACKAY

Crop Genetics Department, Scottish Crop Research Institute, Invergowrie, Dundee DD2 5DA, UK.

Introduction

The rediscovery in 1900 of Mendel's published work of 1865 (Mendel, 1865) marked the birth of modern genetics, and it also opened up the possibility for crop improvement by scientific breeding methods based on a sound knowledge of the inheritance of economically important traits. However, the principal cultivated species of potato, the European or Irish potato, *Solanum tuberosum* subsp. *tuberosum*, is not an ideal species for classical genetic research. It was not until the late 1930s that geneticists recognized that Tuberosum was in fact a tetraploid which displays tetrasomic inheritance (Lunden, 1937; Cadman, 1942). Even then potato breeding continued to remain empirical and unsophisticated, in essence a continuation and intensification of the approach adopted by Knight almost 200 years ago (Knight, 1807). Compared with other crops, potato breeding this century has also appeared relatively less successful and inconsistent with the rapid progress that was made during the 19th century. This has been primarily attributed to its narrow genetic base (Simmonds, 1969; Mendoza and Haynes, 1974), but may also in part be due to the rather conservative, empirical approaches taken by 20th-century breeders.

It is therefore useful to briefly review the evolution of Tuberosum and the crop's genetic base before examining the breeding strategies for clonally propagated potatoes during the 20th century. (Breeding strategies for true potato seed (TPS) are reviewed in the next chapter.) Consideration can then be given to how the breeding of tetraploid potatoes could become less empirical and placed on more scientifically sound, genetically based methods. For recent reviews of breeding *per se*, the reader is referred to Mackay (1987a), Caligari (1992), Tarn *et al.* (1992) and Douches and Jastrzebski (1993), and for technological advances to Ross (1986) and Jellis and Richardson (1987). Caligari (1992) and Douches and Jastrzebski (1993) can also be consulted for details on hybridization techniques.

The Genetic Base of Modern Breeding

Evolution of the crop

The origins of cultivated potatoes and species relationships have been dealt with in detail by Hawkes in Chapter 1. The subject is further reviewed in Simmonds (1976) and Hawkes (1990). However, certain essential features will be reiterated here because of their importance to the understanding of 20th-century breeding strategies.

Three main stages in the evolution of the modern potato crop can be recognized: the domestication of wild tuber-bearing, diploid species in South America c. 7000 years ago; the emergence of a cultivated tetraploid form in South America and its introduction into Europe in 1570; the transformation of this tetraploid from a botanic curiosity to an important staple food crop, adapted to the long-day growing conditions of northern Europe by the end of the 18th century, and its dispersal from Europe to many other parts of the world then and since (Figure 21.1). Throughout these stages, naturally occurring genetic variation, which arose via mutation and sexual recombination, allied to conscious and unconscious selection by humans, together with natural selection in new environments, led to the evolution of the modern potato.

Relatively few of the 235 tuber-bearing *Solanum* species were involved in the early domestication process, probably just several closely related and interfertile members of series Tuberosa, similar to, perhaps identical with, the present-day *S. leptophyes* or *S. canasense* (Hawkes, Chapter 1). As diploid outbreeders with a single S-locus, multiallelic, gametophytic self-incompatibility system (Dodds, 1965), their constituent populations will have consisted of highly heterozygous individuals, intolerant of inbreeding. Human selection for less bitter-tasting, hence less toxic, alkaloid-free tubers from amongst these heterogeneous populations will have resulted in clones which could be planted to reproduce edible tubers in quantity. Thus evolved primitive, cultivated diploid species such as *S. stenotomum*, which were also very variable, self-incompatible outbreeders (Dodds and Paxman, 1962; Cipar *et al.*, 1964; Dodds, 1965). In fact, with respect to tuber shapes, tuber skin and flesh colours, these primitive cultivated forms are more variable than their wild progenitors, a probable consequence of preferential selection and retention of attractive mutations by man (Simmonds, 1976; Glendinning, 1983). The lack of tuber dormancy and faster tuber development in the cultivated diploid species *S. phureja*, which evolved from *S. stenotomum*, enables it to produce up to three crops per year in the lower, warmer, eastern valleys of the Andes, and provides another example of artificial selection by Andean farmers (see chapter 1 by Hawkes). It should perhaps be pointed out that irregularities in the self-incompatibility system which require further investigation, have been reported in *S. phureja* and *S. stenotomum* (Abdalla and Hermsen, 1971).

It has been suggested that the cultivated tetraploid species, *S. tuberosum*

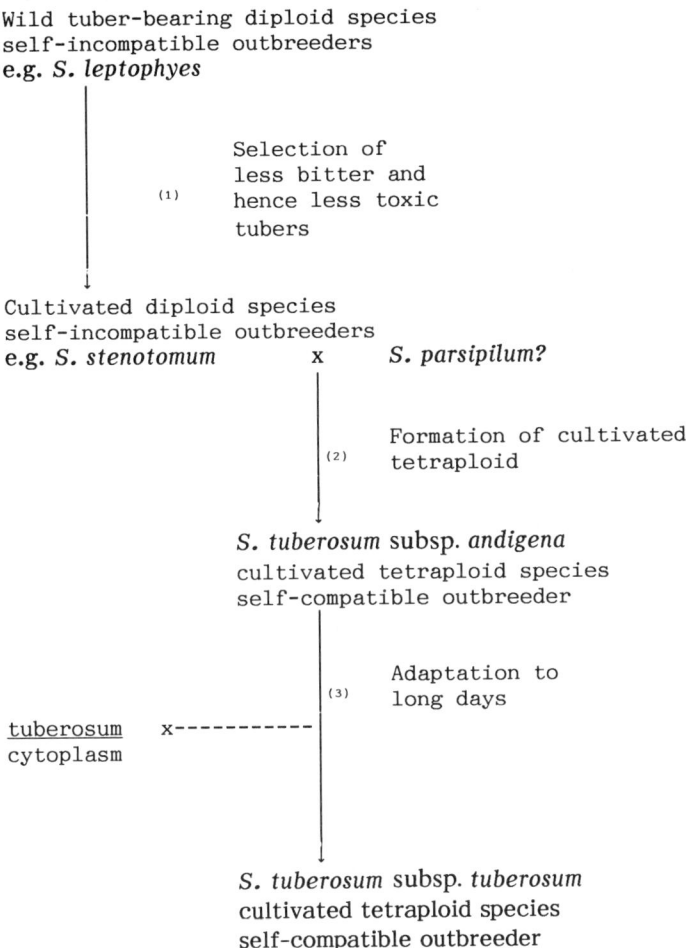

Fig. 21.1. Evolution of the European potato.

subsp. *andigena* most probably evolved, in the Andes of southern Peru and northern Bolivia, as a hybrid between *S. stenotomum* and the wild weed species *S. sparsipilum*, either via $2n$, gametes (Ortiz and Ehlenfeldt, 1992) or from normal haploid, n, gametes followed by subsequent chromosome doubling of the hybrid (see chapter 1 by Hawkes). Recent molecular evidence has, however, raised doubts about *S. sparsipilum*'s role and Grun (1990) refers simply to an unidentified wild species with actinomorphic calyces. The hypothesis that Andigena is a simple autotetraploid of *S. stenotomum* is unlikely to be true, despite its cytological behaviour and tetrasomic inheritance. Andigena is, however, self-compatible, because the gametophytic self-incompatibility system breaks down in tetraploids (Lewis, 1943, 1954). None the less, Andigena populations consist of highly heterozygous individuals which display inbreeding depression on selfing – in contrast with

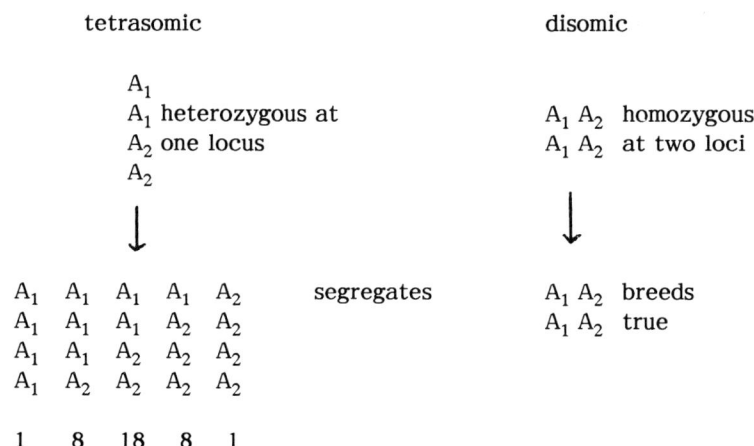

Fig. 21.2. Comparison of selfing an autotetraploid with tetrasomic inheritance and an allotetraploid with disomic inheritance.

allopolyploid *Solanum* species, which show disomic inheritance and are tolerant of inbreeding (Simmonds, 1976). This contrast may in part be due to the fact that allopolyploids can fix different alleles in their constituent parental genomes (Figure 21.2). However, recent estimates of 80% natural selfing in an artificially constructed Andigena population (Glendinning, 1976) pose the questions; why did a greater tolerance of inbreeding not evolve in a crop which has evolved by a similar route to Neotuberosum, or, alternatively, if outbred progeny were at a selective advantage, why did a greater degree of outcrossing not evolve?

The original tetraploid populations of Andigena would have been genetically variable, as collections of primitive cultivars of this species appear to confirm by reference to their range of tuber morphology (Glendinning, 1983). These populations largely superseded the diploid cultivated species, but the reason why the tetraploid status proved superior in agriculture is still a matter of conjecture. However, of significance to this discussion is the fact that these primitive tetraploid forms were adapted to the short days of their centres of origin proximal to the equator. An essential, prerequisite step in the evolution of Andigena to Tuberosum was the species adaptation to the long-day summer growing conditions of northern Europe. This adaptation was presaged by the evolution of the Chilean forms of Tuberosum in prehistoric times, a fact which once suggested that the European form had originated from southern Chile, a hypothesis now largely discounted (Hawkes, Chapter 1). There is also now little doubt that Tuberosum was introduced from Europe into North America and not direct from South America (Hawkes, Chapter 1).

This progressive evolution in Europe of Andigena to a staple food crop, Tuberosum, is therefore presumed to have arisen through selection by the early cultivators for earlier-tubering, higher-yielding clones derived from seedlings from naturally occurring berries, the consequence of open, largely self-pollination (Simmonds, 1969). It also seems likely that it occurred from a narrow genetic base of two introductions from South America to Europe in the 16th century and possible further casual introductions in the 17th and 18th centuries (Glendinning, 1983). Having occurred, it also meant that further introductions of Andigena origin were unadapted in comparison and hence did not enter the European and North American gene pool.

Breeding during the 19th century

As the potato crop assumed its role of a staple food crop, so too did problems associated with its clonal, vegetative means of reproduction – for example, the degeneration of seed stocks due to 'the curl'. This was initially believed to be due to natural degeneration following generations of clonal reproduction and was 'cured' by growing new stocks from true seed and selecting new healthy clones, a practice which led to the selection of many new 'improved' varieties (for review see Salaman, 1926). It also led to a trade in seed tubers from areas less troubled by these problems; for example, seed from Scotland to England (Glendinning, 1983). When this 'degeneration' of clonal stocks was eventually shown to be due to systemic infection by viruses, the reproduction of seed tubers in areas geographically and climatically less favourable to the aphid vectors of those viruses became enshrined in, and is reflected today in, statutory seed certification schemes and protected regions (e.g. for Great Britain, see HMSO, 1991).

The 'great murrain', the pan-European devastation of the potato crop in the mid-19th century (Large, 1940), intensified breeding efforts, albeit from a reduced genetic base. The discovery that certain cultivars were immune to wart, *Synchytrium endobioticum*, and that this immunity was heritable provided added impetus in the early 20th century (Simmonds, 1969; chapter 18 by Wastie).

The first reported deliberate crossing, by artificial pollination, between different potato varieties was by Knight (1807). This practice, however, did not become widespread, in Britain at least, until the latter half of the 19th century. Even then, the raising of seedlings from seed of self-set berries remained a common practice, and, with one reported exception, no American breeders employed artificial cross-pollination in their breeding programmes (Douches *et al.*, 1991). At the same time, breeding in the 19th century was paralleled by a decline in sexual fertility, thus restricting the breeders' choices of parents. This may in part have been due to deliberate selection for non-berrying varieties, as advocated by Knight (1807), who thought this would lead to higher tuber yields. Another possible reason may have been the eventual consequence of inbreeding – due to repeated use of berries produced in the field by open, i.e. self-pollination. The introduction into the USA in 1851 of a particular Chilean Tuberosum variety, Rough Purple Chili (Goodrich, 1863), whose descendants were widely employed as parents at the turn of the century, also cannot be ignored as a contributive factor in the decline in sexual fertility of Tuberosum *sensu lato* at this time. There is now substantial evidence that cytoplasmic and cytoplasmic/nuclear interactions are causative factors in Tuberosum sexual infertility (Grun, 1990), and that the T cytoplasm of Rough Purple Chili and its female descendants was likely to have diverged sufficiently from the A cytoplasm of Andigena to be an important factor. The origin of the T cytoplasm of Chilean subsp. *tuberosum* is still a matter of controversy (Grun, 1979, 1990; Hawkes, Chapter 1).

In essence there can be little doubt that Tuberosum, the European (and North American) cultivated form of the potato, is founded on a narrow genetic base relative to that available in its centres of origin and that, during its evolution in Europe up until the 20th century, this genetic base will have been further eroded. It is therefore somewhat extraordinary that such a crop species has been able to succeed in such a wide range of environments to become the fourth most important food crop in the world.

Breeding Strategies in the 20th Century

Introgression from wild species

Recognition of Central and South America as the centres of origin and diversity of the tuber-bearing members of the genus *Solanum*, and hence the primary sources of genes for disease and pest resistances lacking in European cultivars, has led to numerous collecting expeditions during the early 20th century. Pioneered by the Russians in the 1920s (Hawkes, Chapter 1), several potato germplasm collections have been established worldwide; the Vavilov collection in Russia, the Dutch–German gene bank in Braunsweig, Germany, the Gross-Lusewitz in Germany, the Commonwealth Potato Collection (CPC) in the UK, Sturgeon Bay in the USA, the Balcarce collection in Argentina, the Valdivia in Chile and, of course, the 'world' collection at the International Potato Center (CIP) in Peru.

Whilst the introgression of desirable genes from wild species into Tuberosum is discussed in detail by Hermsen (chapter 23), no discussion on potato breeding could be complete without reference to this strategy, for few modern-bred cultivars lack some genetic input from wild species or primitive cultivated forms. As early as 1909, Salaman identified resistance to late blight in a hybrid of Tuberosum with *S. demissum* and backcrossed this into Tuberosum (Muller and Black, 1951). The blight-resistant 'W races' of Tuberosum were produced in Germany at about the same time, by the same method (Ross, 1979). By manipulating ploidy and with due regard to the endosperm balance number (EBN), virtually any potato species can be utilized in introgression of its genes into Tuberosum (Hawkes and Jackson, 1992; Ortiz and Ehlenfeldt, 1992). The advent of somatic fusion may widen the scope still further (Chapter 23). Nevertheless, despite this apparent access to an extraordinarily wide gene pool, it is also important to recognize that relatively few species have been used to any extent in the breeding of successful modern cultivars (Ross, 1986; Plaisted and Hoopes, 1989; Hawkes in chapter

1). In practice, introgression tends to be targeted at rather few specific traits and, whilst other genetic variation may be introgressed, Glendinning (1983) concluded that the gene pool of modern cultivars was still rather limited and inbred due to close relationships between their parents and ancestors. Mendoza and Haynes (1974) had already reached similar conclusions for North American cultivars.

Neotuberosum

In 1959 Simmonds in the UK began an experiment designed to broaden the genetic base of Tuberosum, and to test the hypothesis that the Tuberosum group had in fact evolved by selection from Andigena, by breeding and selecting an Andigena-based population which might produce parents capable of direct incorporation into modern cultivar breeding programmes (Simmonds, 1969). A similar programme was started in the USA by Plaisted in 1963 with material supplied by Simmonds (Rasco *et al.*, 1980; Plaisted, 1987), by Maris (1989) in 1967 in The Netherlands, and by Tarn in 1968 in Canada (Glendinning, 1987). These populations have so diverged from their original source material that they have now earned the name Neotuberosum.

Starting with unimproved, non-adapted Andigena accessions from the Commonwealth Potato Collection (CPC), Simmonds applied simple phenotypic, recurrent, mass selection methods broadly similar to those employed by 19th-century breeders. Large populations of seedlings were grown in the field and mass selection was applied on those which produced the highest yields of tubers of acceptable sizes, shapes and colours, etc. Open-pollinated berries were harvested from isolated plots of selected plants in the next clonal generation, from which another cycle of seedlings was grown, thus reproducing a sexual generation every 2 years. Late blight resistance was also enhanced by growing and selecting subpopulations each year under severe natural blight attack. In so far as R genes are unknown in Andigena, this resistance was presumed to be of a polygenically inherited, horizontal nature.

Plaisted (1987) modified his scheme in two ways in 1970. He began deliberate pollination of selected clones with bulk collections of pollen, also from selected clones, to avoid inbreeding depression from selfing. He also began raising seedlings in pots to reproduce seed tubers for planting field trials. This resulted in better plants for selection and allowed glasshouse seedling screens for late blight and virus resistance, although it also increased the length of each cycle to 3 years. Maris (1989) used hand pollination between individual selected clones in his scheme.

Within four generations, Simmonds (1969) reported good progress, with the better Andigena clones comparable in yield and maturity to Tuberosum cultivars and better on average in terms of blight resistance (see also Simmonds and Malcolmson, 1967); but these clones were still inferior in tuber shape to modern cultivars, though not in cooking quality. Furthermore, yield heterosis was demonstrated in crosses between the better Neotuberosum clones and

Tuberosum cultivars. Recurrent mass selection was discontinued in 1978, when the emphasis changed to demonstrating the potential of this improved material in breeding cultivars. The cultivar Shelagh resulted from a cross in 1974 between a Neotuberosum clone and a Tuberosum breeding clone, the first such hybrid to enter the UK National List (Anon., 1986). Further details, including information on the processing qualities and disease and pest resistances found amongst the SCRI Neotuberosum population, have been reported elsewhere (Glendinning, 1975a, b, c, 1976, 1981, 1987). One of the more recent findings has been the identification of Neotuberosum clones with improved low-temperature-induced sweetening properties (Brown et al., 1990; Mackay et al., 1990).

Plaisted (1987) made similar progress with his Neotuberosum population and has also found useful variation for several pest and disease resistances. He has, however, reported that, whilst Neotuberosum × Tuberosum hybrids backcrossed to Tuberosum were comparable with Tuberosum × Tuberosum progenies, the hybrids themselves tended to produce smaller-sized tubers with poorer conformity (deep eyes). Tarn and Tai (1983) concurred with these findings in agreeing that first-generation hybrids were not the best parental material. Nevertheless, the first cultivar from Plaisted's programme, cv. Rosa, was a Neotuberosum × Tuberosum hybrid (Plaisted et al., 1981). Maris (1989) reported reciprocal differences, suggesting that progenies with Tuberosum cytoplasm were higher-yielding, whilst, in contrast, the Andigena (Neotuberosum) cytoplasm usually resulted in higher male fertility.

These programmes have therefore demonstrated that through simple mass selection under northern-latitude, long-day summer conditions, Andigena will adapt and produce parents suitable for direct incorporation into modern potato breeding programmes. However, as is discussed in Chapter 3, the formula for response to selection under mixed selfing and random mating is very complex for diploids and not yet derived for autotetraploids. It is therefore not possible to conclude which method of pollination, open or controlled, or of mass selection is best, and Glendinning's discussion (1989) has not resolved the issue.

More sophisticated population improvement schemes, involving recurrent selection and progeny testing for general combining ability (GCA) (Mendoza, 1987) or the use of composite crosses involving Andigena and Tuberosum (Maris, 1989), might be superior; but there is clearly still great scope for more theoretical and practical research on the 'Neotuberosum' approach to potato improvement.

Breeding at the diploid level

Most of the evolutionarily more advanced tuber-bearing wild species are diploid with an EBN of 2, the same as dihaploids of *S. tuberosum* with which they will cross readily. These wild species are sources of genetic resistances to diseases, pests and abiotic stresses, as well as genetic diversity in general. Their hybrids with dihaploids are usually adapted to long-day conditions and will form tubers. If they also produce 2n gametes by FDR, the genetic diversity of the wild species can be efficiently transferred to the tetraploid cultivated form and results in about 25% of the wild species genes in the final product. This has led to Peloquin and coworkers proposing a novel breeding strategy designed to both introgress specific characteristics and broaden the genetic base (Hermundstad and Peloquin, 1987; Jansky *et al.*, 1990), in a way similar to that envisaged by Chase (1963) in his analytic breeding scheme. The genetic aspects of these procedures are discussed in the chapters by Tai and by Ortiz and Peloquin. Population improvement by recurrent selection can precede return to the tetraploid level, a recent example being the work of Rousselle-Bourgeois and Rousselle (1992).

During the period 1962–1979, Carroll (1982, 1987) employed a mass selection method to produce a population of group Phureja/Stenotomum adapted to long-day north European conditions. Carroll's scheme relied on natural seed set, by insects, and included both seedling and tuber populations with a minimum generation cycle of 2 years. The self-incompatibility of these diploids was presumed to ensure cross- rather than self-fertilization. The population rapidly adapted to long-day conditions and yield improved over several generations, mainly as a result of increase in tuber size without a reduction in tuber numbers. The proportions of oval/long oval, regular-shaped tubers increased, but increased selection was required for shallow eyes and improved dormancy. It was also possible to demonstrate variation for late blight resistance and to select for this in the field. Direct hybridization of members of this improved diploid population with tetraploid Tuberosum cultivars via unreduced pollen grains, ensured that the diploids made a 50% contribution to the tetraploid hybrids, rather than 25% had they first been crossed with dihaploids. Some of these tetraploid hybrids were superior to standard tetraploid cultivars in both total and marketable yield, generally producing more tubers per plant with slightly lower mean tuber weights (Carroll and De Maine, 1989). Haynes (1980) initiated a similar diploid population based on Phureja/Stenotomum in 1966 and this is mentioned in the chapter by Ortiz and Peloquin.

The principal aim of much of the work described has been to improve unadapted diploid germplasm to a point where it can be hybridized with tetraploid Tuberosum and produce material of cultivar status. If it was possible to produce large numbers of fully fertile true dihaploids of Tuberosum, then population improvement at the diploid level could be contemplated as a breeding method for further improvement of adapted germplasm.

There is clearly much scope for further theoretical comparisons of the different strategies and methods for the employment of diploids in tetraploid potato improvement. However, for the time being, breeding at the tetraploid level remains the most practicable way forward in the immediate future.

Breeding at the tetraploid level

Recent molecular studies confirm that good progress has been made during the 20th century in introducing new germplasm into the tetraploid Tuberosum gene pool. There is substantial diversity in chloroplast DNA of more recently introduced European cultivars (Waugh et al., 1990) and nuclear DNA polymorphisms detected by RFLP procedures demonstrate the importance of wild species and primitive forms in the development of modern cultivars (Powell et al., 1991).

Nevertheless, the breeding of varieties *per se* has changed little in essence except in its scale and technology. It remains by and large empirical and genetically unsophisticated, relying primarily on phenotypic recurrent selection as exemplified by potato breeding programmes worldwide: Canada (Tarn et al., 1992), England (Howard, 1978), Germany (Fitschen, 1984, in Ross, 1986), Ireland (Kehoe, 1982), Northern Ireland (Lee, 1985), The Netherlands (Neele et al., 1988) and the USA (Douches and Jastrzebski, 1993). The Scottish Crop Research Institute (SCRI) programme as it operated prior to 1982 has been described in detail by Mackay (1987a) and Caligari (1992) and can be viewed in its essential features as a typical breeding programme (Figure 21.3). Pair crosses were made each year between clones with complementary features, usually chosen on the basis of their phenotypes, although over time, of course, breeders did get to know which clones tended to perform as better parents than others in practice. Caligari (1992) has described various techniques used by breeders to encourage flowering and seed set, such as shading in glasshouses to reduce temperatures below 22°C, removing daughter tubers, placing flowering stems in jars of water with an antibacterial agent and grafting potato stems on to tomato stocks (Figure 21.4). The year after crossing, approximately 100,000 seedlings were raised in pots in aphid-screened glasshouses in order to produce tubers (clones) for field evaluation (Figure 21.5). A single tuber was taken from each 'selected' seedling clone for field planting; the subsequent plants were grown at wide spacing under a high-grade seed regime – rather atypical of normal ware production – and the 'best' clones selected visually at harvest. By the second clonal generation, the original population of 100,000 seedling genotypes would have been reduced to about 4000 clones, thus eliminating 96% of the genetic variation. In all of the examples cited earlier, between 85% and 98% of genotypes would have been discarded by this same stage, although in the Canadian programme there is no selection in the seedling generation in the glasshouse.

Details again vary between programmes, but at SCRI, which was fairly typical, the selected clones were replanted as unreplicated three to four plant plots and selected again at harvest, in the field, by visual inspection of yields, tuber morphology, etc., reducing the population to about 1% of the original 100,000 seedlings (the figures for the other programmes cited are between 1% and 4%). Several recent reviews have clearly argued that this type of early generation selection is very ineffective (Tai and Young, 1984; Caligari, 1992; Tarn et al., 1992), although some screening for disease and pest resistances

Year			
0		DECIDE OBJECTIVES AND CHOOSE PARENTS	
		GLASSHOUSE	
1		c. 200 pair crosses	
2		c. 100,000 seedlings in pots visual selection	
		SEED SITE visual selection	
3		40,000 single plants (first clonal year)	
4		4,000 4-plant plots	

ASSESSMENT FOR YIELD AND QUALITY AND SPECIAL DISEASE TESTS

Year	SEED SITE number of plants	CLONES number	WARE SITE number of plots and plants per plot
5	6	1000	2 x 5
6	20	360	2 x 10
7	100	120	2 harvests of 2 x 10
8	300	40	WARE TRIALS at a number of sites
9	700	20	WARE TRIALS at a number of sites
10	2000	4	OFFICIAL TRIALS
11	2000	2	OFFICIAL TRIALS

MULTIPLICATION FROM VIRUS TESTED STEM CUTTINGS (= NUCLEAR STOCKS) AND COMMERCIALISATION OF BEST CLONE

Fig. 21.3. SCRI Potato Breeding Programme in the days before progeny tests were used.

can be employed effectively at these stages, particularly where that resistance is governed by major dominant genes (Plaisted et al., 1984; Swiezynski, 1984; Jellis, 1992).

The relatively slow rate of natural vegetative reproduction (approximately eightfold per year under SCRI conditions), allied to the complicated logistics of accurately assessing 1000 or more clones for a very large number of traits, meant that another 6 years elapsed, at SCRI, before one or a few potential cultivars could be entered with confidence into official statutory (National List) trials. During this period, decreasing numbers of selected clones were

Fig. 21.4. Pollination glasshouse showing plants on bricks with daughter tubers removed.

Fig. 21.5. Seedlings in glasshouse.

grown in increasingly sophisticated trials over as wide a geographical range as economics permitted and breeding objectives demanded. During these intermediate and final stages of selection, the production of seed tubers was separated from the trials which were grown under ware conditions, designed as far as was possible to approximate to those of good commercial practice. In addition to yield and agronomic performance, clones undergoing selection were assessed for their cooking and processing characteristics and tested for their resistances to numerous pests and diseases. The selection criteria and testing procedures in such a multitrait, multistage scheme have largely been governed by practical considerations and experience of the reliability of the various tests used, rather than by genetical knowledge, such as heritabilities or genetic correlations between traits.

In such a breeding scheme it could take 9 years or more to acquire sufficient information on selected clones before using them in a further cycle of crossing and selection. If viewed overall as a recurrent phenotypic selection programme, such a breeding strategy is clearly inefficient and in theory at least there is much scope for improvement. In the next section, consideration is given as to how such breeding work can be made genetically more sophisticated and selection efficiency improved.

Strategies for Increasing Selection Efficiency Based on Genetic Principles

Multiplex parents

There are several well-documented instances in the potato where resistances to diseases of major economic importance are governed by major dominant genes, inherited in a Mendelian fashion. Amongst them are the H_1 gene for resistance to pathotypes RO_1 and RO_4 of *Globodera rostochiensis* (see chapter 14 by Phillips), numerous 'R' genes conferring race-specific resistance to *Phytophthora infestans* (Malcolmson and Black, 1966), as well as genes conferring race-specific and comprehensive resistance to the common viruses PVX and PVY, such as the *Nx* and *Ny* genes or Rx_{adg} and Ry_{sto} genes, respectively, amongst others (Cockerham, 1970). The capacity of the blight fungus to rapidly evolve resistance-breaking strains soon led breeders to abandon the use of R genes (Wastie, 1991a), but in contrast the H_1, *Nx*, *Rx*, *Ny* and *Ry* genes have maintained their effectiveness for decades, and, particularly in the case of the viruses, are likely to do so for the foreseeable future, because of the nature of the host/pathogen relationship. The probability of a resistant clone becoming infected naturally with a mutant, resistance-breaking, strain of a virus is very low and there is no selection pressure to favour such mutants in a susceptible host. Indeed, to date, no strains of PVY have yet been found that are capable of overcoming Ry_{sto} (C. Brown, pers. comm.).

However, most cultivated varieties possessing one or other of these genes, and most breeders' clones, tend to be simplex at their resistance gene loci. Consequently, when used as parents with susceptible (nulliplex) partners, only 50% of their progeny are likely to inherit the resistance gene. Screening out the unwanted susceptible phenotypes is wasteful of time and resources, but not to do so leaves an unacceptably high probability that ensuing cultivars will be susceptible. Some prebreeding to produce multiplex parents for some of these resistances is an attractive alternative (Figure 21.6), as first discussed by Cadman (1942) with respect to the *Nx* gene for race-specific resistance to PVX, and explained in detail by Toxopeus (1953). The use of parents that are quadruplex for a major resistance gene would ensure that all of their progeny would be resistant, as too would triplex parents, in the absence of double reduction; and even duplex parents would have a much enhanced likelihood of passing their resistance on, reducing the need for screening but enhancing the probability of producing resistant cultivars. This approach has been discussed by SCRI researchers and implemented as part of their breeding strategy (Mackay, 1987b). The most difficult problem associated with this approach is the need to progeny-test putative multiplex parents by hybridizing them with susceptible testers, but, as Toxopeus (1953) argued, it should only be necessary to test 26 seedlings of a progeny to distinguish a 5:1 ratio (duplex × nulliplex) from a 1:1 ratio (simplex × nulliplex) with a probability of 97.5%, or to be reasonably certain that the parent is quadruplex or triplex

Rrrr x Rrrr   Simplex genotype x simplex genotype
↓

¼rrrr   duplex genotype recognized in test cross to rrrr by 5 to 1
½Rrrr   (11 to 3 with chromatid segregation) ratio of resistant to susceptible
¼RRrr   progeny compared with 1 to 1 (13:15) ratio if simplex genotype

RRrr x RRrr   duplex genotype x duplex genotype
↓

$\frac{1}{36}$rrrr   triplex or quadruplex genotype recognized in test cross to rrrr as gives all
$\frac{8}{36}$Rrrr   resistant progeny (or 27 to 1 susceptible with chromatid segregation in
$\frac{18}{36}$RRrr   triplex)
$\frac{8}{36}$RRRr
$\frac{1}{36}$RRRR

Fig. 21.6. Production of parents multiplex for a major dominant resistance gene R with chromosomal segregation.

Table 21.1. Segregation of H_1 gene in cross 15205 between clone 10341 ab 18 (duplex) and Cara (simplex). (Data kindly supplied by M.F.B. Dale and M.S. Phillips, SCRI.)

Clone	Susceptible tester	Progeny		Conclusion about clone
		Resistant	Susceptible	
10341 ab 18	Dr Mackintosh	58	12	Duplex
15205 ab 3	Torridon	1	31	Nulliplex with one escape?
15205 ab 6	Torridon	33	1	Triplex with double reduction?
15205 ab 12	Torridon	17	18	Simplex
15205 ab 14	Torridon	30	6	Duplex

and not duplex or simplex. An example from the SCRI programme is shown in Table 21.1.

Bearing in mind the many characteristics required in a successful cultivar, it is important that multiplex parents will have been selected for these too. There is clearly a lot of work required in producing parents multiplex for several major resistance genes, but the process will gain in momentum as the frequency of multiplex parents increases in a breeding programme, and there are techniques to accelerate the whole process. For example, duplex and quadruplex tetraploid stocks can be synthesized from doubling appropriate dihaploids, which themselves can be produced from duplex and triplex sources. The use of monohaploids also offers exciting possibilities (see chapter 7 by Jacobsen and Ramanna).

Accepting the desirability of parents multiplex for resistance genes, most economically important traits in the tetraploid potato display continuous variation where individual genes cannot usually be recognized, except through linkage to markers. It is therefore necessary to resort to biometrically based methods in order to select parents with good general combining ability. First, however, Tarn *et al.* (1992) have advocated using coancestry analysis to avoid closely related parents and hence inbreeding depression, and Powell *et al.* (1991) have suggested using genetic distance based on molecular markers to select diverse parents capable of producing high-performing progeny.

Identifying parents with good general combining ability

Although it may not be possible to identify individual genes and to deduce the actual genotype of a parental clone from its phenotype, it is possible to estimate the extent to which clones are likely to transmit their desirable properties to their offspring. The general combining ability (GCA) of a parental clone provides an assessment of its gametic input, as judged by the mean performance of its progenies from crosses with other clones. In order to make valid inferences, these latter clones should be either all those of interest or a random sample of them. Two types of factorial crossing designs are commonly used to determine combining abilities (see chapter by Bradshaw on quantitative genetics theory). In the first, one set of clones is crossed with another set which complements it for desirable traits. In the second, a diallel set of crosses is made amongst clones showing a range of values for the trait or traits of interest. Where possible, reciprocal crosses should be made and ideally the subsequent progenies assessed in more than one environment in order to study genotype–environment interactions.

The departure of a progeny mean from that expected on the basis of the GCAs of its parents is called specific combining ability (SCA). When all of the variation between progenies proves to be due to SCA, it is not possible to predict or anticipate the performance of a progeny from any pair cross. The breeder is then obliged to hybridize all possible parents in all possible combinations and identify the best combinations by examination of their progenies. At the other extreme, if all the variation is attributable to GCA, the breeder can accurately predict the outcome of a pair cross if the parental GCAs are known from other crosses. Furthermore, if parental GCAs are highly correlated with their phenotypes (which approximate to their genotypes if accurately assessed), the midparental value should provide a good prediction of mean performance of the progeny. After all the parent–offspring covariance and the variance of GCAs both largely depend on the additive genetic variance, in a tetraploid random-mating population in equilibrium. Obviously, in the real world both GCA and SCA will often be found to contribute in varying proportions. Neele *et al.* (1991a) have suggested that SCA tends to be more important than GCA in crossing schemes involving closely related parents (e.g. Killick, 1977), and Tai (1976) suggested the same for traits which have been subjected to continuous directional selection, such as tuber yield. Table 21.2 provides a representative sample of GCA/SCA results for some economically important traits, but these must be interpreted with caution. Table 21.2 also provides reports of GCA–parent and midparent–progeny correlations. It can be seen that, where GCA is more important than SCA these correlations range from 0.69 to 0.97 and from 0.41 to 0.85 respectively. Hence they are usually high enough to be useful. As SCA increases in importance, so the correlations tend to fall in value, as expected, until GCA and midparent values have little predictive power. Neele *et al.* (1991a) found that, whilst neither the selfed nor the dihaploid progenies of tetraploid clones could generally be used to improve on midparent predictions, test crosses with diploids which produced diploid pollen via first-division restitution (FDR) did give better results, particularly for ware yield (GCA = SCA). Ortiz *et al.* (1991) also reported larger differences in the GCAs of tetraploid parents when they were pollinated with diploids producing diploid pollen via FDR, rather than with other tetraploid parents. They attributed this to diploid pollen via FDR being more heterozygous, but more homogeneous than normal diploid pollen from tetraploid parents. Neele (1990) reported that, unlike midparent value for tuber yield, differences in harvest index between parents could be used to predict the tuber yield of their progeny. The use of such predictive methods in the presence of SCA is worthy of further investigation.

Early-generation selection

It is clear that in all breeding programmes the maximum genetic variation is exposed and available for selection in the early generations, i.e. the seedling (usually glasshouse) and first clonal (usually field) stages. Any increases in efficiencies of selection in these stages are likely to result in major improvements in the quality of material advancing to the later stages of selection and to increase the likelihood of genetic improvement in cultivar production. During the 1980s a number of experimental studies confirmed that visual selection, as traditionally practised in the seedling (glasshouse (GH)) and first and second clonal years (FCY, SCY), is very ineffective. But there is still no general consensus as to how to address this problem. The essential feature of these experiments was to assess random samples of clones (progenies) from a number of crosses in replicated trials over a number of early generations without imposing selection on the population(s) under study. This meant that the ratio of genetic variation (σ_G^2) to total variation (σ_P^2), and the extent of genotype (clone) × environment interactions could be determined and expressed in terms of broad-sense heritability ($h_b^2 = \sigma_G^2/\sigma_P^2$) or intraclass correlation (see chapter by Bradshaw on quantitative genetics theory). Selection could then be applied to these populations retrospectively, and predicted and observed responses (R) compared in terms of changes in the mean of the population through the classical formula $R = h_b^2 S$, where S is the selection differential. Finally, a multistage selection strategy could be developed to maximize the chance of retaining and identifying the best clones. Simmonds

Table 21.2. General and specific combining ability (>, < and = used where components of variance estimated; recip = reciprocal difference).

Trait and reference	Combining ability	GCA-parental value correlation	Midparent-progeny correlation
After-cooking blackening			
Dalianis et al., 1966	GCA		
Killick, 1977	SCA		
Bacterial wilt			
Tung, 1992	GCA < SCA (both × location interaction)		
Breeders' visual preference			
Brown and Caligari, 1989	GCA		0.41
Maris, 1989	GCA (>recip)	0.97	0.72
Neele et al., 1991a	GCA > SCA		0.64
Cracked tubers (%)			
Killick, 1977	SCA		
Date of emergence			
Maris, 1989	GCA (<recip) < SCA	0.30	0.16
Eye depth			
Neele et al., 1991a	GCA > SCA		0.78
Haulm type			
Maris, 1989	GCA < SCA	0.77	0.55
Hollow heart			
Veilleux and Lauer, 1981	GCA and SCA		
Late blight in foliage			
Killick and Malcolmson, 1973	GCA < SCA		
Malcolmson and Killick, 1980	GCA		
Stewart et al., 1992	GCA (SCA × year)	0.97	
Late blight in tubers			
Stewart et al., 1992	GCA and SCA	0.67	
Maturity			
Killick, 1977	GCA		
Maris, 1989	GCA (>recip)	0.84	0.72
Mechanical damage			
De Maine et al., 1992	GCA	0.91	
Number of main stems			
Maris, 1989	GCA (>recip)	0.97	0.85

Potato cyst nematode-Pallida
  Phillips et al., 1979   GCA = SCA   0.46
  Phillips and Dale, 1982   1) GCA
                               2) GCA = SCA

Plant height
  Maris, 1989   GCA (= recip) < SCA   0.26   0.21

Regularity of tuber shape
  Neele et al., 1991a   GCA > SCA   0.85

Shape of tuber
  Neele et al., 1991a   GCA > SCA   0.50

Specific gravity
  Tai, 1976   GCA > SCA
  Killick, 1977   GCA and SCA
  Maris, 1989   GCA (> recip)   0.85   0.81
  Neele et al., 1991a   GCA > SCA   0.74

Texture
  Killick, 1977   SCA

Tuber number
  Tai, 1976   SCA
  Killick, 1977   GCA and SCA
  Veilleux and Lauer, 1981   GCA
  Brown and Caligari, 1989   GCA   0.71
  Maris, 1989   GCA (> recip)   0.69   0.59
  Neele et al., 1991a   GCA > SCA   0.57

Tuber weight (mean)
  Tai, 1976   GCA (> GCA x year)
  Killick, 1977   GCA and SCA
  Brown and Caligari, 1989   GCA and SCA   0.40
                                         (and GCA x sites)
  Maris, 1989   GCA (> recip)   0.96   0.78
  Neele et al., 1991a   GCA > SCA (> recip)   0.81

Tubers < 30 mm by weight (%)
  Neele et al., 1991a   GCA = SCA (> recip)   0.73

Tuber yield
  Plaisted et al., 1962   GCA < SCA (both
                                         interact with locations)
  Tai, 1976   SCA
  Killick, 1977   SCA
  Veilleux and Lauer, 1981   GCA and SCA
  Brown and Caligari, 1989   GCA   0.55
                                         (and GCA x sites)
  Maris, 1989   GCA (= recip)   0.78   0.51
  Neele et al., 1991a   GCA = SCA (> recip)   0.14

(1985) presented a theory for two-stage selection and Neele et al. (1989) showed how to incorporate economic considerations, but a complete theory for multistage, multitrait selection is not yet available, although various strategies can be compared by computer simulation, as Cornish (1990a, b) has done for a diploid selfing series.

At SCRI, Brown et al. (1987a) assessed the progenies from eight pair crosses through three generations, initially as seedlings in the glasshouse and subsequently as clones in the field. A random sample of 571 genotypes from the eight crosses contributed data to all three generations, although the initial population was larger. They showed that selection in both the glasshouse (GH) and first clonal year (FCY) produced a positive response, as determined in the second clonal year (SCY), (GH v. SCY, $r = 0.36$; FCY v. SCY, $r = 0.64$); but it resulted in a substantial loss of potentially superior clones. Only 27% and 56% deemed acceptable in the SCY would in fact have been retained had selection actually been practised in the GH plus FCY and FCY generations, respectively. Moreover, a comparison of a random sample of clones with others from the same crosses, which had been selected in the then commercial breeding programme, showed that GH and FCY selection was at best at random. Brown et al. (1987b), however, were able to demonstrate that the ranking of the mean of a cross remained relatively constant over five diverse environments, GH, FCY $\times$ 2 (seed and ware sites) and SCY $\times$ 2 (seed and ware sites). The rank correlations ranged from 0.69 to 0.95 when based on 70 clones per progeny, and not much lower (0.53 to 0.92) when based on 25 clones. They concluded that progeny evaluation, by breeders' visual preference scores, could be used to identify crosses with the highest potential for producing commercial cultivars, and these results were confirmed from a study on a further 52 progenies (Brown et al., 1988).

Tai and Young (1984) rejected seedling (GH) selection and considered cross-selection in FCY, but concluded that it could not be used to replace mass selection, because the increase in heritability did not in their experience translate into an increase in response to selection. They recommended that a moderate intensity of selection over several generations gave a better balance between genetic advance and loss of valuable genotypes.

Neele et al. (1989) quantified this conclusion from an experiment with 600 clones chosen at random from 20 progenies, in the equivalent to FCY and SCY, over 2 years. Their correlations were FCY v. SCY (seed site), 0.69 and 0.66; FCY v. SCY (ware site), 0.54 and 0.45; and SCY (seed) v. SCY (ware), 0.66 and 0.61. They concluded that the most economical selection procedure would be to select (retain) approximately one-third (between 20 and 50%) of FCY clones for trial in SCY, followed by 20% of the SCY clones, to give 100 clones for trial in the third clonal year (see also Neele et al., 1988). They concluded that this mild selection regime in The Netherlands would result in a higher proportion of clones in late-maturity classes, with higher-specific-gravity tubers surviving. They also concluded that selection for breeders' visual preference (their variant, 'plant appearance') could not be improved upon by independently selecting for its principal components, which in their material were tuber yield, stolon length and tuber appearance (Neele et al., 1991b). This supports Maris's (1988) earlier conclusion that selection for individual components of visual preference (his 'general impression') was usually no more reliable than selection for visual preference itself. He did conclude, however, that negative selection should be applied to component characters according to their heritabilities, in order to avoid rejection of too many valuable genotypes at an early stage.

Gopal et al. (1992) argued that individual clonal selection could be more efficient if based on specific characters and that seedlings should be raised in the field rather than glasshouse to permit a fuller expression of yield potential and tuber characters. They also concluded that seedlings of poor vigour can be discarded prior to transplanting to the field. This does rather presuppose that a suitable site and resources are available, particularly as the transplantation of thousands of potato seedlings in the field, as opposed to aphid-screened glasshouses, would be fraught with practical difficulties under most normal conditions, as such seedlings are delicate and very susceptible to late frosts and aphid-transmitted viruses, in addition to weed control problems.

What conclusions can be drawn from these experimental results? Selection on the basis of breeders' 'visual preference' in GH, FCY or SCY is clearly attractive. Large numbers of genotypes can be quickly assessed and a decision, reject or select, taken immediately. It provides a means of integrating the shortcomings of every visual character into the final selection decision. As a subjective selection index, it will only elicit the optimum response when its component characters are weighted by their heritabilities and due account is taken of any genotypic correlations. Some knowledge of these factors would allow breeders to take account of them in their subjective index. Theoretically, the optimum response to clonal selection is expected from some combined index of progeny mean and individual clone deviation, but this would remove the simplicity of selection on breeders' visual preference. There is, however, convincing evidence that the heritability of 'visual preference' is sufficiently low to warrant using cross means to reject entire crosses, on the grounds that they are less likely than others to contain outstanding individual clones, and that this can be done on the basis of glasshouse-raised seedling progenies (Brown et al., 1988; Brown and Caligari, 1989). The evidence also supports the application of mild clonal selection within the selected progenies in the FCY and then the assessment of the SCY generation under ware conditions because of strong clone–harvest date interactions.

In the next section other progeny tests are reviewed which can also be used to select between progenies at the seedling stage, thus reducing the total number of clones to be assessed in the next generation and hence permitting mild selection to be practised.

Development and use of progeny tests

At SCRI, in recent years, much effort has been expended in the development of progeny tests which can be used to select between progenies at the earliest opportunity, as well as to study the inheritance of quantitative traits and to identify the best parents for future breeding. Some of these tests can be applied to seedling populations, from the sowing of true seed, and others to tuber progenies. The overriding criterion for their use is that they must adequately reflect the subsequent performance of clones grown in the field, under normal agricultural conditions. The test based on visual preference, for agronomic potential and tuber characteristics, was described in detail in the preceding section. Tests have been developed and applied for quantitative resistance to a range of pests and diseases, including potato cyst nematode (Phillips, 1981), late blight in the foliage (Stewart et al., 1983; Caligari et al., 1984, 1985) and in tubers (Wastie et al., 1987), gangrene (Wastie et al., 1990) and powdery scab (Wastie, 1991b).

In practice, 80 seedlings (genotypes) per progeny, grown as four replicate samples of 20, are considered desirable; but, where large numbers of progenies are screened or disease testing facilities are restricted, experience has shown that as few as 40 or even 20 can suffice.

Other researchers, Neele and Louwes (1989), have carried out similar research and have shown that quality characters of glasshouse-raised seedling tuber progenies, such as dry matter and chip (i.e. crisp) quality, can be related to subsequent performance of the same progenies clonally reproduced in the field. However, there are conflicting results from attempts to relate tuber yields between glasshouse seedlings and their subsequent field performance as clones. Brown and Caligari (1989) found a high correlation ($r = 0.9$) between seedling and second clonal years, but Neele et al. (1991a) were unable to demonstrate any relevant relationship between tuber yield, numbers of tubers and mean tuber weight of seedlings and their field performance as clones. Nevertheless, once progeny tests are validated they can be used to assess the mean performance of a cross for a number of traits and thus select the best crosses. This is usually done by independent culling, but would be more efficient if selection indices could be derived. Where individual genotypes of progenies are assigned a score, e.g. breeders' preference, it is possible to practise multivariate cross prediction, using multivariate probabilities based on the progeny means and within progeny variances, and the correlations between variates (Brown and Caligari, 1988). By such a technique the probability of a clone within a progeny exceeding or achieving a given target can be calculated for each cross. Ideally, genotypic variances and correlations should be used, but they are not available in the seedling generation. However, in the second clonal year where there was replication, Caligari and Brown (1986) showed with univariate cross prediction for breeders' preference that the phenotypic variance gave a similar set of predictions to the genotypic variance. Furthermore, although the square root of the phenotypic variance added increasingly to the accuracy of the prediction as the target value increased, it never became a major component in such predictions. Hence, with multiple traits, a selection index which takes account of heritabilities of progeny means and correlations between traits should prove adequate.

Having identified the best crosses from progeny tests from limited screenings, larger samples of true seed of these can be sown. From these resowings, clones containing all or most of the desirable traits can be sought over a number of clonal generations in a similar conventional selection programme to that already described, but with a much improved likelihood of success.

It is, of course, extremely unlikely that major improvements in resistance to every pest and disease and in yield and quality can be achieved in a single round of hybridization and selection. The numbers of genes involved and genetic linkages alone would render this highly improbable, even if all the desired traits were available in a single pair cross. Consequently, whilst new improved cultivars will be produced by pair crosses of improved parent clones, the longer-term success of a potato breeding programme will be determined by the overall efficiency of germplasm enhancement by recurrent selection, i.e. by increasing the frequency of desirable genes in the population as a whole, so that future pair crosses have a greater likelihood of producing yet higher frequencies of desirable combinations. Progeny tests offer the means to replace phenotypic recurrent selection with a much more efficient, multitrait, genotypic, recurrent selection programme, in which the generation cycle time can be reduced by several years, because parents with good general combining ability can be recognized shortly after each round of hybridization. Then their progeny can be used for subsequent crossing cycles and selection, whilst cultivars are being produced from resowings of the best progenies.

Maximizing the response to clonal selection with finite resources

Seedling or tuber progeny tests provide the means to rapidly and efficiently identify those progenies with the greatest likelihood of containing desirable clones, but in order to select the best clones from the better progenies larger-scale, clonally replicated trials are necessary. This is particularly so for complex traits such as yield and polygenic disease resistance. The number of clones that can be assessed is more likely to be determined by the resources, manpower and finance available rather than biological constraints, and is probably a more accurate guide to the size of a breeding programme than the number of seedlings raised per annum. Vermeer (1991) showed how to maximize the response to selection with finite resources (e.g. with a fixed total number of plots), for single-stage, single-trait selection, using tuber yield as an example. The classic formula for the response to selection (R) is:

$$R = i\sigma_G^2 \Big/ \left(\sigma_G^2 + \frac{\sigma_{GE}^2}{n} + \frac{\sigma_e^2}{nr}\right)^{\frac{1}{2}}$$

The intensity of selection (i) is determined by the proportion of clones (genotypes) selected, but unreliable selection of too few clones will reduce the ultimate response achieved over generations. The genotypic variance between clones (σ_G^2) is fixed at any given stage, but changes with selection. The number of environments (n) is most commonly a function of the number of locations (sites) and seasons (years); σ^2_{GE} is the genotype–environment interaction variance and σ^2_e the plot to plot variation within the replicates (r) in each environment. The latter will largely be a reflection of soil heterogeneity and is influenced by plot size and shape as well as the efficiency of blocking in the trial design. Ideally, in maximizing the response to selection, it would also be useful to take costs of resources into consideration.

The breeder can obviously manipulate i, n, σ^2_e and r in order to maximize R in a single-stage, single-trait selection scheme, providing σ^2_G and σ^2_{GE} and the soil heterogeneity are known. If intergenotypic competition is a problem, then the expected correlated response in monoculture should also ideally be computed. By accommodating changes in σ^2_G with selection as well as changes in other parameters, the theory can be extended to multistage selection. Cornish (1990a, b) has shown how this can be achieved for selection in a diploid selfing programme, and a similar approach could be adopted for multistage clonal selection. Ideally, a selection index would accommodate multitrait selection, but at present independent culling levels are the normal practice.

Vermeer (1990) computed coefficients of variation ($\sqrt{\sigma^2}/\mu$, where μ is the overall mean) for genotypic differences and for genotype–location, genotype–year and genotype–location–year interactions from reported data from potato trials for yield and other traits. These not only varied considerably between traits, but also between trials within traits. Thus breeders need to estimate values for their own breeding programmes; Vermeer's are not generally applicable and therefore not reported here. However, his paper does provide a useful set of references to genotype–environment interaction. It is also clear that genotype–environment interaction components in potato trials are commonly the same order of magnitude as the genotypic differences. This clearly limits the gain in response achieved through increased heritability in a single environment. Thus testing at more than one location, for more than one year, is desirable.

As replicated trials in many environments may not be feasible in the early stages of a breeding programme, it raises the question as to how selection in one set of environments for performance in other environments compares with direct selection in these other environments. It thus poses the strategic question as to whether or not more than one selection programme is required. The theoretical aspects of this are dealt with in the chapter on quantitative genetics theory, where the point is made that the potato's vegetative means of reproduction lends itself to selection experiments in contrasting environments. Whilst extensive studies have not, to our knowledge, been carried out, there are reports comparing the performance of a limited number of genotypes selected in one environment and trialled in others. Brown *et al.* (1992), for example, found a greater correlation for total marketable yield between the Scottish ware site where clones were selected and sites in England (0.43 to 0.70) than with sites in the Mediterranean (0.00 to 0.67). Although the very best clones from the Scottish ware site performed reasonably well in the Mediterranean, the results support the idea that selection would be optimized by selecting in environments more similar to those in which the cultivars are to be grown.

A final strategic question seeks to consider the balance between maximizing immediate or short-term gains versus longer-term progress. As potato breeding consists of repeated cycles of crossing and selection, maximizing the short-term gains by intensive selection will result in the loss of desirable genes by random genetic drift, and hence in a lower selection limit than with mild selection. It therefore requires continuing input in the early stages of new genetical variation, preferably as adapted germplasm. Fortunately, the need for a broad genetic base of adapted potato germplasm has been recognized and sought during the 20th century.

Future Prospects

In the foreseeable future it is unlikely that lack of suitable genetic variation *per se* will limit progress in potato breeding. Indeed, as the 20th century draws to its close, there appears to be tremendous scope for faster conventional progress as genotypic recurrent selection replaces phenotypic recurrent selection. More use of wild species can be anticipated as barriers to sexual hybridization are circumvented by techniques such as somatic fusion and as desirable genes are transferred more quickly into adapted germplasm by introgression and breeding at the diploid level. Derivatives of protoplast fusion and tissue culture technology may lead to limited chromosome transfer, and the use of *Agrobacterium tumifaciens Ti* plasmid-mediated transformation for the transfer of specific genes is already available, and has been used to produce transgenic potato plants resistant to viruses (Barker *et al.*, 1992; Huisman *et al.*, 1992). Other advances in biotechnology described in this book will no doubt play a role in accelerating progress in potato breeding. However, most economically important traits display continuous variation. In this respect the use of molecular markers to locate quantitative trait loci (QTLs) offers an exciting prospect. If these QTLs can be manipulated in the same way as Mendelian factors by following the segregation patterns of closely linked markers, this could substantially impact upon the number of cycles of hybridization and selection – and avoid the problems associated with selection for traits which are substantially modified in expression by environmental factors. Much will depend on the numbers of loci segregating and how useful molecular markers obtained from widely divergent diploid crosses, chosen to maximize genetic polymorphisms, will prove to be when applied to less polymorphic tetraploids of different genetic constitution.

In the real world of practical potato breeding, economics will ultimately determine the rate of progress and the means of achieving it. In this respect the continuing development and application of more genetically soundly based scientific methods to 'conventional' potato breeding have much to offer. Most authorities agree that the biotechnologically based methods will supplement, but not replace, traditional breeding methods, by adding more tools to the breeders' armoury.

References

Abdalla, M.M.F. and Hermsen, Th.J.G. (1971) A two-locus system of gametophytic incompatibility in *Solanum phureja* and *S. stenotomum*. *Euphytica* 20, 345-350.

Anon. (1986) *Plant Varieties and Seeds Gazette* 259, August.

Barker, H., Reavy, B., Kumar, A., Webster, K.D. and Mayo, M.A. (1992) Restricted virus multiplication in potatoes transformed with the coat protein gene of potato leafroll luteovirus: similarities with a type of host gene-mediated resistance. *Annals of Applied Biology* 120, 55-64.

Brown, J. and Caligari, P.D.S. (1988) The use of multivariate cross prediction methods in the breeding of a clonally reproduced crop (*Solanum tuberosum*). *Heredity* 60, 147-153.

Brown, J. and Caligari, P.D.S. (1989) Cross prediction in a potato breeding programme by evaluation of parental material. *Theoretical and Applied Genetics* 77, 246-252.

Brown, J., Caligari, P.D.S., Mackay, G.R. and Swan, G.E.L. (1987a) The efficiency of visual selection in early generations of a potato breeding programme. *Annals of Applied Biology* 110, 357-363.

Brown, J., Caligari, P.D.S. and Mackay, G.R. (1987b) The repeatability of progeny means in the early generations of a potato breeding programme. *Annals of Applied Biology* 110, 365-370.

Brown, J., Caligari, P.D.S., Dale, M.F.B., Swan, G.E.L. and Mackay, G.R. (1988) The use of cross prediction methods in a practical potato breeding programme. *Theoretical and Applied Genetics* 76, 33-38.

Brown, J., Mackay, G.R., Bain, H., Griffith, D.W. and Allison, M.J. (1990) The processing potential of tubers of the cultivated potato, *Solanum tuberosum* L., after storage at low temperatures. 2. Sugar concentration. *Potato Research* 33, 219-227.

Brown, J., Dale, M.F.B. and Mackay, G.R. (1992) Selecting potato clones in Scotland for general adaptability in England and countries around the Mediterranean. In: Rousselle-Bourgeois, F. and Rousselle, P. (eds) *Proceedings of the Joint Conference of the EAPR Breeding and Varietal Assessment Section and the Eucarpia Potato Section*. INRA-Ploudaniel, Landerneau, France, pp. 43-48.

Cadman, C.H. (1942) Autotetraploid inheritance in the potato: some new evidence. *Journal of Genetics* 44, 33-52.

Caligari, P.D.S. (1992) Breeding new varieties. In: Harris, P. (ed.) *The Potato Crop*, 2nd edn. Chapman and Hall, London, pp. 334-372.

Caligari, P.D.S. and Brown, J. (1986) The use of univariate cross prediction methods in the breeding of a clonally reproduced crop (*Solanum tuberosum*). *Heredity* 57, 395-401.

Caligari, P.D.S., Mackay, G.R., Stewart, H.E. and Wastie, R.L. (1984) A seedling progeny test for resistance to potato foliage blight (*Phytophthora infestans* (Mont.) de Bary). *Potato Research* 27, 43-50.

Caligari, P.D.S., Mackay, G.R., Stewart, H.E. and Wastie, R.L. (1985) Confirmatory evidence for the efficacy of a seedling progeny test for resistance to potato foliage blight (*Phytophthora infestans* (Mont.) de Bary). *Potato Research* 28, 439-442.

Carroll, C.P. (1982) A mass-selection method for the acclimatization and improvement of edible diploid potatoes in the United Kingdom. *Journal of Agricultural Science* 99, 631-640.

Carroll, C.P. (1987) The use of diploid *Solanum phureja* germplasm. In: Jellis, G.J.

and Richardson, D.E. (eds) *The Production of New Potato Varieties*. Cambridge University Press, Cambridge, pp. 231–234.

Carroll, C.P. and De Maine, M.J. (1989) The agronomic value of tetraploid F_1 hybrids between potatoes of group Tuberosum and group Phureja/Stenotomum. *Potato Research* 32, 447–456.

Chase, S.S. (1963) Analytic breeding in *Solanum tuberosum* L. – a scheme utilizing parthenotes and other diploid stocks. *Canadian Journal of Genetics and Cytology* 5, 359–363.

Cipar, M.S., Peloquin, S.J. and Hougas, R.W. (1964) Inheritance of incompatibility in hybrids between *Solanum tuberosum* haploids and diploid species. *Euphytica* 13, 163–172.

Cockerham, G. (1970) Genetical studies on resistance to potato viruses X and Y. *Heredity* 25, 309–348.

Cornish, M.A. (1990a) Selection during a selfing programme. I. The effects of a single round of selection. *Heredity* 65, 201–211.

Cornish, M.A. (1990b) Selection during a selfing programme. II. The effects of two or more rounds of selection. *Heredity* 65, 213–220.

Dalianis, C.D., Plaisted, R.L. and Peterson, L.C. (1966) Selection for freedom from after cooking darkening in a potato breeding program. *American Potato Journal* 43, 207–215.

De Maine, M.J., Bradshaw, J.E. and Caligari, P.D.S. (1992) Inheritance of external mechanical damage resistance of potato cultivars. *Annals of Applied Biology* 121, 379–384.

Dodds, K.S. (1965) The history and relationships of cultivated potatoes. In: Hutchinson, J.B. (ed.) *Essays in Crop Plant Evolution*. Cambridge University Press, Cambridge, pp. 123–141.

Dodds, K.S. and Paxman, G.J. (1962) The genetic system of cultivated diploid potatoes. *Evolution* 16, 154–167.

Douches, D.S. and Jastrzebski, K. (1993) Potato. In: Kalloo, G. and Bergh, B.O. (eds) *Genetic Improvement of Vegetable Crops*. Pergamon Press, Oxford, pp. 605–644.

Douches, D.S., Ludlam, K. and Freyre, R. (1991) Isozyme and plastid DNA assessment of pedigrees of nineteenth century potato cultivars. *Theoretical and Applied Genetics* 82, 195–200.

Glendinning, D.R. (1975a) Neo-Tuberosum: new potato breeding material. 1. The origin, composition, and development of the Tuberosum and Neo-Tuberosum gene pools. *Potato Research* 18, 256–261.

Glendinning, D.R. (1975b) Neo-Tuberosum: new potato breeding material. 2. A comparison of Neo-Tuberosum with unselected Andigena and with Tuberosum. *Potato Research* 18, 343–350.

Glendinning, D.R. (1975c) Neo-Tuberosum: new potato breeding material. 3. Characteristics and variability of Neo-Tuberosum, and its potential value in breeding. *Potato Research* 18, 351–362.

Glendinning, D.R. (1976) Neo-Tuberosum: new potato breeding material. 4. The breeding system of Neo-Tuberosum, and the structure and composition of the Neo-Tuberosum gene pool. *Potato Research* 19, 27–36.

Glendinning, D.R. (1981) Evaluate Neo-Tuberosum potatoes as parental material for use in breeding cultivars. *Scottish Crop Research Institute Annual Report 1981*, pp. 181–183.

Glendinning, D.R. (1983) Potato introductions and breeding up to the early 20th century. *New Phytologist* 94, 479–505.

Glendinning, D.R. (1987) Neo-tuberosum. *Scottish Crop Research Institute Annual Report 1987*, pp. 77–78.

Glendinning, D.R. (1989) Some aspects of autotetraploid population dynamics. *Theoretical and Applied Genetics* 78, 233–242.

Goodrich, C.E. (1863) The origination and test culture of seedling potatoes. *Transactions of New York State Agricultural Society* 23, 89–134.

Gopal, J., Gaur, P.C. and Rana, M.S. (1992) Early generation selection for agronomic characters in a potato breeding programme. *Theoretical and Applied Genetics* 84, 709–713.

Grun, P. (1979) Evolution of the cultivated potato: a cytoplasmic analysis. In: Hawkes, J.G., Lester, R.N. and Skelding, A.D. (eds) *The Biology and Taxonomy of the Solanaceae*. Academic Press, London, pp. 655–665.

Grun, P. (1990) The evolution of cultivated potatoes. *Economic Botany* 44(Suppl. 3), 39–55.

Hawkes, J.G. (1990) *The Potato: Evolution, Biodiversity and Genetic Resources*. Belhaven Press, London.

Hawkes, J.G. and Jackson, M.T. (1992) Taxonomic and evolutionary implications of the endosperm balance number hypothesis in potatoes. *Theoretical and Applied Genetics* 84, 180–185.

Haynes, F.L. (1980) Progress and future plans for the use of Phureja–Stenotomum populations. In: *Report of Planning Conference 1980. Utilization of the Genetic Resources of the Potato III*. International Potato Center, Lima, Peru, pp. 80–88.

Hermundstad, S.A. and Peloquin, S.J. (1987) Breeding at the $2x$ level and sexual polyploidization. In: Jellis, G.J. and Richardson, D.E. (eds) *The Production of New Potato Varieties*. Cambridge University Press, Cambridge, pp. 197–210.

HMSO (1991) The Seed Potatoes Regulations 1991, 1991 No. 2206 and 1992 No. 1031. The Seed Potatoes (Amendment) Regulations 1992. HMSO, Edinburgh.

Howard, H.W. (1978) The production of new varieties. In: Harris, P.M. (ed.) *The Potato Crop*, 1st edn. Chapman and Hall, London, pp. 607–646.

Huisman, M.J., Cornelissen, B.J.C. and Jongedijk, E. (1992) Transgenic potato plants resistant to viruses. *Euphytica* 63, 187–197.

Jansky, S.H., Yerk, G.L. and Peloquin, S.J. (1990) The use of potato haploids to put $2x$ wild species germplasm into a usable form. *Plant Breeding* 104, 290–294.

Jellis, G.J. (1992) Multiple resistance to diseases and pests in potatoes. *Euphytica* 63, 51–58.

Jellis, G.J. and Richardson, D.E. (eds) (1987) *The Production of New Potato Varieties*. Cambridge University Press, Cambridge.

Kehoe, H.W. (1982) Potato breeding. *An Foras Talun Research Report 1982*, pp. 42–48.

Killick, R.J. (1977) Genetic analysis of several traits in potatoes by means of a diallel cross. *Annals of Applied Biology* 86, 279–289.

Killick, R.J. and Malcolmson, J.F. (1973) Inheritance in potatoes of field resistance to late blight (*Phytophthora infestans* (Mont.) de Bary). *Physiological Plant Pathology* 3, 121–131.

Knight, T.A. (1807) On raising of new and early varieties of the potato (*Solanum tuberosum*). *Transactions of the Horticultural Society of London* 1, 57–59.

Large, E.C. (1940) *Advance of the Fungi*. Jonathan Cape, London.

Lee, H.C. (1985) Breeding new export potato varieties. *Agriculture in Northern Ireland* 60, 198–201.

Lewis, D. (1943) The physiology of incompatibility in plants. III. Autopolyploids. *Journal of Genetics* 44, 171–185.

Lewis, D. (1954) Comparative incompatibility in angiosperms and fungi. *Advances in Genetics* 6, 235-285.

Lunden, A.P. (1937) Arvelighetsundersokelser i potet. (Inheritance studies in the potato.) *Meldinger fra Norges Landbrukshøishkole* 17, 1-156.

Mackay, G.R. (1987a) Selecting and breeding for better potato cultivars. In: Abbott, A.J. and Atkin, R.K. (eds) *Improving Vegetatively Propagated Crops*. Academic Press, London, pp. 181-196.

Mackay, G.R. (1987b) Screening for resistance to diseases and pests. In: Jellis, G.J. and Richardson, D.E. (eds) *The Production of New Potato Varieties*. Cambridge University Press, Cambridge, pp. 88-90.

Mackay, G.R., Brown, J. and Torrance, C.J.W. (1990) The processing potential of tubers of the cultivated potato, *Solanum tuberosum* L., after storage at low temperature. I. Fry colour. *Potato Research* 33, 211-218.

Malcolmson, J.F. and Black, W. (1966) New R-genes in *Solanum demissum* Lindl. and their complementary races of *Phytophthora infestans* (Mont.) de Bary. *Euphytica* 15, 199-203.

Malcolmson, J.F. and Killick, R.J. (1980) The breeding values of potato parents for field resistance to late blight measured by whole seedlings. *Euphytica* 29, 489-495.

Maris, B. (1988) Correlations within and between characters between and within generations as a measure for the early generation selection in potato breeding. *Euphytica* 37, 205-224.

Maris, B. (1989) Analysis of an incomplete diallel cross among three ssp. *tuberosum* varieties and seven long-day adapted ssp. *andigena* clones of the potato (*Solanum tuberosum* L.). *Euphytica* 41, 163-182.

Mendel, G. (1865) *Experiments in Plant Hybridisation*, with an introduction by R.A. Fisher. 1965 centennial reprint, Oliver & Boyd, Edinburgh and London.

Mendoza, H.A. (1987) Advances in population breeding and its potential impact on the efficiency of breeding potatoes for developing countries. In: Jellis, G.J. and Richardson, D.E. (eds) *The Production of New Potato Varieties*. Cambridge University Press, Cambridge, pp. 235-245.

Mendoza, H.A. and Haynes, F.L. (1974) Genetic relationship among potato cultivars grown in the United States. *HortScience* 9, 328-330.

Muller, K.O. and Black, W. (1951) Potato breeding for resistance to blight and virus diseases during the last hundred years. *Zeitschrift für Pflanzenzuchtung* 31, 305-318.

Neele, A.E.F. (1990) Study on the inheritance of potato tuber yield by means of harvest index components and its consequences for choice of parental material. *Euphytica* 48, 159-166.

Neele, A.E.F. and Louwes, K.M. (1989) Early selection for chip quality and dry matter content in potato seedling populations in greenhouse or screenhouse. *Potato Research* 32, 293-300.

Neele, A.E.F., Barten, J.H.M. and Louwes, K.M. (1988) Effects of plot size and selection intensity on efficiency of selection in the first clonal generation of potato. *Euphytica* Suppl., 27-35.

Neele, A.E.F., Nab, H.J., de Jongh de Leeuw, M.J., Vroegop, A.P. and Louwes, K.M. (1989) Optimising visual selection in early clonal generations of potato based on genetic and economic considerations. *Theoretical and Applied Genetics* 78, 665-671.

Neele, A.E.F., Nab, H.J. and Louwes, K.M. (1991a) Identification of superior parents in a potato breeding programme. *Theoretical and Applied Genetics* 82, 264-272.

Neele, A.E.F., Nab, H.J. and Louwes, K.M. (1991b) Components of visual selection in early clonal generations of a potato breeding programme. *Plant Breeding* 106, 89–98.

Ortiz, R. and Ehlenfeldt, M.K. (1992) The importance of endosperm balance number in potato breeding and the evolution of tuber-bearing *Solanum* species. *Euphytica* 60, 105–113.

Ortiz, R., Peloquin, S.J., Freyre, R. and Iwanaga, M. (1991) Efficiency of potato breeding using FDR in gametes for multitrait selection and progeny testing. *Theoretical and Applied Genetics* 82, 602–608.

Phillips, M.S. (1981) A method of assessing potato seedling progenies for resistance to the white potato cyst nematode. *Potato Research* 24, 101–103.

Phillips, M.S. and Dale, M.F.B. (1982) Assessing potato seedling progenies for resistance to the white potato cyst nematode. *Journal of Agricultural Science* 99, 67–70.

Phillips, M.S., Wilson, L.A. and Forrest, J.M.S. (1979) General and specific combining ability of potato parents for resistance to the white potato cyst nematode (*Globodera pallida*). *Journal of Agricultural Science* 92, 255–256.

Plaisted, R.L. (1987) Advances and limitations in the utilization of Neotuberosum in potato breeding. In: Jellis, G.J. and Richardson, D.E. (eds) *The Production of New Potato Varieties*. Cambridge University Press, Cambridge, pp. 186–196.

Plaisted, R.L. and Hoopes, R.W. (1989) The past record and future prospects for the use of exotic potato germplasm. *American Potato Journal* 66, 603–627.

Plaisted, R.L., Sanford, L., Federer, W.T., Kehr, A.E. and Peterson, L.C. (1962) Specific and general combining ability for yield in potatoes. *American Potato Journal* 39, 185–197.

Plaisted, R.L., Thurston, H.D., Sieczka, J.B., Brodie, B.B., Jones, E.D. and Cetas, R.C. (1981) Rosa: a new golden nematode resistant variety for chipping and tablestock. *American Potato Journal* 58, 451–455.

Plaisted, R.L., Thurston, H.D., Brodie, B.B. and Hoopes, R.W. (1984) Selecting for resistance to diseases in early generations. *American Potato Journal* 61, 395–399.

Powell, W., Phillips, M.S., McNicol, J.W. and Waugh, R. (1991) The use of DNA markers to estimate the extent and nature of genetic variability in *Solanum tuberosum* cultivars. *Annals of Applied Biology* 118, 423–432.

Rasco, E.T., Jr, Plaisted, R.L. and Ewing, E.E. (1980) Photoperiod response and earliness of *S. tuberosum* ssp. *andigena* after six cycle of recurrent selection for adaptation to long days. *American Potato Journal* 57, 435–447.

Ross, H. (1979) Wild species and primitive cultivars as ancestors of potato varieties. *Proceedings of the Conference: Broadening the Genetic Base of Crops, 1978*, pp. 237–243.

Ross, H. (1986) *Potato Breeding – Problems and Perspectives*. Advances in Plant Breeding 13, Paul Parey, Berlin and Hamburg.

Rousselle-Bourgeois, F. and Rousselle, P. (1992) Création et sélection de populations diploide de pomme de terre (*Solanum tuberosum* L). *Agronomie* 12, 59–67.

Salaman, R.N. (1926) *Potato Varieties*. Cambridge University Press, Cambridge.

Simmonds, N.W. (1969) Prospects for potato improvement. *Scottish Plant Breeding Station Forty-Eighth Annual Report 1968–69*, pp. 18–38.

Simmonds, N.W. (1976) Potatoes. In: Simmonds, N.W. (ed.) *Evolution of Crop Plants*. Longman, London and New York, pp. 279–283.

Simmonds, N.W. (1985) Two-stage selection strategy in plant breeding. *Heredity* 55, 393–399.

Simmonds, N.W. and Malcolmson, J.F. (1967) Resistance to late blight in Andigena potatoes. *European Potato Journal* 10, 161–166.

Stewart, H.E., Taylor, K. and Wastie, R.L. (1983) Resistance to late blight in foliage (*Phytophthora infestans*) of potatoes assessed as true seedlings and as adult plants in the glasshouse. *Potato Research* 26, 363-366.

Stewart, H.E., Wastie, R.L., Bradshaw, J.E. and Brown, J. (1992) Inheritance of resistance to late blight in foliage and tubers of progenies from parents differing in resistance. *Potato Research* 35, 313-319.

Swieżyński, K.M. (1984) Early generation selection methods used in Polish potato breeding. *American Potato Journal* 61, 385-394.

Tai, G.C.C. (1976) Estimation of general and specific combining abilities in potato. *Canadian Journal of Genetics and Cytology* 18, 463-470.

Tai, G.C.C. and Young, D.A. (1984) Early generation selection for important agronomic characteristics in a potato breeding population. *American Potato Journal* 61, 419-434.

Tarn, T.R. and Tai, T.C.C. (1983) Tuberosum × Tuberosum and Tuberosum × Andigena potato hybrids: comparisons of families and parents, and breeding strategies for Andigena potatoes in long-day temperate environments. *Theoretical and Applied Genetics* 66, 87-91.

Tarn, T.R., Tai, G.C.C., De Jong, H., Murphy, A.M. and Seabrook, J.E.A. (1992) Breeding potatoes for long-day, temperate climates. In: Janick, J. (ed.) *Plant Breeding Reviews 9*. John Wiley & Sons, New York, pp. 217-332.

Toxopeus, H.J. (1953) On the significance of multiplex parental material in breeding for resistance to some diseases in the potato. *Euphytica* 2, 139-146.

Tung, P.X. (1992) Genetic variation for bacterial wilt resistance in a population of tetraploid potato. *Euphytica* 61, 73-80.

Veilleux, R.E. and Lauer, F.I. (1981) Breeding behaviour of yield components and hollow heart in tetraploid-diploid vs. conventionally derived potato hybrids. *Euphytica* 30, 547-561.

Vermeer, H. (1990) Optimising potato breeding. I. The genotypic, environmental and genotype-environmental coefficients of variation for tuber yield and other traits in potato (*Solanum tuberosum* L.) under different experimental conditions. *Euphytica* 49, 229-236.

Vermeer, H. (1991) Optimising potato breeding. II. A model for optimising single-stage and single-trait selection. *Euphytica* 53, 151-157.

Wastie, R.L. (1991a) Breeding for resistance. In: Ingram, D.S. and Williams, P.H. (eds) *Phytophthora infestans, the Cause of Late Blight of Potato. Advances in Plant Pathology* 7, 193-224.

Wastie, R.L. (1991b) Resistance to powdery scab of seedling progenies of *Solanum tuberosum*. *Potato Research* 34, 249-252.

Wastie, R.L., Caligari, P.D.S., Stewart, H.E. and Mackay, G.R. (1987) A glasshouse progeny test for resistance to tuber blight. *Potato Research* 30, 533-538.

Wastie, R.L., Mackay, G.R., Caligari, P.D.S. and Stewart, H.E. (1990) A glasshouse progeny test for resistance to gangrene (*Phoma foveata*). *Potato Research* 33, 131-133.

Waugh, R., Glendinning, D.R., Duncan, N. and Powell, W. (1990) Chloroplast DNA variation in European potato cultivars. *Potato Research* 33, 505-513.

22 Breeding Potatoes Based on True Seed Propagation

A.M. GOLMIRZAIE, P. MALAGAMBA AND
N. PALLAIS

International Potato Center (CIP), Apartado Postal 5969, Lima, Peru.

Introduction

The potato was domesticated for vegetative propagation in the high Andes of South America (Ochoa, 1990). This method of propagation proved to be quite adaptable to the environmental conditions of temperate latitudes, where potatoes soon became a staple food, but not for the conditions of developing countries in the torrid zone, in which potatoes are still mostly consumed as a luxury food (Horton, 1987). Although about 30% of the world's total potato volume produced is grown in tropical regions, yields and per capita consumption are usually less than half those obtained in the developed countries located in the temperate zone (Horton, 1987). Various alternatives for increasing the production and consumption of this nutritious food in the tropics and subtropics have been investigated at the International Potato Center (CIP) during the last 20 years (Accatino and Malagamba, 1982), including the use of true potato seed (TPS) from sexual reproduction.

The principal advantage of using TPS over seed tubers lies in the considerable reduction of seed costs due to the much smaller amounts of planting material needed. Depending on the method used, from 50 to 250 g of TPS could be used to replace the 2 tonnes or so of seed tubers that are needed to grow 1 hectare of potatoes. Another very important advantage of TPS is its flexibility of planting time, which is especially appropriate to the conditions of the high-altitude regions of the tropics. Although potatoes could often be planted throughout most of the year in many of these regions, farmers must plant all of their imported seed tubers at the same time, as dictated by the physiological age of the tubers. On the other hand, TPS can be sown successively at intervals and in amounts that would best suit the specific needs of farmers. This would tend to decrease storage costs and stabilize prices by providing a more continuous supply of potatoes in the market-place.

Table 22.1. Comparison of TPS and seed tuber systems (adapted from Golmirzaie et al., 1990).

	TPS	Seed tuber
Seed or requirements according to tuber size and planting density	50-250 g of seed ha^{-1}	1000 to 2000 kg ha^{-1}
Disease situation	Free of fungi, bacteria, nematodes, insects and most viruses except potato spindle tuber viroid (PSTV) and PVT	May carry nematodes, insects, fungi, bacteria, viruses and viroids
Labour requirement during establishment	High during the initial phase of the crop	Low requirements, mechanized
Early growth	More susceptible to weed competition, pests, diseases and abiotic stresses; irrigation is required	Less susceptible to stresses due to high vigour and uniformity
Tuber maturation	Late and variable (15-20 days later than clonal material)	Early and uniform
Tuber uniformity	Low, tubers less suitable for some industrial processing (potato chips, french fries, etc.)	High, tubers well adapted for industrial processing
Cost of production	Reduced cost due to elimination of seed tuber storage and transportation costs	Higher cost due to seed tuber transportation and storage costs
Fitness to cropping systems	Easier fitness to the cropping systems, because the date of sowing does not depend on tuber ageing	Narrow margin of fitness to the cropping systems

Worldwide interest in the potential of TPS technology has grown very fast among CIP's client countries. Currently, more than 50 national potato programmes are involved in collaborative on-farm research to test and adapt TPS technology to local conditions. Commercial potato production via TPS is increasing significantly in a number of countries (Malagamba, 1988). Private commercial production of TPS varieties is also expanding fast. TPS technology is at the starting-gate of the race to develop a new green revolution, designed specifically to improve the production and consumption of potatoes in the developing countries of the torrid zone.

Propagating potatoes sexually offers a tremendous advantage over existing systems of potato production where tubers are used as the means of propagation. Growing potatoes from botanical seed can release 2 tonnes ha^{-1} for planting for human consumption. Table 22.1 shows a comparison of these two potato propagation systems. Clearly, propagation from TPS may be more advantageous than from tubers in some areas of the tropics, particularly because TPS costs less and offers flexibility in any agricultural system.

CIP began breeding for potato production from sexual seed in 1972. In 1975, genetic variances and heritabilities for various criteria unique to sexual propagation were estimated for populations of *Solanum tuberosum* at CIP (Thompson et al., 1983). In 1977, the selection of TPS adapted to tropical and subtropical environments became a research priority at CIP (Mendoza, 1984). The strategy was to focus on the selection of TPS progenies capable of producing high potato yields with acceptable uniformity. By 1984, several promising progenies adapted to short-day conditions and with moderate to high levels of disease resistances had been selected at CIP (Mendoza, 1983). In an effort to make TPS potato production a reality, in this chapter we are presenting some theoretical and practical approaches developed by scientists at CIP and other institutions around the world.

Genetic Aspects

The cultivated potato is a complex tetraploid which exhibits tetrasomic inheritance (Swaminathan and Howard, 1953; Howard, 1970). The theory of inbreeding and crossbreeding in such tetraploids is more complicated than in diploids. For example, interactions can occur between more than two alleles at a locus and gametes can contain two alleles which are identical by descent (see chapter 3 by Bradshaw for details). These differences have consequences for breeding strategies aimed at exploiting heterosis (hybrid vigour), particularly where sexual rather than asexual reproduction is being used.

Effects of inbreeding

The primary effect of inbreeding is to reduce the frequency of heterozygous genotypes. The frequency of tetra-allelic genotypes declines rapidly under selfing, being multiplied by $\frac{1}{6}$ in each generation, in the absence of double reduction. The frequency of triallelic genotypes also declines rapidly (Bennett, 1976). Busbice and Wilsie (1966) suggested that, when there is a high frequency of loci containing three or four different alleles, a rapid loss of vigour might occur on inbreeding as a result of these decreases. In their genetic model, the genotypic value of a locus carrying four different alleles is the sum of four additive gene values plus 11 interaction values, namely six possible two-gene interactions (first-order), four three-gene interactions (second order) and one four-gene interaction (third-order). The value of manoallelic genotypes is simply the sum of four identical additive gene values. With one generation of selfing ($F = 0.167$), approximately 83% of the third-order, 67% of the second-order and 47% of the first-order interactions are lost at tetra-allelic loci, without any increase in homozygosity.

According to Jackson *et al.* (1984), the commonest explanation of inbreeding depression is that it is due to the fixation of unfavourable or deleterious recessives. Inbreeding increases the frequency of homozygous loci, some of which will become homozygous for these deleterious recessives. Jackson *et al.* (1984) cited Jinks and Lawrence (1983), who indicated that inbreeding depression is due to the fact that the characters of interest are determined by genes with non-additive effects, including epistasis, and those genes with deleterious effects are only a subset. These genes will segregate upon inbreeding. The association between inbreeding depression and homozygosity is explained by two rival hypotheses: partial/complete dominance and overdominance (Krebs and Hancock, 1990). In the first hypothesis, inbreeding depression results from the fixation of recessive or partially recessive deleterious alleles at loci encoding the trait of interest (mutational load). The overdominance hypothesis attributes inbreeding depression to the loss of favourable allelic interactions at these loci (segregational or balanced load).

Experimental results in the potato and other crops have shown yield and fertility depression through different generations of inbreeding (Krantz and Hutchins, 1929; Krantz, 1951; Busbice and Wilsie, 1966; Howard, 1970; Hecker, 1972; Atlin, 1985b; Mendoza and Marca, 1986). However, several researchers have reported differences between the responses of lines to inbreeding (Krantz and Hutchins, 1929; Krantz, 1951; Jackson *et al.*, 1984). Apparently some lines might be able to 'resist' the inbreeding effect, thus offering us the possibility of using them in a TPS breeding programme with expected favourable results.

Heterosis

Heterosis or hybrid vigour is the converse of inbreeding depression. The general theory of heterosis was first developed for diploid species (Jinks, 1983). Two hypotheses have been put forward in order to explain heterosis, namely the dominance/complementary epistasis and the overdominance hypothesis. Mendoza and Haynes (1974) analysed the genetic basis of heterosis for yield in potato. They suggested that the overdominant model, which implies a close positive correlation between heterozygosity and yield, better explains this phenomenon. Amoros and Mendoza (1979) later demonstrated this theory. They worked with populations with different levels of heterozygosity and found higher heterotic levels with the most heterozygous materials. The same explanation was also suggested by Chase (1963) and Mendiburu *et al.* (1974). They proposed the use of four divergent diploids (including dihaploids of *S. tuberosum* group *Andigena* and group *Tuberosum*) in order to create a tetraploid with genomes from the four different parents that maximizes the number of tetra-allelic loci. In the proposed schemes, $2n$ gametes formed by first-division restitution (FDR) in meiosis play an important role, because FDR results in about 83% of the potential heterozygosity being transmitted to the hybrid. However, Sanford and Hanneman (1982) suggested the existence of a heterotic threshold. They synthesized complex hybrids containing genomes from three different *S. tuberosum* groups (three-way hybrids), utilizing $2n$ gametes in $4x$–$2x$ crosses, and compared them with two-way and one-way hybrid families. The three-way hybrids were never significantly superior to the two-way hybrids for vigour, yield or tuber type. They concluded that a simple and direct two-way hybridization approach may be optimal. Crossing unrelated genetic material as a means of exploiting heterosis in a TPS programme offers promising perspectives for rapid progress.

Breeding Strategies

The main objective in a TPS breeding programme is to obtain uniform progenies for agronomic and reproductive characters with resistance to important diseases (Golmirzaie and Mendoza, 1988). Several different methods can be used to reach this goal: inbreeding, using diploid parental lines that produce $2n$ gametes, or using tetraploid parental lines (Mendoza, 1980).

Inbreeding

With inbreeding, the highest level of gametic uniformity can be achieved. However, when two inbred and non-related autotetraploid potato genotypes are crossed, the hybrid has an inbreeding coefficient of $F = 0.33$ (Mendoza, 1980). This is because the effect of parental inbreeding partially remains in the progenies and does not disappear as in a diploid, where the inbreeding coefficient in the progenies will be zero.

While inbreeding gives the highest level of gametic uniformity and homogeneity, resulting in completely uniform progenies, it also causes significant yield reductions and poor performance stability. This was confirmed by Golmirzaie *et al.* (1987), who studied the inbreeding effect on yield and agronomic traits in different inbred generations of TPS. The individual potato families expressing inbreeding depression varied between the measured characters. The results showed that it is possible to select tetraploid families that are less depressed by inbreeding for different agronomical and reproductive characters. Other researchers found the same response (Atlin, 1985b; Mendoza and Marca, 1986). These results would indicate that inbreeding is a feasible strategy for producing improved TPS genetic materials (or progenies).

Use of diploid parental lines

Peloquin *et al.* (1985) found mutants in diploid potatoes which display abnormalities during the meiotic process, such that gametes with the somatic chromosome number $2n$ are produced. The mechanisms for the production of these gametes have been identified in microsporogenesis (Ramanna, 1974; Mok and Peloquin, 1975; Iwanaga, 1985) and in megasporogenesis (Iwanaga and Peloquin, 1979; Jongedijk, 1985). The genetic control of $2n$ egg formation is not known, but that of $2n$ pollen has been well established (Watanabe and Peloquin, 1988). In diploid potatoes $2n$ gametes are produced by modified meiosis, genetically equivalent to first-division restitution (FDR) or second-division restitution (SDR) (Iwanaga et al., 1989). FDR gametes transmit about 83% of their heterozygosity and are more genetically uniform than SDR gametes (Peloquin, 1979). The importance of $2n$ gametes is not only to achieve maximum heterozygosity in progenies of $4x-2x$ matings, but also to produce adequate homogeneity in the progenies by an appropriate choice of tetraploid parent (Golmirzaie and Mendoza, 1988; Carroll and De Maine, 1989). Peloquin (1983) presented some breeding schemes aimed at producing TPS. One of these schemes involved obtaining $4x$ progeny from $4x-2x$ FDR, using selected $2x$ hybrids (with good agronomic and reproductive characters) from matings between dihaploids ($2n = 2x = 24$) of tetraploid cultivars or advanced selections (selected for fertility and $2n$ gametes) and introductions of cultivated or wild 24-chromosomes species (selected for $2n$ gamete formation and desired traits, such as disease resistance). These selected $2x$ hybrids are crossed to cultivars which possess good adaptation and are unrelated to the dihaploids involved in the $2x$ hybrid. The $4x$ hybrids from the $4x-2x$ crosses obtain their adaptation and horticultural type from the cultivar, and the desired traits and allelic variation from the dihaploid-species hybrid. Several researchers (Mendoza, 1980; Macaso-Khwaja and Peloquin, 1983; Ortiz and Iwanaga, 1986) have confirmed the utility of FDR ($2n$) pollen in $4x-2x$ crosses as an appropriate mechanism to obtain uniformity and vigour, two characteristics that are required for the successful utilization of TPS. Superior yields of $4x-2x$ progenies have been reported (Mok and Peloquin, 1975; De Jong and Tai, 1977; Mendiburu and Peloquin, 1977; McHale and Lauer, 1981; Veilleux and Lauer, 1981; Masson and Peloquin, 1987). This method is therefore superior to the use of an inbreeding system for producing TPS. There are, however, some limitations in the use of diploids that need to be considered before they can be proposed for use as TPS parental lines: a maturity period that is too long, and agronomic traits that need attention in the breeding of diploid clones. These clones can best be used to transfer resistance and other specific attributes, often found only in diploid species, to the tetraploid progenies.

Use of tetraploid parental lines

Tetraploid parental materials are commonly utilized to produce TPS (Golmirzaie and Mendoza, 1988). Although the meiotic process produces heterogeneous gametes that will result in heterogeneous progenies, there are tetraploid progenies from $4x-4x$ crosses with acceptable uniformity and high yield; the latter trait is particularly common in those combinations of parental lines having a wide genetic background (Golmirzaie and Mendoza,

1985). Research on the utilization of tetraploid parents carried out at CIP includes: (i) selection of the best cultivars for TPS use from available breeding material; (ii) evaluation of different clones for TPS parental line selection; and (iii) development of different types of cultivars for use by farmers either as hybrids, as synthetic lines or for open pollination (OP). The production of cybrids (cytoplasmic hybrids) for transferring male sterility into TPS female parents is also being explored (Galun et al., 1988).

Selection of best cultivars

Many progenies were evaluated in different mating designs for a variety of traits at several locations in Peru to identify the progenies as well as the progenitors to be used in TPS production. The type of genic action was evaluated for each trait: germination; plant vigour, shape and uniformity; earliness; and tuber uniformity for colour, shape, and size. Thompson et al. (1983) found non-additive effects for tuber yield in a Neotuberosum population but the heritability estimates for yield components and number and size of tubers were still high. Their results indicate that yield can be increased by selecting parents with a high-yielding capacity. Golmirzaie and Mendoza (1985) reported that heritability estimates for plant vigour and tuber yield, uniformity and shape were also high. They remarked that recurrent selection cycles do not decrease genetic variability in the short term and that steady gains are expected from further selection for these traits.

Selection for TPS parental lines

To remove the constraints on the expansion of TPS production programmes, TPS breeders should select clones with good agronomic and reproductive characters. A breeding strategy based on these priorities would therefore concentrate on the selection of parental lines that produce flowers and set berries under tropical conditions. A CIP project is currently under way to evaluate and select parental lines for agronomic and reproductive characters under normal short-day length and high temperatures at San Ramon, Peru (800 m altitude, 11°08′ latitude). The steps for this strategy are presented in Table 22.2. By using this strategy, TPS parental lines that flower well and set berries under warm tropical conditions can be selected (Table 22.3). These TPS parental lines can be easily used by the farmer to obtain open-pollinated TPS or by any other programme to produce hybrids or synthetic progenies (these progenies are obtained from a limited number of parental lines planted together under isolation in the field to allow free intermating).

Development of different types of cultivars

Farmers in developing countries can receive TPS either from a commercial producer or from a national public programme that produces primarily hybrid

Table 22.2. Factors included in developing parental lines for TPS production.

Selecting clones from different sources

Evaluating selected clones for agronomic characters
    Plant type
    Earliness
    Tuber colour, shape and size

Evaluating selected clones for TPS characters
    Flower initiation
    Flowering intensity
    Duration of flowering
    Number of flowers per inflorescence
    Style length
    Type of anther
    Flowers' attraction for bees
    Pollen production
    Fruit set
    Seed set
    Seed weight

Screening clones for disease and insect resistance

Crossing clones by using different mating systems

Evaluating progenies from selected clones
    Germination
    Plant vigour, shape and uniformity
    Earliness
    TPS characters of segregation
    Tuber colour, shape and uniformity
    Progeny evaluation for disease and insect resistance

Table 22.3. Performance of the ten highest-yielding TPS parental lines in San Ramon, 1987.

Parental line	Weight per plant (kg)	No. of tubers per plant	DF	FI	FD	SL	AT	BA	PP
GS-72	1.363	10.5	42	3	3	3	2	2	1
GS-251	1.069	21.2	32	3	5	5	2	2	5
GS-593	1.060	12.5	47	3	5	5	2	2	5
GS-8	1.044	10.0	37	7	7	4	2	2	5
GS-620	1.019	14.0	40	3	3	4	2	2	5
GS-215	1.011	13.3	40	5	3	5	1	1	1
GS-697	1.033	15.2	26	5	3	4	2	2	3
GS-723	1.000	6.2	31	5	7	5	2	2	3
GS-227	0.998	16.5	29	5	7	4	2	2	5
GS-477	0.992	20.7	40	3	3	5	2	2	3

DF = days to flowering; FI = flowering intensity; FD = flowering duration; SL = style length; AT = anther type; BA = bees attraction; PP = pollen production.
Scale: 1 to 9 for DF, FI, FD, SL (1 = low, 9 = high); 1 to 5 for PP (1 = low, 5 = high); for AT (1 = distorted, 2 = normal); for BA (1 = yes, 2 = no).

TPS. Farmers can also produce their own TPS by collecting the seed resulting from natural pollination in the field (OP).

Hybrids

These progenies are produced, by controlled pollination, from parents that have good general combining ability (GCA) for agronomic and reproductive characters. Hybrid progenies have been evaluated at different locations in Peru as well as in CIP's regions. Golmirzaie and Mendoza (1985) have reported on the necessity of testing the parental clones for different traits in many locations before they are recommended as potential TPS parental lines. Mendoza (1985) stated that, for the use of TPS in commercial potato production, it is important to identify parental clones with a high GCA for yield and tuber uniformity. However, one of the most important traits for these lines is earliness in their progenies. A potato crop grown from TPS has a tendency to have late maturity due to the heterogeneity of the parental lines. Since tropical areas have the best potential for adopting TPS, selected hybrids with earliness will make the use of TPS economically feasible in these areas (Golmirzaie, 1987).

Synthetics

A promising method for producing low-cost TPS is to create a synthetic population. This is achieved by identifying parents with good GCA for agronomic and reproductive characters and creating different synthetics, depending on the number of parents originally included. Selected TPS parental lines are planted together and isolated from other potato crops in the field. A variable number of parental lines can be used (2, 4, 6 or more) to produce natural synthetics.

In another approach, synthetic population (controlled) crosses are made to produce a two-parent population and the mixture concept used to obtain four- and six-parent populations (a mixture is an agronomically uniform TPS population, produced by bulking equal amounts of seed from various cross combinations). Farmers would be provided with a mixture of TPS of non-related parents with good male fertility and good adaptation.

The performance of the synthetic TPS lines depends on the rate of selfing and the number of parental lines in the initial population (Golmirzaie, 1987). By studying different types of synthetic lines (natural as well as controlled) in relation to the number of parents (2, 4, 6 or more) in the base population, Golmirzaie (1987) showed that the performance of the progeny from using two parents to produce natural or controlled synthetic lines was as good as or better than using four or more parents. Therefore, by selecting the right parental lines with high GCA, it is possible to obtain a high performance of TPS in a naturally produced synthetic population with only two parents (Table 22.4).

Open pollinated (OP) progenies

This is the simplest type of genetic material used in TPS production and it can be used by farmers in developing countries. The cost of OP seed is significantly lower than that of hybrid seed, and it does not require special

Table 22.4. Results from testing different combinations of parental lines in natural synthetic (N) and control synthetic (C) populations. San Ramon, 1986.

No. of parents	No. of plants at harvest		No. of tubers		Total weight (kg)	
	N	C	N	C	N	C
Two	19	24	241	321	11	12
Four	21	25	277	308	11	11
Six	19	26	234	343	10	12

skills or large investments in land and facilities. Therefore, selection of relatively high-yielding progenies produced by OP is the first step in establishing a TPS production scheme. Because the yield of OP progenies tends to be lower than that of hybrid progenies, the yielding ability of progenies needs to be increased. Open-pollinated seed could be obtained from two different sources; one is to harvest seed from F_1 hybrid families and to continue with OP generations; and the other is to harvest seed from adapted clones and to continue with OP generations. Several studies have shown that, in general, hybrid TPS progenies perform better than OP progenies (Mendoza, 1984; Kidane-Mariam et al., 1985). OP progenies also show decreased performance through successive generations. Atlin (1985b) suggested that these progenies are substantially inbred because seed in the first generation and subsequent generations results from selfing or matings between relatives. However, there are certain genotypes whose OP progenies perform just as well and which do not show any yield and/or vigour reduction when taken from the OP_1 generation to other successive generations of OP. A possible explanation of these results is that inbred individuals are less viable and fertile than heterozygous ones, and they therefore contribute less to the next generation. The OP seed obtained for the next generation comes from individuals that are the most heterozygous among the selfed group, or they result from an outcross. This type of genotype can be selected and given to farmers in developing countries, who could then use OP seed from their own crops (Golmirzaie, 1987). By selecting early at the seedling stage for the most vigorous individuals, it is possible to increase the performance of OP progenies for agronomic and reproductive characters by more than 35% (Golmirzaie and Mendoza, 1986). Experiments were carried out in 1985 in three locations of Peru - Huancayo, Lima and San Ramon - to confirm this result. Open-pollinated seeds were collected from ten clones, and five populations of 150, 300, 450, 600 and 750 seeds were obtained from each clone and grown in trays. A total of 120 seedlings were taken from each combination of clone and population. These seedlings were transplanted into the field with selection pressures of 16%, 20%, 27%, 40%, and 80%. The rate of improvement for agronomic and reproductive characters in the parents was linearly related to the rate of selection for seedling vigour in the population prior to transplanting to the field. It is possible that farmers could conduct such a selection process and maintain their stock of seeds through repeated cycles of natural pollination (OP) without any reduction in yield or other important characters.

CYBRIDS

Around 50% to 60% of the TPS production cost originates from the emasculation of the flowers of the female parent. One method of producing low-cost seeds is by using cybrid TPS populations from protoplast fusion. A cybrid plant is made up of cells of hybrid origin that presumably contain nuclei from one (or both) mother cell types, as well as some portion of the cytoplasm of the donor cells (Maliga, 1980). With this method, cytoplasmic male sterility (CMS) of the tetrad type is transferred from a clone (donor) into a TPS parental line (recipient) (Galun et al., 1988). An important aspect of this method is that tetrad sterile pollen attracts bees and therefore natural cross-pollination is encouraged. This type of cybrid population can be produced in the same way that F_1 hybrid maize is produced, by interplanting female and male rows of potato in various combinations. Berries should be collected from the female rows where they have formed by natural pollination, since their sterile pollen attracts bumble-bees. When feasible, hand pollination can be carried out without emasculation in this type of population (Golmirzaie et al., 1991).

Summary and Conclusions

The autotetraploid nature of the potato makes its manipulation difficult when aiming at sexual propagation. Heterosis, the converse of inbreeding depression, should, however, be exploited in a TPS programme. Breeding strategies have the purpose of producing high-yielding and uniform progenies. Inbreeding is not a feasible strategy for hybrid TPS production because of the remnant level of inbreeding present in the progenies coming from parental lines obtained by this methodology. Diploid parental lines that produce unreduced gametes ($2n$ gametes) transmit high heterotic levels and resistance to pest and diseases to their progeny in $4x$–$2x$ crosses. For the best results, however, agronomic characters in the $2x$ parental material should be improved before their use (prebreeding work), and well-adapted $4x$ material (advanced lines or varieties) should be used as female parent. Use of suitable tetraploid parental lines have given good results in the $4x$–$2x$ breeding scheme. Parents should have high GCA for yield, and good agronomic and reproductive characters. Among the types of genetic materials (hybrids, open-pollinated, synthetics) to be given to the farmers, F_1 hybrids have performed better than open-pollinated material, although some OP progenies have given similar or even better yield than hybrids. Synthetic varieties composed of two progenitors and produced under controlled or natural conditions have behaved equal to or better than those with four progenitors. The possibility of using cybrids is emerging as an effective method of producing low-cost seeds. Cytoplasmic male sterility (CMS) of the tetrad type may be transferred from a clone (donor) into a TPS parental line (recipient) without any modification of the agronomic attributes of the recipient. In this way, emasculation can be avoided and, since tetrad sterile pollen attracts bees, they can be used to perform the pollination, or it may be done by hand.

References

Accatino, P. and Malagamba, P. (1982) *Potato Production from True Seed*. International Potato Center, Lima, Peru.

Amoros, W.R. and Mendoza, H.A. (1979) Relationship between heterozygosity and yield in autotetraploid potatoes. *American Potato Journal* 56, 455. (Abstract).

Atlin, G. (1985a) Farmer maintenance in TPS varieties. In: *Report of Planning Conference on Innovative Methods for Propagating Potatoes, 10-14 December 1984, Lima, Peru*. International Potato Center, Lima, Peru, pp. 39-62.

Atlin, G. (1985b) Inbreeding in TPS progenies: implications for breeding and seed production strategy. In: *Present and Future Strategies for Potato Breeding and Improvement. Report of the XXVI Planning Conference, 12-14 December 1983, Lima, Peru*. International Potato Center, Lima, Peru, pp. 71-85.

Bennett, J.H. (1976) Expectations for inbreeding depression on self-fertilization of tetraploids. *Biometrics* 32, 449-452.

Busbice, T.H. and Wilsie, C.P. (1966) Inbreeding depression and heterosis in autotetraploids with application to *Medicago sativa* L. *Euphytica*, 15, 52-67.

Carroll, C.P. and De Maine, M.J. (1989) The agronomic value of tetraploid F_1 hybrids between potatoes of group Tuberosum and group Phureja/Stenotomum. *Potato Research* 32, 447-456.

Chase, S.S. (1963) Analytical breeding of *Solanum tuberosum*. *Canadian Journal of Genetics and Cytology* 5, 359-363.

De Jong, H. and Tai, G.C.C. (1977) Analysis of tetraploid-diploid hybrids in cultivated potatoes. *Potato Research* 20, 111-121.

Galun, E., Perl, A. and Aviv, D. (1988) Protoplast fusion-mediated transfer of male sterility and other plasmone-controlled traits. In: *Applications of Plant Cell and Tissue Culture*. Ciba Foundation Symposium 137, Wiley, Chichester, pp. 97-112.

Golmirzaie, A.M. (1987) Performance of TPS synthetic populations. *American Potato Journal* 64, 440 (Abstract).

Golmirzaie, A.M. and Mendoza, H.A. (1985) Earliness in TPS progenies. *True Potato Seed (TPS) Letter* 6(2), 1.

Golmirzaie, A.M. and Mendoza, H.A. (1986) Effect of early selection for seedling vigor on open-pollinated true potato seed. *American Potato Journal*, 63, 426 (Abstract).

Golmirzaie, A.M. and Mendoza, H.A. (1988) Breeding strategies for true potato seed production. *CIP Circular* 16(4), 1-8.

Golmirzaie, A.M., Bretschneider, K. and Ortiz, R. (1987) Inbreeding effect on the production and agronomical characters of different true potato seed generations. In: *Abstracts of the Tenth Triennial Conference of the European Association for Potato Research*, Aalborg, Denmark, 26-31 July, pp. 294-295.

Golmirzaie, A.M., Ortiz, R. and Serquen, F. (1990) Genetica y Mejoramiento de la Papa mediante semilla (sexual). International Potato Center, Lima, Peru.

Golmirzaie, A.M., Tenorio, J. and Serquen, F. (1991) Organelle transfer (mitochondrias) for the production of cybrid potatoes and their possible use as TPS parental lines. *Report of the Planning Conference on Application of Molecular Techniques* to *Potato Germplasm Enhancement*. International Potato Center (CIP), Lima, Peru, pp. 121-128.

Hecker, R.J. (1972) Inbreeding depression in diploid and autotetraploid sugarbeet, *Beta vulgaris* L. *Euphytica* 21, 106-111.

Horton, D.G. (1987) *Potatoes, Production, Marketing and Programs in Developing Countries*. Westview Press, Boulder, Colorado.

Howard, H.W. (1970) *Genetics of the Potato Solanum tuberosum*. Logos Press, London.

Iwanaga, M. (1985) Ploidy level manipulation approach: development of diploid populations with specific resistance and FDR $2n$ pollen production. In: *Present and Future Strategies for Potato Breeding and Improvement. Report of the XXVI Planning Conference, 12-14 December 1983*. International Potato Center, Lima, Peru.

Iwanaga, M. and Peloquin, S.J. (1979) Synaptic mutant affecting only megasporogenesis in potatoes. *Journal of Heredity* 70, 385-389.

Iwanaga, M., Jatala, P., Ortiz, R. and Guevara, E. (1989) Use of FDR $2n$ pollen to transfer resistance to root-knot nematodes into cultivated $4x$ potatoes. *Journal of the American Society of Horticultural Sciences* 114, 1008-1013.

Jackson, M.T., Taylor, L. and Thompson, A.J. (1984) Inbreeding and true potato seed production. In: *Report of Planning Conference on Innovative Methods for Propagating Potatoes, 10-14 December, Lima, Peru*. International Potato Center, Lima, Peru.

Jinks, J.L. (1983) Biometrical genetics of heterosis. In: Frankel, R. (ed.) *Heterosis. Monographs on Theoretical and Applied Genetics 6*. Springer-Verlag, Berlin, pp. 1-46.

Jinks, J.L. and Lawrence, M.J. (1983) The genetical basis of inbreeding depression and heterosis; its implications for plant and animal breeding. *PORIM Bulletin*, Kuala Lumpur, Malaysia.

Jongedijk, E. (1985) The pattern of megasporogenesis and megagametogenesis in diploid *Solanum* species hybrids: its relevance to the origin of $2n$-eggs and the induction of apomixis. *Euphytica* 34, 599-611.

Kidane-Mariam, H.M., Arndt, G.C., Macaso-Khwaja, A.C. and Peloquin, S.J. (1985) Comparisons between $4x \times 2x$ hybrid and open pollinated true potato seed families. *Potato Research* 28, 35-42.

Krantz, F.A. (1951) Potato breeding in the United States. *Zeitschrift für Pflanzenzüchtung* 29(3), 388-393.

Krantz, F.A. and Hutchins, A.E. (1929) Potato breeding methods II. Selection in inbred lines. *Minnesota Agricultural Experimental Station, Technical Bulletin* 58.

Krebs, S.L. and Hancock, J.F. (1990) Early-acting inbreeding depression and reproductive success in the highbush blueberry, *Vaccinium corymbosum* L. *Theoretical and Applied Genetics* 79, 825-832.

Macaso-Khwaja, A.C. and Peloquin, S.J. (1983) Tuber yields of families from openpollinated and hybrid true potato seed. *American Potato Journal* 60, 645-651.

McHale, N.A. and Lauer, F.J. (1981) Breeding value of $2n$ pollen from diploid hybrids and *Phureja* in $4x$-$2x$ crosses in potatoes. *American Potato Journal* 58, 365-375.

Malagamba, P. (1988) Potato production from true seed in tropical climates. *Hort Science* 23, 495-500.

Maliga, P. (1980) The need and the search for genetic markers in plant cell cultures. In: Salov, F. *et al.* (eds) *Plant Cell Cultures: Results and Perspectives*. Elsevier/North Holland Biomedical Press, Amsterdam, pp. 107-114.

Masson, M.F. and Peloquin, S.J. (1987) Heterosis for tuber yields and total tuber solids content in $4x \times 2x$ FDR-CO crosses in potato. In: Jellis, G.J. and Richardson, D.E. (eds) *The Production of New Potato Varieties: Technological Advances*. Cambridge University Press, Cambridge, pp. 213-217.

Mendiburu, A.O. and Peloquin, S.J. (1977) The significance of $2n$ gametes in potato breeding. *Theoretical and Applied Genetics* 49, 53-61.

Mendiburu, A.O., Peloquin, S.J. and Mok, D.W.S. (1974) Potato breeding with haploids and $2n$ gametes. In: Kasha, K. (ed.) *Haploids in Higher Plants*. University of Guelph, Ontario, Canada, pp. 249-258.

Mendoza, H.A. (1980) Preliminary results on yield and uniformity of potatoes grown from true seed. In: *Report of Planning Conference on the Production of Potatoes from True Seed, 13-15 September 1979, Manila, Philippines*. International Potato Center, Lima, Peru, pp. 156-172.

Mendoza, H.A. (1983) Selection of uniform progenies to use TPS in commercial potato production. In: *Report 16 of Planning Conference on Present and Future Strategies for Potato Breeding and Improvement*. International Potato Center (CIP), Lima, Peru, pp. 87-97.

Mendoza, H.A. (1984) Selection of uniform progenies to use in TPS commercial potato production: innovative methods for propagating potatoes. *Report of 28th Planning Conference*, International Potato Center, Lima, Peru, pp. 5-16.

Mendoza, H.A. (1985) Selection of uniform progenies to use TPS in commercial potato production. In: *Report of Planning Conference on Innovative Methods for Propagating Potatoes, 10-14 December, Lima, Peru*. International Potato Center, Lima, Peru.

Mendoza, H.A. and Haynes, F.L. (1974) Genetic basis of heterosis for yield in the autotetraploid potato. *Theoretical and Applied Genetics* 45, 21-25.

Mendoza, H.M. and Marca, J.L. (1986) Performance of selfed, open pollinated, and hybrid progenies in a *Solanum tuberosum* ssp. *andigena* population. *American Potato Journal* 63, 444 (Abstract).

Mok, D.W.S. and Peloquin, S.J. (1975) Breeding value of $2n$ pollen (diplandroids) in tetraploid $\times$ diploid crosses in potatoes. *Theoretical and Applied Genetics* 45, 21-25.

Ochoa, C. (1990) *The Potato of South America: Bolivia*. Cambridge University Press, Cambridge.

Ortiz, R. and Iwanaga, M. (1986) Manipulacion de niveles de ploidia en papa: Evaluacion agronomica de progenies tetraploides derivadas de cruzamientos $4x-2x$. *Primer Congreso Peruano de Genetica*, Lima, Peru.

Peloquin, S.J. (1979) Breeding methods for achieving phenotypic uniformity. In: *Report of Planning Conference on the Production of Potatoes from True Seed, 13-15 September 1979, Manila, Philippines*. International Potato Center, Lima, Peru. pp. 151-155.

Peloquin, S.J. (1983) New approaches to breeding for the potato for the year 2000. In: Hooker, W.J. (ed.) *Research for the Potato in the Year 2000. Proceedings International Congress, Lima, Peru, 22-27 February 1982*. International Potato Center (CIP) Lima, Peru, pp. 32-34.

Peloquin, S.J., Okwuagwu, C.O., Leue, E.F., Hermundstad, S.A., Stelly, D.M., Schroeder, S.H. and Chujoy, J. E. (1985) Use of meiotic mutations in breeding. In: *Present and Future Strategies for Potato Breeding and Improvement. Report of the XXVI Planning Conference, 12-14 December 1983*. International Potato Center, Lima, Peru.

Ramanna, M.S. (1974) The origin of unreduced microspores due to aberrant cytogenetics in the meiocytes of potatoes and its significance. *Euphytica* 23, 20-39.

Sanford, J.C. and Hanneman, R.E, Jr (1982) A possible heterotic threshold in the potato and its implications for breeding. *Theoretical and Applied Genetics* 61, 151–159.

Swaminathan, M.S. and Howard, H.W. (1953) The cytology and genetics of the potato (*Solanum tuberosum*) and related species. *Bibliographia Genetica* 30(11), 271–281.

Thompson, P.G., Mendoza, H.A. and Plaisted, R.L. (1983) Estimation of genetic parameters for characters related to potato propagation of true seed (TPS) in an andigena population. *American Potato Journal* 60, 393–401.

Veilleux, R.E. and Lauer, F.I. (1981) Breeding behaviour of yield components and hollow heart in tetraploid–diploid vs. conventional derived potato hybrids. *Euphytica* 30, 547–561.

Watanabe, K. and Peloquin, S.J. (1988) Occurrence of 2n pollen and *ps* gene frequencies in cultivated groups and their related wild species in tuber-bearing Solanums. *Theoretical and Applied Genetics* 78, 329–336.

23 Introgression of Genes from Wild Species, Including Molecular and Cellular Approaches

J.G.TH. HERMSEN

Department of Plant Breeding (IVP), Wageningen Agricultural University, PO Box 386, 6700 AJ Wageningen, The Netherlands.

Introduction

Introgression in the classical sense is the incorporation of genes from one individual or population (=donor) into the gene complex or gene pool of another (=recipient). It usually consists of hybridization between the donor and recipient and repeated backcrossing to the recipient (recurrent parent) under selection in successive generations for the genes to be introgressed. Donor and recipient may belong either to the same taxon, or to different taxons with various degrees of relationship. In a classical backcrossing procedure the amount of recipient genome being substituted may vary from a hybrid with initially equal proportions of donor and recipient genes, to a genotype that is nearly isogenic with the recurrent parent (self-fertilizing species) or with its selfed progeny (cross-fertilizing species), the degree of isogenicity being dependent on the number of backcrosses and on the method and intensity of selection in F_1 and successive backcross generations.

Recent developments in cell and molecular biology have greatly extended the breeder's opportunities for hybridization and introgression, and have provided more sophisticated methods of indirect selection using molecular markers linked to the traits to be introgressed. Introgression can thus be defined in a broader sense as the incorporation of DNA from whatever source (even synthetic genes) into the nuclear, chloroplast or mitochondrial DNA complex of a living organism. It is this distinction and this broadened concept of introgression which provides the framework of this chapter.

It is meaningful to distinguish between the natural hybridization, introgression and selection that are so important in the evolution of crop species, and the man-mediated hybridization, introgression and selection that are so important for improving crop species by breeding. This chapter will focus only on man-aided gene transfer and introgression in potato breeding.

Table 23.1. List of 40 tuber-bearing *Solanum* species occurring in this chapter and arranged in alphabetical order of their abbreviations.

Abbrev.	*Solanum* species	Abbrev.	*Solanum* species
acl	*acaule*	ifd	*infundibuliforme*
agf	*agrimonifolium*	jam	*jamesii*
ber	*berthaultii*	lgl	*lignicaule*
blb	*bulbocastanum*	mrf	*morelliforme*
brd	*brevidens*	opl	*oplocense*
cap	*capsicibaccatum*	phu	*phureja*
chc	*chacoense*	pld	*polyadenium*
chm	*chomatophilum*	plt	*polytrichon*
chn	*chancayense*	pnt	*pinnatisectum*
clr	*clarum*	pta	*papita*
cmm	*commersonii*	scr	*sucrense*
col	*colombianum*	sto	*stoloniferum*
cph	*cardiophyllum*	tar	*tarijense*
crc	*circaeifolium*	tbr	*tuberosum*
dms	*demissum*	subsp. adg	subsp. *andigena*
etb	*etuberosum*	subsp. tbr	subsp. *tuberosum*
fen	*fendleri*	trf	*trifidum*
frn	*fernandezianum*	tuq	*tuquerrense*
gon	*goniocalyx*	ver	*verrucosum*
hjt	*hjertingii*	vrn	*vernei*

In the centres of diversity from Chile to the state of Nebraska (USA), 225 wild tuber-bearing *Solanum* species have been recognized according to the latest taxonomic interpretation by Hawkes (1990) and partitioned in 21 taxonomic series. Out of this huge reservoir 40 *Solanum* species occur in this chapter and are listed in Table 23.1 in alphabetical order of their abbreviations as proposed by Simmonds (1963) and updated by Huaman and Ross (1985). Only these abbreviations of species will be used in the text.

Introgression in classical breeding is the result of hybridization or (back) crossing, genetic recombination related to interspecific relationship, and direct or indirect selection in F_1 and backcross generations. These components will be discussed under the headings:

1. Barriers to interspecific hybridization and ways to overcome them.
2. Interspecific relationship, meiotic chromosome pairing and genetic recombination.
3. Traditional and marker-aided selection, and rate of recovery of the recipient genotype.

In a final section gene transfer via transformation will be treated as an alternative means of introgression in potato breeding.

Barriers to Interspecific Hybridization and Ways to Overcome Them

Crossing or hybridization in a broad sense can be defined as any natural or artificial fusion of two genetically different cells leading to hybrid progeny. Hermsen (1992) distinguishes between sexual and somatic crossability, where sexual crossability is the probability of obtaining hybrid progeny from the fusion in pairs of different specialized haploid gametes, and somatic crossability is the probability of obtaining hybrid progeny from the fusion in pairs of different naked unspecialized somatic cells or protoplasts. When the probability is zero, the parents concerned display non-crossability or incongruity.

When two species have a normal pollen-pistil relationship, then, after intercrossing, an unimpeded chain of processes is initiated, based on a perfect interaction between pollen genes (or gene complexes) and matching pistil genes (or gene complexes). Incomplete matching means one or more missing links in the chain and thus partial or complete failure of hybridization and introgression.

Prezygotic barriers

Prezygotic barriers are expressed between pollination and fertilization as non-germination of pollen on stigma; germination, but no penetration into stigma; penetration, but inhibition of pollen tube growth to different degrees and at different sites in style, ovary or ovulum. Two concepts describe in a general way the genetics of such barriers: 'penetration capacity' of the pollen donor, including all genes or gene complexes in the pollen controlling its capacity to overcome barriers to hybridization of alien females, and 'barrier capacity' of the female parent, including the genes or gene complexes that control all barriers in the pistil against hybridization by alien pollen. This genetic model, first proposed for *Lycopersicon* species by Hogenboom (1973), was further investigated in *Solanum* by Hermsen *et al.* (1977). These authors intercrossed separate F_1 plants from a diploid tbr × tbr population as females with separate plants from a selfed ver population. Surprisingly, four types of ver plants were detected, discriminating four matching types of tbr plants. The differences were of a qualitative nature as is apparent from Table 23.2. Table 23.2 shows the pattern of acceptance/non-acceptance (+/−) that was found, and a tentative genetic interpretation of the results. The central part of the pattern (bold type) suggests a gene-for-gene relationship between penetration and barrier genes. A similar model was suggested by Hermsen (1977) to explain results from intraspecific and reciprocal interspecific crosses between four *Solanum* species with quantitatively different penetration and barrier capacities (Table 23.3).

Unilateral incompatibility is non-acceptance of pollen in only one direction of a reciprocal cross. It was first investigated by Lewis and Crowe (1958) and described as a function of the *S* locus for incompatibility. It is very

Table 23.2. Pattern of acceptance (+) and non-acceptance (−) from pollinating four 2x tbr plants differing in barrier capacity with four ver plants differing in penetration capacity. Typical gene-for-gene interaction pattern in the centre (bold type).

Functional barrier genes from tbr	Functional penetration genes from ver			
	$p_1 + p_2$	p_1 only	p_2 only	None
$B_1 + B_2$	+	−	−	−
B_1 only	+	**+**	**−**	−
B_2 only	+	**−**	**+**	−
None	+	+	+	+

Table 23.3. Crossability pattern and pollen tube growth in a diallel involving four *Solanum* species - ver, phu, pnt and blb - in increasing order of penetration and barrier capacity. + = fruits obtained: $\frac{1}{4}$ etc. = pollen tubes arrested at $\frac{1}{4}$ etc. of the style length from the stigma. Gene-for-gene interaction pattern in the centre (bold type).

♀ \ ♂	ver	phu	pnt	blb
ver	+1/1	+1/1	+1/1	+1/1
phu	−1/4	**+1/1**	**−2/3**	−1/4
pnt	−1/5	**−1/2**	**+1/1**	+2/3
blb	−1/5	−1/6	−1/4	+1/1

common in plant families comprising both self-incompatible (si) and self-compatible (sc) species, such as the Solanaceae. Most common is inhibition of sc pollen in an si style, but also si × si and sc × sc inhibition occurs (review Abdalla and Hermsen, 1972a). The genetic basis of acceptance of pollen from sc ver was studied in styles of si chc by Grun and Aubertin (1966) and in styles of si diploid tbr by Hermsen *et al.* (1974). Grun and Aubertin postulated two to four unlinked recessive genes controlling acceptance. The data obtained by Hermsen *et al.* could be explained by two genetic models:

1. The first one, with two unlinked recessive genes (only $a_1a_1a_2a_2$ being acceptor), was able to explain 12 out of 14 ratios observed in selfed progeny from two acceptors and five non-acceptors, and in seven F_1s from crosses between non-acceptors and acceptors.

2. The second one, with one dominant acceptor gene *A* and one dominant inhibitor *I* (*iiAA* and *iiAa* being acceptor) could explain all 14 ratios.

Breeders may overcome or circumvent natural prezygotic barriers between species by:

1. Selecting matching genotypes within parent species (see Table 23.2).

2. Pollinating a large number of female genotypes with a pollen mixture of many male genotypes in each interspecific cross.

3. Choosing the right direction of the cross in case of unilateral incompatibility.
4. Applying mechanical means such as cutting styles and direct pollination of ovules.
5. Using mixtures of incompatible and compatible (= mentor or rescue) pollen (Adiwilaga and Brown, 1991).
6. Double pollination, first by genetically marked compatible (= pioneer) pollen, and then by the incompatible pollen (Watanabe et al., 1990; Singsit and Hanneman, 1991a), or the other way around (Iwanaga et al., 1991).
7. Applying somatic hybridization, as will be explained later on.

It should be clear that breeders can sometimes force interspecific hybridization and introgression, where these are not likely to occur in nature.

Postzygotic barriers

Postzygotic barriers are expressed during and after fertilization in ovules, or more specifically in embryo sacs, and also during growth and flowering of F_1 plants or even later in segregating F_2 or BC_1 generations. These barriers, like the prezygotic ones, may cause partial or complete failure of hybridization and introgression.

Embryo development may be arrested in its early stages due to the action of deleterious gene(s), to degeneration of the suspensor of the embryo cutting off the nutritive supply (Sangduen et al., 1983), to evolutionary differences between parental genomes in timing and duration of cell division, to somatoplastic sterility associated with prolific growth of nucellus and integuments (Cooper and Brink, 1940) and, most frequently, to endosperm degeneration.

Normal endosperm is indispensable for nourishing the embryo. Endosperm failure is most frequent with interspecific crosses, but is also associated with intraspecific interploidy crosses. Balanced endosperm normally originates from the fusion of two haploid cells from the female parent and one haploid cell from the male parent. In intraspecific crosses between parents with equal ploidy the endosperm is balanced. In intraspecific interploidy crosses (e.g. $4x \times 2x$) the $5x$ endosperm is unbalanced and $3x$ embryos will rarely survive.

In interspecific crosses the situation may be more complicated. In order to explain the crossing behaviour of different species, each species is assigned a hypothetical value in the endosperm. Ploidy and hypothetical value are equal in tbr and in most diploids and their derived autotetraploids, but they are different in allopolyploids and in most Mexican and some South American diploid species (Johnston et al., 1980; Hanneman, 1984). The hypothetical value is named endosperm balance number (EBN), and an interspecific hybrid embryo is embedded in balanced endosperm if the ratio of female EBN and male EBN in the endosperm is 2:1. This implies that balanced endosperm is obtained when both parent species have the same EBN. In the following survey

Table 23.4. Grouping of *Solanum* species (and haploid and doubled derivatives) based on ploidy and EBN number. Abbreviations of species are according to Simmonds (1963) and Huaman and Ross (1985).

Ploidy, EBN	*Solanum* species (and haploid and doubled derivatives)
2x, 1 EBN	blb brd cap chn clr cmm cph crc etb frn jam lgl mrf pld pnt trf; amphimonoploids from allotetraploid species
2x, 2 EBN	chc chm ifd tar; Megistacroloba 2x species, Tuberosa 2x species except chn, dihaploids from natural and man-made autotetraploids
4x, 2 EBN	All allotetraploids tested so far: acl agf col fen hjt opl plt pta scr sto tuq; all doubled 2x, 1 EBN species
4x, 4 EBN	tbr subsp. tbr and adg; all doubled 2x, 2 EBN species (autotetraploids)
6x, 4 EBN	Demissa species; acl subsp. *albicans*

(Table 23.4) *Solanum* species and haploid and doubled derivatives are grouped according to their ploidy–EBN ratio, as far as known from crosses with standard species.

EBNs can be manipulated by meiotic (2*n* gametes) or mitotic chromosome doubling and by haploidization, as indicated in Table 23.4. Species with different EBNs can be intercrossed only when crosses are made on a large scale (Table 23.5). The few hybrids thus obtained are ascribed by Ehlenfeldt and Hanneman (1988) to a kind of natural restoration of the EBN ratio to 2:1 by rare events such as endomitosis and proliferation of nuclei in the secondary embryo sac nucleus, or to multiple fertilization of the central nucleus.

Ehlenfeldt and Hanneman (1988) obtained 11 diploid F_1 hybrids from the cross cmm × chc (1 EBN × 2 EBN; see Table 23.7). Four of them were selected on the basis of fertility and fruit set, crossed in a complete diallel, and further backcrossed reciprocally with each parent. Table 23.6 summarizes the experiment and pooled data obtained. Analysis of the populations led to the following tentative genetic model being proposed:

1. The EBN system in cmm and chc is based on three unlinked loci with additive effects.
2. Within a species the three genes have an identical EBN value, but the chc genes are twice as effective as the cmm genes.
3. The genes are homozygous in both species; the F_1 hybrids thus contain six alleles, three from each species. So in the segregating population the genic EBN values vary from a minimum of three to a maximum of six, due to recombination.
4. Surplus of female EBN dosage results in 45% small but viable seeds, whereas surplus of male EBN dosage gives 12.4% viable seeds that are relatively large (Table 23.6). So there is a better tolerance of female surplus.

The pooled observations (Table 23.6) for the backcrosses (45% and 12% viable seeds) approximate the expectations from the model (50% and 12.5%

Table 23.5. Crossability of *Solanum* species with different EBNs. All crosses reciprocal to the ones mentioned in column 1 failed completely; — = figure not reported; p. fl. = pollinated flowers; b = berries; s = seeds; s/b = seeds per berry.

		Number of					
Cross	EBN	p. fl.	b	s	s/b	Hybrid plants	Reference
ver × blb	2 × 1	309	168	48	0.3	7	1
acl × blb	2 × 1	4826	523	383	0.7	103	2
etb × ver	1 × 2	1282	29	356	12.3	—	3
brd × ver	1 × 2	1082	11	13	1.2	—	3
cmm × chc	1 × 2	900	—	54	—	45	4
crc × 2x-tbr	1 × 2	400	75	49	0.7	47	5
crc × phu	1 × 2	10	5	2	0.4	2	5
crc × ber	1 × 2	16	5	6	1.2	6	5
sto × 4x-tbr*	2 × 4	94	49	—	—	38	6
fen × blb	2 × 1	235	5	3	0.6	—	7
hjt × cmm	2 × 1	303	29	2	0.07	—	7
hjt × cph	2 × 1	526	60	3	0.05	—	7
sto × cph	2 × 1	240	62	11	0.18	—	7
pta × cmm	2 × 1	263	65	70	1.08	—	7

* = double pollination and embryo rescue.
1 Hermsen and Ramanna (1976); 2 Hermsen (1966); 3 Hermsen (1983); 4 Ehlenfeldt and Hanneman (1988); 5 Louwes *et al.* (1992); 6 Singsit and Hanneman (1991a); 7 Singsit and Hanneman (1991b).

Table 23.6. Pooled results from a complete diallel and reciprocal backcrosses involving four selected F_1 hybrids from cmm × chc. Data from Ehlenfeldt and Hanneman (1988).

	EBN		Surplus EBN	Pooled numbers of seeds		Variation in seed size
Crosses	♀	♂		Total	Viable (%)	
$F_1 \times F_1$	$1\frac{1}{2}$	$1\frac{1}{2}$	None	5100	70.6	Large to small
$F_1 \times$ cmm	$1\frac{1}{2}$	1	Female	2797	42.9	Average to small
chc $\times F_1$	2	$1\frac{1}{2}$	Female	1480	47.3	Average to small
$F_1 \times$ chc	$1\frac{1}{2}$	2	Male	2987	13.4	Average to large
cmm $\times F_1$	1	$1\frac{1}{2}$	Male	381	11.3	Average to large

respectively), but discrepancies exist when considering individual crosses. Furthermore, the pooled data of the $F_1 \times F_1$ crosses (71% viable seeds) do not fit the expectations from the model (55% viable seeds). In order to explain the biased ratios the authors point to a general underestimation of the number of aborted seeds and to seed abortion due to factors other than endosperm failure. When they converted the ratios of viable : aborted seeds of the model to ratios of viable seeds : total number of developing seeds (aborted + viable), the predictions of the model showed a good overall agreement with the original gross data on seed set.

This genetic analysis is the first attempt to give a biological basis to the hypothetical EBN concept.

A hypothesis basically similar to EBN was developed independently by Nishiyama and Yabuno (1979). Their concept of activation index (AI) is the quotient of activation value (AV) assigned to the male nucleus and response value (RV) assigned to the egg and to each of the polar nuclei, such that $AI = AV/2 RV$ for the endosperm (in fact the reverse of the EBN ratio). Both AV and RV are supposed to be controlled by one or more genes. In each species $AV = RV$, and thus AI of the endosperm from an intraspecific cross between plants of equal ploidy is 0.5. AV values are assigned to different species on the basis of results from interspecific crosses, given an AV value = 1, to a randomly chosen control species. They may have any value, including fractions, as opposed to EBNs. This makes this hypothesis less simple, but more realistic and flexible than the EBN hypothesis.

The significance of EBN and AI for introgression breeding is in their predictive value of results from interspecific crosses and of backcrossing, and in their potential to guide the breeder in overcoming barriers to introgression.

The role of ploidy manipulation

When considering the grouping of *Solanum* species based on EBN and ploidy (Table 23.4), it is apparent that, for introgression of desirable characters from wild species (except hexaploid) directly into tetraploid potato cultivars, the first step is always an inter-EBN cross. Such crosses usually fail, due to embryo abortion following endosperm collapse. As stated before, this barrier can be overcome by extensive crossing, with or without embryo rescue (Table 23.5), or circumvented by equalizing the parental EBNs. This can be done in two ways. The first is by doubling the chromosome number of the lower-EBN wild species, making use of natural $2n$ gametes (meiotic doubling), or by colchicine treatment or *in vitro* explant culture (mitotic doubling). The second is by haploidization of the polyploid cultivars through pseudogamy, for which excellent phu clones are available which are homozygous for embryo spot (Hermsen and Verdenius, 1973). Doubling is the obvious approach for introgression at the tetraploid level and haploidization for introgression at the diploid level. However, as a rule selected dihaploids cross well with $2x$, 2 EBN, wild species, so that selected F_1 hybrids can be raised directly to the tetraploid level via $2n$ gametes in crosses with $4x$ cultivars and a good agronomic performance achieved, as shown by Peloquin and his associates (Hermundstad and Peloquin, 1985; Peloquin *et al.*, 1989).

Besides a low berry and seed set (Table 23.5), inter-EBN hybridization has some striking features:

1. The hybrid plants obtained usually consist of two groups with different ploidy (Table 23.7), when the lower-EBN parents produce both n and $2n$

Table 23.7. Different ploidy in F_1 from interspecific inter-EBN crosses. The higher ploidy in each F_1 is due to $2n$ gametes from the lower-EBN parent.

Cross	EBN	Number of hybrids				Reference
		$2x$	$3x$	$4x$	$5x$	
$4x$ acl $\times$ $2x$ blb	2×1	0	88	15	0	Hermsen (1966)
$2x$ cmm $\times$ $2x$ chc	1×2	11	34	0	0	Ehlenfeldt and Hanneman (1988)
$2x$ crc $\times$ $2x$ tbr	1×2	8	39	0	0	Louwes et al. (1992)
$2x$ ver $\times$ $6x$ dms	2×4	0	0	2	1	J.G.Th. Hermsen (unpublished)
$2x$ tbr $\times$ $6x$ dms	2×4	0	0	7	5	J.G.Th. Hermsen (unpublished)
$4x$ hjt $\times$ $2x$ cmm	2×1	0	1	1	0	Singsit and Hanneman (1991b)
$4x$ pta $\times$ $2x$ cmm	2×1	0	1	17	0	Singsit and Hanneman (1991b)

gametes. The frequency of $2n$ gametes is usually very low, but, because $2n$ gametes functionally equalize the parental EBNs, they have a strong competitive advantage in fertilization.

2. If inter-EBN crosses are successful, this occurs in one direction only, not in the reciprocal one. This holds true for all the crosses included in Table 23.5, where only the successful direction of the crosses is given in column 1. In the failing direction, pollen tube inhibition usually prevents fertilization.

The role of $2n$ *gametes in evolution and introgression*

The natural occurrence of $2n$ gametes in many tuber-bearing *Solanum* species is well established (Den Nijs and Peloquin, 1977a; Watanabe and Peloquin, 1991), although their frequency may vary greatly within and between species owing to genetic, environmental and physiological factors. Such $2n$ gametes may originate in different ways, but genetically the basic distinction is between $2n$ FDR gametes (reduction of $2n$ to n chromosomes followed by restitution to $2n$ through incomplete first division) and $2n$ SDR gametes (reduction of $2n$ to n chromosomes followed by restitution to $2n$ through incomplete second division). The mechanisms involved in the origin of FDR and SDR gametes and their genetic consequences have been described by several authors (Mok and Peloquin, 1975; Ramanna, 1979; Hermsen, 1984; see also chapter 5 by Tai). It can be calculated that FDR gametes introgress 80% of the intact parental genotype to the hybrid progeny and SDR gametes 40%. Desynaptic mutants (Jongedijk and Ramanna, 1988) produce completely sterile n gametes, but, when $2n$ gametes occur, they are $2n$ FDR gametes which originated from a pseudohomoeotypic division (Ramanna, 1983; Jongedijk et al., 1991), and which introgress about 95% of the intact parental genotype to hybrid progeny. These high percentages of intact gene transfer through FDR gametes are especially important when the parent plant carries many desirable genes for qualitative as well as quantitative characters.

Besides their value in breeding and introgression, $2n$ gametes have also played a major role in the evolution of polyploid species of *Solanum* (Den Nijs and Peloquin, 1977b; Iwanaga and Peloquin, 1982), as well as in many other plant families (review Harlan and De Wet, 1975).

The role of $2n$ gametes in overcoming critical stages in introgression breeding schemes has been well documented by examples of realized introgression, in which embryo rescue and the use of bridging species have also been exploited to bring about introgression from distantly related species to potato cultivars (see below, section on examples of successful introgression breeding).

Embryo rescue during the first steps of introgression

It should be mentioned that embryo rescue is becoming more and more sophisticated and powerful, where embryo abortion is due to defective endosperm, such as in inter-EBN crosses. Embryo rescue is also frequently used in distant hybridization. Like EBN – or ploidy – manipulation, embryo rescue can only be successful if prezygotic barriers do not inhibit fertilization.

Embryo rescue may be applied both in the first interspecific cross and in the first backcross. Adiwalaga and Brown (1991), Iwanaga *et al.* (1991) and Singsit and Hanneman (1991a) used embryo culture to circumvent inter-EBN barriers.

Louwes *et al.* (1992) used sexual hybridization and Mattheij *et al.* (1992) somatic hybridization to include crc subsp. crc, a good source of late blight and nematode (*Globodera pallida*) resistance, in their introgression breeding programme. The species ($2x$, 1 EBN) was successfully crossed as a female with $2x$ tbr, gon × tbr, ber and phu (all $2x$, 2 EBN) by applying *in vitro* cultivation of young seeds in a nutrient medium after gibberellic acid treatment. Both diploid and triploid hybrids were obtained, which could be successfully backcrossed to diploid tbr. The crosses with crc as male were not successful, and neither were the reciprocal crosses of crc with pnt and pld, in spite of both species being 1 EBN, like crc. This may be due to prezygotic barriers. By using embryo rescue techniques, Eijlander and Stiekema (1994) have even been successful in bridging the crossability gap between $6x$ *S. nigrum* and tbr cv. Désirée, and expect success from backcrossing the hybrids with tbr using the same techniques.

Somatic hybridization: common barriers with sexual hybridization

Somatic hybridization may be an important first step to introgression breeding. The pre- and postzygotic crossability barriers to sexual hybridization treated so far do not apply to somatic hybridization. Here, however, finding the right conditions may meet with difficulties, i.e. the right procedure for protoplast culture, fusion and selection of hybrids and for regeneration to plants is crucial for success. The rate of success is also very much dependent on plant family, genus, species, and even cultivar. Because there is virtually no barrier to the fusion of protoplasts, the scope of hybridization and introgression is very much widened by somatic hybridization.

In this connection the success of somatic hybridization of some recalcitrant species (brd, blb, pld, crc) directly with tbr cultivars (Helgeson, 1992; Mattheij *et al.*, 1992) demonstrates its potential, the more so because some somatic hybrids could be sexually backcrossed to cultivars, and introgression of resistance to PLRV and *Erwinia* soft rot could be demonstrated in the backcross generation (Helgeson and Williams, 1991). The classical way in these cases (Hermsen, 1983; Louwes *et al.*, 1992) took more time and effort (see below, section on examples of successful introgression breeding).

Sexual and somatic hybridization have a number of barriers in common which are a consequence of disharmony between parental idiotypes or of deleterious genes or gene combinations:

1. Chromosome elimination, e.g. phu chromosomes in sexual hybrids (Ramanna and Hermsen, 1971; Clulow *et al.*, 1991) and in somatic hybrids (Pijnacker *et al.*, 1989).
2. Recessive lethal genes causing embryo death just before or after germination, e.g. albino, yellow cotyledon, tiny dwarf, defective xylem (Simmonds, 1965; Hermsen *et al.*, 1978).
3. Recessive genes for abnormal plant growth, e.g. virescens, droopy, yellow margin, dark green dwarfs (Simmonds, 1965; Hermsen *et al.*, 1978).
4. The presence of a large proportion of weak and unthrifty plants in the F_2 of interspecific crosses due to disharmonious gene combinations from the two parents ('hybrid breakdown'), especially when they are distantly related.
5. Finally, sterility may occur both in sexual and in somatic hybrids, although they may differ in this respect, as will be shown in the next section.

Male Sterility and Introgression of Plasmon-coded Traits into Potato

Male sterility in F_1 plants due to reduced homologous pairing is characteristic for sexual hybridization of certain remotely related species, such as Tuberosa species on the one hand and Morelliformia, Bulbocastana, Pinnatisecta, Polyadenia and Circaeifolia on the other. Etuberosa species, for example, when crossed with Pinnatisecta species give rise to completely male- and female-sterile F_1 progeny. Such sterility may be overcome by mitotic chromosome doubling, giving each chromosome an identical partner.

Anorthoploid species or species hybrids ($3x$, $5x$) are usually highly sterile or, if somewhat fertile, cannot breed true due to aneuploid gametes. Female fertility is usually sufficient, however, for further (back)crossing and thus for introgressive hybridization. Doubling the chromosome number has little effect on fertility.

The causes of sterility are sometimes unknown, e.g. in early varieties of tbr subsp. tbr, as well as in many accessions of tbr subsp. adg. The genetic basis of male sterility in most $2x$, 2 EBN species × phu male, and of male fertility in the reciprocals has not yet been analysed as far as the author is aware.

Cytoplasmic male sterility (CMS) is based on an interaction between cytoplasmic (mitochondrial) genes and matching, mostly dominant, nuclear genes. Interspecific hybrids from crosses of ver ♀ with diploid species from series Tuberosa and Commersoniana are well investigated. Depending on the accession of ver, different types of CMS have been detected, such as 'tetrad' CMS (also induced by *Longipedicellata* cytoplasm), 'eclipse' CMS, 'striped vacuolar' CMS and 'undivided microsporocyte' CMS (Buck, 1960; Grun *et al.*, 1962; Abdalla and Hermsen, 1972b; Ramanna and Hermsen, 1974). Doubling the chromosome number has no effect on CMS, e.g. $2x$ (ver × blb) and $4x$ (ver × blb) had an average pollen stainability of 10.6% and 11.4% respectively (Hermsen and Ramanna, 1976). The only way to circumvent the CMS barrier is by plasmon substitution, either through repeated backcrossing or via fusion of pretreated somatic cells.

Using a tbr acceptor of ver pollen, J.G.Th. Hermsen (unpublished) produced ver genotypes with tbr cytoplasm via the substitution backcrossing scheme (tbr acceptor × ver) × ver, etc. The clones $(t)V^5$ and $(t)V^6$ thus obtained were used as bridging species in introgression schemes (Hermsen, 1983), where CMS could thus be avoided.

A more recent and quick *in vitro* method to substitute ver cytoplasm for tbr cytoplasm is by fusion of an enucleated somatic tbr cell with a somatic ver cell with inactivated cytoplasm (iodoacetate treatment); the fusion product is a cell with a ver nucleus and tbr cytoplasm, comparable to the aforementioned backcross product $(t)V^6$. Similarly, this method can be used to introduce CMS into a cultivar of tbr by using an enucleated ver cell with an iodoacetate-treated somatic cell of the tbr cultivar. An informative paper on this approach to introgression of advantageous plasmon-coded traits into potato, with its phylogenetic limitations, is by Perl *et al.* (1990), where one can find examples and further details. These authors demonstrated the feasibility of this so-called donor–recipient protoplast fusion technique for potato.

Mariani *et al.* (1990) fused a natural ribonuclease gene ('barnase') from *Bacillus amyloliquefaciens* with a tapetum-specific promoter of a tobacco gene (TA29). This chimeric construct selectively destroyed the tapetal cell lineage during anther development and prevented pollen formation. This male sterility could be restored using a construct bearing the bacterial gene for 'barstar', a protein inhibitor of 'barnase', linked to the tapetum-specific promoter (Mariani *et al.*, 1992). This is a novel genetic method for introgressing pollen sterility and fertility restoration into crop plants by transformation. It has been successfully applied in oil seed rape and tobacco, and is expected to be successful in other crops as well.

To conclude this section it should be emphasized that male sterility may greatly influence natural hybridization, while CMS may promote unilateral introgression into the CMS parent. Furthermore, other cytoplasmically controlled characters besides CMS may be transferred and introgressed into crop species, using the donor–recipient protoplast fusion technique (Perl *et al.*, 1990, including more references).

Interspecific Relationship, Meiotic Chromosome Pairing and Genetic Recombination

The occurrence of introgression is associated with phylogenetic relationships between *Solanum* species via crossability and genetic recombination. Closely related species can as a rule be intercrossed easily, although barriers may occasionally occur in this category (see previous sections). With increasing genetic distance, the number and strength of crossability barriers also increase, and genetic recombination is restricted or even impeded. This implies a need for large-scale crossing and growing large progenies to recover desirable recombinants. It may be relevant to mention here that gene recombination in F_1 hybrids of tbr × ber could be promoted by a short cycle of culturing petiole callus, thus accelerating introgression of insect resistance from ber into tbr (Lentini *et al.*, 1990).

Breeders using wild species need cytogenetical studies to gain an insight into the meiotic processes and types of segregants that can be expected. Matsubayashi (1991) reviewed extensive data on meiotic pairing and pollen fertility in *Solanum* species and interspecific F_1 hybrids. He reached the following general conclusion: species of all taxonomic series, both diploid (except Etuberosa) and polyploid, are related by sharing one basic genome, coded A. A number of structural modifications of the A genome have evolved following geographical isolation, leading to reduced chromosome pairing and recombination between A and modified A genomes in F_1 hybrids. Polyploid species, besides A or modified A, carry additional genomes which are also differentiated and are given different symbols (B, C, D, P) by Matsubayashi.

Matsubayashi (1991) listed 59 *Solanum* species on the basis of ploidy and genome composition. The data are summarized in Table 23.8. The genome A and/or modified A occurs in 95% of the species listed and E only

Table 23.8. Numbers of species carrying the genome(s) indicated in column 1. A^1, C^1, D^1 = modified A, C and D genomes respectively. Classification based on Table 1 in Matsubayashi (1991).

Genome(s)	Number of species			Remarks
	2x	4x	Total	
AA	24	16	40	Basic genome
A^1A^1	12	12	24	Nine modifications of A
EE	3	0	3	In Etuberosa
AA + A^1A^1	—	8	8	In Tuberosa, Acaulia
A^1A^1 + C^1C^1	—	3	3	In Conicibaccata
A^1A^1 + PP	—	1	1	In *S. tuquerrense*
AA + BB	—	4	4	In Longipedicellata
AA + DD + D^1D^1	—	4	4	In Demissa

in diploid Etuberosa species (Ramanna and Hermsen, 1981). The genomes B, C, P and D or their modifications have not been found in diploid species. However, such diploid species may: (i) still exist, but have remained undiscovered or not yet been analysed (Hawkes, 1978); (ii) be extinct after their incorporation into allopolyploids (Hawkes, 1978); or (iii) never have existed, but the originally homologous genomes have changed after their incorporation into polyploids.

Dvorak (1983), a wheat cytogeneticist, reviewed literature data on pairing behaviour in diploid and allopolyploid *Solanum* species, in F_1 hybrids from 36 diploid × diploid, 27 allopolyploid × diploid and 20 allopolyploid × allopolyploid crosses and finally in haploids from sto, plt and dms. He stated that there is nearly complete pairing in diploid species and F_1 hybrids, a considerable amount of homoeologous pairing (besides the 12 homologous bivalents) in the allopolyploid × diploid F_1s, clearly less homoeologous pairing in allopolyploid × allopolyploid F_1s, low homoeologous pairing in haploid sto (4.0 II), plt (7.9 II) and dms (6.1 II) and no homoeologous pairing at all in the allopolyploid species. From these data he concluded that the potential for homoeologous pairing of allopolyploids is greater than the actual amount of homoeologous pairing in allopolyploids, their haploids and the allopolyploid × allopolyploid hybrids. He also concluded that the most obvious explanation is that homoeologous pairing is suppressed by genes present in allopolyploid, but not in diploid, *Solanum* species, which are similar to the well-established *Ph* and other genes in allopolyploid wheats and their relatives. So Dvorak (1983) favoured genetic control of homoeologous pairing and further assumed non-structural chromosome differentiation diffused over non-coding sequences in all chromosomes to explain the exclusive bivalent pairing in allopolyploids. On the other hand, Matsubayashi (1991) did not think that such a gene control system of homoeologous pairing is applicable to most of the potato species, but rather that lack of homoeologous pairing in allopolyploid *Solanum* is due to structural chromosome differentiation after the origin of the allopolyploid. In the cytogenetic literature on other allopolyploids, the same two ideas are encountered. Much research is needed to gain more insight into the level of generalization of either concept.

Traditional and Marker-aided Selection, and Rate of Recovery of the Recipient Genotype

Selection in traditional backcross breeding is usually direct selection for desirable, and against undesirable, donor traits in F_1 and backcrosses, aimed at recovering the recipient genotype while retaining one or a few donor traits. With a closely related donor and recipient, optimal gene recombination and efficient detection methods for the target donor traits, the backcross method is very suitable and has been applied on a large scale in potato breeding. Therefore 80% of all potato cultivars bred in Europe and North America carry genes from related *Solanum* species, especially dms, adg and vrn, the main target traits being resistance to late blight, cyst nematodes and viruses (Hermsen, 1989, for more details).

The most frequent application in the potato is indeed introgression of resistances from wild species by sorting through F_1 and backcross generations using pathogen, pest and abiotic stress screens to aid selection. The duration of nobilization of an F_1 from a wild × cultivated cross depends on the genetic distance of the parents and the selection efficiency. When crossability is poor and/or gene exchange at meiosis in hybrids is reduced, many pollinations have to be made and large populations screened to find the right recombinants for backcrossing. Direct selection may then be a labour-intensive and time-consuming procedure. Recovery of the recipient genotype near to the donor target genes is a special problem, because even after several backcrosses these genes are flanked by donor parent fragments of considerable length. This was theoretically calculated by Stam and Zeven (1981) and also experimentally demonstrated with RFLP markers by Young and Tanksley (1989).

Traditional indirect selection for desirable traits, using easily scored closely linked traits with a high heritability, has been practised and has proved useful when target traits are difficult to handle. However, suitable linked traits are rarely available.

The recent discovery of huge numbers of molecular DNA markers (RFLPs and RAPDs) is a breakthrough in the introgression breeding of several crops, including the potato. Saturated linkage maps of the potato (e.g. Bonierbale *et al.*, 1988; Gebhardt *et al.*, 1989) and several other crops are now available. An exhaustive review of the use of molecular markers in breeding for disease resistance was written by Melchinger (1990), and several applications of RFLPs in potato breeding have been given by Bonierbale *et al.* (1990). RFLPs are the most widely used markers at present. They are very useful (generally occurring, stable Mendelian inheritance, codominant and neutral), convenient (detectable in nearly all tissues, at all stages and ages) and numerous in exons as well as in introns. At present, RFLPs require expensive laboratory supplies, but detection methods are generally the same irrespective of species or marker. Selection for target traits can be replaced by indirect selection for RFLP(s) that are tightly linked with them. In addition, RFLPs which are specific to donor species can be used to select against undesired donor DNA. These possibilities may speed up introgression and reduce the number of backcrosses needed to recover the recipient genotype. Marker-aided selection is especially advantageous for introgressing disease resistance (Melchinger, 1990):

1. No inoculum problems such as uniformity of spread, the right races or isolates.
2. No exotic or quarantined pathogens needed; testing possible in off-season nurseries when no inoculum is available.
3. No trouble with environments that strongly influence the expression of resistance genes.
4. Testing by markers feasible at all stages of development, whereas some resistance genes are not expressed before the full-grown stage.

5. If resistance is recessive, an extra progeny test is normally needed after every backcross, but not with marker-aided selection because of the codominant expression of the marker alleles.

Marker-aided selection is highly promising, but practical applicability is dependent on the amount of time saved, on the relative costs of direct versus marker-aided selection and, last but not least, on the degree of linkage between marker and target gene. A saturated RFLP linkage map is required for efficient utilization of marker-aided selection techniques, and the significant costs and time required for RFLP assays still limit the application.

Michelmore *et al.* (1991), working with RAPD markers in lettuce, recently published a very elegant and rapid method to detect markers tightly linked to target genes. They used a population segregating for a target gene, for example monogenic resistance, and subdivided it into the two classes, resistant and susceptible. The individuals of a class are identical for the target gene, and similar for the marker genes linked to it, such that they are more similar as the linkage gets closer. In other words, the two classes are different at the target locus and its linkage block (± 25 cM on either side of the locus), but on average similar for all other chromosome regions. So screening two bulks of DNA samples, one from each class of individuals, for differences in molecular markers enables the identification of the markers belonging to the linkage block of the target locus: they are those for which the classes are distinct. The method is called 'bulked segregant analysis' and is suitable for both self- and cross-fertilizing crops.

Examples of Successful Introgression Breeding

The first example elaborated in this section deals with the introgression of late blight resistance from diploid blb (coded B in the scheme) into tetraploid tbr (coded T), by using two late-blight-susceptible bridging species, tetraploid acl (coded A) and diploid phu (coded P), and amphidiploidy. Successive steps and results are given in Tables 23.9 and 23.10.

The observed and expected average chromosome pairing at meiosis and the genome composition of the blb hybrids at various stages of development are presented in Table 23.11. Expected pairing is based on two assumptions:

1. The chromosomes of blb (B) do not pair with any other chromosome.
2. The A and C chromosomes of acl pair with each other when single, but pair preferentially (A with A, C with C) when double.

Comparing the observed and expected configurations in Table 23.11 it is evident that in the ABP and ABPT hybrids the blb chromosomes show considerable pairing with phu chromosomes, and also some pairing with the chromosomes of acl (multivalents). Phu is therefore a crucial link in the introgression of late blight resistance. This is also clear when the late blight scores of $(AB)^2 \times$ tbr (step 4 in Table 23.9) are compared with those of ABP and ABPT (steps 5 and 7). Clones from $(AB)^2 \times$ tbr were highly susceptible

Table 23.9. Introgression scheme leading to ABPT.

1.	Cross	$4x$ acl × $2x$ blb: crossability genetically determined in both parent species
2.	Result	$3x$ AB (88 plants): pollen stainability <5%
		$4x$ AB (15 plants): pollen stainability 12%
		Origin $4x$ AB: $2n$ gametes from blb
3.	Doubling	$6x$ (AB)^2: pollen stainability 85%
		$8x$ (AB)^2: pollen stainability 29%; crosses and selfings unsuccessful; plants discarded
4.	Cross	$6x$ (AB)^2 × $4x$ tbr cultivars: successful, but late blight resistance lost (explanation in text); plants discarded
5.	Cross	$6x$ (AB)^2 × $2x$ phu: successful and resistance maintained (explanation in text)
6.	Result	$4x$ (AB)^{2}P: chromosome mosaicism in root tips; $2n = 42$-48; pollen stainability 8-37%
7.	Cross	$4x$ (AB)^{2}P × $4x$ tbr cultivars: strong barriers; berry set 7.8%, 0.12 seeds per berry, 21% seed germination leading to 36 hybrid plants per 18.616 pollinated flowers
8.	Result	(AB)^{2}PT, later on coded as ABPT: pollen stainability 39.3% (range 5-93%); late blight reaction and ploidy, see Table 23.10

Table 23.10. Ploidy and late blight reaction of 32 ABPT clones. SS-RR = from extremely susceptible to extremely resistant.

	Number of clones				
$2n$	Total	SS	S	R	RR
48	24	8	5	8	3
72	4	—	—	—	4
49	1	—	—	—	1
46	1	—	1	—	—
65	1	—	—	—	1
66	1	—	—	—	1
Total	32	8	6	8	10

both at Wageningen and in the Toluca valley (Mexico), indicating no recombination, in accordance with the expected meiotic configurations (unfortunately no analysis made) in the last line of Table 23.11. ABPTs segregated from very susceptible to highly resistant (Table 23.10), as expected with recombination.

Further backcrosses of ABPT with tbr cultivars are not hampered any more and late blight resistance is amply transferred to later generations.

A second introgression scheme, leading to the quadruple hybrid VEPT, displayed barriers, steps and results parallel to those in the ABPT scheme. PLRV resistance was introgressed from Etuberosa (E) species (brd, etb, frn) and late blight resistance from both the bridging Pinnatisecta (P) species and ver (V with tbr cytoplasm: see section on male sterility and introgression of

Table 23.11. Observed and expected average chromosome pairing at meiosis and genome composition of hybrids assuming the genome codes AACC for acl, BB for blb, PP for phu and TTTT for tbr. Pairing genomes underlined.

		Configurations					
		Expected		Observed			
Hybrids	Genome composition	II	I	IV	III	II	I
3x AB	<u>AC</u> B	12	12	0	0.44	12.07	10.45
4x AB	AC <u>BB</u>	24	0	0.12	0.11	23.10	0.38
6x AB	<u>AA</u> <u>CC</u> <u>BB</u>	36	0	0.19	0.14	35.19	0.56
8x AB	<u>AA</u> <u>CC</u> <u>BBBB</u>	–	–	Not analysable (stickiness)			
4x ABP	<u>AC</u> BP	12	24		0.85	18.93	3.43
4x ABPT	½ (<u>AC</u> BP) <u>TT</u>	18	12	0.39	0.33	22.06	1.28
6x ABPT	<u>AC</u> BP <u>TT</u>	24	24	0.50	0.40	33.28	2.12
5x ABT	<u>AC</u> B <u>TT</u>	24	12	Not analysed			

plasmon-coded traits) (two-way introgression). There are no barriers to backcrossing VEPT with potato cultivars. Briefly the VEPT scheme is as follows:

2x etb (E) × 2x pnt (P) → 2x EP (sterile) → 4x EP (fertile)
2x ver (V) × 4x EP → 3x VEP (sterile) → 6x VEP (fertile)
6x VEP × 4x tbr cultivars (bottleneck) → 5x VEPT with strong multiplication resistance to PLRV and intermediate resistance to late blight

Parallelism to both previous introgression schemes also occurred, when three sources of late blight resistance were introgressed into a tbr cultivar background, the resistance sources being 2x blb, 2x ver (with tbr cytoplasm: see section on male sterility and introgression of plasmon-coded traits) and 4x sto. The quadruple hybrids thus obtained were coded VBST and the scheme can be briefly summarized as follows:

2x ver (V) × 2x blb (B) → 2x VB (sterile) → 4x VB (fertile)
4x VB × 4x sto (S) → 4x VBS
4x VBS × 2x tbr (T) → 3x VBST (sterile) → 6x VBST (fertile)
6x VBST × 4x tbr cultivars (bottleneck) → 5x VBSTT, showing late blight resistance varying continuously from intermediate to extreme

The quadruple hybrids required more backcrossing for nobilization than ABPT. By using early tbr cultivars as the recurrent parent, the nobilization process could be accelerated.

Details of these introgression breeding schemes can be found in Hermsen (1983).

Similarities with the above extensive introgression schemes can be observed in later attempts to introgress genes from allotetraploid Longipedicellata and Acaulia species and from diploid Etuberosa species into tbr cultivars (Hermsen, 1983; Chavez et al., 1988; Bamberg, 1990; Camadro and Espinillo, 1990; Adiwilaga and Brown, 1991).

Characteristic features of classical introgression breeding schemes involving distantly related sources of desirable donor genes can be summarized as follows:

1. Sterility occurring once (ABPT) or twice (VEPT, VBST).
2. Fertility restored by mitotic or meiotic doubling giving rise to amphidiploids.
3. Some steps only possible with $2n$ gametes from the lower-EBN parent or by haploidization of the other.
4. One or two bridging species serving as a channel for desirable genes to the final recipient, and sometimes also contributing desirable genes themselves (VBST).
5. A severe bottleneck occurring mostly at the stage where tbr cultivars are included for the first time; bottlenecks can be overcome by large-scale crosses over a long period of time using different male and female genotypes, or by adjusting the parental EBNs; overcoming the bottlenecks is also possible by applying ovule culture. Once the bottleneck has been overcome, there are no more problems in further (back)crosses with cultivars.

In this connection the success of somatic hybridization of some recalcitrant species (brd, blb, pld, crc) directly with tbr cultivars (see subsection on somatic hybridization) is promising, the more so because some somatic F_1 hybrids could be sexually backcrossed with cultivars and introgression demonstrated.

Transformation: a Way of Rapid Introgression

Transformation is the transfer of alien genes into a plant and their incorporation into its genome. For dicotyledonous species such as potato, *Agrobacterium*-mediated transformation is the most obvious method at present. Basically, genes from different plant families and from animals, insects, bacteria and viruses, as well as synthetic genes, can be incorporated into the plant genome, and expressed if provided with the right regulatory sequences. Because isolated genes are transferred and not whole donor genomes, no backcrosses are needed and the recipient genotype remains largely intact. Therefore, transformation is potentially the shortest procedure for introgressing alien genes into a crop species, and hybridization barriers are not relevant.

However, the number of desirable genes that have been pin-pointed and isolated, a requirement for transformation, is still limited. This also holds true for the potato, although the number of genes that are available for transforming the potato is increasing and comprises:

1. Resistance genes from potato species.
2. Virus genes encoding coat proteins of PVX, PVY and PLRV, and conferring resistance in the potato to the corresponding viruses.

3. Bacterial genes changing the starch composition.
4. Plant genes encoding the patatin proteins in potato.
5. Genes from *Bacillus thuringiensis* (*Bt* genes) conferring resistance to insects.
6. Plant genes encoding pathogenesis-related (PR) proteins, which may bring about resistance to fungi.
7. Plant or insect genes encoding small proteins (thionins, cecropins, apidecins) which are active against bacteria.

Without genetic engineering, such genes could not be made available for transformation of the potato.

Transformed cells must have the ability to regenerate into plants. This is a genetically controlled character and different crops or even varieties of one crop have different optimum requirements for regeneration, which are often insufficiently known. The same is true of the ability to be transformed, for which varieties may also differ greatly. The potato in general has a good regeneration and transformation ability, but the genetic differences mentioned may be limiting factors for making progress with certain genotypes. Another drawback may be somaclonal variation among the regenerated transgenic plants, and here also there are varietal differences in the amount and nature of somaclonal variation.

It is certainly not true that transformed clones of a cultivar differ from the corresponding untransformed cultivar only by the incorporated alien DNA. However, such an ideal result might be achieved by selection among the independent transformants obtained, if a sufficient number of such transformants is available. Recently, Jongedijk *et al.* (1992) reported stability of expression of PVX resistance based on the PVX coat protein gene introduced into the cultivars Bintje and Escort. Furthermore, the authors found 17.9% of the Bintje- and 81.8% of the Escort-derived transgenic clones to be true to type for 50 defined morphological traits, tuber yield and grading. Finally, the true-to-type clones could be distinguished unambiguously from the untransformed cultivars, using the polymerase chain reaction with suitable primers. These promising results were obtained from the first systematic field experiments of this kind known to the author.

References

Abdalla, M.M.F. and Hermsen, J.G.Th. (1972a) Unilateral incompatibility: hypotheses, debate and its implications for plant breeding. *Euphytica* 21, 32–47.

Abdalla, M.M.F. and Hermsen, J.G.Th. (1972b) Plasmons and male sterility types in *Solanum verrucosum* and its interspecific hybrid derivatives. *Euphytica* 21, 209–220.

Adiwilaga, K.D. and Brown, C.R. (1991) Use of 2n pollen-producing triploid hybrids to introduce tetraploid Mexican wild species germ plasm to cultivated tetraploid potato gene pool. *Theoretical and Applied Genetics* 81, 645–652.

Bamberg, J.B. (1990) A practical method for making wild potato germplasm from

Mexico accessible for evaluation and utilization. *American Potato Journal* 67, 539 (Abstract).

Bonierbale, M.W., Plaisted, R.L. and Tanksley, S.D. (1988) RLFP maps based on a common set of clones reveal modes of chromosomal evolution in potato and tomato. *Genetics* 120, 1095–1103.

Bonierbale, M.W., Ganal, M.W. and Tanksley, S.D. (1990) Applications of restriction fragment length polymorphisms and genetic mapping in potato breeding and molecular genetics. In: Vayda, M.E. and Park, W.D. (eds) *The Molecular and Cellular Biology of the Potato*. CAB International, Wallingford, UK, pp. 13–24.

Buck, R.W. (1960) Male sterility in interspecific hybrids of *Solanum*. *Journal of Heredity* 51, 13–14.

Camadro, E.L. and Espinillo, J.C. (1990) Germ plasm transfer from the wild tetraploid species *Solanum acaule* Bitt. to the cultivated potato, *S. tuberosum* L. using 2n eggs. *American Potato Journal* 67, 737–749.

Chavez, R., Brown, C.R. and Iwanaga, M. (1988) Application of interspecific sesquiploidy to introgression of PLRV resistance from non-tuberbearing *Solanum etuberosum* to cultivated potato germplasm. *Theoretical and Applied Genetics* 76, 497–500.

Clulow, S.A., Wilkinson, M.J., Waugh, R., Baird, E., De Maine, M.J. and Powell, W. (1991) Cytological and molecular observations on *Solanum phureja*-induced dihaploid potatoes. *Theoretical and Applied Genetics* 82, 545–551.

Cooper, D.C. and Brink, R.A. (1940) Somatoplastic sterility as a cause of seed failure after interspecific hybridization. *Genetics* 25, 593–617.

Den Nijs, A.P.M. and Peloquin, S.J. (1977a) 2n-gametes in potato species and their function in sexual polyploidization. *Euphytica* 25, 585–600.

Den Nijs, A.P.M. and Peloquin, S.J. (1977b) Polyploid evolution via 2n gametes. *American Potato Journal* 54, 377–386.

Dvorak, J. (1983) Evidence for genetic suppression of heterogenetic chromosome pairing in polyploid species of *Solanum*, sect. Petota. *Canadian Journal of Genetics and Cytology* 25, 530–539.

Ehlenfeldt, M.K. and Hanneman, R.J. (1988) Genetic control of endosperm balance number (EBN): three additive loci in a threshold-like system. *Theoretical and Applied Genetics* 75, 825–832.

Eijlander, R. and Stiekema, W.J. (1994) Biological containment of potato (*Solanum tuberosum*): outcrossing to its related wild species black nightshade (*Solanum nigrum*) and bittersweet (*Solanum dulcamara*). *Sexual Plant Reproduction* 7(1) (in press).

Gebhardt, C., Ritter, D., Debener, T., Schachtschabel, U., Walkmeier, B., Uhrig, H. and Salamini, F. (1989) RFLP analysis and linkage mapping in *Solanum tuberosum*. *Theoretical and Applied Genetics* 78, 65–75.

Grun, P. and Aubertin, M. (1966) The inheritance and expression of unilateral incompatibility in *Solanum*. *Heredity* 21, 131–138.

Grun, P., Aubertin, M. and Radlow, A. (1962) Multiple differentiation of plasmons of diploid species of *Solanum*. *Genetics* 47, 1321–1333.

Hanneman, R.E. (1984) Assignment of endosperm balance numbers (EBN) to tuberbearing *Solanum* species. In: *Report to the NCR-84 Potato Genetics Technical Committee*, Des Plaines, Illinois, USA.

Harlan, J.R. and De Wet, J.M.J. (1975) On O. Winge and a prayer: the origins of polyploidy. *Botanical Review* 41, 362–390.

Hawkes, J.G. (1978) Biosystematics of the potato. In: Harris, P.M. (ed.) *The Potato*

Crop – the Scientific Basis for Improvement. Chapman and Hall, London, pp. 15–69.

Hawkes, J.G. (1990) *The Potato: Evolution, Biodiversity and Genetic Resources.* Belhaven Press, London.

Helgeson, J.P. (1992) New genes for resistances to viruses, fungi and bacteria through somatic hybridization. *Proceedings International Symposium on Advances in Potato Crop Protection*, Wageningen, Netherlands. Abstract, p. 43.

Helgeson, J.P. and Williams, C. (1991) Somatic hybrids of potato and *Solanum brevidens*: RFLP analysis of DNA introgression in sexual progeny. In: *Proceedings Second International Potato Molecular Biology Symposium*, St Andrews, Scotland. (Abstract).

Hermsen, J.G.Th. (1966) Crossability, fertility and cytogenetic studies in *Solanum acaule* × *Solanum bulbocastanum. Euphytica* 15, 149–155.

Hermsen, J.G.Th. (1977) General considerations on interspecific hybridization. In: Sanchez-Monge, E. and Garcia-Olmedo, F. (eds) *Proceedings 8th Congress of Eucarpia*, Madrid, Spain, 1977, pp. 299–304.

Hermsen, J.G.Th. (1983) Utilization of wide crosses in potato breeding. *Report of a Planning Conference on Present and Future Strategies for Potato Breeding and Improvement.* International Potato Center, Lima, Peru, pp. 115–132.

Hermsen, J.G.Th. (1984) Mechanisms and genetic implications of $2n$-gamete formation. *Iowa State Journal of Research* 58, 421–434.

Hermsen, J.G.Th. (1989) Current use of potato collections. In: Brown A.H.D., Frankel, O.H., Marshall, D.R. and Williams, J.T. (eds) *The Use of Plant Genetic Resources.* Cambridge University Press, Melbourne, pp. 68–87.

Hermsen, J.G.Th. (1992) Introductory considerations on distant hybridization. In: Kalloo, G. and Choudhury, J.B. (eds) *Distant Hybridization in Crop Plants.* Monographs on Theoretical and Applied Genetics 16. Springer-Verlag, Berlin, 1–14.

Hermsen, J.G.Th. and Ramanna, M.S. (1976) Barriers to hybridization of *Solanum bulbocastanum* Dun. and *S. verrucosum* Schlechtd. and structural hybridity in their F_1 plants. *Euphytica* 25, 1–10.

Hermsen, J.G.Th. and Verdenius, J. (1973) Selection from *Solanum tuberosum* group Phureja of genotypes combining high-frequency haploid induction with homozygosity for embryo-spot. *Euphytica* 22, 244–259.

Hermsen, J.G.Th., Olsder, J., Jansen, P. and Hoving, E. (1974) Acceptance of self-compatible pollen from *Solanum verrucosum* in dihaploids from *S. tuberosum.* In: Linskens, H.F. (ed.) *Fertilization in Higher Plants.* North-Holland Publishing Company, Amsterdam, pp. 37–40.

Hermsen, J.G.Th., Govaert, I., Hoekstra, S., Loon, C. van, and Neefjes, C. (1977) Analysis of the effect of parental genotypes on crossability of diploid *Solanum tuberosum* with *S. verrucosum*: a gene-for-gene relationship? In: Sanchez-Monge, E. and Garcia-Olmedo, F. (eds) *Proceedings of the 8th Congress of Eucarpia*, Madrid, Spain, 1977, pp. 305–312.

Hermsen, J.G.Th., Taylor, L.M., Breukelen, E.W.M. van, and Lipski, A. (1978) Inheritance of genetic markers from two potato dihaploids and their respective parent cultivars. *Euphytica* 27, 681–687.

Hermundstad, S.A. and Peloquin, S.J. (1985) Germplasm enhancement with potato haploids. *American Potato Journal* 62, 479–487.

Hogenboom, N.G. (1973) A model for incongruity in intimate partner relationships. *Euphytica* 22, 219–233.

Huaman, Z. and Ross, R.W. (1985) Updated listing of potato species names, abbreviations and taxonomic status. *American Potato Journal* 62, 629–641.

Iwanaga, M. and Peloquin, S.J. (1982) Origin and evolution of cultivated tetraploid potatoes via 2n-gametes. *Theoretical and Applied Genetics* 61, 161-169.

Iwanaga, M., Freyre, R. and Watanabe, K. (1991) Breaking the crossability barriers between disomic tetraploid *Solanum acaule* and tetrasomic tetraploid *S. tuberosum*. *Euphytica* 52, 183-191.

Johnston, S.A., den Nijs, A.P.M., Peloquin, S.J. and Hanneman, R.E. (1980) The significance of genetic balance to endosperm development in interspecific crosses. *Theoretical and Applied Genetics* 57, 5-9.

Jongedijk, E. and Ramanna, M.S. (1988) Synaptic mutants in potato, *Solanum tuberosum* L. I. Expression and identity of genes for desynapsis. *Genome* 30, 664-670.

Jongedijk, E., Ramanna, M.S., Sawor, Z. and Hermsen, J.G.Th. (1991) Formation of first division restitution (FDR) 2n-megaspores through pseudohomotypic division in *ds-1* (desynapsis) mutants of diploid potato: routine production of tetraploid progeny from $2x$ FDR $\times 2x$ FDR crosses. *Theoretical and Applied Genetics* 82, 645-656.

Jongedijk, E., De Schutter, A.A.J.M., Stolte, T., Van den Elzen, P.J.M. and Cornelissen, B.J.C. (1992) Increased resistance to potato virus X and preservation of cultivar properties in transgenic potato under field conditions. *Bio/Technology* 10, 422-429.

Lentini, Z., Earle, E.D. and Plaisted, R.L. (1990) Insect-resistant plants with improved horticultural traits from interspecific potato hybrids grown *in vitro*. *Theoretical and Applied Genetics* 80, 95-104.

Lewis, D. and Crowe, L.K. (1958) Unilateral interspecific incompatibility in flowering plants. *Heredity* 12, 233-256.

Louwes, K.M., Hoekstra, R. and Mattheij, W.M. (1992) Interspecific hybridization between the cultivated potato *Solanum tuberosum* subspecies *tuberosum* L. and the wild species *S. circaeifolium* subsp. *circaeifolium* Bitter exhibiting resistance to *Phytophthora infestans* (Mont) de Bary and *Globodera pallida* (Stone) Behrens. 2. Sexual hybrids. *Theoretical and Applied Genetics* 84, 362-370.

Mariani, C., Beuckeleer, M. de, Truettner, J., Leemans, J. and Goldberg, R.B. (1990) Induction of male sterility in plants by a chimeric ribonuclease gene. *Nature* 347, 737-741.

Mariani, C., Gossele, V., Beuckeleer, M. de, Block, M. de, Goldberg, R.B., Greef, W. de and Leemans, J. (1992) A chimaeric ribonuclease-inhibitor gene restores fertility to male sterile plants. *Nature* 357, 384-387.

Matsubayashi, M. (1991) Phylogenetic relationship in the potato and its related species. In: Tsuchija, T. and Gupta, P.K. (eds) *Chromosome Engineering in Plants: Genetics, Breeding, Evolution*, Part B. Elsevier, Amsterdam, Oxford, New York, Tokyo, pp. 93-118.

Mattheij, W.M., Eijlander, R., De Koning, J.R.A. and Louwes, K.M. (1992) Interspecific hybridization between the cultivated potato *Solanum tuberosum* subspecies *tuberosum* L. and the wild species *S. circaeifolium* subspecies *circaeifolium* Bitter exhibiting resistance to *Phytophthora infestans* (Mont.) de Bary and *Globodera pallida* (Stone) Behrens. 1. Somatic hybrids. *Theoretical and Applied Genetics* 83, 459-466.

Melchinger, A.E. (1990) Use of molecular markers in breeding for oligogenic disease resistance. *Plant Breeding* 104, 1-19.

Michelmore, R.W., Paran, I. and Kesseli, R.V. (1991) Identification of markers linked to disease-resistance genes by bulked segregant analysis: a rapid method to detect markers in specific chromosomal regions by using segregating populations. *Proceedings National Academy of Science* 88, 9828-9832.

Mok, D.W.S. and Peloquin, S.J. (1975) Three mechanisms of 2n pollen formation in diploid potatoes. *Canadian Journal of Genetics and Cytology* 17, 217–225.

Nishiyama, I. and Yabuno, T. (1979) Triple fusion of the primary endosperm nucleus as a cause of interspecific cross-incompatibility in *Avena*. *Euphytica* 28, 57–65.

Peloquin, S.J., Yerk, G.L., Werner, J.E. and Darmo, E. (1989). Potato breeding with diploids and 2n gametes. *Genome* 31, 1000–1004.

Perl, A., Aviv, D. and Galun, E. (1990) Protoplast-fusion derived *Solanum* cybrids: application and phylogenetic limitations. *Theoretical and Applied Genetics* 79, 632–640.

Pijnacker, L.P., Ferwerda, M.A., Puite, K.J. and Schaart, J.G. (1989) Chromosome elimination and mutation in tetraploid somatic hybrids of *Solanum tuberosum* and *S. phureja*. *Plant Cell Reporter* 8, 82–85.

Ramanna, M.S. (1979) A re-examination of the mechanisms of 2n gamete formation in potato and its implications for breeding. *Euphytica* 28, 537–561.

Ramanna, M.S. (1983) First division restitution gametes through fertile desynaptic mutants of potato. *Euphytica* 32, 337–350.

Ramanna, M.S. and Hermsen, J.G.Th. (1971) Somatic chromosome elimination and meiotic chromosome pairing in the triple hybrid 6x (*Solanum acaule* × *S. bulbocastanum*) × 2x *S. phureja*. *Euphytica* 20, 470–481.

Ramanna, M.S. and Hermsen, J.G.Th. (1974) Unilateral 'eclipse sterility' in reciprocal crosses between *Solanum verrucosum* Schlechtd. and diploid *S. tuberosum* L. *Euphytica* 23, 417–421.

Ramanna, M.S. and Hermsen, J.G.Th. (1981) Structural hybridity in the series Etuberosa of the genus *Solanum* and its bearing on crossability. *Euphytica* 30, 15–31.

Sangduen, N., Kreitner, G.L. and Sovensen, E.L. (1983) Light and electron microscopy of embryo development in annual × perennial *Medicago* species cross. *Canadian Journal of Botany* 61, 1241–1257.

Simmonds, N.W. (1963) Abbreviations of potato names. *European Potato Journal* 6, 186–190.

Simmonds, N.W. (1965) Mutant expression in diploid potatoes. *Heredity* 30, 65–72.

Singsit, C. and Hanneman, R.E. (1991a) Rescuing abortive inter-EBN potato hybrids through double pollination and embryo culture. *Plant Cell Reporter* 9, 475–478.

Singsit, C. and Hanneman, R.E. (1991b) Haploid induction in Mexican polyploid species and colchicine-doubled derivates. *American Potato Journal* 68, 551–556.

Stam, P. and Zeven A.C. (1981) The theoretical proportion of the donor genome in near-isogenic lines of self-fertilizers bred by backcrossing. *Euphytica* 30, 227–238.

Watanabe, K. and Peloquin, S.J. (1991) The occurrence and frequency of 2n pollen in 2x, 4x, and 6x wild, tuber-bearing, *Solanum* species from Mexico, and Central and South America. *Theoretical and Applied Genetics* 82, 621–626.

Watanabe, K., Arbizu, C., Schmiediche, P. and Jackson, M.T. (1990) Germplasm enhancement methods for disomic tetraploid species of *Solanum* with special reference to *S. acaule*. *American Potato Journal* 67, 586 (Abstract).

Young, N.D. and Tanksley, S.D. (1989) RFLP analysis of the size of chromosomal segments retained around the *Tm-2* locus of tomato during backcross breeding. *Theoretical and Applied Genetics* 77, 353–359.

Index

Acaulia 19
  classification and chromosome numbers 12
  endosperm balance number 17
  evolution 17
  genome formula 22
  hexaploidy 16
  tetraploidy 16
ACB 296-8
activation index 522
activation value 522
ae mutant 161
aewx mutant 161
after-cooking blackening 296-8
AI 522
albinism 266
allozyme analysis of wild species 24-5
alpha taxonomy 15
Alternaria solani, resistance to 290, 403, 406-7
amf mutant 160, 162-3, 164
amylopectin, debranched 161-2
amylose extender mutant 161
amylose-free mutant 160, 162-3, 164
Andean potato weevil *see Premnotrypes* spp.
aneuploids 53, 58
  production from dihaploid-wild species crosses 150
  production for genetic analysis 144
  regenerated plants 199

anther culture in production of mono- and dihaploids 135, 157-8, 178-9, 180
anthocyanin pigmentation 263-5
aphids, resistance to 354
ascorbic acid in tubers 296
autotriploidy in wild species 15
auxins and somaclonal variation 206
AV 522

B chromosomes 53
bacterial disease *see specific organisms*
bacterial wilt *see Pseudomonas solanacearum*
barnase 526
barrier capacity 517
Basarthrum 21
base number 43-4, 48-9
  evidence from karyotype analysis 45-7
  molecular evidence 47-8
  and pairing behaviour of monoploids 45
  use of monohaploids in determination 158-9
Belonolaimus longicaudatus 319
black scurf 418-19
blackleg *see Erwinia* spp.
breeding programmes
  2*n* gametes in 126-8

539

breeding programmes *contd*
   based on true seed propagation 499–510
   clonally propagated potatoes 467–91
   dihaploid-wild species hybrids in 147–8
   diploids in 145–6, 474–5
   early-generation selection 483, 486–7
   identification of parents with good GCA 482–3, 484–5
   for insect resistance 448–9, 457–60
   introgression of genes from wild species 35–8, 472–3, 515–34
   maximizing the response to selection 489–91
   molecular markers in 223–6
   monohaploids in 166–7
   19th century 471–2
   progeny tests 487–9
   somaclonal variation in 206–8
   strategies for increasing selection efficiency 480–2
   tetraploids in 476–9
   tissue culture in 177–90
   20th century 472–9
   for virus resistance 354–7
broad-sense heritability 72
brown rot *see Pseudomonas solanacearum*
Bulbocastana 19
   classification and chromosome numbers 10
   endosperm balance number 52
   isozyme analysis 24
   leaf flavonoids 23
   morphology 16

calyx
   shape 271
   tube length 272
α-chaconine 290
chaperonins 242
chimeras, periclinal 56
chlorophyll synthesis in tubers 291, 292–3
chloroplast DNA
   changes in regenerated plants 202
   distribution 33–4
   variations in 222

chromosome instability
   applications 61–2
   in vitro somatic 57–60
   in vivo somatic 53–7
   in regenerated plants 198–204, 205
   structural 60–1
Circaeifolia 19
   classification and chromosome numbers 10
   endosperm balance number 52
   isozyme analysis 24
   leaf flavonoids 23
   morphology 16
citric acid 296–7
Clavibacter michiganensis subsp. *sepedonicus* 429
   resistance to
     future prospects 441
     in modern cultivars 432
     and somaclonal variation 439–40
     in wild species and primitive cultivars 434
CMS 509, 526
coefficient of coancestry 77
   determination for dihaploids 137–9
coefficient of double reduction 77, 134
   estimation using dihaploids 139–42
   relationship with gene–centromere map distances 142–3
cold acclimatization 249–50
cold-induced sweetening 250–1, 300–1
cold stress 248–51
Colorado beetle, resistance to 35, 36, 290
Commersoniana 19
   classification and chromosome numbers 10
   endosperm balance number 52
   isozyme analysis 24
   morphology 16
common scab, resistance to 36, 419–21
Commonwealth Potato Collection 35
Conicibaccata 19
   classification and chromosome numbers 11
   endosperm balance number 17, 52
   genome formula 22, 23
   hexaploidy in 16
   isozyme analysis 24
   leaf flavonoids 23
   migration routes 17, 20
   tetraploidy in 16

corky scab 421
corolla characteristics 272
corolla index 272
correlated responses 75
covariance of half sibs 86
covariances between relatives 84–7
crosses between pure lines
    disomic inheritance 92–4
    tetrasomic inheritance 94–6
cryopreservation 177
cultivation
    and polyploidy 52
    in South America 5–7, 8–9
culture medium, and somaclonal variation 206
Cuneoalata 19
    classification and chromosome numbers 11
    endosperm balance number 17, 52
    genome formula 22
    isozyme analysis 24
    leaf flavonoids 23
cyanidin 265
cybrids
    genomic changes in 201
    in production of true potato seed 509
    and resistance to *Phytophthora infestans* 386
cytokinins and somaclonal variation 206
cytoplasmic male sterility 509, 526

dehydrins 248
delphinidin 265
Demissa 19
    classification and chromosome numbers 12
    distribution 20–1
    endosperm balance number 52
    genome formula 22
    hexaploidy in 16
    immunology 25, 26
dihaploid induction 54, 109
dihaploids 53–4, 56, 109
    agronomically useful somaclonal variation in 207
    analysis of frequency distribution of populations 143
    in combination with protoplast fusion 189–90

determination of coefficient of coancestry 137–9
determination of coefficient of inbreeding 137–9
disadvantages of use in genetic analyses 155
and estimation of coefficient of double reduction 139–42
hybridization with wild species 146–50
production 135, 156–8, 178–9
utilization 136–7, 181–2
diplandroid gametes 109, 112
diplogynoid gametes 109, 112
diploids
    advantages of genetical analysis at level of 133–5
    in breeding programmes and genetic analysis 145–6, 474–5
    genetic system 144
    as parental lines in true potato seed production 503–4
disomic inheritance 92–4
    comparison with tetrasomic inheritance 133, 134, 470
Ditylenchus spp. 319
DNA markers *see* molecular markers
domestication
    archaeological evidence for 3–4
    historical evidence 4
double reduction 76–7, 78, 133–4
drop tests 289
drought resistance 37, 244–8
dry matter of tubers 304–6
dry rot, resistance to 290, 415–18, 433
dwarfism 265–6

early blight, resistance to 290, 403, 406–7
early-generation selection 483, 486–7
EBN 16–21, 51–2, 519–22
eelworm *see Globodera pallida*; *Globodera rostochiensis*
embryo rescue, and introgression of genes from wild species 524
embryo spot 54, 156, 273–4
endoreduplication 59, 199
endosperm balance number 16–21, 51–2, 519–22
endosperm failure 519

environmental stress, impact on yield 239–55
Erwinia spp. 429
   resistance to 251–4
      applications of genetic engineering 440
      future prospects 440
      in modern cultivars 431
      molecular factors in 437, 438
      and somaclonal variation 439–40
      in somatic hybrids 436
      transfer from primitive to modern cultivars 434–5
      in wild species 36, 432–3
Estolonifera 7, 10, 19
Etuberosa 7, 19
   classification and chromosome numbers 10
   endosperm balance numbers 21, 52
   genome formula 22
   isozyme analysis 25
Europe, introduction of potato to 4–5

F_2 triple-test cross 93
Fa gene 326
Fb gene 326
FDR 109–12
FDR-NCO 112
ferridichlorogenic acid 296–7
field immunity 340
first division restitution 109–12
flanking marker models 103–4
flavour of tubers 306
flower characters 271–4
folic acid in tubers 296
frost tolerance 35, 37, 249–50
fry colour 301–3
fungal diseases *see specific fungi*
Fusarium spp. *see* dry rot; Fusarium wilt
Fusarium wilt, resistance to 403, 407

gangrene 421–2
gene banks 35
gene-centromere mapping
   and $2n$ gametes 112–14
   with molecular markers 221
   relationship of distances with coefficient of double reduction 142–3

gene transfer, and resistance to *Phytophthora infestans* 387
general combining ability
   identification in parental clones 482–3, 484–5
   in non-equilibrium populations 87–8
   variance 86
generation matrix 79, 80
genetic correlation 75
genetical variation, partitioning 82–3
genome evolution 43–62
genomic *in situ* hybridization 52
genotype–environment interactions 72–5
genotype–microenvironment interaction 73
germplasm collections 35
gibberellins 241
Giemsa C-banding 46–7
GISH 52
glandular trichomes 451, 452, 454, 457
Globodera pallida 319
   pathotype scheme 321–3
   resistance to 332–3
      breeding for 166
      and glycoalkaloids 290
      in haploids 181–2
      major gene inheritance 326–8
      quantitative 328–9
      sources of 323
      in wild species 35, 36
   tolerance to attack by 330–1
Globodera rostochiensis 319
   pathotype scheme 321–3
   resistance to 332–3
      and glycoalkaloids 290
      major gene inheritance 323, 325–6
      mapping gene loci 225
      quantitative 328–9
      in somatic hybrids 188
      sources of 323
      in wild species 36
   tolerance to attack by 330–1
glycoalkaloids in tubers 290–2
Gm gene 351, 352
green peach aphid *see Myzus persicae*
greening of tubers 291, 292–3
Gro1 gene, mapping 225

H1 gene 225, 323, 325, 332, 480
H2 gene 326-7, 332
H3 gene 327-8
heat resistance in wild species 37
heat-shock proteins 163, 242-3
heat stress 240-4
heterosis 502-3, 509
hexaploidy in wild species 16
hollow heart 245, 246, 287-8
    incipient 288
homozygosity, association with
    inbreeding depression 502
hrp genes 438-9
hybrid vigour 502-3, 509
hybridization
    barriers to interspecific 517-25
    monitoring with molecular markers 224
    and production of true potato seed 507
hypoxic stress 251-4

inbreeding 77-80
    decline in heterozygosity on 92, 93
    to produce true potato seed 503
inbreeding coefficient 77, 78, 80
    determination for dihaploids 137-9
inbreeding depression 91-2, 501-2
incongruity 517
inflorescence types 271
Ingifolia 11, 19
insecticides 447
insects 448
    resistance to
        breeding for 457-60
        criteria for breeding programmes 448-9
        screening for 454-7
        status of 449-54
integrated pest management 447
intercropping 240-1
interdihaploids 189-90
internal rust spot 288
International Potato Center 35
interval mapping 104
introgression
    definition 515
    monitoring with molecular markers 224
IPM 447

IRS 288
isozyme analysis of wild species 24-5
ivy leaf mutant 270

Juglandifolia 7, 10, 19

K1 gene 325-6

late blight *see Phytophthora infestans*
leaf
    characters 268-70
    discs in rapid propagation 175
    flavonoids in wild species 23
leafminer flies *see Liriomyza huidobrensis*
Leptinotarsa decemlineata, resistance to 35, 36, 290
Lignicaulia 19
    classification and chromosome numbers 10
    endosperm balance number 52
    morphology 16
Liriomyza huidobrensis 448
    resistance to 454
        screening for 456
    status of 452
Longipedicellata 19
    classification and chromosome numbers 12
    endosperm balance number 17, 52
    genome formula 22
    immunology 25, 26
    migration routes 17, 20
    tetraploidy in 16
lubimin 378

Ma gene 351
Macrosiphum euphorbiae 36
Maglia 11
Maillard reaction 250, 300
male sterility in F_1 plants 525-6
malvidin 263
maximal equational segregation 134
Megistacroloba 19
    classification and chromosome numbers 11

Megistacroloba *contd*
  endosperm balance number 17, 52
  evolution 17
  genome formula 22
  isozyme analysis 24
Meloidogyne spp. 319
  resistance to 37, 331-2, 435
meristem culture 176
microspores, *in vitro* selection on 183-4
microtubers 177
minitubers 177
mites 448-9
mixture 507
molecular genetics 213-29
molecular mapping
  monohaploids in 165-6
  use of molecular markers 214-21
molecular markers 213-14
  in genetic and physical mapping 214-21
  in genetic resources management and biosystematics 222-3
  in germplasm utilization and breeding programmes 223-6
  use in selection 529-30
monohaploids 156
  in analysis of multigene families 163, 165
  in breeding programmes 166-7
  in combination with protoplast fusion 189-90
  in comparison of mutations with wild-type versions 165
  in determination of base number 158-9
  in molecular biology 163-6
  in molecular mapping 165-6
  in mutation studies 159-63
  in potato genetics 158-63
  production 156-8, 178-9, 180
  utilization 181-3
Morelliformia 19
  classification and chromosome numbers 10
  immunology 25, 26
  isozyme analysis 24
morphological characteristics 263-74
multigene families, monohaploids in analysis 163, 165
multiplex parents 480-2

mutations
  comparison with wild-type version 165
  studies using monohaploids 159-63
Myzus persicae 448
  resistance to 36, 454

Na gene 346, 348, 350
Nacobuss aberrans 319
narrow-sense heritability 89
Nb_{tbr} gene 348, 349, 350
Nc gene 344
nematodes *see individual species*
Neotuberosum 473-4
  nematode resistance 330
  resistance to *Phytophthora infestans* 384
niacin in tubers 296
N_L gene 343, 355
nodal band pigmentation 269
non-crossability 517
Ns gene 352
Nx_{chc} gene 346, 348-9, 349, 350, 480
Nx_{tbr} gene 346, 348-9, 350, 480
Ny_{chc} gene 344-5, 346, 480
Ny_{dms} gene 344-5, 346, 480

Olmosiana 19
  classification and chromosome numbers 11
  morphology 16
omega taxonomy 15
open pollinated progenies in production of true potato seed 507-9
ovary wall pigmentation 273
overdominance 502

pantothenic acid in tubers 296
Paratrichodorus spp. 319
parental line breeding 356-7
parthenogenesis in production of mono- and dihaploids 135, 156-7, 178
partial/complete dominance 502
patatin 295-6
  multigene family 163, 165, 221
pathogenicity-related proteins 374, 378-9
PCN *see Globodera pallida*; *Globodera rostochiensis*

PCR 185, 214
pelargonidin 265
penetration capacity 517
pesticides 447
Petota 7, 19
petunidin 263, 265
PFK 301
Phomea foveata 421-2
phosphofructokinase 301
photosystem II, effects of heat stress on 241-2
Phthorimaea operculella 448
   resistance to 435, 454
     screening for 456
     status of 450-1
phytoalexins 377-8, 417
Phytophthora infestans 365, 366
   adaptation 371-2
   changes in 389
   genetic markers 370
   host-pathogen interaction
     biochemistry 375-9
     histology and cytology 374-5
   mating types 367-8
   oospore production 369-70
   parasexuality 370
   ploidy 368-9
   resistance to 389-91
     breeding for 166, 382-8
     general 373, 381
     genetic basis 379-81
     in haploids 182
     induced 373-4
     methods of incorporation 385-7
     non-host 373
     origin of 381-2
     race-specific 372-3, 381
     selection for 387-8
     in somatic hybrids 186-7
     sources of 383-5
     types of 372-4
     in wild species 36
   sexual reproduction 365, 367-70
   transformation 371
   variability in 365
phytuberin 378
pigmented whorl 265, 268
Pinnatisecta 19
   classification and chromosome numbers 10
   endosperm balance number 52

   immunology 25, 26
   isozyme analysis 24
   leaf flavonoids 23
   morphology 16
Piurana 19
   classification and chromosome numbers 11
   endosperm balance number 17, 52
   genome formula 22
   tetraploidy in 16
plant type 265-7
ploidy manipulation in introgression of genes from wild species 522-3
PLRV *see* potato leaf roll virus
Polyadenia 19
   classification and chromosome numbers 10
   isozyme analysis 24
   morphology 16
polymerase chain reaction 185, 214
polyploidy 49-52, 57-8
   in regenerated plants 199
postharvest losses 251-4
potato cyst nematode *see Globodera pallida*; *Globodera rostochiensis*
potato leaf roll virus (PLRV)
   resistance to 342-3
     breeding for 166
     in haploids 181-2
     in wild species 36
   strains 342
potato mop-top furovirus 353
potato spindle tuber viroid 278, 354
potato tuber moth *see Phthorimaea operculella*
potato unmottled curly dwarf virus 354
potato virus A 346, 348
potato virus M 350, 351
potato virus S 351-2
potato virus T 353
potato virus V 353
potato virus X (PVX)
   resistance to 348-50
     breeding for 166
     in haploids 181-2
     mapping gene loci 224-5
     in somatic hybrids 186
     in wild species 36
   strains 348

potato virus Y (PVY)
    resistance to 344–6, 347
        breeding for 166
        in haploids 181–2
        in somatic hybrids 186
        in wild species 36
    strains 344
Potatoe 10–11, 19
powdery scab 421
PR proteins 374, 378–9
Pratylenchus spp. 319
premature cytokinesis 111
Premnotrypes spp. 448
    resistance to 453–4, 457
processing qualities, inheritance 285–306
progeny tests 487–9
protein content of tubers 294–6
protein electrophoresis on wild species 23–4
proteinase inhibitor 163
protoclones 182
protoplasts
    fusion 184–90
    genomic changes in 200, 204
    in vitro selection on 182–3
    in rapid propagation 175–6
PS II, effects of heat stress on 241–2
pseudogamy 135
Pseudomonas solanacearum 429
    resistance to
        applications of genetic engineering 440
        future prospects 440
        in modern cultivars 432
        molecular factors in 437–9
        transfer from primitive to modern cultivars 435
        in wild species 36, 433–4
PVA 346, 348
PVM 350, 351
PVS 351–2
PVX *see* potato virus X
PVY *see* potato virus Y
pyridoxine in tubers 296

QTLs *see* quantitative trait loci
quality characteristics, inheritance 285–306

quantitative genetics 71–2
quantitative trait loci (QTLs) 101–5, 167, 491
    location with $2n$ gametes 71–2
    molecular markers 225–6

R genes 372–3, 379, 380–1, 390, 480
random mating populations
    in equilibrium 81–7
    response to selection 88–91
randomly amplified polymorphic DNA (RAPD) markers 213–26
rapid propagation 174–6
recombinant inbred lines 166
regeneration capacity, genetics of 179–81
response to selection, maximizing 489–91
response value 522
restriction length fraction polymorphism (RFLP) 239
    molecular markers 213–26
    wild species 27–34
Rhizoctonia solani 418–19
RI lines 166
riboflavine in tubers 296
ring rot *see Clavibacter michiganensis* subsp. *sepedonicus*
rishitin 378, 417
RKN *see Meloidogyne* spp.
Rm gene 350–1, 355
root-knot nematode *see Meloidogyne* spp.
Rotata 17, 19, 20
Rough Purple Chili 471–2
RV 522
Rx1 gene, mapping 225
Rx2 gene, mapping 225
Rx_{ac1} 349–50
Rx_{adg} 349–50, 480
$Rx_{(scr)}$ 348, 349
Rx_{tbr} 349
Ry_{adg} gene 346
Ry_{hou} gene 345
Ry_{sto} gene 344–6, 355, 356, 480

s gene 352
second division restitution (SDR) 109–12

selection
    marker-aided 529-31
    response of random mating
        populations to 88-91
self-incompatibility 228-9
    in wild species 22
serine 248
sexual crossability 517
sexual polyploidization, genetic
        consequences of 2*n* gametes in
        120-6
shaker tests 289
shoots 174
sloughing 298
soft rot *see Erwinia* spp.
solanidine 290
α-solanine 290
Solanum abancayense 18
S. *acaule* 6
    chromosome number 12
    disease resistance 37
    endosperm balance number 18
    frost resistance 7, 35, 249
S. *acroglossum* 18
S. *acroscopicum* 331
S. *agrimonifolium* 17
    chromosome number 11
    endosperm balance number 18
    genome formula 22
S. *ajanhuiri*
    chromosome number 12
    evolution 6
    isozyme analysis 25
    nematode resistance 331
S. *alandiae* 11
S. *albicans* 12, 17, 18
S. *amabile* 18
S. *ambosinum* 18
S. *berthaultii*
    adaptation 7
    chromosome number 11
    disease resistance 37
    endosperm balance number 18
    insect resistance 451
    thermotolerance 241
S. *boliviense*
    chromosome number 11
    endosperm balance number 18
    frost tolerance 249
S. *brachistotrichum*
    chromosome number 10
    endosperm balance number 18
    isozyme analysis 25
S. *brachycarpum*
    adaptation 13
    chromosome number 12
    endosperm balance number 18
S. *brevicaule*
    chromosome number 11
    endosperm balance number 18
    nematode resistance 329
S. *brevidens*
    chromosome number 10
    endosperm balance number 21
    isozyme analysis 25
    self-compatibility 22
S. *bukasovii* 11, 18
S. *bulbocastanum*
    chromosome number 10
    endosperm balance number 18
    immunological studies 26, 27
    isozyme analysis 25
    leaf flavonoids 23
    nematode resistance 331
S. *canasense* 5, 6, 468
    chromosome number 11
    endosperm balance number 18
    isozyme analysis 24
S. *capsibaccatum*
    chromosome number 10
    endosperm balance number 18
    isozyme analysis 25
    nematode resistance 331
S. *cardiophyllum* 10, 18
S. *chacoense* 6, 20, 24
    adaptation 13
    base number 43
    chromosome number 11
    disease resistance 35, 37
    endosperm balance number 18
    resistance to bacterial disease
        435
    thermotolerance 241
S. *chancayense* 18
S. *chaucha*
    chromosome number 12
    evolution 6, 52
S. *chomatophilum*
    chromosome number 11
    endosperm balance number 18
    frost tolerance 249
S. *circaeifolium* 10, 18

S. clarum
    chromosome number 10
    immunological studies 27
    isozyme analysis 25
    leaf flavonoids 23
S. colombianum
    adaptation 13
    chromosome number 11
    endosperm balance number 18
S. commersonii
    adaptation 13
    chromosome number 10
    disease resistance 37
    endosperm balance number 18
    frost tolerance 249
    insect resistance 450
S. curtilobum 6, 12, 52
S. demissum 21
    adaptation 13
    base number 43
    chromosome number 12
    crossing with 17
    disease resistance 35, 37, 385
    endosperm balance number 18
    genome formula 22
    isozyme analysis 24
    thermotolerance 241
S. edinense 12, 35, 382
S. etuberosum
    chromosome number 10
    endosperm balance number 18, 21
    isozyme analysis 25
    self-compatibility 22
S. fendleri
    base number 43
    chromosome number 12
    endosperm balance number 18
    isozyme analysis 24
S. fernandeziamum 18, 21, 25
S. gandarillasii
    chromosome number 11
    endosperm balance number 18
    nematode resistance 331
S. gourlayi
    chromosome number 11, 16
    endosperm balance number 17, 18
    nematode resistance 323, 329, 331
    protein electrophoresis 24
S. guerreroense 12, 18
S. hjertingii 12, 18
S. hondelmannii 11
S. hougasii 12, 18
S. huancabambense 18
S. incamayoense 24
S. infundibuliforme
    chromosome number 11
    endosperm balance number 18
    protein electrophoresis 24
    topiary mutant 265
S. ingifolium 11
S. iopetalum 12, 18
S. jamesii
    adaptation 13
    base number 43
    chromosome number 10
    endosperm balance number 18
    thermotolerance 241
S. juzepczukii 6, 12, 52
S. kurtzianum
    chromosome number 11
    endosperm balance number 18
    nematode resistance 323, 324, 325
    thermotolerance 241
S. laxissimum 18
S. leptophyes 5, 6, 468
    chromosome number 11
    endosperm balance number 18
    nematode resistance 329
S. lesteri 10
S. lignicaule
    chromosome number 10
    endosperm balance number 18
    nematode resistance 331
    protein electrophoresis 23
S. longiconicum 11, 17, 22
S. lycopersicoides
    chromosome number 10
    relationship to *Lycopersicon* 27, 30
S. maglia 11, 24, 37
S. marinasense
    chromosome number 11
    endosperm balance number 18
    protein electrophoresis 23
S. medians 18
S. megistacrolobum 6
    chromosome number 11
    endosperm balance number 18
    frost tolerance 7, 249
    isozyme analysis 24
    nematode resistance 331
S. microdontum
    adaptation 13

S. microdontum contd
   chromosome number 11
   disease resistance 37, 435
   endosperm balance number 18
   nematode resistance 331

S. mochiquense, endosperm balance number 18

S. morelliforme
   chromosome number 10
   isozyme analysis 25
   leaf flavonoids 23
   primitive characteristics 16
   self-compatibility 22

S. moscopanum 11, 18, 23

S. multidissectum
   chromosome number 11
   endosperm balance number 18
   frost tolerance 249
   nematode resistance 323, 324, 326–7

S. multiinterruptum 18

S. neocardenasii 7, 11

S. olmosense 17

S. oplocense 17
   chromosome number 11, 16
   cold acclimatization 249
   endosperm balance number 18
   nematode resistance 329

S. oxycarpum 17
   chromosome number 11
   endosperm balance number 18
   genome formula 22

S. pampasense 18

S. papita
   chromosome number 12
   endosperm balance number 18
   thermotolerance 241

S. pascoense 18

S. phureja 6, 7, 468
   chromosome number 12
   cold resistance 251
   disease resistance 37, 433–4, 435, 438
   endosperm balance number 18
   isozyme analysis 24
   protein electrophoresis 23

S. pinnatisectum
   chromosome number 10
   endosperm balance number 18
   insect resistance 450
   isozyme analysis 24

S. piurae 11

S. polyadenium
   chromosome number 10
   immunological studies 26, 27
   isozyme analysis 25
   self-compatibility 22

S. polytrichon
   chromosome number 12
   cold acclimatization 249
   endosperm balance number 18

S. raphanifolium 33
   endosperm balance number 18
   resistance to bacterial diseases 435

S. × sambucinum 27

S. sanctae-rosae
   chromosome number 11
   endosperm balance number 18
   frost tolerance 249
   nematode resistance 323

S. santolallae 11

S. schenkii 12

S. semidemissum 12

S. sogarandinum 18

S. sparsipilum 5
   chromosome number 12
   disease resistance 37, 329, 435
   endosperm balance number 18
   in evolution of *S. tuberosum* 155
   insect resistance 450
   isozyme analysis 24
   nematode resistance 323, 329, 331–2
   protein electrophoresis 23–4

S. spegazzinii
   chromosome number 12
   endosperm balance number 18
   nematode resistance 37, 323, 324, 326, 329, 331
   thermotolerance 241

S. stenotomum
   endosperm balance number 18
   evolution 5–6
   in evolution of *S. tuberosum* 155, 468
   isozyme analysis 25
   protein electrophoresis 23–4
   resistance to bacterial disease 435

S. stoloniferum
   adaptation 13
   chromosome number 12
   disease resistance 37
   endosperm balance number 18
   thermotolerance 241

S. sucrense
   chromosome number 12
   endosperm balance number 17, 18
   insect resistance 450
   nematode resistance 329
   tetraploidy 16
   thermotolerance 241
S. tarijense
   adaptation 7
   chromosome number 11
   disease resistance 37
   endosperm balance number 18
   insect resistance 450
S. toralapanum 11
S. trifidum 10, 18
S. tuberosum
   chloroplast DNA distribution 33–4
   nematode resistance 330
S. tuberosum cv. Adretta, aneuploidy 53
S. tuberosum cv. Alpha, sensitivity to drought 247
S. tuberosum cv. BelRus, resistance to soft rot 251, 253
S. tuberosum cv. Cara, drought tolerance 247
S. tuberosum cv. Desiree
   aneuploidy 58
   drought tolerance 245, 247
S. tuberosum cv. Duke of York, karyotype analysis 46
S. tuberosum cv. Fetwell, agronomically useful somaclonal variation in 207
S. tuberosum cv. Fortyfold, chromosome instability 59
S. tuberosum cv. Foxton, agronomically useful somaclonal variation in 207
S. tuberosum cv. Lemhi, drought tolerance 245
S. tuberosum cv. Majestic, structural chromosome instability 60–1
S. tuberosum cv. Maris Bard, chromosome instability 59
S. tuberosum cv. Maris Piper, agronomically useful somaclonal variation in 207
S. tuberosum cv. Nooksack, drought tolerance 245
S. tuberosum cv. Russet Burbank agronomically useful somaclonal variation in 207

sensitivity to drought 245, 246
S. tuberosum cv. Spunta, drought tolerance 245, 247
S. tuberosum cv. Torridon
   aneuploidy 53
   chromosome instability 57
S. tuberosum cv. Up-to-Date, drought tolerance 247
S. tuberosum cv. Veenster, sensitivity to drought 245
S. tuberosum subsp. *andigena*
   chromosome number 12
   day-length adaptation 4–5
   dwarfism in 265
   endosperm balance number 18
   evolution 5–6, 468–70
   nematode resistance 37, 323, 324, 327–8, 331
   protein electrophoresis 23–4
   resistance to bacterial disease 432–3
   resistance to *Phytophthora infestans* 383–4
S. tuberosum subsp. *tuberosum*
   chromosome number 12
   endosperm balance number 18
   evolution 5–6, 468–71
   protein electrophoresis 23–4
   resistance to *Phytophthora infestans* 383–4
S. tuquerrense 11, 18
S. vallis-mexici 12
S. venturii 18
S. vernei
   adaptation 13
   chromosome number 12
   nematode resistance 35, 37, 323, 324, 327, 328–9, 331
S. verrucosum 21
   adaptation 13
   chromosome number 12
   disease resistance 37
   endosperm balance number 18
   self-compatibility 22
S. vidaurrei 24
S. violaceimarmoratum
   adaptation 13
   chromosome number 11
   endosperm balance number 18
S. weberbaueri 18
S. yungasense 11
solavetivone 378

somaclonal variation 197-8
  in breeding programmes and biotechnology 206-8
  factors influencing 204, 206
  origin and nature 198-204, 205
  and resistance to bacterial diseases 439-40
  and resistance to *Phytophthora infestans* 383-4
somatic crossability 517
somatic hybridization 184-8
  and genomic changes 201
  and introgression of genes from wild species 524-5
  and resistance to bacterial diseases 436
  and resistance to *Phytophthora infestans* 386
South America, potato cultivation 5-7, 8-9
specific combining ability 482-3
  variance 86
spindle tuber viroid, resistance to in wild species 36
Spongospora subterranea 421
SPS 250-1
starch
  alterations through mutation and selection 160-3
  amylose-free 160-1
Stellata 17, 19, 20
stem
  canker 418-19
  characters 267
  explants in rapid propagation 175
stigma type 273
stolbur, resistance to 353-4
stolon length 279
Streptomyces scabies, resistance to 36, 419-21
subsoiling 247
sucrose-6-phosphate synthase 250-1
sugar content of tubers 300-4
Synchytrium endobioticum, resistance to 36, 412-15, 416, 471
synthetic populations, and production of true potato seed 507, 508

table qualities, inheritance 285-306
terminal leaflet shape 269

Tetranychus urticae, resistance to 454, 455
tetraploids
  breeding at level of 476-9
  as parental lines in true potato seed production 504-9
  wild species 15-16
tetrasomic inheritance 76-80, 94-6
  biometrical genetical analysis 115-18
  comparison with disomic inheritance 133, 134, 470
Thanatephorus cucumeris 418-19
thermotolerance 240-4
thiamine in tubers 296
threonine 248
tissue culture 173-4
  in breeding programmes 177-90
  and genomic changes 200-1
  in germplasm maintenance and propagation 174-7
  in vitro selection on single cells 182-4
  and living collections 177
tobacco rattle tobravirus, resistance to 353
transformation 533-4
transgenic plants
  genomic changes in 201
  and virus resistance 357
transposon mutagenesis 162-3, 164
transposons 227-8
Trichodorus spp. 319
trisomic analysis 150
true potato seed, breeding programmes based on 499-510
TRV, resistance to 353
tuberization, effects of heat stress on 243-4
Tuberosa 19
  classification and chromosome numbers 11-12
  endosperm balance number 17, 52
  evolution 17
  genome formula 22
  hexaploidy in 16
  isozyme analysis 24
  leaf flavonoids 23
  polyploidy in 52
  tetraploidy in 16

tubers
    after-cooking blackening 296–8
    blackspot 289
    brown centre 288
    cold-induced sweetening 250–1, 300–1
    comparison of propagation from with true potato seed 500–1
    cooked texture 298–9
    dry matter 304–6
    enzymic browning 299–300
    eye colour 276
    eye depth 279
    flavour 306
    flesh colour 275
    glycoalkaloids in 290–2
    greening 291, 292–3
    growth cracking 286–7
    hollow heart 245, 246, 287–8
    internal rust spot 288
    lack of blackening in wild species 37
    nutritional value 294–6
    periderm pigmentation 274–5, 276
    resistance to mechanical damage 289–90
    shape 276–9
    shattering 289
    spraing 353
    sprout colour 276
    sugar content 300–4
$2n$ gametes
    in analysis of tetrasomic inheritance 115–18
    in breeding programmes 126–8
    and gene-centromere mapping 112–14
    in introgression of genes from wild species 523–4
    in location of quantitative trait loci 118–20
    modes of formation 109–12
    in unilateral and bilateral sexual polyploidization 120–6

UDP-glucose pyrophosphorylase 250–1
Ulluco 3
unilateral incompatibility 517–18
UPPL 250–1

Verticillium wilt, resistance to 403, 404–6
virescens 266–7
viruses
    aggressiveness 340
    pathogenicity 340
    resistance to 339
        associated with hypersensitivity 340
        associated with tolerance 341
        breeding for 354–7
        extreme 341
        terminology 339–40
    resistance to infection by 341–2
    resistance to vectors of 342, 354
    virulence 340
    see also specific types
visual preference 486–7
vitamin content of tubers 296

wart, resistance to 36, 412–15, 416, 471
water stress 244–8
waxy mutant 160
weighted least-squares procedure 113–14
wild species
    adaptive ranges 7, 10–14
    allozyme analysis 24–5
    biochemical systematics 23–7
    comparative immunology 25–7
    crossability and compatibility 21–2
    evolutionary relationships 16–34
    genome relationships 22–3
    hybridization with dihaploids 146–50
    isozyme analysis 24–5
    leaf flavonoids 23
    molecular systematics 27–33
    morphology 16–21
    protein electrophoresis 23–4
    resistance to *Phtophthora infestans* 385
    systematics 15–16
    value in potato breeding 35–8, 472–3, 515–34

Yungasensa 19
    classification and chromosome numbers 11
    endosperm balance number 17, 52
    and evolution of wild potato species 17
    genome formula 22